Sustainable Development Goals Series

The **Sustainable Development Goals Series** is Springer Nature's inaugural cross-imprint book series that addresses and supports the United Nations' seventeen Sustainable Development Goals. The series fosters comprehensive research focused on these global targets and endeavours to address some of society's greatest grand challenges. The SDGs are inherently multidisciplinary, and they bring people working across different fields together and working towards a common goal. In this spirit, the Sustainable Development Goals series is the first at Springer Nature to publish books under both the Springer and Palgrave Macmillan imprints, bringing the strengths of our imprints together.

The Sustainable Development Goals Series is organized into eighteen subseries: one subseries based around each of the seventeen respective Sustainable Development Goals, and an eighteenth subseries, "Connecting the Goals," which serves as a home for volumes addressing multiple goals or studying the SDGs as a whole. Each subseries is guided by an expert Subseries Advisor with years or decades of experience studying and addressing core components of their respective Goal.

The SDG Series has a remit as broad as the SDGs themselves, and contributions are welcome from scientists, academics, policymakers, and researchers working in fields related to any of the seventeen goals. If you are interested in contributing a monograph or curated volume to the series, please contact the Publishers: Zachary Romano [Springer; zachary.romano@springer.com] and Rachael Ballard [Palgrave Macmillan; rachael.ballard@palgrave.com].

Sandi Hilal · Merve Bedir ·
Mette Ramsgaard Thomsen ·
Martin Tamke
Editors

Design for Partnerships for Change

Proceedings of the UIA World Congress of Architects Copenhagen 2023

Springer

Editors
Sandi Hilal
Department of Architecture
and Built Environment
Lund University
Lund, Sweden

DAAR (Decolonizing Architecture Art
Research)
Beit Sahur, Palestine

Mette Ramsgaard Thomsen
CITA (Centre for Information
Technology and Architecture)
The Royal Danish Academy
Architecture, Design, Conservation
Copenhagen, Denmark

Merve Bedir
BAK (basis voor actuele kunst)
Utrecht, The Netherlands

Martin Tamke
CITA (Centre for Information
Technology and Architecture)
The Royal Danish Academy
Architecture, Design, Conservation
Copenhagen, Denmark

ISSN 2523-3084 ISSN 2523-3092 (electronic)
Sustainable Development Goals Series
ISBN 978-3-031-36992-6 ISBN 978-3-031-36993-3 (eBook)
https://doi.org/10.1007/978-3-031-36993-3

This Springer imprint is published by the registered company Springer Nature Switzerland AG
The registered company address is: Gewerbestrasse 11, 6330 Cham, Switzerland

Scientific Committee

General Reporter, Alternate General Reporter

Mette Ramsgaard Thomsen, Professor and Head of CITA (Centre for Information Technology and Architecture), Royal Danish Academy—Architecture, Design, Conservation
Martin Tamke, Associate Professor, CITA (Centre for Information Technology and Architecture), Royal Danish Academy—Architecture, Design, Conservation

Panel Chairs

Panel 1: Design for Climate Adaptation
Billie Faircloth, Research Director, Kieran Timberlake, Adjunct Professor University of Pennsylvania
Maibritt Pedersen Zari, Educator and Researcher, Associate Professor, Auckland University of Technology
Panel 2: Design for Rethinking Resources
Carlo Ratti, Professor and Director of the Senseable Lab, MIT, Founding Partner of Carlo Ratti Associati
Mette Ramsgaard Thomsen, Professor and Head of CITA (Centre for Information Technology and Architecture), Royal Danish Academy—Architecture, Design, Conservation
Panel 3: Design for Resilient Communities
Juan Du, Professor and Dean of the John H. Daniels Faculty of Architecture, Landscape and Design, University of Toronto
Anna Rubbo, Senior Researcher, Center for Sustainable Urban Development (CSUD), The Climate School, Columbia University
Panel 4: Design for Health
Arif Hasan, Chairperson Urban Resource Center Karachi, former Visiting Professor NED University Karachi, former member of UNs Advisory Group on Forced Evictions (PA)
Christian Benimana, Co-Executive Director and Senior Principal MASS Design Group

Panel 5: Design for Inclusivity
Magda Mostafa, Autism Design Principal Progressive Architects Professor of Design Department of Architecture The American University in Cairo Cairo, Egypt
Ruth Baumeister, Associate Professor of Theory and History of Architecture Aarhus School of Architecture
Panel 6: Design for Partnerships for Change
Sandi Hilal, Co-Director, Decolonizing Architecture Art Research (DAAR), Lise Meinert Visiting Professor, Lund University
Merve Bedir, Founding member, Center for Spatial Justice and Kitchen Workshop; Co-director, Aformal Academy

Special Advisors

Katherine Richardson, Professor in Biological Oceanography and Leader of Sustainability Science Centre, University of Copenhagen
Chris Luebkeman, Leader of the Strategic Foresight Hub, Office of the President, ETH Zürich
Thomas Bo Jensen, Head of Research, Aarhus School of Architecture
Camilla Ryhl, Research Director, Bevica Fonden

Peer Review Committee

We would like to thank all the members of the Peer Review Committee for this volume for their enduring effort and valuable advice.

Alejandro Jaramillo Quintero
Angelo Bucci
Athanasios Spanomaridis
Binita Magaiya
Carolina Vasilikou
Catalina Ortiz
Charlotte Malterre-Barthes
Chensi Shen
Chirag Sharma
Chrisna du Plessis
Cid Blanco Junior
Dale Sinclair
Demetra Kourri
Fadi Shayya
Elena Parnisari
Erika Henriksson
Esther Charlesworth
Filipa Corais
Francisco Díaz
Gisle Løkken
Gizem Deniz Güneri
Haohao Xu
Hou Yuming
Isabella Rusconi
Jacopo Gresleri
Jean-Pierre Chupin
Joel Chan
Judy Chen
Kamala Visweswaran
Kathleen Stokes
Kristian Kreiner
Kweku Addo-Atuah
Li Brian Zhang
Ma Haoyi

Marcella Arruda
Mark Linder
Matxalen Acasuso Atutxa
Merve Bedir
Mia Heide
Michael Alexander Ulfstjerne
Michael K. Jenson
Milton Garavito
Nadine Samaha
Neha Korde
Runa Johannessen
S. M. Kaikobad
Sabine Storp
Samir Harb
Sandi Hilal
Satria A. Permana
Sayon Pramanik
Sol Pérez-Martínez
Sony Devabkhatuni
Sowmya Raman
Tanjina Khan
Teodora Comanita
Vanessa Ma
Xiaohui Zhang
Yael Padan
Yanyu Cui
Zhang Tong
Zhang Yue

Preface

Introduction

In the autumn of 2022, as part of the preparations for the UIA World Congress of Architects 2023 Copenhagen, we invited Panel Chair and MASS Design Group architect Christian Benimana to Copenhagen to speak to our collegiate and students. In his introduction, he outlined the dramatic land use change in Rwanda following the country's population growth over the last 50 years. Pointing to the maps, he argued that we have passed a tipping point and that our given societal infrastructures cannot simply be extended or optimised to support this new situation. We cannot build 500 universities or 600 hospitals, he said, instead we need systemic change to rethink what a university is and what a hospital can be. We need to question how our institutions, infrastructures and communities can change in the way they address those in need and what access can be, and therefore also how architecture, its practices, embedded knowledge and products can be methods of instigating change.

The present proceedings presents six volumes examining the knowledge foundation for such change. As proceedings for the Science Track of the UIA World Congress of Architects 2023 Copenhagen Sustainable Futures—Leave no one behind, they contain a total of 296 papers investigating, showcasing and arguing for how change can be imagined across the built environment. By asking how architecture can help achieving the UN Sustainable Development Goals (SDGs), the presented papers collect the research- and practice-based results of a global community. Together they ask what the future of the built environment can be and how design as action and as knowledge can create new roles for architecture and the communities it serves.

This Preface starts with the articulation of our profound gratitude to the Scientific Committee and the community of submitting authors and peer reviewers that have been part of this effort. During the last two and half years, we have worked together with the Scientific Committee's Panel Chairs and Special Advisors to form a vision for the Science Track. The process has been an education, not only in our understanding of the SDGs, the transformative power of design creation or the wider societal role of the built environment, but also in keeping our minds open to the many positions that architecture can be thought through and its critical role in engaging different knowledge cultures and perspectives. We therefore start by thanking the 17 members of the Scientific Committee, the contributing 656 authors of the 296 accepted papers, the 1486 authors of the more than 750 paper submissions and the 536 peer reviewers that have all made this project possible.

Platform

The UIA World Congress 2023 Copenhagen starts with an ambition. Pitched in 2017, only one year after the launch of the UN Sustainable Development Goals, the central nerve is the articulation of the profound agency of architecture and how it plays an acute role in achieving the SDGs. In the congress, the Science Track is given a particular role. Initiated early in the planning process, the aim has been to place the Science Track at the heart of the congress in order to collect its underpinning knowledge foundation and shape its criticality through a broad outreach to a global community. Sustainability, like architecture, is a wicked problem. Its solutions are dependent on the way we ask, the methods we use and the contexts in which we work. To ask how architecture can be part of the dynamic fulfilment of the UN SDGs is to ask: who are the communities we design with and for, what is the knowledge we draw upon and how can its sharing change how we think about what our built environment can be.

One of the central drivers in our preparatory work for the Science Track has been the realisations of the blindness of the UN SDGs to the agency of architecture. The SDGs seek to steer behaviour both through impacting legislation and wider societal value sets. They establish priorities and galvanise efforts across communities by identifying targets and providing shared yard sticks in the form of indicators. In doing so, they inscribe a world view of its defining actors; the governmental bodies, industries and communities that can be leveraged upon to instigate change. And in this world view, architecture is strangely absent. At present, none of the UN SDGs declare targets that directly articulate architecture as a driver for change nor are there any indicators that evaluate its role. The built environment is only mentioned as a driver for resilient communities but without real value setting of the role of planning and design. This despite the extensive and complex impact architecture holds on human and non-human well-being; the way we live our lives, shape equity and use our resources.

For us, this realisation has led to the overarching aim of using the congress to build awareness. To argue for and demonstrate how architecture has the ability to afford change in the way we understand and construct the world around us and therefore how it as a situated practice engaging directly with both legislation, industry and the communities in which architecture *takes place* can become a direct way of effecting change.

Vision

The Science Track is formed around six panels of which this volume is one. The vision of the six panels is to articulate six differentiated perspectives onto how architecture can be part of achieving the SDGs while reinforcing their interconnectedness. The panels are in part mapped to existing fields while at the same time suggesting new. By bringing together otherwise fragmented knowledge across the breadth of architecture's research and practices, the aim is to bring together knowledge across research, practice and education to provoke new perspectives, new alliances and concrete action. In articulating the panels, the Scientific Committee asks pertinent and provocative questions that challenge the field and position the SDGs as active goal posts. These questions form the chapters of each volume asking how architectural knowledge creation can innovate the thinking, design and making of architecture.

- ***Design for Climate Adaptation***: With profound urgency, global communities are acting and adapting to the earth's changing climate. Our built environment, the most common habitat of humans, should interact with the earth's ecosystems and climates in a sustainable and regenerative way. 'Design for Climate Adaptation' emphasises people, multiple forms of research, knowledges and action for high and low-tech solutions that make buildings, neighbourhoods, landscapes, cities and regions regenerative, resilient and adaptive to climate change impacts.
- ***Design for Rethinking Resources***: Design shapes our world, from the places we live in to objects we use every day. As we grow more aware of the limits of our planet's resources shifting from an exploitative to a restorative, regenerative and circular design ideology becomes fundamental. 'Design for Rethinking Resources' examines approaches to resourcefulness in architecture; how sustainability challenges the foundations of our material practices and how they can change with it.
- ***Design for Resilient Communities***: A resilient community anticipates, adapts to and recovers from adversity. Climate change, the global pandemic and political upheavals in many countries have revealed social, economic and environmental inequalities that threaten communities worldwide. These fault lines disproportionately impact the poor, people of colour, the racially or ethnically marginalised and women. 'Design for Resilient Communities' encourages innovative solutions and facilitates the development of knowledge and skills necessary for adaptation and recovery.
- ***Design for Health***: Architecture and health are inseparable. From the direct design of hospitals and places for healing to the strategic design of infrastructures and city planning, architecture affects physical and mental health of individuals and communities. 'Design for Health' asks how architecture can reconceive health as a design issue, how land rights impact healthy living, how legislation, planning and building impact inequality and access to water and how single buildings and the civic

construction of hospitals, health clinics and community buildings can operate in unison with local environments and ecologies to create a safe and healthy space for all.

- ***Design for Inclusivity***: No individual deserves to experience space in a manner that is less safe, less comfortable or less accessible as a result of their identity or challenges. Sustainability, in its most holistic definition, cannot be achieved without a collective act. 'Design for Inclusivity' aims to critically define the constructs and categories of who exactly we are excluding, and why, in order to mindfully develop strategies to mitigate this exclusion.
- ***Design for Partnerships for Change***: 'Design for Partnerships' is about recognising the asymmetrical relationships between states, public spaces, civil societies and private domains to find new balances for the existing power structures. By challenging the ontology of universalism, it examines how architecture and the built environment can play an essential role in creating a ground for care through local governance, space making practices, imaginaries and scenarios of plural(istic) political, socially and ecologically sustainable futures.

Critical Positions

The two and half years of preparation has been an inspiring experience through which we have witnessed the power of architectural thinking in action—its interweaving of the critical and the creative ideation as well as its inherent inventiveness orientation towards the future. As part of the curation of this work, we have defined a series of critical positions by which to understand the correlation between architectural thinking and the UN SDGs. A first position has been to challenge the inherent anthropocentrism and perceived lack of hierarchy between the goals; the Tabula Rasa effect as Johan Rockström names it Rockström 2016. The SDGs have been criticised for failing to recognise that planetary, people and prosperity concerns are interconnected (Kotzé et al. 2022). In forming the six panels of the Science Track, we seek to position a rupture to the modernist axiom that the environment is situated outside of us. Instead, we understand the SDGs as a balancing between planetary and human needs which needs to be holistically addressed.

A second position is the critical appreciation that the SDGs retain an adherence to an underlying model of growth. The Science Track asks what the future practices of architecture can be, what the ethical roles of architectural design are and how architecture knowledge can create change in how architecture is produced both within and without of models of growth. It seeks to identify who the partners of architecture practice can be both through grassroot community action and through industry-based models.

A third position is the challenge of the embedded universalism within the SDGs. The SDGs maintain a universalism that is common to the UN system and underlies much of UN's work. However, this fundamentally modernist position of understanding sustainability as 'a problem to be solved' and placing agency with legislation leaves questions of agency, voice and power unchallenged. The Science Track seeks to incorporate this criticism through the panel calls and their associated sub-questions by provoking reflection on the perceived neutrality of architecture's own humanist traditions and insist on the query of how architecture is produced, by people and for people.

The challenge to universalism has also led to a review of the scientific practice of knowledge dissemination. The call for papers deliberately encourages exchanges and learnings across different knowledge and practice silos. This is effected through differentiated publication formats that include scientific knowledge production as well as design-based knowledge production, narrative formats such as oral history, visual essays, as well as dialogue-based exchanges and argumentative essays. The aim of these formats is to expand the possibility of transdisciplinary knowledge exchange and include voices that are not commonly part of academic and professional discourse.

The fourth and final position is to understand the SDGs as part of a changing world. The SDGs set out a 14-year-long project. Any project of that length needs to build in methods of reviewing its own fundamental value sets and core conceptual foundation. The intensifying and accumulating effects of climate change, the aftermath of the COVID-19 pandemic, the continued stress on the world's resources and the increasingly multi-partisan war in Ukraine have deep and unequal repercussions on global communities. To engage with the SDGs is to correlate the goals to a changeable understanding of both needs and means. It is to commit to a continual address of both the contexts and instruments of change-making. In the Science Track, our focus on the concrete and the actionable through presentations of cutting-edge research, real-world case studies and near future focussed arguments argue for a situated understanding of the SDGs. This emphasis contextualises the SDGs within the multiple and diverse practices of architecture as well as the disparate places in which architecture takes place. The perspectives, methods and means are purposefully broad. They seek to represent the breadth of the solution space needed for the systemic change needed. They also purposefully include different voices and different styles to make present the different actors, different knowledge streams and different institutions that create this change.

Perspective

The result is six volume proceedings tracking a wide and multifarious interpretation on how architecture can be part of achieving the SDGs. Across their individual chapters, we see a breadth of enquiries asking who the communities are, who the actors are and what the means of architectural production are. They ask how we can shape the methods of architectural thinking as well as their associated technologies, how they can be distributed and what is the consequence of their sharing.

The proceedings instantiates a moment in time. As research strands, they are part of larger trajectories of knowledge creation. Where our aim for the

World Congress is to facilitate new discussions and exchange enabling synergy across silos and geographies, it is clear that the full potential of this conversation is only just beginning. The World Congress coincides with the half-way mark of the SDGs. Launched in 2016 and with a projected completion date of 2030, we need to transition from a place of planning and speculating to one of action. The work of the Science Track is therefore marked by a sense of urgency. The desire is to define the effort of this work not in terms of their individual results, but more as a launch pad for future exchange and collaboration. We hope that what is created here is a community of dedicated actors all with a shared stake in the well-being of future generations. Our hope is that the legacy of this project will be that we can retain this commitment and grow its stakeholders to mature these propositions into actionable change.

We profoundly thank the Scientific Committee for their immense effort and profound engagement in shaping the Science Track. Thank you to Billie Faircloth, Maibritt Pedersen Zari, Carlo Ratti, Anna Rubbo, Juan Du, Arif Hasan, Christian Benimana, Magda Mostafa, Ruth Baumeister, Sandi Hilal, Merve Bedir, Katherine Richardson, Chris Luebkeman, Thomas Bo Jensen and Camilla Ryhl.

Copenhagen, Denmark Mette Ramsgaard Thomsen, General Reporters
Martin Tamke, Alternate General Reporters

References

Kotzé LJ, Kim RE, Burdon P, du Toit L, Glass L-M, Kashwan P, Liverman D et al (2022) Chapter 6: planetary integrity. In: Sénit C-A, Biermann F, Hickmann T (eds) The political impact of the sustainable development goals: transforming governance through global goals? Cambridge University Press, Cambridge, pp 140–171

Rockström J, Sukhdev P (2016) The SDGs wedding cake—Stockholm resilience centre. Retrieved from https://www.stockholmresilience.org/research/research-news/2016-06-14-the-sdgs-wedding-cake.html. Accessed on 04 Apr 2023

Partnerships for Change

Collective Thinking and Writing

Husam Abusalem, Kweku Addo-Atuah, Wiebke Ahues, Syarifah Ismailiyah Alathas, Mª Angeles López Amado, Matxalen Acasuso Atutxa, Andrea Terceros Barron, Pernille Maria Bärnheim, Merve Bedir,Nezih Burak Bican, Camillo Boano, Edoardo Bruno, Angelo Bucci, Esther Charlesworth, Wei Chunyu, Irina Teodora Comanita, Brandon Cuffy, Geetanjali Deshmukh, Francisco Díaz, Billie Faircloth, Valeria Federighi, John Fien, Camilla Forina, Maria Camila Gomez, Gizem Guneri, Magdalena Haggärde, Xu Haohao, Karin Hartmann, Erika Henriksson, Sandi Hilal, Yuming Hou, Marie Frier Hvejsel, Ariel Hernán Jacubovich, S M Kaikobad, Akshatha Ravi Kumar, Sonia Puente Landázuri, Felipe Lanuza, Sofia Leoni, Joanna Lesna, Joel Peter Weber Letkemann, Tommaso Listo, Leslie Louie, Gisle Løkken, Vanessa Ma, Natalia Brener Maceiras, Binita Magaiya, Ilya Fadjar Maharika, Kasper Martlev, Luis Carlos Mestrinho, Vasimalai Mp, Daniel Huertas Nadal, Sarah Naarden, Yiorgos Papamanousakis, Elena Parnisari, M Elvira Carregado Pazos, Laura Otero Pedraza, Satria Agung Permana, Elamuhil S, Arsh Saluja, Nadine Samaha, Mary Kristine Segovia, Chensi Shen, Ouyang Sheng, Manindra Shrestha, Caroline Sohie, Keqing Tang, Carolina Vasilikou, Aurea Soto Vázquez, Teresa Táboas Veleiro, Chunyu Wang, Yifan Wang, Ryan Welch

It all started when...

Sandi and Merve were invited to chair a panel with the title "Partnerships for Change" at the UIA congress in Copenhagen in 2023. This introduction takes its point of departure from this same invitation.

Sandi

is Living in Europe, and was raised in Palestine she is constantly faced and challenged by the description of integration, which is portrayed as if it is something that will simply arrive, sooner or later. But in her experience, the way integration is understood within the framework of European Nation States is more like a race towards behaving as an ideal guest; an ideal that few can ever attain. When someone is asked to behave as a perfect guest in Europe, their power to act as a host is taken away. She decided that she was no longer willing to accept the role of the "eternal guest" that she have been allocated; she decided not to run that race. Instead, she wanted to explore other ways of understanding, defining, and challenging the dominant notion of integration, and look for alternatives to the binary of inclusion and exclusion by asking the questions:

Who has the right to host? Who is told they need to behave like a perfect guest? **How can we analyze the power of hosting as a means of becoming a visible and demanding agency?** *When she was invited to be a member of the UIA World Congress 2023 CPH Scientific Committee she was determined to act upon this* **right to be a host** *rather than accept to be included as an eternal guest to add to the dominant frames of other needed voices and different backgrounds.*

has moved between the Netherlands, Turkey, and Hong Kong during this project. Her continuous question is about belonging and ownership. She accepts herself as the inevitable result of several imperfect but invasive institutions of nation-state, religion, tradition, and family, and realizes that she has evaded or negated those frameworks through collectivity. She studied architecture at Middle East Technical University, a place and time that has taught her to practice unlearning and relearning, and asking the question of herself, discipline, work, and what constitutes life.

Merve

In accepting to chair the UIA Congress Merve was interested in ***deforming the current academic and scientific frameworks of architecture.***

Sandi and Merve ask:

What does it mean to accept to chair a panel in a universalist Western Eurocentric platform such as the UIA? What role can the two of them play in such a platform? Are they feeding on the glory of one dominant universal narration of the world, or do ***they have the possibility of creating ruptures in this scientific congress system by questioning its very foundations?***

The first thoughts..

They both believe in the power of storytelling and think that if they would only turn this invitation into a story that they are living, with all its complexity and contradiction, then this might be one possible way of acting on such a platform rather than passively declining the invitation. The beauty of storytelling is that you do not have to control the result and create a pre-framing so nothing would go wrong. Storytelling has the ability to open up for the unknown and to simply narrate it. To narrate the story is for them a way to escape the frame. By narrating the story, they are sure that they would already be reframing the frame and creating another story, and hope that this story will create other stories.

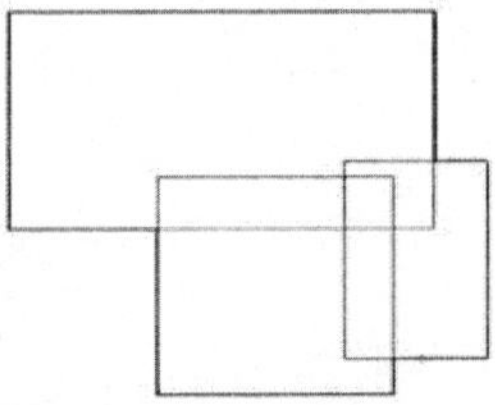

As Merve and Sandi prepared for the congress call, they questioned the same foundations of how architecture is taught and the way it is perceived today by proposing to reframe, rethink and reimagine different meanings for some words that are heavily used in the environment in which they both operate, words such as community participation, land, agency, commons and design. They thought that these words are usually used to frame and many times to even colonise the discourse. They thought that many authors from different places can propose different meanings to these words, creating multiple frames in constant transformation and that this might be one of the ways to reframe and possibly to decolonise these words dominating the discourse. They indeed hoped that the partnerships for change would potentially meet at the threshold of the frames rather than inside these frames. By doing so, many authors could feel the agency to begin to redefine such foundational words with multiple frames and act upon them, rather than aiming to make the universally dominant definitions of these words more inclusive.

Time passed...

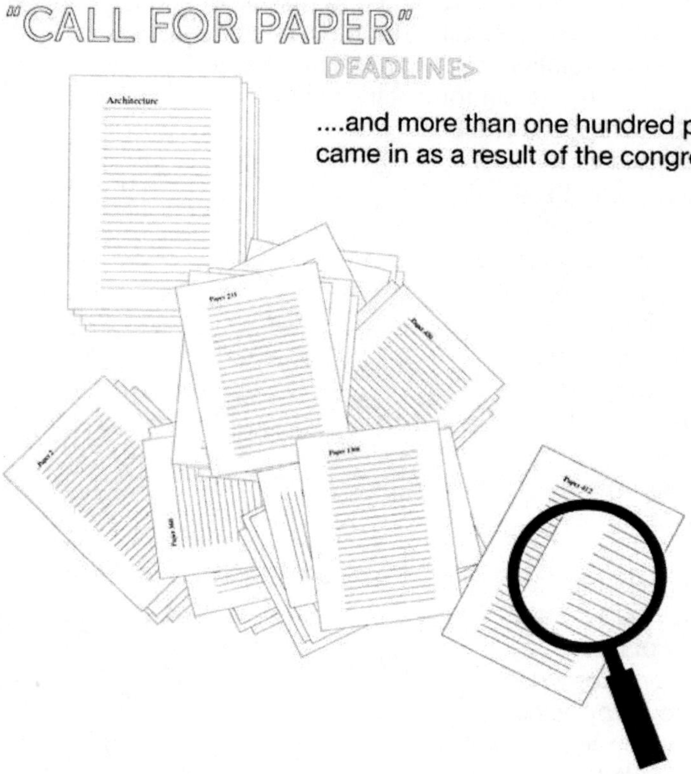

....and more than one hundred papers came in as a result of the congress call.

These papers needed to be peer reviewed as requested by the system. Both Sandi and Merve were very troubled by the need for peer reviewing. The review for them was the surveillance system to maintain that one frame and keep that frame dominant and in control.

What to do? Were they willing to do it? Were there other ways to still work from within and against the system? Were they naive to think they can do it? It was a difficult moment when they both thought that it might indeed have been a very bad, naive idea.

A moment of crisis....

The papers arrived **in the middle of this crisis** and they decided to go through these papers... This same crisis turns into ***euphoria***. This same dominant frame brought together

so many people that are eager to build as many frames as they can but they never **found a platform that can bring them together.**

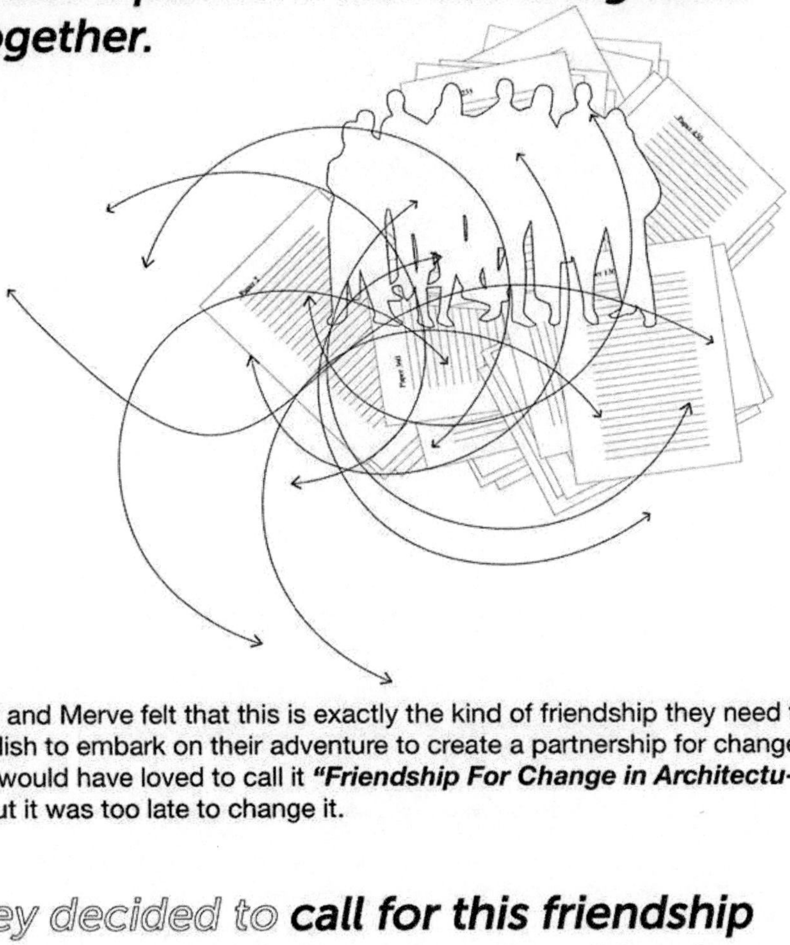

Sandi and Merve felt that this is exactly the kind of friendship they need to establish to embark on their adventure to create a partnership for change. They would have loved to call it ***"Friendship For Change in Architecture"*** but it was too late to change it.

They decided to **call for this friendship** though, and this is where the story began.

Merve and Sandi invited everyone in the group for a first meeting to express their frustration with the practice of both chairs as gatekeepers and peer-reviewed congresss, and they proposed to write the introduction to this book collectively.

Dear paper author(s),

We want to thank you for submitting your paper to Panel 6: Design for Partnerships for Change. We are glad to let you know that your paper has been Accepted with Changes.

As part of the peer review and revision process we want to invite you to participate in an online meeting on

Tuesday December 20
at
Colombia / US (east coast): 7am – 10am
Bolivia: 8am – 11am
Chile / Argentina: 9am – 12am
UK: noon – 3pm
CET / CAT: 1pm – 4pm
Turkey: 3pm – 6pm
India: 5:30pm – 8:30pm
Bangladesh: 6pm – 9pm
Chongqing / Philippines: 8pm – 11pm
Australia: 11pm – 2am

On this link
https://kglakademi.zoom.us/j/63816986344?pwd=OGN6MWREVzJZbFB-GT1ppRXFGbkROQT09

We have scheduled three hours but may be able to end earlier.

The aim of the meeting is for all authors of papers that have been Accepted with Changes to collectively discuss the intention of Design for Partnerships for Change with Panel Chairs Merve Bedir and Sandi Hilal.

This collective dialogue will help inform the forthcoming revision of your paper, that must be finalised and submitted by January 15, 2023. We want to explain the motivation for our decisions in selecting papers to be invited for participation and revision. We also want to encourage the revision process to be a collective act shared amongst the authors of papers Accepted with Changes.

Furthermore, we want to invite all authors of papers Accepted with Changes to partake in the formulation of the introduction to the Springer proceedings publication.

They explained to the group that during the process of reading the papers, they took into consideration the question of inclusion. Who produces knowledge and who has access to that knowledge and the constituents of the knowledge produced? The bodies that are intended to be included in academia are expected and trained to learn the language and codes in order to be included in the frame. Instead, they wanted to open the door for papers that prioritise life experiences, storytelling, and oral knowledge.

One of the door codes Sandi and Merve questioned was the practice of quotations. The other one was language, which suffers from translation and neutrality. They were looking for writing that was open to the unknown. Their priority was about the clear positioning of the story and the location where the story takes place. With this first meeting, they invited the group to write the introduction to the panel collectively. They insisted that they were not going for perfection, they were not going for writing a book with no errors and encased in perfection, but rather for building a community that will help change and challenge the systems, that will dare to take a position and reframe the frame in order to provide access to knowledge.

One aspect Sandi and Merve realised that they had never thought about before was that this same UIA platform brings together not only people that are interested in belonging to the frame but many that feel the urge to disrupt by changing the frame rather than completely abandoning it.

A moment of euphoria....

They began to feel the desire to cultivate this kind of partnership, to be able to change while understanding the need to disrupt as a means to create an agency of belonging.

Pernille Maria Bärnheim
Copenhagen, Denmark
12.00 p.m.

Daniel Huertas Nadal
Bogotá, Colombia
6.00 a.m.

Elamuhil S Madurai
India
4.00 p.m.

Angelo Bucci,
Pescara, Italy
12.00 p.m.

Ariel Hernán Jacubovich,
Buenos Aires, Argentina
8:00 a.m.

Carolina Vasilikou
London, United Kingdom
11.00 p.m.

Chensi Shen
Tianjin, China
7.00 p.m.

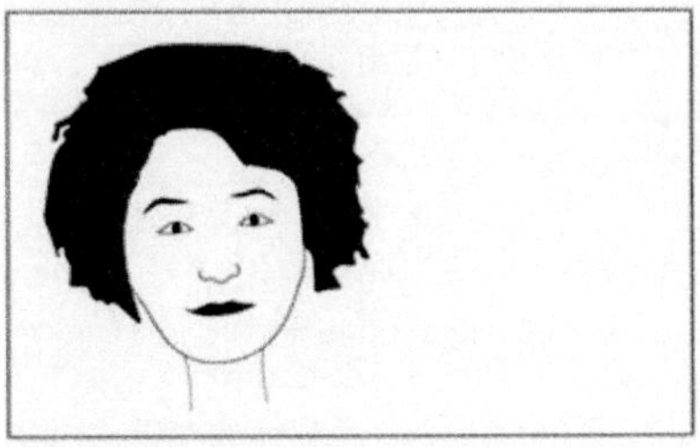

Merve Bedir
Hong Kong
7.00 p.m.

Elena Parnisari
Porto, Portugal
11.00 p.m.

Hou Yuming
Shanghai, China
7.00 p.m.

Erika Henriksson,
Gothenburg, Sweden
12.00 p.m.

Kweku Addo Atuah
United States
6.00 a.m.

Keqing Tang
Shanghai, China
7.00 p.m.

Luis Carlos Mestrinho
Guimarães, Portugal
11.00 p.m.

Mary Kristine Argao Segovia
Pasig, Philippines
7.00 p.m.

Nadine Samaha
Beirut, Lebanon
Melbourne, Australia
10.00 p.m.

Satria A. Permana
Indonesia
12.00 p.m.

S M Kaikobad
Dhaka, Bangladesh
12.00 p.m.

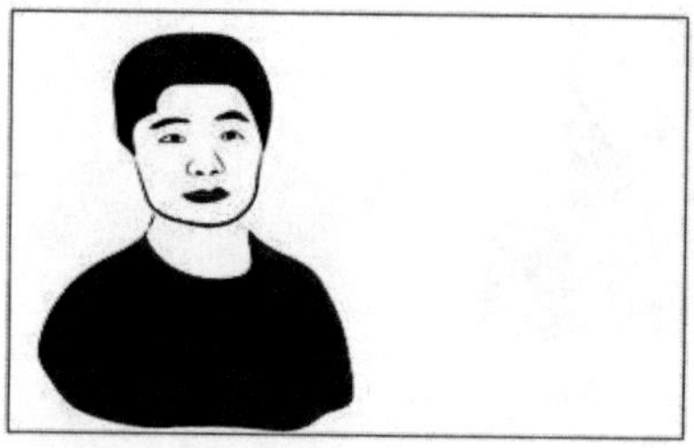

XU Haohao
Changsha,China
7.00 p.m.

Yifan Wang
Zürich, Switzerland
12.00 p.m.

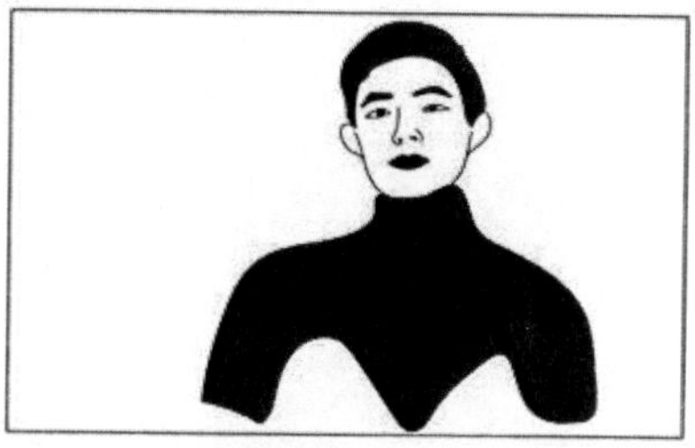

Chunyu WANG
Chongqing, China
Beijing, China (home)
7.00 p.m.

Irina Teodora Comanita
Copenhagen, Denmark
Bucharest, Romania
12.00 p.m.

Sandi Hilal
??
12.00 p.m.

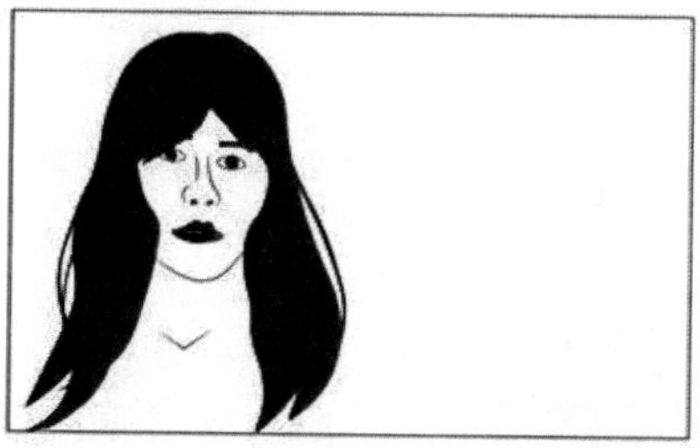

Sofia Leoni
Turin, Italy
12.00 p.m.

Why do we write collectively?

Angelo:

The challenge of a panel entitled "Partnership for Change" lies precisely in the word 'change'. What kind of change? Who is entitled to choose which change?

As architects, we have to avoid believing that we can easily define the change to be implemented, to think that we can find a solution, and that this solution can be defined by a small group of people, even by a single person. Searching for the solution is to define a precise point at which to arrive, but changes need time and the future is not predictable. The only thing we can do is to define a direction, a plausible scenario that will be used as a compass rather than a map.

The ambition of this experimentation is to start tracing a path toward a future scenario based on comparison, engagement, plural voices, decision-making, shared objectives and collaboration. A scenario in which the project is no longer the work of a single person but derives from an open process that different voices can access, each with its own cultural, perceptive or empathetic contribution, helping the process and even becoming a fundamental part of it.

Merve:

The reasons for writing the introduction collectively are to meet and learn from each other, to problematize the prevailing narrative and the context, and to think and build knowledge collectively. There is confidence to gain in writing collectively, helping each other, nerding, perfecting and finding ways to help each other. By writing collectively, we test reframing and its adaptability to continue our endeavours in the long term.

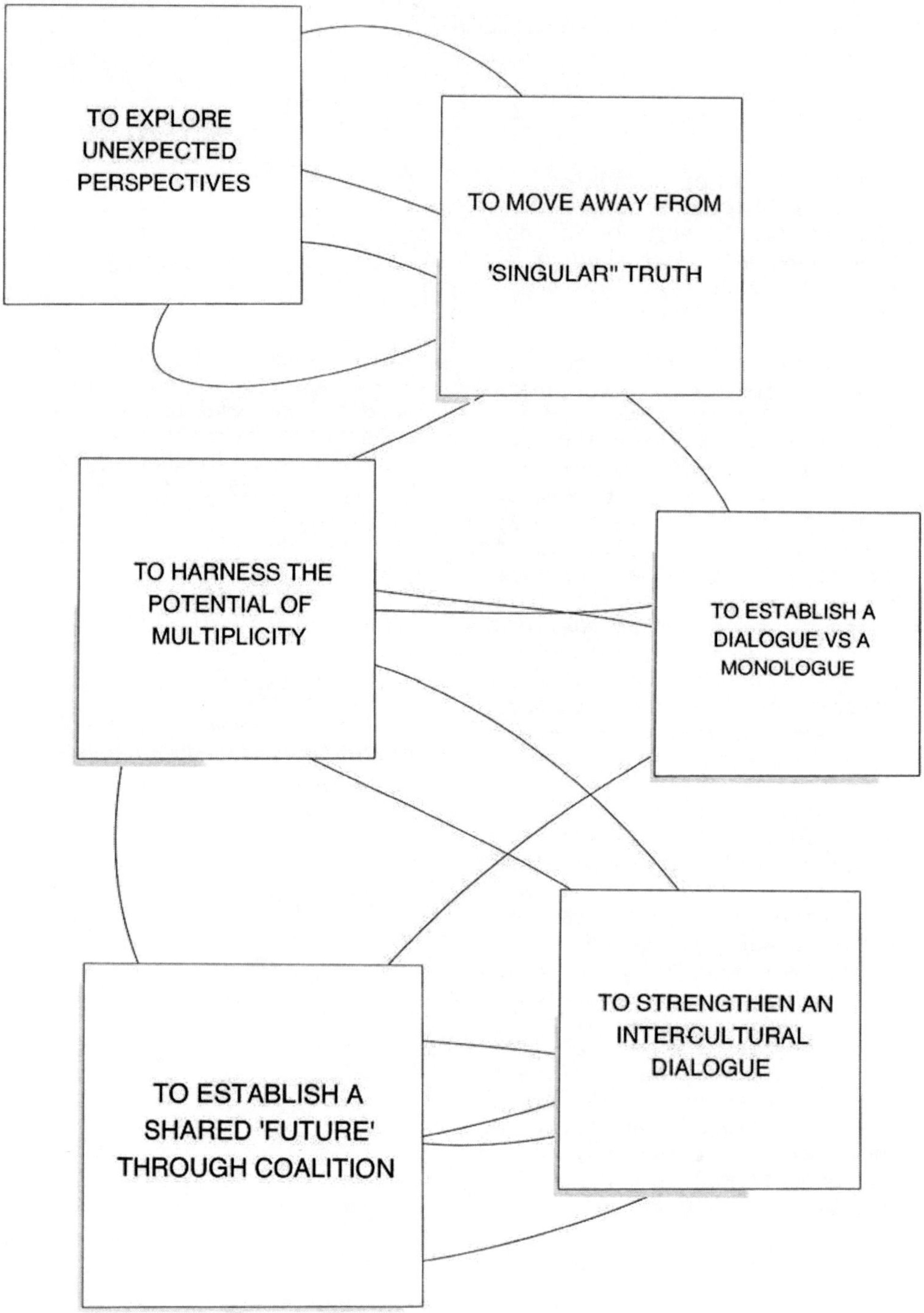
TO EXPLORE UNEXPECTED PERSPECTIVES
TO MOVE AWAY FROM 'SINGULAR" TRUTH
TO HARNESS THE POTENTIAL OF MULTIPLICITY
TO ESTABLISH A DIALOGUE VS A MONOLOGUE
TO STRENGTHEN AN INTERCULTURAL DIALOGUE
TO ESTABLISH A SHARED 'FUTURE' THROUGH COALITION

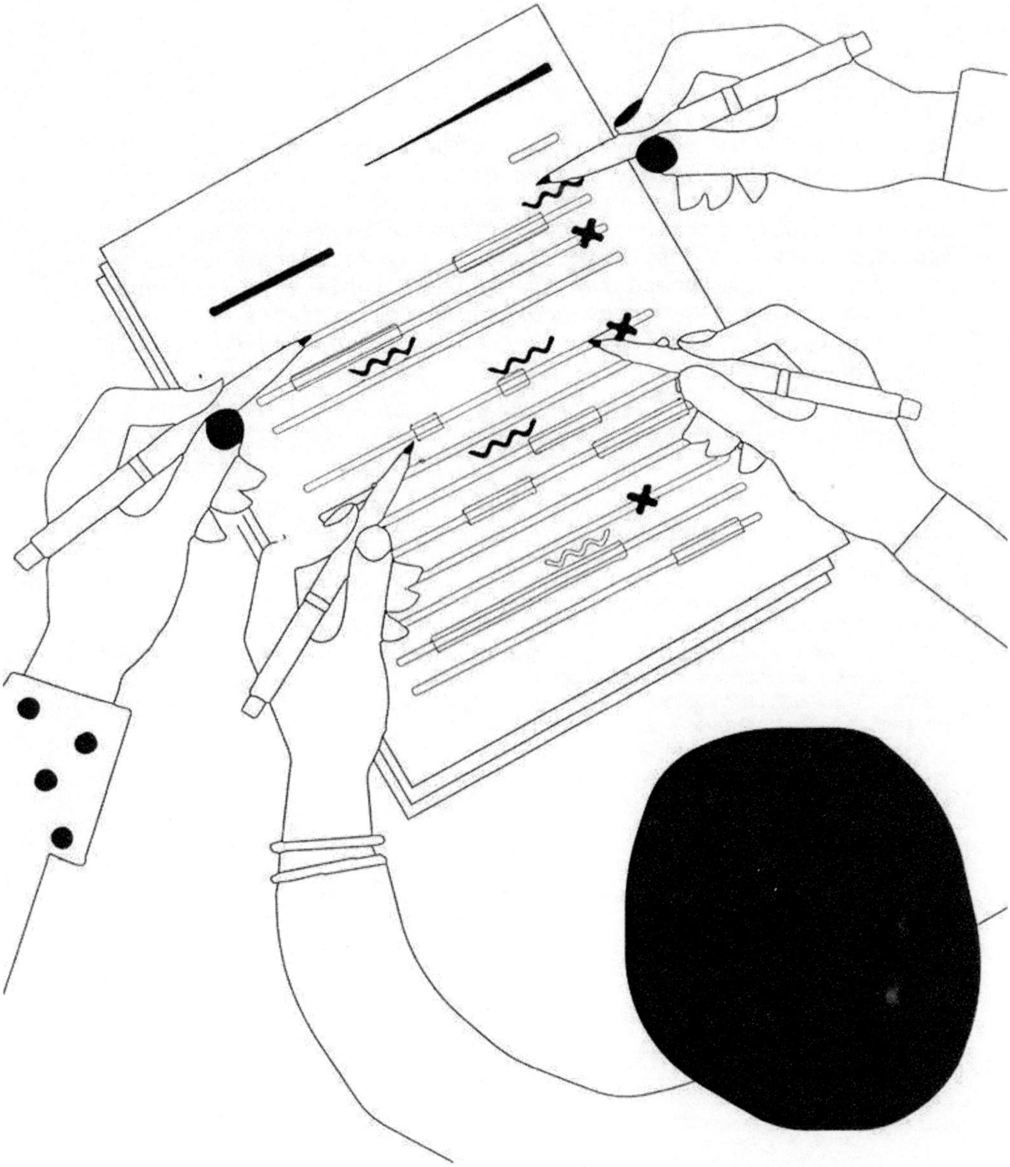

What are our rituals

Sandi:
We decided to create our own ritual of writing, the ritual of waking up every morning and acting upon the document, reading each other's words and reacting to each other's concepts.
We believe rituals give room for collective writing, that unpredictability and unplanning forge a way for escaping the preplanning and give space for the unpredictable to take place and help us transgress our boundaries in collectivity. It helped us to dare together, feeling protected by these morning rituals that we had created.

from: Sandi
to: ALL
object: COLLECTIVE RITUALS
monday, january 30th, 10:26 AM/CET

Good morning dears!
This morning I waked up with low energy, and thought that I had no energy to share my thoughts in our collective document, I though decided to go and read the document again and it filled me with energy and joy! it felt very good. This is what community sometimes is able to do .. when your own energy is not enough it provides you with some needed one.
Thank you all for being part of this genuine process! if nothing it is bringing me a great hope in a moment where news from Palestines is deeply depressing.
Very warm regards
Sandi

from: Laura
to: ALL
object: reply: COLLECTIVE RITUALS
monday, january 30th, 17:26 AM/CET

Hi from Bolivia!

I am only catching up with everything now! I haven't been able to join any of the previous meetings, I apologize.
I was getting married the first week of January, that's why.
Hosting everyone in the jungle hasn't been easy!
I have been reading the shared document, I have to say, what an amazing group of humans !
I've just added the personal bio to the doc to begin with.

See you on the 1st of February!

Laura

NS: As per Maslow's hierarchy we should maybe be self actualising, and spirit transcending understanding our essence, values and potential. Similarly Carol Sanford the great regenerative practitioner states that it's crucial to focus on inner work before expanding th router work as developing and nurturing our relationships, understanding that we are nested in wholes than starting intervening in the fields and nodes.

(AB): Seneca: "There is no favorable wind for the sailor who does not know where to go."

~~The challenge of a panel entitled "Partnership for change" lies precisely in the word change.~~
~~What kind of change? Who is entitled to choose which change?~~
~~Coming from a culture based on the precepts of Functionalism, Rationalism, and the Modern Movement, a As architects, we have to avoid believing that we can easily define the change to be implemented, to think that we can find a solution, the only solution and that this solution can be defined by a small group of people or, even, by a single person.~~
~~This approach, based on problem-solving, has had to deal, over time, with the complex dynamic system we live in and in which this type of approach fails to bring about real changes. In these highly complex systems, if you want to implement a change, you have to implement open processes that generate "quality events" that begin to direct the system towards a "preferable state".~~

Searching for "the" solution is to define a precise point where to arrive, but changes need time and the future is not predictable. The only thing we can do to gain a change is to define a direction, a plausible scenario that will be ~~used as a compass, instead of "the" solution as a map.~~usable
~~Furthermore, defining a scenario is impossible for a single person because each one of us has his/her own bias - due to the culture he or she comes from, due to the educational path, etc. Working in collectives helps to avoid mistakes in defining present and future scenarios and, moreover, helps to have a broader view of the topic we are focusing on.~~
~~Although a group has always boundaries, to be a multitude of cultures, backgrounds, and points of view, might allow for avoiding mistakes usually made by single persons.~~
-
~~The idea to structure a group, involved in a col~~
~~process, arises precisely from this reflection.~~

Nad ne Samaha
11 57 24 feb

The context in this nstance s a conversation on how to think and write collectively in a global virtual space Yet our experience in this context has been transcended by our exchange of ideas and sharing of our stories

Merve Bedir
08 29 23 feb

Angelo I would suggest to take out the individual quotes and keep the voice in the collective that we form altogether My other suggestion would be to shorten th s part a little bit if it s possible

Angelo Bucci
09 26 23 feb

Ok I can work on t in the afternoon feel free to suggest changes and I will check later

Who are we?

Merve:

We acknowledge that we do not easily fit into the established physical, economic, social and gender patterns. We think with not just humans but with the environment too. We are those in migration, creating our own community of practice. We are the grassroots voices, people on the ground, are engaged with the land. Hardly any of us have their main language as English. We are the migrant, woman, LGBTQAI+, marginalised person that keep the door open.

Sandi and Merve decided to initiate the discussion by thinking of one of the most used words in architecture: community. Indeed, ours was a very intense, heated and extensive, discussion. We spent many hours debating the way we use the word 'community' in architecture. We never finished debating and we felt the impossibility of agreeing on one definition to describe what the word community might mean. We might have agreed that a question in a congress room such as "Where is the voice of the community?" might be highly problematic.

Collective thinking through oral and written conversations

Conversation on partnership and communities

Sandi:
Working within a UN agency (UNRWA) for long time, I learned from many refugees that when you speak of community in an abstract manner, you put yourself in a powerful space of hearing the voices of the people and hearing the unheard. The moment we think of community as one entity, we almost overcome the complexity of seeing them as humans. "Where is the involvement of the community?" is a typical question that shows the inability of recognising the power structures and the power imbalances within that same community.

Teresa:
Community as a term (where is the community, the power battles?) creates new colonialism. By mentioning a community from a superior position, do we perpetuate the same power structures?

Nadine:
Beauty creates community. Engaging and acting is important and community arises and develops. You start with action and energy. We don't need to define or seek it absolutely.

Karin:
When we talk about community, it's maybe not important to define it, but to hear every voice and accept that it changes. In Germany, the participation starts but then the loudest voices reflect the community, and it's usually the voices of men who have a lot of time. It's important to be aware that every voice is heard.

Geetanjali:
Data about communities empowers them. Data about slum communities, by collecting data and holding talks and focus-group discussions, when we do that (ourselves in the community), people really come forward and voice their opinion.

Nadine and Luis:
We need to change our anthropocentric view of the world, which is leading to profound changes in Earth systems and significant loss of biodiversity, and replace it with a more eco-centric focus, where humanity is no longer exclusively exercising forms of dominion over nature, but exploring other forms of relationships on different terms. In this sense, the idea of partnership conceived as the state of being a partner would include not only people and organisations as 'partners', but also non-human actors. As animistic as this proposal may seem at first sight for Western thought, we recall that several cultures take this position as the cornerstone of their relationship with the planet. Thus, one possible path is to broaden the notion of collectivity to include humans and non-humans, which will eventually require the decentralisation of humans in design.

Teresa:
What defines a community? Sense of belonging? Territory? A common goal? A group of people? How do we establish the basis?

Merve:
If there is a community, it is never the same from one time to the next.

Sandi:
Community is always framed. There is a demand to define the community, and those who demand its definition don't imagine themselves within the community.

Gisle & Magdalena:
If there is an inside, there is also an outside. By defining a community you also define who is not part of it. Our concern is more about actor-network thinking, which is more open and beyond the immediate perception of community.

Burak:
Who are the members of the community? When are they together and when are they not? What are their aims and motivations to gather? How do they participate in coming up with targets and end goals? If we ask such simple questions, maybe there are different definitions of the framing. In our case, the means of participation are important. We talk about power structures and how to engage and be active in a community.

Teresa:
Community has positive and negative meanings. When words are overused, they lose their meaning. Currently, community means nothing to me.

Sandi:
Why are we disturbed by this term? People use it thinking that it is still necessary.

Merve:
Why don't we pluralise the meanings, definitions. We don't have to consolidate a single definition at a certain time.

Teresa:
Collective then is what I prefer to use, instead of community, because community always includes a powerful group.

Ariel:
Community is not homogenous, as if they know what they want. Groups are always changing, and usually, they don't know what they need. Collective groups, what they want—we work around controversy of territory, and around controversies, actors will gather. Then we introduce ourselves as architects to bring architectural capacity to the network. Community doesn't work as a concept. We don't limitate ourselves to an idea of community. A group of actors join with non-human actors, like things, and have the capacity to make other actors act, to transform and intervene in the network.

Sarah:
Community is about a territory. Community of practice refers to a position of learning, and agreement/disagreement, friction points, could be confronting and memorable.

Teresa:
Agreeing with @Gisle & Magdelena; the question is whether the word 'community' means inclusion or exclusion.

Angelo:

Following up on @Burak's question, the difference between a community and a crowd is about their idea of the future, of the development to achieve. From an evolutionary point of view, communities start from common perspectives of the future. To feel ourselves part of the construction and development of this future engages people, makes them aware and propositive, ensuring participation in defining policies and future changes.

Binita:
Community is continuously changing based on the future that the community wants. People involved will feel themselves part of the community and the future scenario. The mistake we are making is trying to define a community, as if it is locked in one position. Let's think of the future; this is the best way to create community.

Karin:
We all belong to different communities. Sometimes, we feel like a host; sometimes we feel we are guests, depending on the sense of belonging we might have in the specific situation. However, in architecture, the term 'community' is used differently and with more distance, merely as a good argument for a planning decision.

Gisle & Magdalena:
The concept of community presupposes some kind of common interest within a group, and in our opinion, it is rather ambiguous or even dubious because it can easily become the means of controlling the individual, also because someone will inevitably define the purpose and values of the group or 'community', which will not necessarily be in the interest of everyone. Rather than defining things and people in closed categories, we believe there is a need for finding ways to work with possibilities in open and unbiased ways.

Daniel:
The word 'community' is always suggestive but never fully defined, holding out the promise of containing the values of interaction, mutual support and commonality. Community is often spoken of as an absolute value when what is truly profound is the collective, the set of actions that defines those values. It is more interesting to think about the collective than the community.

Teresa:
Maybe if we have common goals and challenges in the world, we should shatter the term 'community' because this term is limited by its own boundaries.

During this conversation, Merve and Sandi realised that conceptualising the unknown will be imperative for such a panel. They understood that framing the future work should not be an abstraction that represents what people can't reach. The idea of future should not be a limitation to externalise ourselves from the questions we deal with. Instead, future should be something to embrace as a productive and generative unpredictability. Therefore, it is a fertile terrain for creative research in the present.

Future is a way to talk about doubt, failure and our uncertainties, rather than being framed in an expected output that looks safe and certain.

Conversation on what we would like to change collectively

Sandi:
When I studied architecture, I was encouraged to shine as the only star in the sky. One name, one building, one architect is the aim of many in the world of architects. As a teacher, I inform my students at the beginning of our experience together that they will be graded collectively, and that the collectivity we are building in our class is one that will validate them based on their ability to create collectivities rather than to compete, one against the other. In the course of the collectivity building in the class, driven by the grading system, students begin to work with each other, support each other's projects, each contributing with their knowledge. Most of the time, the final result is very complex and inclusive. I would like to live in a society where my work is validated and acknowledged based on my ability to contribute to the collectivity rather than on my ability to compete and alienate other members in my community in order to shine alone.

Nadine:
How can we change our mindset from working competitively to working collectively? This could be the departure point. It is interesting how First Nations people gathered in the past, always in a circle so everyone could participate. The Romans broke that circle and created the amphitheatre in a semi-circle. Our modern society created parallel lines that can never meet with a rectangular stage where spectators and presenters are separated. Collective engagement vanished and the idea of stars and heroes started to rise.

Nadine:
Yearning to connect with a wider architecture community, Sarah and I tried to put words to what we are currently working on and what we wanted to achieve in the future. It's that need for deeper connection that made me become involved in this collective writing. As an architect practitioner, where we are choked with constraints and regulations, I was becoming used to being framed and having a box-ticking formula. The first meetings were frustrating for me as I was keen for a structure, or a framework. As the meetings progressed, I started to feel more in tune with the exploration of the meaning of collective writing as more of a virtual chatting connection. I started to sense the shedding of some of my colonised skin, trying to unlearn but feeling happier in a wider conversation and letting this exploration evolve.

Merve:
I have also been coming to this question myself for a while now, from a perspective of the discipline. For a while now, I have worked at the edge of institutions, along the boundaries of communities, at the borders of countries and cities. I have always questioned and tried to get out of the vocabulary that is conceptualised by the formal and the discipline; I have collaborated with non-designers. What I realise though is that I often need or want to come back to my discipline to borrow material(s) and craft(s) from it. If we are talking about a community of architects/-alike, then I am curious to know what are the materials and crafts that we can borrow from our discipline(s), as well as those that we can contribute from our non-disciplinary belongings?

Leslie:
@Sandi, I agree with your sentiments. In the paper we submitted, we focused on trying to highlight the value of leadership and expertise in a population that is normally ignored in supply chain governance. In the US at this time, many are trying to address this issue but the main approach is sort of adopting a heroic consumerist attitude, which I think is detrimental to the cause. I worked for a fair-trade non-profit that focused on creating opportunities for artisans in impoverished communities. That professional experience, as well as my personal history as a second-generation Filipino American, has led me to believe that the key to addressing social issues and driving change is community. Rather than focusing on a heroic narrative when addressing an issue of justice, I think our industry needs to learn how to be better members of the global community. In this type of work, we are not the cause of change but the supporters of change. Growing up, I learned the stories of my family, who experienced extreme poverty. It was not the Peace Corps that came to our family's village that changed our lives; it was our family, and it was our community that allowed my mother to escape poverty. Many of my friends have similar stories that reveal that it isn't heroism that saves people from modern slavery or child labour. Resilient communities are what change people's lives. I also want to live in a world that values collaboration over competition. I believe that will be the key to driving true change.

Nadine:
Thank you, Leslie, for sharing your story. I went to a healing ceremony on Australia Day as a few Australians feel that this day is more a day of mourning for the First Nations people: the Aboriginals. They call it 'Invasion Day' as they mourn their history. During the ceremony, Aunty/Professor Dianne Kerr, a Senior Wurundjeri Elder of the Wurundjeri people and knowledge keeper from the Ganun William Barak clan, spoke eloquently. Instead of being bitter, despite all the trauma that she and her ancestors suffered during colonialism, she was more interested in seeing how to move forward and be more inclusive. Her words resonated with me as she feels as an elder she is responsible not only for her family and clan but for all newcomers to her land.

We don't need more heroes or supposed saviours; we as individuals need to assume responsibility to create a better environment for our children and their offspring. One has to take the lead from the bottom up and never get trapped in being the victim.

Francisco:
I think the question of what kind of change is key, and it's at the root of the discussion this panel is trying to raise. The underlying assumption is that we are not comfortable with the world as it is now or with our role in it. Thus, we are looking for ways to redesign or reframe it. However, we are doing so with the few tools we have in academia: gathering thoughts and writing. That's a much-needed first step, and we must take it. Yet, we can also take the opportunity to envision what comes next: what worlds can we imagine from our vulnerable yet privileged position, what alternative worlds can our conversations open and how can this partnership help design new worlds?

Angelo Bucci:
@Francisco, I totally agree with you. It's a matter of fact that to change something, we need to understand what kind of change we are pursuing and what this change will generate. This kind of collective experiment might be helpful in trying to find out the changes that are possible to achieve in our small community, first of all, and to show how collective thinking may be useful to break some rules and reach a higher target, a broader view of the problems and a wiser approach to issues, and, most of all, to understand that we need collective thinking and collective designing as a tool to face the dynamic, complex system. In my opinion, our attempt should be seen as an attempt to influence a sub-system, the academic one, to achieve a first step toward the progress of the entire system, and we are doing so by trying to create a high-quality intervention also recognised as controversial and antagonistic.

Nadine:
How can we explain to our readers that they will not find answers in this story but more questions and doubts? We believe that to relearn how to share doubt and think about it collectively is what we need as a scientific community in this moment of history to productively and collectively perform against the frame as ways to continue transforming and reframing. A river flows and forges its path though new grounds without following a specific route structure. As in Khalil Gibran's poem about the river, it is only when the river reaches the ocean that she realises that she is the ocean.

Daniel:
I like to think that we are waters of the same river, claiming to join and travel together.

Nadine:
Khalil Gilbran's river poem called "Fear" suggests that the river is a metaphor for surrendering to the flow of currents and embracing the unknown. The loss of identity when disappearing into the sea of collective flow is also an expansive process and not necessarily a dissolving or diluting act. Indigenous Australians observe that the most robust and diverse species are at the interface of freshwater and saltwater ecosystem challenges.

It is said that before entering the sea
a river trembles with fear.
She looks back at the path she has traveled,
from the peaks of the mountains,
the long winding road crossing forests and villages.

And in front of her,
she sees an ocean so vast,
that to enter
there seems nothing more than to disappear forever.

But there is no other way.
The river can not go back.

Nobody can go back.
To go back is impossible in existence.

The river needs to take the risk
of entering the ocean
because only then will fear disappear,
because that's where the river will know
it's not about disappearing into the ocean,
but of becoming the ocean.

Khalil Gilbran, Fear

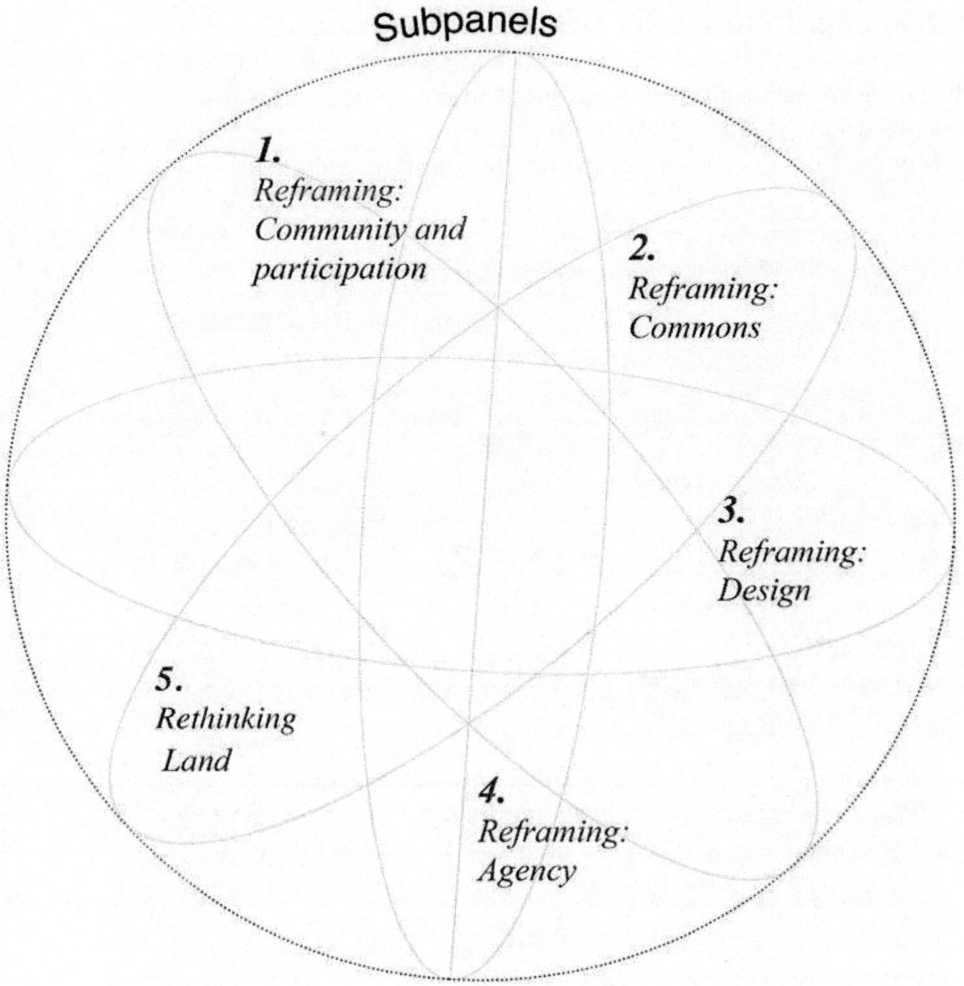
Subpanels
1.
Reframing:
Community and
participation
2.
Reframing:
Commons
3.
Reframing:
Design
4.
Reframing:
Agency
5.
Rethinking
Land

1. Reframing: Community and Participation

Sandi and Merve asked a series of questions to prompt responses from different researchers, practices, and places.
In Re-framing Community – Participation the questions were:

How to reframe the understanding of communities beyond who is the care-giver and who is care-receiver, who is the guest and who is the host, who is the saviour and who needs to be saved today?
How to re-frame the understanding of the concept of 'communities' in their role of political subjects able to create alliances and contribute actively in more inclusive and open collective domains?
How to understand participation beyond the asymmetrical relationship between "organizers" and the participating communities that are often reduced to "relief recipients", who should gratefully endorse the attempt of those who are there to help them?
How to re-frame participation not only as a performative tool that makes organisers feel good but in its ability to challenge dominant power structures and to find new balances between existing structures and how not to hesitate to understand participation as a tool of operation to negotiate conflict?

2. Reframing: Commons

Sandi and Merve thought that the modernist assumptions of public space and natural habitat was about their design and management that eventually allow them to exist without people and communities.
In Re-framing Commons the question was:

> *How to re-frame the understanding of the commons as spaces near and far, created by the interaction of people, where commons can exist only if people and more-than-human agencies of habitat are constantly producing them?*

3. Reframing: Design

Merve and Sandi thought that the discursive and disciplinary aspects of design sustain the dominance of a unique modernist universalism through the skills acquired at school and the knowledge produced in academia, which fail to respond to the needs and desires across particular situations. The references of expertise that inform the roles of designers within the communities and environments diminish the potential and impact of design both as intervention and as critical imagination.
In Re-framing Commons the question was:

> *How to re-frame design in its role of challenging dominant structures by strengthening and amplifying alliance building and people-centered space making? How to set the architectural language free from a unique universalism, towards a language of plurality?*

4. Reframing: Agency

Sandi and Merve thought that the decision-making power and the proximities that different roles, people, and communities have amongst each other, the ways and dynamics of negotiation within and across communities relate to the degrees of autonomy, collectivity and representation of agencies.
In Re-framing Agency the questions were:

How to link and design agency in the absence of the state?
How to re-frame agency when aiming for design for change, and design for partnerships?
What are the potentials of designers' agency?

5. Rethinking: Land

Merve and Sandi thought that the relationship of people with land, soil, technologies, and environment needs to be re-thought, and land needs to be considered as a totality of the ground including what is stood on, what is below as well as above.
In Re-thinking Land the questions were:

Who are those that look after land?
What are the ways of maintaining, managing, and caring for land as a space of connecting and sharing?
What are the ways of owning and belonging?
How to expand this re-framing to practices of de-growth and healthy growth?
What are the meanings of infrastructure as the physical and ephemeral glue of land?
How to think of the alliances needed for just transitions of land?

Contents

Part I
Framing Community Participation

1 Retreat in Order to Advance: How Sharing Economy Shapes Co-living Rental Community Design in China

Chunyu Wang

Abstract

In China, the rise of the sharing economy has brought opportunities for co-living rental communities, particularly through the redevelopment of abandoned properties. These properties, often consisting of non-residential functions such as industrial buildings, spare commercial spaces, and spare office spaces, pose a unique challenge in terms of converting the problems into positive traits by design. The sharing economy as a business model gives inspirations to these issues. This study aims to examine how does the sharing economy shape the design of co-living communities by exploring nine typical co-living rental communities that have been transformed from non-residential properties. The research methodology includes thematic visits, focus groups, and typological studies. Four major transformation typologies are discussed, including the transformation of ground floor spaces, rooftop platforms, underground spaces, and private living spaces. The findings indicate that the sharing economy has led to a design approach of "retreat in order to advance", highlighting that sharing the private spaces with the public results in a better utilization of the community resources and promoting the shift from physical spaces to social spaces.

Keywords

Rental community · Sharing economy · Co-living · Community design · China

1.1 Introduction

In China, the growth of co-living rental communities is being driven by the "Internet plus" and sharing economy trend. Construction is being speeded up by the concept of property reconstruction in the sharing economy (Botsman and Rogers 2011). The challenge in reconstructing these existing buildings into shared communities lies in that many of these stock or abandoned properties are industrial buildings, spare commercial spaces, spare office spaces, etc. (Mayer-Schenberger and Cukier 2013). These properties pose a unique design challenge, such as high floor heights, deep floor depths, and insufficient communication spaces. Therefore, it is especially important in the reconstruction design to adapt to new functions and turn these problems into positive traits in a shared way. For newly built communities, another key issue in the design is to form a shared relationship

C. Wang (✉)
School of Architecture and Urban Planning, Chongqing University, Chongqing, China
e-mail: chunyu@cqu.edu.cn

S. Hilal et al. (eds.), *Design for Partnerships for Change*, Sustainable Development Goals Series,
https://doi.org/10.1007/978-3-031-36993-3_1

between communities and cities under intensive land use (Felson and Spaeth 1978).

The functional transformation of physical space is an effective solution to these challenges. This means that part of the community's space is opened up to the public, and the relatively private space is transformed to the space for group use. This type of transformation often occurs at the porch, open ground floor, arcade, rooftop platform, small square, and other intersecting interfaces with the city. The target of sharing can be either the renters within the community or the broader citizenry. The transformation allows the original structures that are difficult to adapt for residential use to be altered for shared use, thus making the entire community space more fully utilized and increasing the level of sharing.

1.2 Methods

Co-living rental communities have been developing for several years, and many cases have emerged. As the economic center of mainland China, Shanghai has the largest concentration of shared rental communities, and most community brands have properties in Shanghai. This study first determined the scope of the representative cases of the largest brands in Shanghai (n = 35), all of which were transformed from existing properties. Then, all of the cases were screened through a six-dimensional sharing evaluation, including functional adaptation, public–private dialogue, group symbiosis, community interaction, urban integration, and network interconnection. Typical cases (n = 9) were found to have a high degree of sharing and were mainly discussed in this study (Table 1.1).

This study utilized a thematic visit method to investigate and evaluate the attributes of the cases, including the surrounding environment, internal building space, and transformation methods. Relevant professionals, residents, and departments were consulted based on preset questions about management and policies. The data collected from the cases was analyzed, including project background, design process, transformation strategy, technique application, and using status. A typological approach was used to sort out the spatial characteristics of the cases. Compared to traditional rental apartments, the design of shared rental communities was shaped by the sharing economy in four major typologies.

1.3 Four Typologies of Design Under Sharing Economy

1.3.1 Transformation of Ground Floors

Transforming the ground floor of a building to a shared space with the general public is a common design technique in modern architecture. Even if the ground floor is not a semi-outdoor space, sharing the entire interior floor space can create an urban shared citizen space. By transforming the ground floor, permeability between the internal shared space and the outdoor public space is ensured through the continuity of the ground floor and the transparency of the interior and exterior.

Norman Foster's HSBC bank in Hong Kong is a case where private space is transformed into semi-private and semi-open urban shared space from the top floor to the bottom floor. In the design, a 12-m-high shared square is created on the ground floor, which is accessible to both the bank's staff and the general public. Moses Safdie and Aedas' Marina Bay Sands Hotel in Singapore also opens its ground floor to all visitors through a large pedestrian corridor that spans the entire three large spatial volumes. Each guest room unit shares the atrium connecting the first floor to the top, increasing the depth of the space in both horizontal and vertical directions. The bottom layer has multiple functions such as business and display, providing visitors with a good experience (Fig. 1.1).

The transformation of existing properties into shared communities often involves transforming the functions of the ground floor. This is due to the fact that many industrial, commercial, and office buildings tend to have higher first floors, some even reaching up to 30 feet in height. Such

Table 1.1 Nine major cases discussed in this study

No	Name	Axonometric	Photo	Address	Original type/ Architect	Reconstruction year	Scale
1	Harbor Apartments North Sichuan Rd	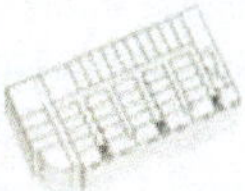		818 North Sichuan Rd, Hongkou, Shanghai	Commerce underground GPT +	2017	3728 m^2
2	Harbor Apartments Shanghai University Rd			3363 Hutai Rd, Baoshan, Shanghai	Commerce GPT +	2016	Approx. 9000 m^2
3	Mofang Apartments Ningguo Rd	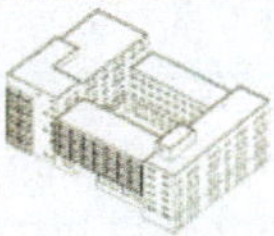		29–7 Longkou Rd, Yangpu, Shanghai	Industrial heritage /	2016	Approx. 6000 m^2
4	Base Complex Zhangjiang			3000–3 Longdong Ave, Pudong, Shanghai	Hotel MASS design	2016	12,630 m^2
5	Harbor Apartments Tiandong Rd	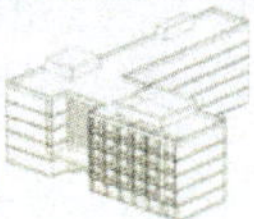		88 Tiandong Rd, Xuhui, Shanghai	Offices GPT +	2016	5745 m^2; 164 rooms
6	Langshi Apartments North Suhe Bay			1000 Qiujiang Rd, Jing'an, Shanghai	Offices /	2018	Floor 18,000 m^2, underground 3500 m^2
7	YOU+ International Community Shilong Rd	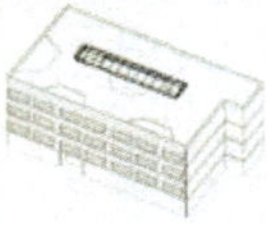		395 Shilong Rd, Xuhui, Shanghai	Industrial heritage Hotel /	2015	Approx. 4000 m^2
8	Vanke Port Apartments Zhangjiang National Innovation			899 Dangui Rd, Pudong, Shanghai	Industrial heritage Atelier Deshaus	2016	/
9	Base Living Shiziwan Rd	/		2280–18 Kaixuan Rd, Xuhui, Shanghai	Industrial heritage MASS Design	2012	2230 m^2

Fig. 1.1 Ground floor and atrium of Marina Bay Sands Hotel. *Source* author

spaces are not suitable for residential use, making it necessary to transform the function in order to adapt the architectural structure for multiple functions.

A typical example of this transformation is the Mofang Apartment Ningguo Road community in Shanghai (Figs. 1.2 and 1.3). The original site was a rubber factory consisting of four enclosed multi-story workshop buildings. The north and south buildings were primarily used for offices and residences; hence, they had a lower floor height, while the east and west buildings had light industrial functions and a higher floor height. From the perspective of elevation, the east and west buildings have significantly higher building heights than the south and north buildings.

The second to fourth floors and partial fifth floor were reconstructed for residential use, primarily designed as single flats on the north and south sides, due to the appropriate floor height. On the other hand, the east and west sides, which had a floor height of over 13 feet, were designed as loft apartments.

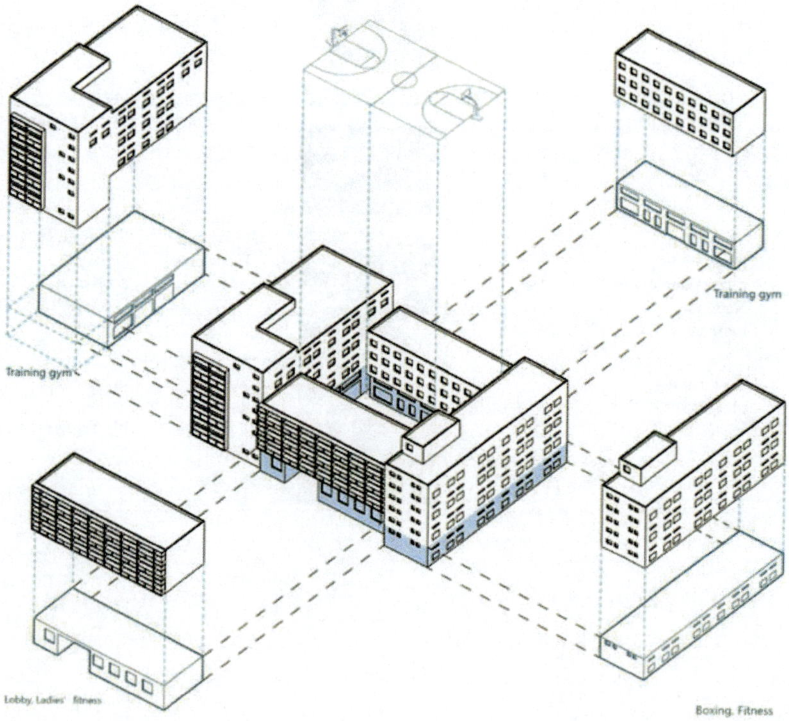

Fig. 1.2 Transformation of the ground floor and the courtyard in Mofang Ningguo. *Source* author

a. Main entrance b. Inner basketball court c. The original factory

Fig. 1.3 Situation photo **a** and **b** and original photo **c** of Mofang Ningguo. *Source* author

In contrast, the ground floors of each building had floor heights exceeding 20 feet, making them unsuitable for residential use. In the transformation process, the ground floors were designated as shared spaces for the community and the general public to be accessed for sports training and fitness. The north side was converted into a sports training hall primarily for indoor competitive sports such as table tennis and badminton, while the west side was transformed into a sports training hall for fencing and other less popular but professional sports. The east side was transformed into a gym and boxing ring, open to all citizens but with a certain discount for community renters. The south side was also used for functions such as reception hall, community office, and women's fitness and yoga. The four buildings enclosed are all for fitness, but they have different and subdivided functions.

In addition to transforming the ground floor, the community also transformed the enclosed inner courtyard square into a basketball court. The size of the inner courtyard square was large enough to accommodate a standard-sized court, and a plastic material floor was added to make it accessible to the public. Basketball is a popular sport among the youth in China, and basketball courts are considered as free urban infrastructure. Hence, the community decided not to charge anyone for using the court. This transformation provided a platform for better communication between renters in the community and the outside world, which is more in line with the needs of young renters, rather than traditional shared office and tea rooms.

Although sharing the ground floor spaces with the public may seem to reduce the living function area and “space efficiency rate” in “retreat”, the appropriate allocation of functions has enriched the shared life in the community and improved the quality of life. This shared space has actually contributed to an overall increase in rent.

1.3.2 Transformation of Rooftop Platforms

As the “fifth facade” of a building, the roof is an important element of modern architecture and can serve as a shared platform for communities. In residential buildings, the transformation of the rooftop platform into a shared space has become a prevalent trend. One such example is the Pinnacle at Duxton in Singapore, designed by ARC Studio Architecture and Urbanism, as part of the Housing and Development Board (HDB)'s affordable housing program. The seven 50-story residential skyscrapers, standing at 538 feet tall, are connected by overpasses and roof terraces, creating a shared space for both the residents and the general public. The roof platform is open for public access from 9 am to 9 pm daily, with an entry fee of 5 Singapore dollars, providing visitors with breathtaking views of the city (Fig. 1.4).

Similarly, Marina Bay Sands Hotel in Singapore, designed by Moshe Safdie and Aedas, features a 130,000-square-foot skypark that connects three 650-foot-high towers. The skypark, which includes runways, infinity swimming

Fig. 1.4 Pinnacle Duxton. *Source* author

Fig. 1.6 Marina Bay Sands Hotel rooftop. *Source* author

pools, pedestrian streets, a coffee bar, and a garden overlook, is open to both hotel guests and the general public. A 20 Singapore dollar voucher for the coffee bar grants visitors access to the shared space, allowing them to enjoy the panoramic views of the city.

These examples demonstrate the trend of transforming rooftop platforms into shared spaces, promoting both public access and group sharing within the community (Figs. 1.5 and 1.6).

Therefore, the transformation of the rooftop platform can foster communication and interaction within the community and even among the broader public through sharing. Many shared rental communities are located in city centers, where the problems of intensive land use and limited space for activities, such as gardens, persist. As a result, the transformation of roof space as a shared platform has become a common and valuable design strategy. The design and reconstruction of rental communities now have numerous examples of transforming the rooftop platform to serve a social purpose, such as the Vanke Port Communities. Through the reconstruction of the original roof structure, the introduction of indoor kitchens, greenhouses, handcraft workshop spaces, and open spaces like dining rooms and sight-seeing platforms, the sense of community belonging is strengthened, conveying an energetic lifestyle and shared values. Based on the spatial reconstruction and utilization, the transformation of rooftop platform functions can be categorized into two modes: the large open platform mode and the functional block mode.

Fig. 1.5 Marina Bay Sands Hotel. *Source* author

Large open platform mode

The large open platform mode is a relatively restrained transformation of the original rooftop platform, usually involving optimization of the paving and the addition of simple seating and rest facilities. This mode offers more adaptability and variability to the community members due to its open space.

The Base Apartment Zhangjiang Community in Shanghai adopts the large open platform mode (Fig. 1.7). The original building was a business hotel with a total area of over 130,000 square feet, surrounded by a large number of IT companies, start-ups, and research institutes. The rooftop platform in this building was designed to serve as a shared platform for various activities. The ground was replaced with a wooden floor to allow for direct seating, and several sets of sunken seats were included. Each set of seats can accommodate over ten people, promoting communication within each group. The simple structure also offers the possibility of converting the seats into other functions. The parapet made up of transparent glass balustrades not only provides renters with elegant views but also enhances the sense of security without hindering the design. Despite its basic renovation, this platform promotes communication among renters and creates a vibrant community culture and life.

The YOU+ International Youth Community Shilong Road, aimed at young entrepreneurs in start-ups, also endeavors to maximize the use of the rooftop platform (Fig. 1.8). The building was originally an industrial factory, then a small hotel, and later an investment institution's office before being reconstructed into the YOU + International Youth Community. The community's investors include Xiaomi Technology, a leading Internet company in China. Given its focus on entrepreneurs, the renters in this community are generally young and have similar living habits and social circles.

The rooftop of the community has undergone significant transformation, with the exception of a few laundry rooms and equipment rooms. The spaces have been transformed into open roof gardens and shared social platforms. The parapet wall and floor have been covered in anti-corrosive wood. On the north side of the platform, four isosceles trapezoidal-shaped areas have been created for the purpose of increasing the green coverage and enhancing the ecological environment of the shared platform. The furniture available for use, such as tables and chairs, is composed of movable coffee tables and chairs. This allows for the flexible utilization of the large space for leisure activities, such as coffee breaks, as well as for hosting events and gatherings. Despite its low cost and simple design, this shared

a. Elevation b. Atrium platform c. Rooftop platform

Fig. 1.7 Photos of Base Zhangjiang. *Source* author

a. Elevation b. Co-working interlayer c. Courtyard

Fig. 1.8 Photos of YOU+ International Youth Community Shilong Road. *Source* author rooftop platform has a high utilization rate, especially in good weather conditions, and regular activities are held there on a weekly basis.

Inserting function blocks mode

Some of the reconstructed shared rental communities have some supplementary functions such as equipment rooms, elevator rooms, and warehouses on the rooftop. These supplementary function blocks, if integrated effectively into the reconstructed community, can form a dialogue with the open rooftop platform.

The Base Apartment Shizi Bay Road Community, facing a shortage of activity space, adopted a sharing solution strategy that involved the use of both the roof platform and functional blocks. The original buildings were two partially five-story clothing textile factories. During the reconstruction process, industrial elements were preserved in the interior design, such as the concrete beams and columns, which were left in their original colors. The facades are simple and white, being harmonious with the surrounding old community. The top floor of the community, which consisted of five previously unused storage rooms, equipment rooms, and staircases, has been transformed into a coffee bar and a shared social kitchen (Figs. 1.9, 1.10 and 1.11).

On the other hand, the Harbor International Shared Community Tiandong Road took a proactive approach to insert new functional blocks in the reconstruction of the roof space (Figs. 1.12 and 1.13). The original site was a six-story office building with an area of nearly 65,000 square feet. The community's L-shaped layout features a large rooftop platform on the top floor, which was previously abandoned. During the reconstruction, a canopy on the platform was transformed into a glass block and gym, coffee, and other functions were added. The gym has good daylighting and ventilation, which has contributed to its high utilization rate. The coffee operates independently and serves both community tenants and visitors. An outdoor rooftop with a shared interactive area, rooftop garden, and coffee seating was also established.

Fig. 1.10 Partial indoor café. *Source* author

Fig. 1.9 Base Shizi Bay Rooftop. *Source* Base-living

Fig. 1.11 Partial outdoors platform. *Source* author

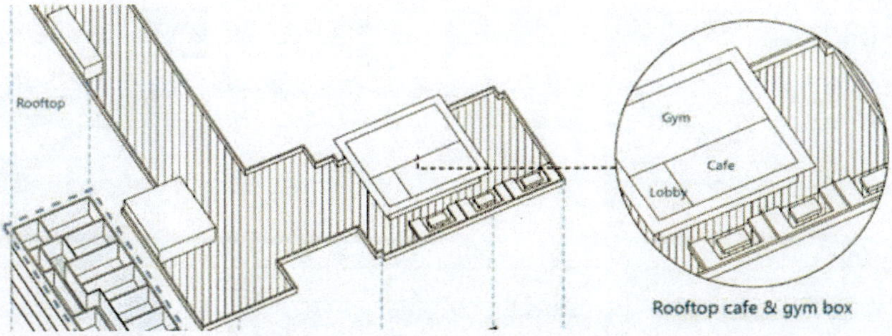

Fig. 1.12 Rooftop platform and function blocks of Harbor Tiandong Road. *Source* author

a. Elevation b. Courtyard c. Rooftop platform

Fig. 1.13 Photos of Harbor Tiandong Road. *Source* author

The transformation of the roof platform into a shared space can effectively address the challenges of intensive land use, insufficient public space, and a lack of shared platforms in the reconstruction of rental communities. Sharing the rooftop space with the public does not take up too much rentable floor space, but brings many benefits of activities. The integration of functional blocks creates an interesting and adequate urban function on the roof terrace, which appeals to both the community and the surrounding population. However, these transformations often lack open access to the roof, such as a stairway or path, which may limit public access and limit the platform's use to primarily serving the community's tenants.

1.3.3 Transformation of Underground Spaces

Shared underground disaster shelter space

Urban disaster shelters, including bomb shelters, flood control levees, and air-raid shelters, are often only functional during specific periods. They are mostly left unused and in standby mode. By integrating these urban spaces and structures with disaster shelter functions and shared space functions, the utilization of resources and land can be optimized and a "combination of peace and war functions" can be achieved.

In Balizhuang Street, Beijing, there was an unused air-raid basement. The original disaster shelter function was not suitable for daily use and had been utilized as an underground rental settlement, which was unsatisfactory and posed security risks. After undergoing transformation, the 16,000-square-foot air-raid basement was converted into a shared space called "Sweet Potato Community", serving nearly 10,000 residents in the surrounding area (Figs. 1.14 and 1.15). The aim of "Sweet Potato Community" was to transform the basement into a community-sharing platform by incorporating functions such as an open salon, a community library, a community college, a gym, and a cinema. Through various activities, the space tried to foster social interaction and break down

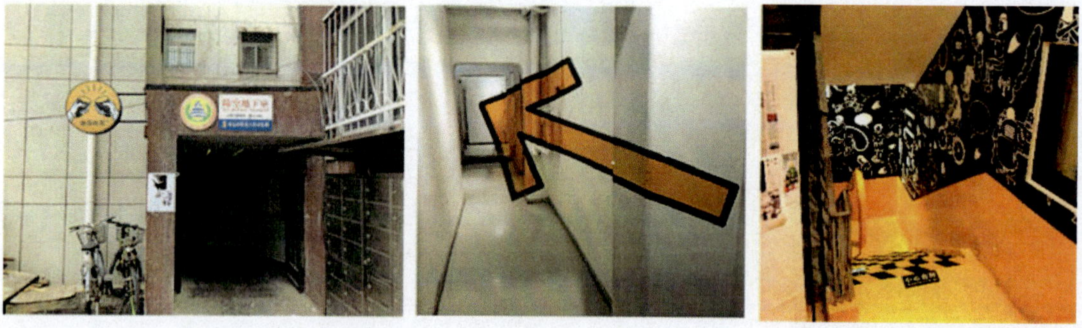

Fig. 1.14 "Sweet Potato Community" entrance, old Balizhuang St, Chaoyang, Beijing. *Source* author

Fig. 1.15 Reading rooms and kids playing together in "Sweet Potato Community". *Source* author

barriers between neighbors, creating a safe and lively urban environment.

Harbor International Shared Community North Sichuan Road is another example of shared air-raid basement in shared rental communities (Fig. 1.16 and 1.17). The basement, which was previously an unused air-raid shelter, was transformed into a comprehensive shared living room and open bar. Upon entering the space, visitors are greeted with a central area consisting of a bar counter and seating. This central area is relatively open and serves as a coffee during the day and a bar at night. To the sides, there are tables and chairs. The outer area contains a small gym and a movie room, while the inner area has a shared library with bookshelves and desks. Although the overall space is not particularly large, it provides a social atmosphere for the community and public and contributes to the shared living experience. Renters within the community feel closely connected to the city through shared underground space, rather than having their own community's property occupied by outsiders.

a. guiding stairs of entrance b. bar counter and lobby c. bar and library

Fig. 1.16 Bar of Harbor International Shared Community North Sichuan Road. *Source* author

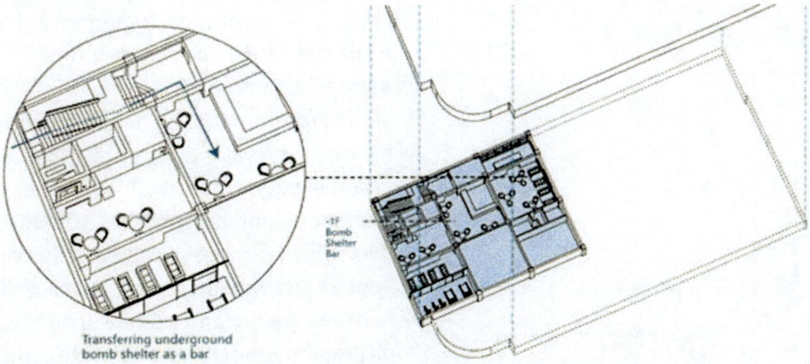

Fig. 1.17 Underground space of Harbor Apartments North Sichuan Road. *Source* author

a. elevation b. Easystorage on -2F c. 1000-liter mini stoage

Fig. 1.18 Langshi Apartment North Suhe Bay and its shared rental storage. *Source* author

Shared underground storage space

The integration of underground storage into high-rise design is a common functional setting. With the growth of the sharing economy, storage space has become a resource that can be shared. Langshi Apartment North Suhe Bay is an example. The original site was a 22-story office building and was transformed into a shared rental community in 2018. During the transformation, the basement space was utilized as a private mini storage service with an intelligent network, thereby making it a "shared private storage" solution for the general public (Figs. 1.18, 1.19, 1.20 and 1.21).

The underground shared storage space provides 24-h access via face recognition technology. It offers two sizes of mini storage units, 1000 L and 2000 L, allowing customers to choose the size that best suits their needs. The process of renewing a lease can be completed through the official online platform, making it simple and convenient. A third of the warehouse space is reserved for the tenants of the Langshi Apartment, who are offered a 30% discount, linking the shared rental housing with shared storage. This new form of shared storage is a novel type of shared space, which makes effective use of the otherwise idle basement space by breaking it into smaller units for usage by others. Sharing the storage space with the public increases the space utilization of the community.

1.3.4 Transformation of Private Living Spaces

The concept of public and private in architecture is a common division of spaces, yet the

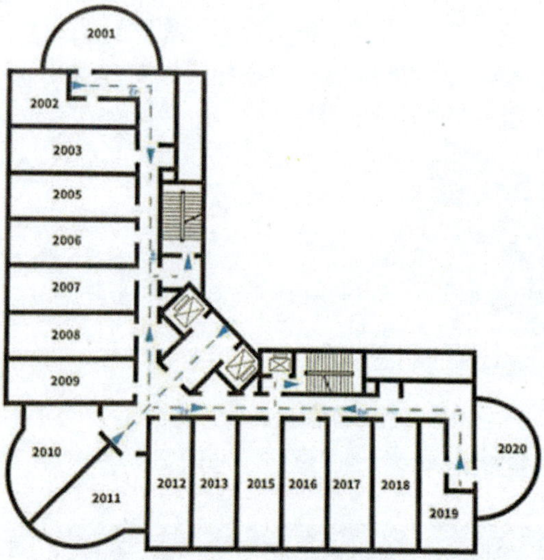

Fig. 1.19 Langshi Apartment North Suhe Bay standard plan. *Source* Langshi, redrawn by author

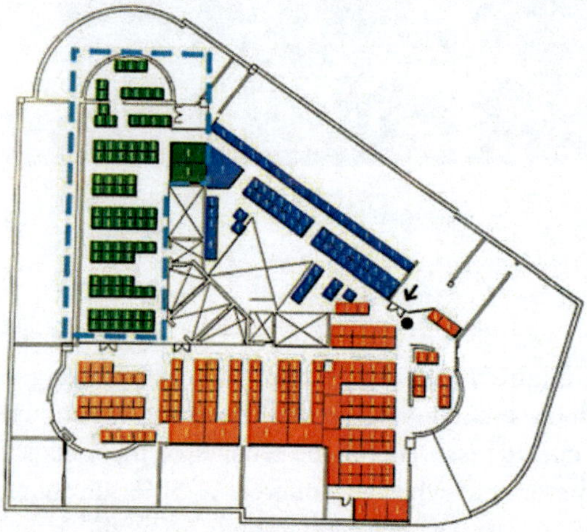

Fig. 1.20 Langshi Apartment North Suhe Bay B2F storage plan. *Source* Easystorage, redrawn by author

boundaries between the two are often blurry. Hermann Herzberg believes that it is not necessary to strictly oppose the concepts of public and private spaces and that there is no fixed distinction between them (Hertzberger 2005). Chamayev and Alexander, on the other hand, emphasize that it is equally important to ensure the privacy of the space and to create a public shared living space (Chermayeff and Alexander 1966). In the context of housing and apartments, privacy is of utmost importance, but finding the right balance between community life and privacy is also crucial.

Therefore, a reasonable transition boundary between private and public spaces should be established, which will determine the level of interaction and information exchange between the two. This boundary is the main expression of the "public–private dialogue" in space. The boundary between public and private spaces requires both a physical boundary and a psychological boundary. Physical boundaries refer to tangible elements such as walls, fences, vegetation, pavement, and differences in elevation that enclose and define a space. A strong physical boundary, such as a fence, separates public and private spaces clearly, while a flexible physical boundary, such as vegetation or height differences, allows for an open space with a permeable boundary. The psychological boundary refers to people's perception of the space on a psychological level. When the physical boundary is not clear, the psychological boundary is used to distinguish between public and private spaces. The most common manifestation of this boundary is the space in front of a door (Figs. 1.22 and 1.23).

When people open the space demarcated by a psychological boundary to the public, they transform privacy to the public sphere, which generates a positive public–private dialogue and sharing. According to C. Alexander's concept of "event pattern", the life and spirit of a place are not dependent on the physical environment, but on the pattern of events experienced by people there (Alexander 1977). The public–private dialogue in a modern rental community can foster community relationships.

In addition to the transformation of ground floor spaces and rooftop spaces, small-scale space transformation from private to sharing also occurs between households within the community. Herman Herzberg's theory of "threshold" has a similar effect, suggesting that residential entrances should not only be seen as entry and exit points, but also as spaces for communication and interaction. The concept of "threshold" serves as both a physical and psychological boundary. In practice, any space in the form of a transition space

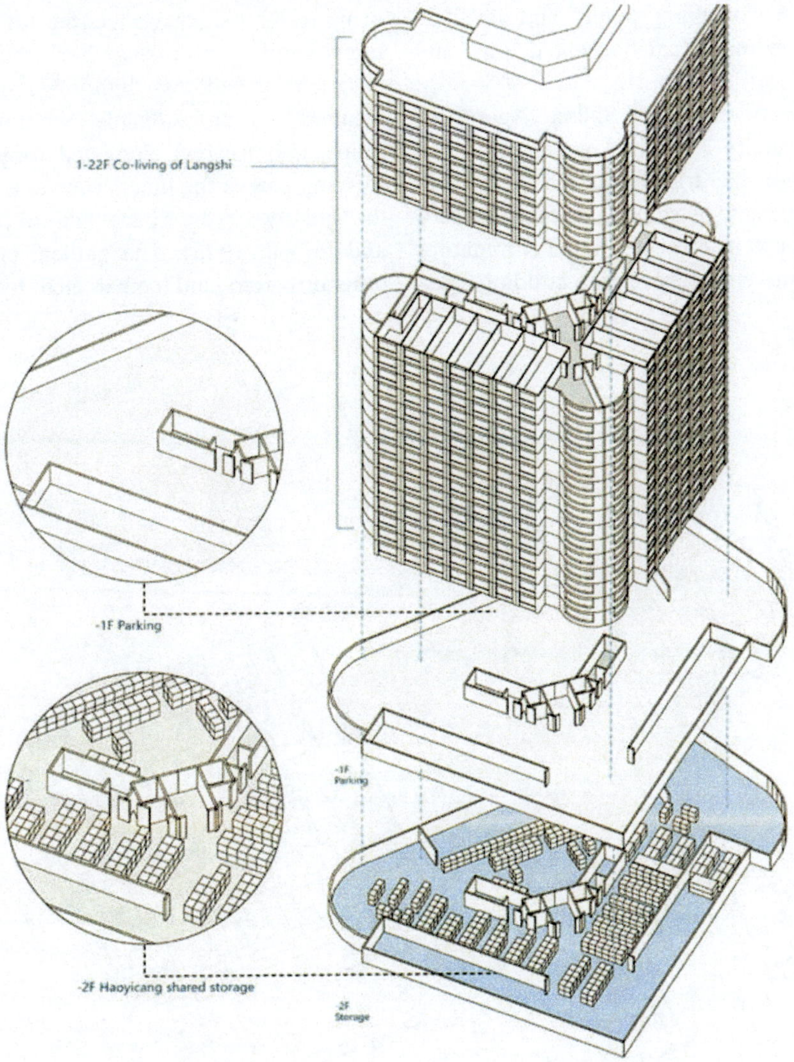

Fig. 1.21 Axonometric drawing of Langshi Apartments North Suhe Bay. *Source* author

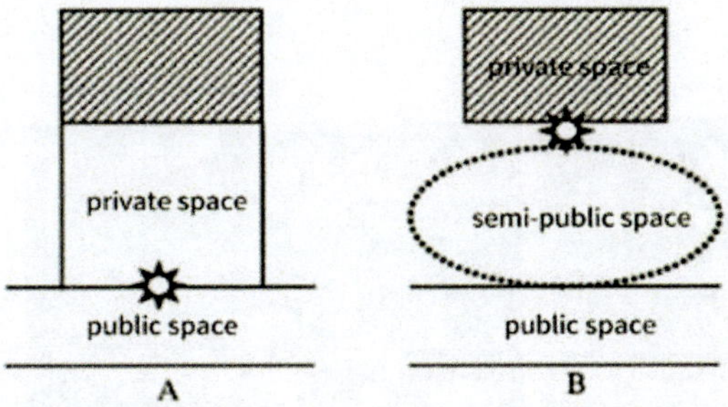

Fig. 1.22 Spatial hierarchy generated by different physical boundaries. *Source* modified from Chermayeff and Alexander 1966

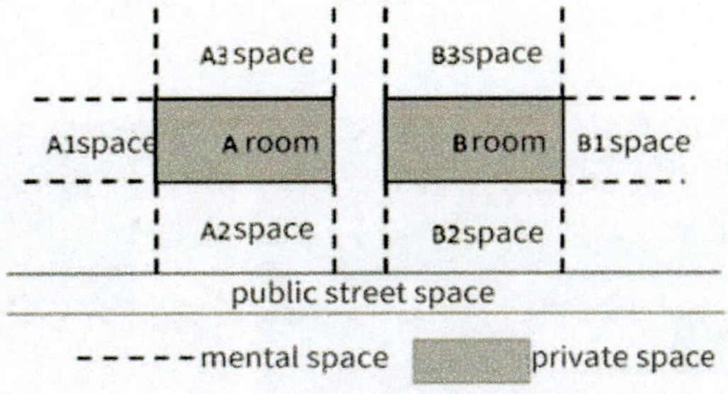

Fig. 1.23 Scope of psychological space domain. *Source* modified from Chermayeff and Alexander 1966

can serve as a "breaking point" and facilitate sharing between individual household units and the community as a whole.

The Harbor International Shared Apartment Shanghai University Road has incorporated this concept through the transformation of private inner room space to a small garden (Fig. 1.24 and 1.25). Prior to its renovation, the community was a large commercial space with building areas of up to 50,000 square feet per floor. Due to the large depth, a group-shared enclosed garden was designed to address lighting issues, but still resulted in unreasonable window–floor area ratios, and narrow, elongated rooms. By transforming part of the inner space to a small garden, the quality of the space was improved (Kahn and Lobell, 1979). The garden, created by the transformation and demarcated by a relatively

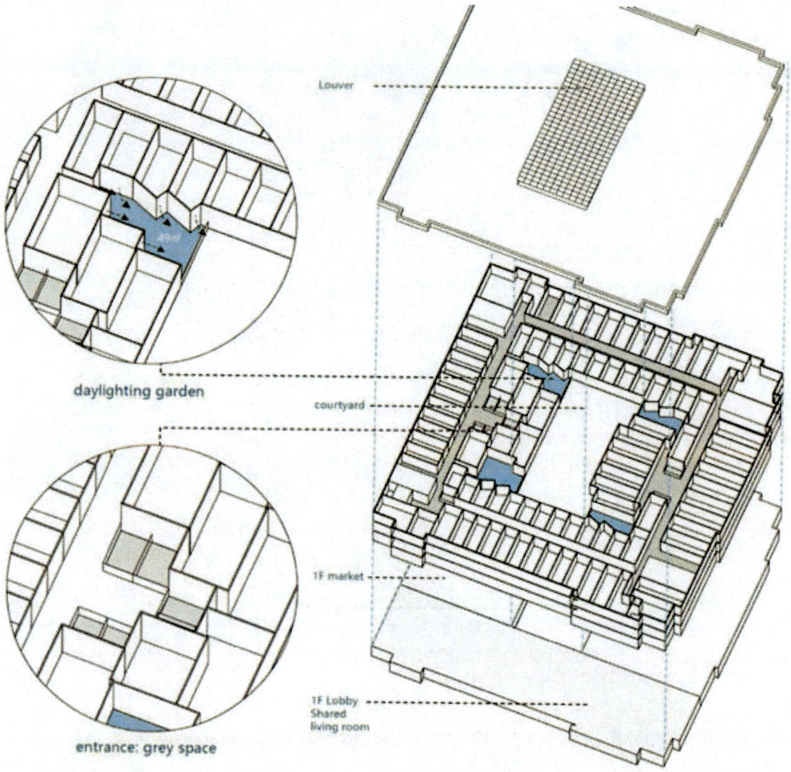

Fig. 1.24 Axonometric drawing of Harbor Shanghai University Road. *Source* author

Fig. 1.25 Current situation of Harbor Shanghai University Road. *Source* author

flexible boundary such as a fence, forms a semi-open space. The convex and concave boundary lines of the garden space avoid issues of window view interference between tenants and broaden both the physical and psychological boundaries of the space. This creates a transition point for public–private dialogue and provides space for communal activities, such as communication and leisure. Some courtyards are equipped with amenities like tables and chairs, storage facilities, and other amenities, further enhancing the functional differentiation of the shared space.

Traditionally, transforming a portion of the interior space of one's home into a shared space may seem like a "retreat", but it is this transformation that enhances the sense of community and instead makes for a better living experience.

1.4 Discussion

1.4.1 Four Major Typologies of Functional Transformation

This study has explored the design in typical co-living rental communities in Shanghai and found four major typologies of functional transformation under the influence of the sharing economy: the transformation of ground floor spaces, rooftop platforms, underground spaces, and private living spaces. The transformation not only allows the original structures that are difficult to adapt for residential use to be altered for shared use, but also makes the entire community space more fully utilized and increasing the level of sharing. This is reflected in various space features:

The transformation of ground floors enhances the permeability between the internal shared space and the outdoor public space. The transformation of rooftop platforms and underground spaces is a way to expand the usable space in the communities, making full use of the available resources. Private living spaces, as the core of the co-living community, have been transformed into shared spaces with more public areas, which foster a better sense of community and increase social interaction.

1.4.2 Rethinking the Space Efficiency of Community Design

Traditionally, it has been believed that a higher "space efficiency rate" is the embodiment of a reasonable and affordable community design. Accordingly, the area inside rooms is maximized to the fullest extent possible. However, this study reveals that the sharing economy has brought about a change of co-living rental communities. On the one hand, tenants do not own the properties, making the calculation of the area of public space less important (Kobayashi 2016; Gentile 2016). On the other hand, the area of rooms within the individual's room is often quite small, and many functions, such as communication, work, and entertainment, take place outside of private spaces. Consequently, tenants place more emphasis on the quality of shared spaces (Goering and Whitehead 2017).

The design approach of "retreat in order to advance" has been highlighted as a result of the sharing economy, emphasizing that sharing private spaces with the public leads to better utilization of community resources. It suggests that it is unnecessary to make "full use" of the space and design as many tenements as possible. At this point, the transformation of private functions to shared platforms represents an individual-collective win–win strategy (Wang 2017).

1.4.3 Sharing Economy Promotes the Transformation of Physical Space into Social Space

Through the analysis of the nine cases, it is found that co-living rental communities under the sharing economy model pay more attention to social attributes and emotional maintenance (Bricocoli and Sabatinelli 2016). While physical space tends to meet functional requirements, shared space is designed to create social relationships. In contrast to traditional rental apartments, co-living shared communities promote neighborhood relations and facilitate social interaction among tenants. Tenants with similar backgrounds are brought together in the same community and can communicate with each other on shared platforms. In this sense, shared communities use space as a means of guiding the evolution of neighborhoods.

1.4.4 Strengths and Limitations

The findings of this study provide valuable insights into the design of co-living rental communities in China and can inspire further research in the field. The design transformations under the influence of the sharing economy can be considered a new trend in the development of co-living rental communities, which can have significant implications for future community design and development.

There are two main limitations to this study. Firstly, while the nine cases selected for the study were representative and diverse in terms of design thinking, they were primarily limited to the Shanghai context. Although the sharing economy has influenced community design thinking throughout China, different regions may have diverged into some different design characteristics depending on their cultural backgrounds. Therefore, future research should investigate the design situations in other regions to further expand the findings of this study. Secondly, the observation method used in this study mainly focused on the point-in-time state, emphasizing activities that are common in the community. Future research should involve a comparison of the before and after changes that the shared design produces. The evidence-based design theoretical framework should be introduced to track the differences in the use of a community before and after the transformation, to further validate the positive effects of shared design practices.

References

Alexander C (1977) A pattern language: towns, buildings, construction. Oxford University Press

Botsman R, Rogers R (2011) What's mine is yours: How collaborative consumption is changing the way we live, vol 5. Collins, London

Bricocoli M, Sabatinelli S (2016) House sharing amongst young adults in the context of Mediterranean welfare: the case of Milan. Int J Hous Policy 16(2):214–231

Chermayeff S, Alexander C (1966) Community and privacy. Kajima Institute Publishing Company

Felson M, Spaeth JL (1978) Community structure and collaborative consumption: a routine activity approach. Am Behav Sci 21:614–624

Gentile A (2016) Rental subsidy and the emancipation of young adults in Spain. Int J Hous Policy 16(2):243–254

Goering J, Whitehead CM (2017) Fiscal austerity and rental housing policy in the United States and United Kingdom, 2010–2016. Hous Policy Debate 27 (6):875–896

Hertzberger H (2005) Lessons for students in architecture (Vol. 1). 010 Publishers

Kahn LI, Lobell J (1979) Between silence and light: spirit in the architecture of Louis I. Kahn. Shambhala

Kobayashi M (2016) Housing policies in Japan. In: The housing challenge in emerging Asia: options and solutions, pp 126–142. Springer

Mayer-Schenberger V, Cukier K (2013) Big data: a revolution that will transform how we live, work, and think. Oxford University Press

Wang C (2017) To improve space utilization efficiency: periodic renting strategies of residential open buildings. UIA 2017 Seoul World Architects Congress, Seoul

2 The Colors of "El Nispero" Reframing Communities, Reframing Territories

Daniel Huertas Nadal
and Maria Camila Gomez

Abstract

This documentary captures the color of the life of the Ma-Majarí of the "el Níspero" Afro-descendant community in the Colombian Caribbean. It shows its people's warmth—so-called "nispereros"—their smiles, dance, houses, and dreams. Thanks to their people, organizations, and leaders, these voices that speak of identity, culture, and ethno-development have made possible new spaces for participation. The research shows its efforts to develop collaborative strategies and habitat construction systems applicable to the community's life plans. From the recognition of the community, research and documentary have been oriented toward the contribution of empowerment and community social participation in constructing a route of cultural identity that allows communities to remain in the territory with dignity. The house, the home, and the construction of its spatiality from everyday life become arguments to discuss a project of cultural interest on a differential approach to habitat. Understanding the importance of orality spaces as an opportunity to find the other implies modifying the practices with which architecture approaches the community. The project uncovers the subtle relationships between intangible cultural and tangible social spaces. Sustainable social development could not be possible if this connection is not protected. Finally, the documentary talks about how to build an alternative development proposal for Afro-rural communities, overcome the social gap, and defend that another development is possible.

Keywords

Intangible cultural heritage · New ethnicities · New territories · Afro-rural housing · Social construction of habitat

2.1 Introduction

I learned to look, taking small steps toward the invisible as if I were an indifferent observer of everyday appearances—a collector of experiences or spaces. In working with the communities, I discovered a sense of complicity. I understood that complicity is entering into what is always seen. It is to make different worlds coincide. It can be from a wood stove or a palm fabric. Once lit, the kitchen becomes a single space. I learned to be part of what I see—being part of the experience, without appropriating the scene, without the violence of the analytical

D. H. Nadal (✉) · M. C. Gomez
School of Architecture and Design, Universidad de Los Andes, Bogotá, Colombia
e-mail: d.huertas@uniandes.edu.co

S. Hilal et al. (eds.), *Design for Partnerships for Change*, Sustainable Development Goals Series,
https://doi.org/10.1007/978-3-031-36993-3_2

Fig. 2.1 Photogram from the documentary. El Nispero, Montes de Maria, Colombia. View of "the down neighborhood." 2021. The diversity of ethnic communities, such as Ma-Majarí from El Nispero in the Colombian Caribbean, calls for the need to make the African diaspora visible in implementing SDGs that admit a differential ethnic approach. Images by the authors

gaze. A look that wants to be more confident, where looking is not to violate; it is not to conquer nor to analyze. To look means finding a complicity state that allows us to discover once and again the reason for being in everyday life. One observes unarmed from intuition. Only in this way can the intangible heritage of Afro-Colombian housing be understood. One has to cross the sandy street, reach the shade of the tree, cross the house to sit in the hut, look at the patio, and talk over coffee. One must listen to the sweet and musical sound of the words in the palenquera language while the breeze refreshes the skin. One has to let oneself be overwhelmed to understand the value of architecture and the meaning of this research (Fig. 2.1).

2.1.1 Considerations

The colors of El Níspero is a documentary, a narrative proposal. A research work on the social construction of the habitat that reflects the color of life of the Afro-descendant community Ma-Majarí of El Níspero in the Colombian Caribbean. This text proposes a contextual framework associated with the documentary film to be projected at the UIA2023 Congress as part of a transmedia communication project.

Communities are not subjects or objects of study. Instead, they are actors who build knowledge. This project is a sample of the shared construction of collective intelligence, an attempt to discover new knowledge transfer systems. With this perspective, it is possible to imagine a real Sustainable Development.

This research focuses on a case study in María La Baja, Montes de María, in the Caribbean area of Colombia, working directly with Ma-Majarí del Níspero Community Council. The Ma-Majarí del Níspero community has led the territorial affirmation processes of Afro-descendants in this area. Hundreds of men and women have suffered the rigors of war by dealing with terror and fear and reaffirming their will of a dignified life through collective efforts. However, violence, in all its expressions, has traditionally affected this territory and its social fabric. This project wants to discard fear and recover the capacity for affirmation, the desire for development, and the redefinition of the colors of community life.

2.1.2 Some Concepts

In the documentary, some concepts appear that are interesting to delimit: the intangible cultural heritage, the spaces associated with intangible culture, and the life and ethno-development plans of the Afro-descendant community councils.

2.1.2.1 The Intangible Cultural Heritage of Afro-Colombian Rurality

The permanent construction of narratives and the oral tradition as an expression of self-development marks a notable difference between black communities and other ethnic groups. Afro-descendant communities dynamically build and transmit their way of living, their traditions, and their way of staying in the territory, giving rise to reflection on the construction of these processes in societies linked to ancestry, identity territorial, and the transmission of knowledge orally. Territory, memory, and culture represent the ethnic perspectives of black communities (Restrepo 2013); for this reason, the construction of new discourses from ancestry bespeaks the traditional debate between the traces of Africanism and the new ethnicization of the Afro-Colombian community councils (Domínguez 2015).

The relevance of intangible cultural heritage lies in its ability to keep ancestry present, reactivate its symbology and connections, and reinforce the collective fabric. Orality time is a dynamic time that modifies past and present relationships. It is the basis of oral transmission in black communities, which permanently updates the tradition and its spatiality (Hampaté Ba 1983). However, the recognition of these values, such as the declaration of San Basilio de Palenque as a Masterpiece of Material and Intangible Heritage of Humanity in 2005 by UNESCO, can mean the recognition of cultural identity or end in an exoticization of cultural practices (Parra-Valencia 2020). In this sense, it is interesting to foresee an approach that is interested in alternative forms of expression and development from a perspective that allows deconstructing the Eurocentric rationality of current policies, developing a proposal of decoloniality both in its cultural approach as well as its development proposals (Walsh 2007). In this sense, communities are not simply trapped in places waiting to be assisted by development (Escobar 2001). Instead, they build networks and narratives of the intangible to question the dominant readings about their territories.

Renewing the perspectives of intangible culture means proposing a new ethnicity as a multiple, diverse, and dynamic process that can constitute a mechanism of subversion against established positions (Gutiérrez et al. 2020). By assuming the renewal of identity, the communities propose a social and political process of ethnic affirmation and cultural strengthening, understood in terms of valuing and legitimizing traditions and cultural practices, to form individuals and communities with a sense of belonging (Rojas and Restrepo 2004).

2.1.2.2 Spaces of the Intangible

The patio, the hut, the tree in front of the house define the Afro-descendant spaces of the intangible (Huertas Nadal and Pinedo Cobos 2021). The house is the only architectural element that can be identified as a spatial manifestation of the Afro-descendant population in the Colombian Caribbean, articulating the practices that allow orality, poetry, and cultural manifestation. From this point of view, housing is the expression of the spaces of intangible culture. A space that must assume the importance of transmitting spatiotemporal experiences on the one hand and ensuring the symbolic transcendence of identity on the other.

The Afro-descendant house of the Caribbean in Colombia represents the link of more than 150 years with Africa (Arteaga 2019). It expresses an image of culture and territory, representing a social and cultural context in permanent development. The articulation between the street, the house, and the patio are essential expressions of daily life that configure a particular projection of heritage, identity, and memory. The house is the intimate space and the expression of the ways of living of this culture (Montoya and Solarte 2016).

2.1.2.3 The Patio

In Colombian Afro-Caribbean housing, the house becomes a threshold that opens onto the patio. The houses have preserved this spatial structure. The patios are associated with women when they link to cooking, while they remain associated with men when they link to farming. The patio in the house strengthens the practice of a solidary coexistence that allows permanence in the territory, food sovereignty, and the expression of own uses and management (Rodríguez 2002).

The patio brings together different traditions, symbolism, and an expression of home that is transmitted orally. The ancestors' stories are told on the patio, hair is cut, cassava is prepared, medicinal plants are cultivated, children play together, and donkeys, pigs, and chickens are raised. The toilet and shower are always on the patio in a small module outside the house. Somehow, the house is due to the patio, where the intimate is shared.

2.1.2.4 The Hut: The "Bohío"

Leaving the house to the patio, we find an articulating space, the hut called bohío: the stove, the hearth. It is the wood stove under a canopy of palm trees. The bohío has represented the family unit since ancient times. It is a space where the word is summoned with fire, and communication and identity arise. Orality keeps the community firm; ancestral wisdom calls to be part of the culture. Through the projection of these identity spaces, Afro-descendant women, communities, and movements are empowered. It is an offering to the family. However, it is also a call to the ancestors to guide the way through their wisdom.

The bohío is the center of identity, the space that conjugates and summons tradition. It is the wood stove. In the bohío, it is vital to recognize the symbolic value of fire, word, and time, which are not related to the functionality of the gas stove but to the tradition of shared communication. Undoubtedly, the bohío constitutes the most representative constructive exponent of pre-Columbian cultures in the Caribbean area (Taboada 2009). The vernacular construction systems of the bohíos have lasted over time, while the housing construction systems have been modified and updated, mainly because only the materiality of the bohío is associated with intangible expressions and manifestations of each community (Larios 2015).

2.1.2.5 The Tree in Front of the House

For African ancestors, the tree is a symbol of memory, of family protection. The tree in front of the house manifests the link with ancestral traditions, materializing the experience toward the outside. The richness of Afro culture includes living inside the patio, around the hut, and outside the tree facing the street. The public space for the Afro-descendant communities of the Colombian Caribbean has a very intense meaning because it represents the collective construction of spatiality that extends from the family space. This appropriation of the public is part of the community fabric, articulating the house with the community. Families fill the houses' doors and the trees' shadows at sunset with chairs. Sitting in the chairs, they can watch people go by and talk. From the street, you can see, in the background, behind the houses, the patios.

The tree in front of the house is an element of urban scale. They are related to a culture where houses do not grow above trees. Instead of developing vertically, the houses multiply their elements horizontally. The roof of the house is only a thermal control element, generally inclined. The tree appears as a reference element, a meeting space, and a public shadow. The tree is also the space of the shared, representative, culturally dependent word.

The hut on the patio and the tree in front of the house are places where life develops identity. The word lives in the patio, feminine, powerful, and alternatively, in the tree at the entrance, masculine and scattered.

2.1.2.6 Life Plans and Ethno-development Plans

The processes of defense of the territory by ethnic and peasant communities in Colombia pose a complex and urgent challenge. Apart from the historical processes of making black communities invisible, and even after the protracted peace process that culminated in the Havana agreements of 2016, difficulties persist in proposing a culturally appropriate development. With the uncertainties of the current post-conflict scenario and the new peace process in Colombia, the regulations and framework initiatives have generated an important debate that affects community life. The particular complexity of the Caribbean region involves historical, ethnic claims in Montes de María, a place of permanent social struggle, where the armed conflict has caused intense forced displacements. This research addresses the scope, impact, and development of life plans by Afro-Colombian community councils, as effective governance mechanisms that allow communities to remain in the territory with dignity. The community councils' life and ethno-development plans are an alternative to the development plans proposed by the government since they suggest territorial affirmation from a differential ethnic approach that enables social renewal.

A Life Plan is not a spontaneous or occasional collective proposal. The LPs are the necessary planning instruments to consolidate a Community Council. In Colombia, this legal figure recognizes an Afro-descendant community due to its ancestral conditions in the territory, with its own identity and culture. The community LP is a document that contains information about the community, resources, and needs. It also includes the changes desired by the community and the projects required for those changes. These community plans examine the collective identity, the future projection of the cultural reality, and the strategic tools to guarantee collective projects' economic, social, and ecosystem sustainability. An LP is one of the fundamental pillars of community strategies when defining a culturally appropriate development perspective.

2.1.3 About Methodologies

The methodological framework of the research proposal is defined from the Participatory Action Research (PAR) models. Since Paulo Freire in the 1970s defending work from the communities, and the construction of Orlando Fals Borda in the 1980s suppressing the critical separations that divided theory and practice, researcher subject and researched object, Latin America has been the emergency scenario and development of participatory investigative methodologies.

PAR is a qualitative research method developed to improve the lives of the communities involved in research projects. It prioritizes qualitative aspects of the processes without intervening or manipulating the information received, such as original ideas, proposals, plans, and dreams that outline the construction of a collective life project. PAR promotes the shared construction of knowledge, allowing the simultaneous formation of communities and research teams (Freire 1970; Fals Borda 2001). Communities identify problems and basic needs to address them through the search for solutions that cause changes in the collective social and cultural reality (Greenwood and Morten 1998). PAR based on community inclusion modifies the research approach and the traceability of the proposed solutions since it requires a commitment to maintaining the exchange over time. In the case of the presented documentary, we have been collaborating throughout four years of continuous exchanges that are currently still in force. Diversifying perspectives, interests, and capacities defines a consensus agenda in which the advancement of research establishes different meeting points for understanding and proposing new agreements (McDonald 2012).

PAR approaches are based on the voices and experiences of the community, guiding research projects toward open solutions where the conclusions seek to define alternatives, strategies, and shared learning. From this perspective, this project focuses on constructing a collective framework defined and based on the sociopolitical, economic, and cultural development context of the Ma-Majarí community of El Nispero in María La Baja. The process avoided the systematization of social cartographies, focusing on those that are not a final product but an essential part of a sequence of scenarios. These social cartographies were not intended to define but to find a place, as Holmes proposes (2005 translation, 19) "*Find a place, find the other or simply find utopia.*"

The principles of the PAR were fundamental in constructing this community project to guarantee work "with" the community and not "for" the community. Each project phase was an approximation, and the results were always the process. The pedagogy is based on objectives and fundamental preconceptions of social visibility, structural social change, and social integration. It was not structured as a teaching or discovery process because that would show that there is something to teach and that there is someone who knows. Instead, he considered the opposite by becoming a collective production. It focused on addressing a process of shared discovery, a participatory exchange that allowed the construction of a comprehensive territory proposal, collectively valuing the opportunities, resources, and difficulties to project a dignified and culturally appropriate habitat.

2.2 Transcription: The Colors of El Níspero

Intervene:

Fidel Barrios. Knower and community leader. Human rights defender. He knows the history of El Nispero from its origin to the present day.

Deyanira Martinez, Luz Elena Romero, Kariné Sanmartin. Leaders of the Community board. They participate in the governing spaces of the community, leading the Corporation of Afro-descendant Women of the Community Council.

Arnulfo Cardosi. Legal representative of the Ma-Majarí Community Council of El Nispero. Adviser at the Governorship of the department of Bolívar for ethnic affairs. Coordinator of the Office of Ethnic Affairs of María la Baja.

Osney Alvarez. Leader of the Afro Youth Corporation of El Nispero.

Alberto Cardosi. Knower and community leader. Displaced by paramilitary violence.

Adriana Santero. Displaced from the department of Santander, resident of the first generations of El Nispero.

Rita Magellan. Community leader, healer, and herbalist. Coordinator of the Association of Midwives Women of El Nispero.

Daniel Huertas. Investigator.

"Construction in bahareque and wooden boards is still widely used. If it is bahareque, everyone helps, kneads and stuffs [the mud]".

Fidel Barrios:

Faustino Palomino said, in a story or a Decima [a ten-verse poem], "When Enrique Wash was killed in Angola, the women ran on land and the men in a nightgown." Enrique Wash lived in Flamenco and made the journey to San Onofre. However, he had an altercation with my grandfather, Manuel Peña, so they challenged each other. Enrique Wash had other enemies, and they hired Benito Mansilla, there in Maria la Baja. Benito Mansilla hid behind some trees that were in what is now a bridge over the Cucal stream, then he shot him and killed him. Since he was a wealthy man with money, his followers conspired with the people of Angola and burned the houses. People were very alarmed because they had not seen that before and left. When they spread out, Mr Blas Audivet came and took over the square where there was a Níspero (Loquat) tree. Where Manuel Camacho lives, on the corner, there was a man named Lucho de la Cruz, who I got to know. Those were the first inhabitants who came here to El Níspero.

Deyanira Martínez:

You live happily. Everything you do, what you remember … in the mornings when you start to get up, people, neighbours, coffee, sweeping at the doors … that is the tranquillity that you experience.

Arnulfo Cardosi:

It generates peace; it generates tranquillity and there is nothing tastier … than cassava with sour cream or rice with coffee. That is part of the essence of each of us, Nispereros (people from El Níspero).

Daniel Huertas:

I found a very beautiful community; that beautiful that it is difficult to define qualitatively or (I suppose) in scientific and anthropological terms. What allows us to better understand it is what they were generating. It is a harmonious community; a community in a process of transition, which was what moved me the most about this situation. They were in a process of cultural transition, still spontaneously practising their ancestry. That is, not forcing it or recovering it but protecting it. I find a young community that therefore looks to the future. I find a community in a process slightly politically disoriented but not because they do not know what they want to do (they have it very clear). The communities are very clear about their processes, but they need a strategic approach to articulate all these realities. So, because of the community's affable and clearly dynamic and experiential character, I got very excited. It is a theoretical challenge to understand, as a professional, what architecture can do in those environments.

Luz Elena Romero:

Here, the houses open at six in the morning, and when it is ten at night, the doors are still open. Here in the village, we generally go to the kitchen [built] with palm leaves since dawn. Everyone who passes by shouts: "Deyaniraaaa", and Deyanira from her kitchen answers: "Whaaaat? here I am, good morning." For example, with the neighbour, we talked from kitchen to kitchen between the patios: "good morning, how are you? Where is so-and-so, where is Yelly?."

Deyanira Martínez/Kariné Sanmartín:

The house may be tiled, and it may have luxuries inside, but since one gets up, after sweeping and making the bed, we go to the kitchen, which is built with palm leaves. There, there is fresh air from the patio … that is something indispensable. There are even people who, after going down to the kitchen, they do not enter the rooms anymore, until going to bed. Something very curious that one has here is that one says: "If the wood stove is not on in the house, the house is sad or ruined." You must always turn on your wood stove, even if you are not going to cook anything, even if you do not have potatoes, but you must turn it on. Yes sir, joy is reflected. We all experience that. Where there is a wood stove, it always must be on, even throwing smoke to make the house look cheerful. If the stove is off, no matter how much joy one has, one says that the house is sad (Fig. 2.2).

Daniel Huertas:

There is a flow, that is, a way to live in those houses, as well as a way to build them. In the research outside the community, these houses are related to Angolan constructions. This is relevant because this whole territory was called Llanos de Angola [Angolan plains], including the constructions in Angola and El Congo but also the indigenous bohíos [huts]. Closed indigenous bohíos in this case. These constructions are usually open, but the Afro, due to their slavery condition and defensive character, always closed them. Curiously, in Africa, they have always been closed, even when there is no defensive condition. However, the model was adopted here because it responded to the defensive conditions needed by the palenques [traditional maroon settlements].

We began to perceive, with the community, that there was something in the houses calling us. We still did not know very well where we were going to end up. One of the workshops regarding the definition of ancestry aimed to identify a series of houses that the community recognized as ancestral. They said: "well … when you talk about ancestral housing, Afro housing, what [type of] housing are you referring to?".

The centre of the cultural articulation for people is the wood stove in the bohío, which is behind the house. This element is absolutely essential. The house is relatively flexible or expendable; the typology is not definitive; it does not define its character; it defines a way of life that can be changed. The updating and recovery

Fig. 2.2 Photogram from the documentary. Luz Elena Romero house, El Nispero, Montes de Maria, Colombia. 2021. The wood stove: Cooking in the "bohío", the heart of Colombian Afro-rural housing. Images by the authors

of spatiality—which normally includes modifying the view of the imaginary with them—provokes a new experience of what is really there. That is the characterization of the house to discover spatial values. It does not necessarily mean that "you all have to build with bahareque and wood". Perhaps the bohío should remain in palm leaves. Perhaps that small construction should remain, which is a very nice thing… The kitchen (that bohío that I refer to as the centre of the research) is found throughout the community. The fact that this is built collectively and not individually; that it is built with the family and with anyone who can help; and that you feed anyone who comes, gives this element a very powerful symbolic value. All this comprises the process of characterization of the house.

Osney Álvarez:

With all the aspects that we have, my dream is that El Níspero is recognized internationally, only for its culture.

(Music not transcribed)

I presume that people always feel emotion about their own culture—when we form Kuagros [a traditional form of resistance used by the enslaved] or when we do activities in the streets, the square, or anywhere—people feel that their territory is valuable. I do not mean the material things but our culture, what we feel being black. Yes, Kuagros are of great importance here.

(Music not transcribed)

Kuagros start from small groups where each one is assigned functions and a specific theme. It can be traditional medicine, sport, dance, or historical memory. Everything is framed by our ancestral customs. Our culture is always reflected (Fig. 2.3).

Rita Magallanes:

For me, being a midwife is very important because it is giving life to children who will come in the future. In the countryside, there is no help for midwives. Sometimes the mothers giving birth do not have time to leave the house, so I feel great pleasure and emotion to receive that child and provide first aid. I want nothing else; I want this to continue. If I die or my colleagues die, I hope there is someone left to represent us, so that the tradition is not lost. It is not so much that it will be lost, but this tradition is very important in a community where there is no medical centre for pregnant women when they urgently need it. That is why we are preparing some new apprentices, so the tradition is not lost.

Luz Elena Romero:

A community council reflects everything regarding our traditions, as the many things encompassed by the word ethnic: practices and customs, dialect, and the daily life that makes us different. Today, the community council is very important, allowing us many spaces, and study opportunities. Through the community council,

Fig. 2.3 Photogram from the documentary. Dancing "bullerengue" on Afro night, El Nispero, Montes de María, 2021. The shared spatiality of the collective and the oral tradition develop fundamental meaning in the struggle to recognize community. Images by the authors

we have received many scholarships to access higher education ... it is our essence.

Daniel Huertas:

Inland rural Afro communities have had to first solve a very powerful conceptual problem, which they mention from the outset: "We do not know if we are Afro or if we are farmers because farmers have always been Afro, as they told us." That leads us to investigate the life model known as the ethno-development plan. This involves diverse debates about whether this is culturalism, especially from an anthropological approach, which comprises an institutional understanding of organizational complexity, or the independent, alternative cultural model that arises from communities.

Arnulfo Cardosi:

One of the greatest satisfactions of creating the ethno-development plan within El Níspero's community council is that each step and guideline was proposed by the community. The community determined the actions and routes to take. The importance of the ethno-development plan is that it leaves the framework for identity processes, safeguarding all cultural and ethnic identities. Present and future generations participate in each of these actions, undoubtedly allowing black people affirmation within the community council.

Daniel Huertas:

What emerges from this strategic diagnosis? It clearly shows the need to respond to the political and governmental situation, the need to generate a future and the need to dialogue with the ancestors. We understand that all this, in principle had to be reflected in their life plan because it is the road map that they must be able to negotiate in a prior consultation, to be able to negotiate a policy, to be able to ask for grants or put together projects.

Arnulfo Cardosi:

I remember Daniel telling us: "Well, tell me how we can do this. Here we have some work teams, the traditional medicine Kuagro and the collective memory Kuagro, the ethno-education Kuagro. Let's give the kids some materials to express how they see themselves in ten years or fifty years from now in El Nispero. And then, let's think about how we managed to contrast that information with the older adults and make that journey within the community council." Every street has a sense that generates permanence in the territory. I believe that the main reason why we let Daniel and the academy entered El Nispero's community council was that it [the process] took absolutely nothing. It was not scripted. Curiously, he has never brought professionals to the community; he brought human beings. Human beings who have shown that they have a profession and want to serve this black community (Fig. 2.4).

Luz Elena Romero:

Here we call him "El gringo". You say [to someone]:

- "Daniel is coming".
- "What Daniel?".
- "The Gringo".

"Ah, yes. We already saw him".

Daniel has been a very important player in these processes. He has accompanied us in the construction of the ethno-development plan.

If you ask the kids up there who Daniel is... The kids welcomed him because in the activities we did, he was always ready. Truthfully, the affection he transmits is very beautiful.

Arnulfo Cardosi:

The empathy that Daniel shows ... he is not a guest, he feels like family.

Daniel Huertas:

My approach, every time I visit a community, makes me feel that I am in the ethical space where I want to be. I understand that there are other realities. I understand that architecture has other spaces and that it is multiple. What fascinates me and makes me fall in love with my profession is that I finally know that I can work in the ethical spaces where I care. Suddenly, putting a tree in front of a house is an architectural action. However, you must understand; you must enter to know if witches fit in the tree; or if they are going to arrive; or if they are going to transform into geese. All that does not create a better project, but makes you love what you do. I do not understand the profession in any other way. That is why working with communities is an exciting world for me. I do not find it in other areas of architecture. That is what I believe.

The most important change ... one of the most beautiful things after being with the community for three years has been to discover (first) that young people (above all) want to stay. Their future plans, of course, may include leaving, but El Nispero has become an opportunity. That is a fascinating thing. They want to build El Níspero and take part in its

Fig. 2.4 Photogram from the documentary. La Victoria neighborhood, El Níspero, Montes de María, 2021. Window in a "bahareque" house, the Afro-Caribbean mud construction typical of the Montes de María region. The self-construction and land management processes are imposed in an institution's absence. Images by the authors

future knowing that they can. That idea: “you can” is fascinating. They want to take part, that is the second aspect that I find exciting. They want to take part in the affirmation of their black community. From their ancestral conditions.

Osney Alvarez:

What I like the most about El Níspero is that, despite the evolution of time and technological advancement, some of the ancestral is always preserved. You see kids on the street, and you tell them something. Let's say you stand on the corners and tell a story.

Alberto Cardosi:

Here, [I like] everything. I have not seen a more beautiful town than this. I have not seen it yet. Because this village is a village where everyone is kind. That is profit being here on El Níspero.

Fidel Barrios:

I like El Níspero in juice [laughter] ... to visit it ... [laughter]

Adriana Santero:

I like that people are very friendly. I have not had problems with them. I like how they are. They treat you well, I also treat them well. I said, if they do not give me this house, let them buy it from me, but here in El Nispero I will buy anywhere.

Arnulfo Cardosi:

El Nispero is life, and as El Nispero is life, my life is anchored to this territory. Every street, every corner, every lady, every gentleman undoubtedly reflects ethnicity and how black we are in this territory. Just thinking about being outside the community council frustrates me. I believe that the greatest fulfilment I can have in my life is to be able to die in my community council; to be buried in the land where I was born. In that land where my belly button was received and planted, here in this territory. I believe that my life is anchored. Here, there are all of those who made history in the creation of this community council and the region. Undoubtedly, this is the best region in Colombia. After El Níspero, there are only mountains and snakes.

Fidel Barrios:

I like El Níspero. I would like to see in El Níspero good lighting, good schools, even some paved streets, drinking water, that we can exercise in a sports centre, that when you come, we can go to the sports centre to ride bicycles [laughter]. That is what I want for El Nispero. I want El Nispero to be Nispero [as the fruit]: sweet, tasty, kind (Fig. 2.5).

2.2.1 Teaser. The Colors of El Nispero

Every street, every corner, every lady, every gentleman undoubtedly reflects ethnicity and how black we are in this territory. Here, there are all of those who made history in the creation of this community council and the region. Undoubtedly,

Fig. 2.5 Photogram from the documentary. New generations in El Nispero, Montes de María, 2021. The visibility of futures where no one is left behind begins with one’s gaze. Images by the authors

this is the best region in Colombia. After El Níspero, there are only mountains and snakes.

If the wood stove is not on in the house, the house is sad or ruined. You must always turn on your wood stove, even if you are not going to cook anything, even if you do not have potatoes, but you must turn it on. Yes sir, joy is reflected. We all experience that. Where there is a wood stove, it always must be on, even throwing smoke to make the house look cheerful. If the stove is off, no matter how much joy one has, one says that the house is sad.

Acknowledgements The University of Los Andes supported this work under the FAPA program (Fondo de Apoyo a Profesores Asistentes—FAPA).The authors would like to thank Ma-Majarí del Níspero Community Council for their trust and love throughout the process.

References

Arteaga R (2019) La vivienda tradicional en el caribe colombiano. Credencial Historia No. 350. Febrero 2019. Arquitectura y tradición. Banrepcultural. Red cultural del Banco de la República. https://www.banrepcultural.org/biblioteca-virtual/credencial-historia/numero-350/la-vivienda-tradicional-en-el-caribe-colombiano

Domínguez MI (2015) Comunidades Negras Rurales de Antioquia: Discursos de Ancestralidad, Titulación Colectiva y Procesos de "Aprendizaje" Del Estado. Estudios Políticos 46:101–123. Instituto de Estudios Políticos, Universidad de Antioquia, Medellín

Escobar A (2001) Culture sits in places: reflections on globalism and subaltern strategies of localization. Polit Geogr 20(2):139–174. Amsterdam, Elsevier

Escobar E (2010) Territorios de Diferencia: Lugar, Movimientos, Vida, Redes, 1st edn. Envión Editors, Bogotá

Fals Borda O (2001) Participatory action research in social theory: origins and challenges. In: Reason P, Bradbury H (eds) Handbook of action research: participatory inquiry and practice. SAGE Publications, Thousand Oaks, CA, pp 27–37

Freire P (1970) Pedagogia do Oprimido, 36ª. Paz e Terra, Rio de Janeiro

Greenwood D, Morten L (1998) Introduction to action research: social research for social change. SAGE Publications, Thousand Oaks, CA

Gutiérrez E, Silva-Tapia A, Garbe S, Cárdenas M (2020) Ethnicities in dispute: new paths, new challenges. Ibero-Am Mag XX(75):207–232. Iberoamericana Vervuert, Madrid-Frankfurt

Hampaté Bá A (1983) A tradição viva. In: História geral da África. Unesco (Org.), Ática

Holmes B (2005) La Comunicación de La Información Cartográfica. In: Conferencia Del Taller de Cartografías Tácticas. FADAIAT, Tunez

Huertas Nadal D, Pinedo Cobos C (2021) Vivienda rural afrocaribe: espacialidad, tradición y futuro. Contesti. Città, Territori, Progetti 2(2):129–158

Larios PM (2015) Vernacular housing in the Colombian Caribbean: diversity within the unit. In: We of the Caribbean, chap 6. Simón Bolívar University Editions, Barranquilla, pp 179–200

McDonald C (2012) Understanding participatory action research: a qualitative research methodology option. Can J Action Res 13(2):34–50. https://doi.org/10.33524/cjar.v13i2.37

Montoya A, Solarte E (2016) San Basilio de Palenque. Patrimonio intangible en riesgo. Rev Uni-Pluri/Versidad 16(2):63–73. Universidad de Antioquia-Facultad de Educación, Medellín

Parra-Valencia L, León EA, Jaramillo LG, Galindo D, Luders S (2020) El lumbalú y las mujeres tejedoras de lo espiritual y comunitario. Psicol Soc 33:1–16. Associação Brasileira de Psicologia Social, Belo Horizonte

Restrepo E (2013) Etnización de La Negridad: La Invención de Las Comunidades Negras Como Grupo Étnico en Colombia. Editorial Universidad del Cauca, Popayán

Rodríguez F (2002) Palenque de San Basilio: masterpiece of the intangible heritage of humanity. Ministry of Culture. Colombian Institute of Anthropology and History, Bogota

Rojas A, Restrepo E (2004) Conflicto e (in)Visibilidad: Retos En Los Estudios de La Gente Negra En Colombia, 1st edn. Universidad del Cauca, Popayán

Taboada DM (2009) The bohío and intangible heritage, dangerous relationships. In: Orality yearbook Nº19, patrimonial sites oral traditions. UNESCO Regional Office for Latin America and the Caribbean, Havana, pp 60–65

Walsh C (2007) Interculturalidad y colonialidad del poder. Un pensamiento y posicionamiento "otro" desde la diferencia colonial. In: Castro-Gómez S, Grosfoguel R (eds) El giro decolonial. Reflexiones para una diversidad epistémica más allá del capitalismo global. Siglo del Hombre, pp 47–62

Download Link

YouTube. https://youtu.be/V7I_MSWGlvE

3 Moving Towards Different Directions —Thoughts About Thinking-with Environments and Each Other

Erika Henriksson

Abstract

'*Moving towards different directions*' is a narrative essay that expose challenges and possibilities of working transdisciplinary and exploratory through a collaborative and participatory, hands-on building project. The essay, that draws from experiences of transforming a building together with residents at a rehabilitation clinic starts with a visit to a building store. There, a situation opens for insights concerning how directions towards different environments has a big impact on both social and material relations. Something that set off a reflective process concerning, how we become different depending on where we are, and what impact environments can have on power relations in a participatory project of making the built environment. But beyond the learnings and insights that are shared through the text, is the essay also an exploration in the format for communicating a relational building project. It is written with the intention to invite the reader to join two builders in their interaction and conversations while moving between different environments and situations.

Keywords

Environments · Participation · Power relations · Hands-on building · Care · Thinking-with

As we approach the store it starts to rain. It's pouring and just before I turn off the main road, I hear the first flash of lightning. The sky is gray blue. It has darkened rapidly. When I left the cottage this morning it was sunny and bright. Like the weather, I've also seen the colors of his sky gravitate towards darker shades during the drive to the building store.

Once we are there, he gets stressed by the store assistant who has another customer at the same time as us and who also can't answer my questions. '*Maybe we should talk to someone else*' he says, looking at the man I asked for help, '*Your boss for example.*' As the assistant walks away, he turns to me and says '*You're so good. I'll support you, but I don't know anything. I don't understand any of this.*' I open my mouth to say that he does know a lot, but just then the store assistant calls out to me, and I turn away from him to answer. While I'm talking, I hear him say, '*I need to smoke, can I have the keys?*' And in a reflexive movement, I hand over the keys to the car and watch him walk away.

It takes some time, but eventually I manage to get more of the leca blocks we're using to wall up a new foundation. I load them on the car, pay, and as I drive out of the building hall, he comes walking towards me from the other corner of the

E. Henriksson (✉)
Architecture and Design, NTNU, Trondheim, Norway
e-mail: erika.henriksson@gu.se

S. Hilal et al. (eds.), *Design for Partnerships for Change*, Sustainable Development Goals Series,
https://doi.org/10.1007/978-3-031-36993-3_3

parking. He gets in the car, and I drive off. It's all quiet.

After a while I feel that I have to say something, '*I'm sorry for turning away, I didn't mean it to be like that. It was just so much happening at once.*'

"*It's okay,*" *he replies,* "*I was the one who didn't know anything. I want to do good, but I didn't understand anything you were talking about, so I just left, even though it felt wrong to do that*"

I can hear in his voice that he is trying to hold back tears and feel uneasy about how the visit turned out, thinking that I should have acted differently.

After driving for a while, the forest ends and wide fields spread out. On the left side of the road stands a large wooden house that looks like a castle, and not far from it a small dirt road leads to the road we are driving on. I turn and follow the road that winds towards the house and ask if we should make a visit.

Three large buildings form a courtyard that flows into a garden. We enter one building which is empty, apart from us, and I can already see how the dark clouds that have colored his existence for the last hour begin to dissipate.

We walk around in different rooms, looking at the furnishings, murals, and building details. At first, we walk together, but we move in different rhythms and drift apart. Sometimes someone calls out to the other to show a discovery, but soon we are completely absorbed in our own experiences.

After a while, we move out of the house and walk towards a garden, where tense drops of water glisten in the green. We follow narrow paths, before deciding to move towards a bench. It's wet, so he wipes off the worst of it with a sweater he's had tied over his shoulders. '*Who do you think has been living in the house?*' He says as we sit down.

'*I don't know, I think they seem to have been clever, but they must have been rich also, don't you think?.*'

He looks out at something far away. '*Sometimes you can almost sense people who aren't there anymore. Do you know what I mean? Energies. It doesn't even have to be people, it can be places or houses that are abandoned, you feel something.* He points to the garden we just walked through, '*Maybe there have been people doing something here. Can you see them among the green? Over there is a little boy sitting on a rock.*'

I manage to catch a glimpse of the boy before he disappears and think that nothing has to be fixed; environments can change in our imagination. Through how we enact and use them, we can transform the meaning of our surroundings, or who we might allow ourselves to be in different situations. At the same time, is it impossible to ignore that what surrounds, has an impact. This became so clear today. How stress can grow in one environment but be calmed in another.

Sometimes it seems to me like the world wants to present us humans as autonomous, stable entities that exist and behave completely independently and without being influenced from the world around, but we are connected to it in every way. We influence and are influenced by environments and things around us, and we influence and are influenced by each other. Something queer feminist cultural theorist Sara Ahmed has highlighted and reflected on through the concept of *orientations*.[1] For me is Ahmed's, discussion on orientations rich and contains both a critical reflection on how forces and norms in society can influence and steer people in different directions. At the same time, as she offers hope and shows that we can choose what we turn towards, and by this take some control over what influences us.

When we started this project, we had to find and understand what the project direction was, but also what we could use to create a movement together. This brings me again to Ahmed, and the concept of bodily horizons, which she describes as 'a 'line' that bodies can reach toward [...]'[2] These can be both physical objects and concepts,

[1] She describes *orientations* as starting points and as directions towards which bodies turn to continue from a "*here*" (Ahmed 2010, p. 236).

[2] Ahmed (2006, p. 552).

which make up all that gathers around an activity, and help us to find our way.[3] Different activities mean that different things gather around people who are acting; a writer needs a desk, paper, and pen to write with. While those who think and test by transforming an old building, need to gather other things and ideas around them. And because we (in the project) have different backgrounds and experiences of life, we needed to spend time to work out what exactly we could reach after and send between each other to create a shared movement. We needed to find what ideas and thoughts could guide us forward, but also what words, concepts, and tools we could use as we built up the direction for our exploratory journey. And by turning to environments around us, we were able to think, discuss and develop the process together with buildings, things, environments, and stories that we came into contact with during spontaneous and planned excursions. In that way, our process developed into a form of situated, embodied, and relational thinking that started from things we shared and had in common because we experienced them together.

Initially, conversations carried out while moving around and visiting different places in the surrounding environment helped us to build a framework and direction. But now that we have started to build and transform the small building, are we also turning to environments to gather materials and tools, such as the visit to the building store a moment ago. Meaning that the function of environments has partly changed. They are no longer just places of experience that can help us build a world of ideas about possible futures. Now our thoughts are also challenged in encounters with the weight and complexity of the material world. And the environments we turn to need to also be materially productive. And as we sit there on the bench, I begin to think about the meaning of the project in relation to environments. I am also paying more and more attention to the store and wondering about the influence that something as specific as a building store can have on interpersonal relationships, on people's perception of themselves and on their relationship to the whole material world around.

I look back on the visit there a moment ago and am reminded of Ahmed's observation that it usually takes a while before one actually arrives at certain activities.[4] Ahmed has also pointed out that we approach activities with different bodies[5] and that different bodies has different allowance to extend in certain spaces.[6] I think that we also approach activities with different knowledge, and expectations of ourselves, for example who we would like to be, or who we feel we should be. Something that points towards that arriving is not just about the moment and the place where we are. The now. The circumstances of our arrival are just as much about what is behind us. What Ahmed calls the background.[7] Each of us has a background and a history. Things that happened earlier in our lives, that influenced and shaped us into who we are able and capable of being at that particular time and place, when we enter.

When I began to direct my interest towards a building approach to architecture, there was a lot that I needed to conquer, grasp, and learn; the craft of building itself, but also the environments and things that went with the building activity. A journey that has taken time, partly because I had to find my own way. But the arrival has also been delayed because of obstacles that I encountered on my way, and initially the building store was such an obstacle.

I think that this was because the building store represents and manifests a knowledge that is part of the building activity and is a place where people can act out and claim to have knowledge. Which at the same time means that non-knowledge can be made visible and obvious to oneself and ones surrounding there.

I remember walking into building stores feeling like I had opened a book and didn't understand half the meaning of the words. Something that made me aware of lots of things that I didn't know and couldn't do. A realization

[3] Ibid., p. 553.

[4] Ibid., p. 549.

[5] Ibid., pp. 552–553.

[6] Ibid., p. 563.

[7] Ahmed (2006, pp. 547, 549, 560).

that also brought feelings of disempowerment, distance, and alienation. For although I was close enough to pick up and feel the things on display in the store, they, and the contexts they could be used in were still very far away from me. I was not participating, and I had not arrived yet.

My journey of arriving has also been affected by having a woman's body in a building world, that still can be described as male coded. I experienced, how a male body more easily were taken seriously, had a voice people choose to listen to over mine, but also had more space to act as a builder. Something that created uncertainty and frustration, where I realized that a path into building was not only about learning the knowledge that inhabits the activity but was also about learning to navigate the various situations and power relations that surround the activity.

Today, the building store and the building world is no longer something alien and uncomfortable, for me.

I have arrived. But that doesn't mean that those I work with have the same feeling about stepping into the store, and therefore not the same opportunity to act. Something that points to another importance that environments have in our project; the role to influence participation and power relationships.

This insight problematizes the idea of the environments as something that we share, as it is not as simple as just because something is physically present, or because we together share the experience in time and space, is it equally accessible to us. I therefore realize that I need to remember to not only think about the material and practical aspects that different environments can offer, but also to reflect on the impact they have on participation, accessibility and how interpersonal relationships are affected and changed depending on where we are.

'*Now, I'm getting tired of this,*' he says, getting up from the bench. '*Why don't we go back and start working on the last plinth?.*'

I snap out of my thoughts and follow him. When we get to the parking, I see the blocks piled on the flatbed of my car and realize that I know very little about them. Then I turn and look around at the landscape and think that the materials humans use to build with come from there. Or not exactly from there. They come from other landscapes around the world where there are rocks, trees, and water. Something that's easy to forget when you're walking around among insulation and bags of concrete inside the building store. There is no connection to the origins of the materials and the consequences their creation have had on the ecological system they were part of and moved through during processes that shaped them into what we see in the store; we know nothing about the hands and bodies that worked on their creation; what materials were combined to shape them into what they are today, or what traces were left in the environments from which they were taken. We miss large parts of their background; everything that happened before they ended up at the store, and therefore select them only on the basis of what they appear to be at that particular time and place.

The anonymity of the blocks makes me realize that the way we currently relate to the materials in the project, allows for a rather limited and reductive relationship with the material world. Something that goes against the ideas we have about what a building process can mean and aspire to do; we have embarked on a journey where we seek to move in a direction where we, by transforming the small house, simultaneously explore how a building process can serve as a tool for practicing caring and responsible relationships with ourselves, each other, and the world around us. And the material world is very much part of that relational fabric we bind together through the process.

With these new thoughts, I drive out onto the road we turned away from a while ago. As the car moves forward, I look at all the little roads leading off the main road and think that it's important to remember to make those turns away from the straight and clear path. That I mustn't get too caught up in the productivity of building but remember our visions and that we can orient ourselves towards something that we don't know what it is. Environments and situations where playfulness and adventurous discovery still can be allowed to be part of the process. And in order to develop ideas based on values other than the current and dominant ones, one might need to enter spaces of exception; small pockets and folds where these values can be nurtured and grow. These 'other' spaces also serve a critical important function in society because they can stand as alternatives to the dominant ones, and therefore can help us in our explorative journey towards other ways of thinking, being, and building.

Our such space for nurturing an alternative desire and approach has been the relationships that have grown between us in the project through the unfolding process. But this togetherness has not happened in a vacuum. It has taken place in various situations that we have created or directed ourselves towards. Environments in which we have talked, thought, speculated, and acted, and of course it matters what values and ideas these environments are made of.

I look at the road in front while my mind is directed on two different theorists that I have been reading lately, which both in different ways helped in deepening and expanding my understanding on how the surrounding matters and has in influence on humans. One of them is feminist thinker Donna Haraway and the other is the philosopher Jaques Rancière.

When reading Rancière have I been especially interested in his discussion on *the distribution of the sensible*,[8] which can be describes as an operation of inclusion and exclusion carried out by the ones in power, a power that is manifested in what is allowed to become visible, sayable, audible, doable, and possible in the (physical) world.[9] Rancière has also pointed out that since this operation determines what have a right to appear in what we share as common, it therefore also determines who and what forms what is part of the common.[10]

For me, does Rancière manage to expose the connection between aesthetics in spatial environments and the politics of inclusion and participation in a society, by pointing towards the meaning and importance that lies within being connected to what is made visible and seen in the common space. At the same time as his writing also makes it very clear for me that human-made environments never are neutral.

This brings me over to Haraway, who in the book '*Staying with the trouble*'[11] repeatedly reminds her reader that it matters what we engage with when forming something new. In the book Haraway states that '*It matters what matters we use to think other matters with; it matters what stories we tell to tell other stories with; it matters what knots knot knots, what thoughts think thought, what descriptions describe descriptions, what ties tie ties. It matters what stories make worlds, what worlds make stories.*'[12]

She builds this argument from the work of anthropologist Marilyn Strathens who once wrote that '*it matters what ideas we use to think new ideas (with).*'[13] And I think that spaces and environments created by people are just that, ideas. They are ideas materialized and manifested in the physical world, created from values, ideologies, and thoughts about how some people, thought that the world should be. Which means that it also matters from which environments one desire, think, and make new environments.

In our project, has working and exploring from a physical place come to mean trying things out in improvisation with the shifting context. Something that has meant that I, and I together with my collaborators, have sometimes realized that we have moved in a direction that is not consistent with the intentions that guide us. It is not surprising that this happens, because each new environment and situation grasps at a multitude of aspects that leads us in different directions, making it difficult to have an overview. However, precisely because we are working in an open and improvisatory process, is it also possible to use insights about misdirection to redirect the movement and explore what can be done differently to get a little closer to something that

[8] Rancière (2000:2004, 2006).

[9] Rancière (2000:2004, p. 85), Smith and Weisser (2011, p. 10).

[10] Rancière (2000:2004, p. 12).

[11] Haraway (2016)

[12] Haraway, p. 12.

[13] Strathen (1992, p. 10).

can give shape to our guiding intentions. When it comes to interpersonal relationships, there are a few things that are particularly important to be aware of in these ever-changing contexts, and that is that there is room for everyone involved in the work to influence the process and their own circumstances in it.

The fact that we who work together have different histories knowledge, skills, and experiences of acting and using power, is something that can have both positive and negative effects; one positive thing is that such a collaboration carries with it the possibility that the process is shaped and influenced by perspectives, values, and voices that are rarely given space in a normative construction process. However, a relationship defined by a power asymmetry can also lead to oppression, be inhibiting, or provide a sense of disempowerment, for those with less power. Which can work to prevent opportunities to actually influence and truly participate.

In our project have I the most power. For example, I do not live in an institution, but am free to move and can control my life to a much greater extent. As a result of this, do I become a link between life in the institution and the society outside; I can provide access to environments and experiences that would otherwise be difficult to access. Which is a circumstance that gives me an insignificant status, which can be present as an invisible force that influences different situations. I also have a power advantage in my role as researcher, architect and as the one who publicly tells our story, and together all this contributes to my strong position within the project, and therefore I have a responsibility to examine how I can use this power with care.

It is hard to escape that I am main responsible for the design of the building process. For example, making the final choice of the techniques we build with, arrange the logistics of construction, or control the stability of the structure, etc. It would even be imprudent of me to pass that responsibility to someone who has no experience of building. But I can ensure that the choices I make and propose grow from shared convictions regarding the direction we are moving towards. I can also make proposals based on a careful consideration of the needs and conditions of others.

'*Can't we drive along the river instead of through the village,*' he says, pointing down towards the valley. I turn off the asphalt road and onto the dirt road that leads us directly to the little house. When we arrive, he quickly gets out and folds down the hatch on the back of the flatbed and starts unloading.

We stack the blocks in a pile against one wall of the house and carry the last of them inside. I sit down on one of the plinths opposite him.

"*Do you remember the tongue in the river?*" I say.

"*Yes, of course. It was nice there. Maybe we can go there sometime? It would be nice to see it again.*"

"*Yeah, maybe we can go there pretty soon. I thought that instead of just working with blocks, we could try to work with stones also. We could for example go to the river and see if there are any stones there that we can work with. What do you think?*"

"*Yes, that would be something. Do you mean that we could collect the stones ourselves, and then build with them?*"

"*Yes, I've been thinking that the building store didn't turn out so well today and maybe we can try something new.*"

"*I think that sounds really good, then we should look for flat stones.*"

The ideas behind the new way of getting materials as well as new material to work with, come from different insights and thoughts. One is to explore how collecting materials can be done in ways where more people than me have the possibility to act. Because, by the river we are no longer facing an environment that requires prior knowledge or sets up for social situations that take a while to arrive at. This could allow the activity of gathering materials to resemble the open movements within the landscape through which we once developed this project. Just like earlier today when we visited the wooden castle. Or like that spring evening when we went between different places along the river and found the long stone tongue that went out into the water.

I also think that collecting materials directly from the places where they been created over long periods of time, can make us more aware of the circumstances of their lives and history. Something that brings us a little closer to some of the backgrounds that are not available to us in the building store.

I see this new orientation as an exercise in responding to circumstances that occur and emerge during the process, where I can train myself to dare to change directions along the way, instead of sticking to predetermined goals and thoughts. At the same time as this is a step towards a process where we turn towards

Fig. 3.1 Turning towards a new environment for gathering materials to build with

Fig. 3.2 Collecting stones down by the river

Fig. 3.3 Mixing mortar

Fig. 3.4 One thing that I have come to find important when working in an exploratory and participatory project is to stay open for allowing approaches and methods to transform and change following insights and learnings that are made through the unfolding work

different environments to explore how relationships and thinking are influenced and played out in different ways depending on what we turn towards. Meaning that we also can see environments as fields for learning about ourselves and the world around us, who we become, and want to become, depending on where we are. That we start to think about the environments from which we create new environments (Figs. 3.1, 3.2, 3.3 and 3.4).

Acknowledgements I want to send out my warmest and deepest thank you to (my co-supervisor) Meike Schalk, who have been a critical and careful reader during the development of this text.

References

Ahmed S (2006) Orientations: toward a queer phenomenology. GLQ J Lesbian Gay Stud 12(4):543–574

Ahmed S (2010) Orientation matters. In: New materialisms—ontology, agency and politics, Coole, Diane & Frost, Samantha. Duke University Press, Durham and London

Haraway D (2016) Staying with the trouble: making kin in the chthulucene. Duke University Press, Durham and London

Rancière J (2000:2004) The politics of aesthetics—the distribution of the sensible. Continuum, New York

Rancière J (2006) Thinking between disciplines: an aesthetics of knowledge. Parrhesia 1:1–12

Smith JR, Weisser A (eds) (2011) Everything is in everything: Jacques Rancière between intellectual emancipation and aesthetic education

Strathen M (1992) Reproducing the future, anthropology, kinship, and the new reproductive technologies. Mancheser University Press, Manchester, UK

4 The Story of the Design Partnership Behind ZeroCityPlus: A Community Game for Participation and Local Governance Beyond Net Zero Goals

Carolina Vasilikou, Yiorgos Papamanousakis, and Felipe Lanuza

Abstract

Can playing a game help us raise awareness of the climate crisis, contribute to behavioural change towards a more sustainable way of life, and create plans and strategies for cities that can regenerate the ecosystems we all depend on? What would this game look like? The paper investigates the design partnership between an academic institution, a small not-for-profit organisation, a research associate, and three local community groups. We researched the development of an urban game that negotiates the conflict between the multiple stakeholders of sustainable urban development, public participation, and climate literacy. The goal is to trace the journey, as citizens, communities, businesses, and local government, together, towards net zero and carbon-negative strategies and action plans, for a more sustainable and resilient city. ZeroCityPlus (ZeroCity +) is one of the fifteen Design Exchange Partnerships participating in the inaugural year of the Future Observatory, a UK-wide research programme coordinated by the Design Museum in partnership with UK Research and Innovation and the Arts and Humanities Research Council. The project methodology was based on design research, prototyping, active community workshops, and a test-learn-improve approach. Successive iterations of gaming workshops were developed with local communities in Nottingham, Reading, and London. The paper presents the findings from the first phase of the project and the story behind the development of a hybrid analogue and digital urban game for public participation beyond net zero.

Keywords

Urban game · Design partnership · Public participation · Climate literacy · Sustainable urban development · Resilient city · Community engagement · Design research

4.1 Introduction

By concentrating the majority of the world's population and being the major source of carbon emissions, cities are a key battlefield against the environmental crisis. The problem of climate change is, to a great extent, an urban problem. How authorities and planners coordinate with communities and other stakeholders is crucial for adapting cities, policies, and collective

C. Vasilikou (✉)
School of Architecture, University of Reading, Reading, UK
e-mail: c.vasilikou@reading.ac.uk

Y. Papamanousakis
Urban Transcripts, London, UK

F. Lanuza
University of Sheffield, Sheffield, UK

S. Hilal et al. (eds.), *Design for Partnerships for Change*, Sustainable Development Goals Series,
https://doi.org/10.1007/978-3-031-36993-3_4

behaviours to reach net zero goals and reduce (if not revert) environmental damage. However, there is a widespread disconnect between urban policymaking and citizens. Often, tokenistic public engagement in local government planning consultations leads to actions and regulations that fail to respond to people's needs and expectations, resulting in scepticism and passive citizenship. Therefore, with the environmental crisis becoming the greatest challenge for cities, a deep and robust citizen-policy consensus appears to be imperative. On such a basis, the mediator of an evidence-based serious game design methodology could contribute to ways that a collective understanding of the environmental problems and sustainable goals opportunities at a local level can be generated and translated into action and community-led policymaking. ZeroCityPlus (ZeroCity + in the initial project; ZC + will be used as an acronym for the rest of the paper) is a design prototype of a community game that aims to involve society as a whole in a meaningful bottom-up vision to produce a much-needed change in the reality of public participation.

4.2 Background

Scientific research has led to the understanding of the environmental crisis becoming increasingly comprehensive and detailed, which has determined the application of several technical solutions at different city levels and scales. Technology can enable the means for "green growth" of cities into environmental sustainability, but it cannot do it alone. There are deeper political transformations that need to occur to change established socio-cultural and economic structures and practices at the level of the local government. These political transformations are less often developed when compared to science and technology, but they are essential to define the framework and complement their advancement. One of them is the relation between local governments and citizens, where distrust and lack of meaningful participation seem the biggest issues.

Within the last decade, there has been a gradually noticeable lack of trust that citizens show to their respective governments and each other. This is inevitably linked with rising levels of loneliness and marginalisation in modern societies. In the rapidly expanding culture of fear and mistrust, currently nurtured in urban environments, playful co-operation and game design can produce opportunities for social interaction and relationship building towards net zero + goals in an innovative, experiential, and playful way. On a political level, games allow information to be shared and discussions to be made, offering a broader perspective on issues at stake and creating empathy. On an institutional level, a gamified interaction can promote experiential opportunities that help with community bonding and encourage people to reflect about the impact of their collective behaviour on the environment. On a personal level, games can train people to deal with certain processes, unexpected difficulties, and surprising situations. ZC + is developed to bridge the gap between local governments and citizens through an evidence-based gamified participatory process. Games have the potential to promote dialogue and trust between different players. Previous experiences in using games for public engagement in urban planning and policymaking contexts have demonstrated their capacity to foster collaboration, integrating collective knowledge (Play the City 2019).

Local governments play a critical role in engaging people in net zero strategies. Their own agency is perceived through tangible impacts due to the processes taking place in the everyday environment and life. Therefore, public engagement could be embedded in governmental net zero strategy, with the potential of both developing it internally at high level, as well as drawing on externally conducted participatory bottom-up processes (Demski 2021; Sasse et al. 2021). This approach fits the strategy of ZeroCityPlus as a gamified participatory process delivered at a council and/or local authority level. This can potentially align with broader strategies to achieve net zero and beyond, as in the case of Bristol's One City Plan (Bristol Council 2021), enhancing public participation in local policy and decision-making, and action for change.

Urban games have already been adopted by cities in piloting projects with great results across the globe, such as Oslo's Trafikkagenten spy game app, which allows young citizens to act as "secret agents" for the city, live-reporting any difficulties or safety hazards that they encounter. They create emotional moments in public space, in an effort to explore how games and new communication technologies create new hybrid social spaces, in which the private and the public are intertwined. Play elements and playful urban spaces have an inverse relationship with fear and mistrust since the prerequisites for a place to be playful are that it should be safe, accessible, and inclusive. Moreover, games, when properly conceived, can act as feedback systems that prescribe certain kinds of player actions, whilst accommodating changes in the system that emerges from play. These case studies, mainly coming from Europe, show few examples of application beyond academic frameworks. Ekim Tan's method of City Gaming (to understand urban planning and design as gaming), developed from her PhD work, stands out as a well-established consultancy practice, platform, and database that integrates multiple games (by different authors) in an open-source platform (Play the City 2019).

Sasse et al. (2021) argue for a wide range of participatory processes, yet in their list of "Methods for Public Engagement in Decision-Making" there is no specific mention of games or gamified processes as possible methods for public participation at a local government level in the UK. The potential of games and gaming in public participation in this context remains largely unexplored, with the ZC + project providing a design research opportunity as well as expanding the practices of community engagement towards addressing environmental concerns from an urban perspective at a local level.

The project is built on existing research experiences of participatory game creation in the UK, where institutions like the Design Council and the British Council have liaised with game experts for creating games as public engagement tools for communities to explore possibilities and innovation in local areas and address the Sustainable Development Goals (Dodig and Groat 2020). For example, the board game "Building Futures" developed by CABE, the RIBA, and the AOC was based on urban planning scenarios to bring communities, authorities, and developers together to encourage debate and consensus reaching in a context simulating a real planning process. The game was trialled in a number of cities in the UK in 2008.

4.3 Methodology

Although some overlaps exist when comparing to some games, the focus on behavioural change and community action towards environmental care and net zero strategies remains unexplored. ZC + used a test-learn-improve approach, to frame within it the agency of an independent mediator that can set a common ground between local authorities and citizens; a space where they can build trust and mutual commitment from the outset.

The design methodology followed that of a "serious game" development (Serious Games Society 2014). The term serious game refers to those games created with the purpose of instructing as well as entertaining, including for example learning and training in political or business settings (Abt 1970); or those that do not have fun or joy as their main purpose (Michael and Chen 2006).

The overall design methodology is related to the non-gaming elements and activities (debrief session, follow-up, guided conversations, etc.) happening prior to, between and after the game stages, framing its relationship with the reality on which it is intended to have an impact. Elements of the envisioned future become an anticipation of a desired common goal upon which stakeholders agree and commit.

A relevant precedent for ZC + , which would now be regarded as a serious game, dates back to 1904, when feminist, activist, and game designer from Maryland Lizzie Magie created "The Landlord's Game" to highlight and illustrate the injustices of the land tenure and showing the social benefits of land tax to discourage speculation. The game later shifted (despite Magie's agency) and became the well-known Monopoly, based on capitalist principles where one landlord is eventually the winner by leading others to bankruptcy. Salen and Zimmerman (2003) point out that being "*the direct progenitor of Monopoly, it is ironic that Magie's game became a parody of exactly what it intended to critique. What began as an earnest attempt to educate the masses about the ills of land monopoly was transformed by Parker Brothers into a rhetorical tool for capitalism itself.*" (Salen and Zimmerman 2003, pp 578–579). Conversely, a number of games intending to counter the principles of Monopoly have appeared. Anti-monopoly, from 1970, which tries to recover the meaning of Magie's "Landlord Game" by showing how harmful monopolies can be for free market (Anspach 1973) has had a number of derivations. Co-opoly tries to counter Monopoly's individualistic drive and puts forward the collaboration between players as a key aspect (Van Slyke et al. 2015).

To a certain degree, ZC + drew its methodological framework from the Landlord's Game mechanics and learnt critically from Monopoly's widespread assimilation in popular culture. Almost everyone has played Monopoly and is likely to quickly learn to play a game following the logic of throwing the dice and moving a token along consecutive squares across a board. Nonetheless, ZC + intended to revert the individualistic success logic that reflects the radical capitalism exploitation that has harmed our societies and the environment.

However, the direct predecessor to ZC + is not Monopoly, but the experimental urban game Remix(c)ity is created by Urban Transcripts in partnership with Konzulat Studios in 2018. ZeroCity + was modelled as a series of design prototypes with a strong community-led input in the design development process, around a collaborative framework of negotiation, empathy building, and solidarity, where players develop coordinated actions and strategies to reduce the carbon footprint of a specific neighbourhood or urban area to net zero. There is not a single winner, only collective winners or collective losers.

ZC + was initially developed as an analogue board game in late 2021, using the ability of this typology to educate players about highly complex systems, and its potential to become consensus-reaching tools in such contexts, as anticipated by Richard Duke nearly five decades ago and multiplied in effect by Play the City in the twenty-first century (Duke 1974; Play the City 2019). That is the case of a recent research experience where a commercial board game (named C02) about the environmental crisis was adapted into a game called Climate Policy to create a model of a complex system for players (representing power companies) to educate citizens about the impact of energy transition at a local and global level (Castronova and Knowles 2015). According to the researchers, when the rules of a board game are clear and explicit, these can be easily changed or adapted, being more flexible in terms of game design and testing than digital games where the rules are "buried in computer code or a game script" (Castronova and Knowles 2015). Another experience of specifically created analogue board games that convey urban planning collective decision-making among participants (civil engineering students) also demonstrates that developing serious games that integrate participatory processes can bring benefits given the nature of analogue forms in fostering collaboration between players (Sousa 2020).

The way in which environmental problems and the climate crisis relate society, culture and the built and natural environment, is perhaps the best example of such complex systems and ZeroCityPlus' development in its first stages and prototyping benefitted from these conditions.

4.4 Underpinnings of an Urban Game that Aims to Mediate Scientific Knowledge with Playful Engagement

Behavioural change at a society level is a vital part of net zero aims, as the majority of emission reductions will require people to adopt low-carbon technologies and lifestyle changes (Demski 2021). As a tool for both action and study, ZC + aimed to enhance the understanding of people's behaviours and attitudes towards net zero + practices in the public context of decision-making and everyday life. This aligns with the strategy of setting common grounds and vocabularies across different stakeholders, for example pointing at the United Nations 17 Sustainable Development Goals and in particular, goals 11 (sustainable cities and communities), 12 (responsible consumption and production), 13 (climate action), and 17 (partnerships to achieve goals (Drew et al. 2021).

Individual behavioural change is often portrayed as the way to spearhead the response to net zero challenges. However, without higher level policy support individuals cannot change systems, nor can climate change be averted by "outsourcing" institutional, governmental, state, and wide societal responsibilities to individuals, who are themselves often reduced to consumers rather than citizens in the participatory process. ZC + 's design framework is about linking individual and collective behaviours to policy change, so essentially saying that for individuals to change, policy needs adapting to support and enable the required changes. The main ingredients of the game's design framework are

1. encompassing diagnosis (sharing of knowledge),
2. action (levers of impact and change discussed collectively),
3. creating opportunities for mutual accountability, and
4. enabling the effective delivery of a common strategy.

As a fast, low-cost prototype, its objective is to support councils in a meaningful communication and co-decision-making with stakeholders and achieve the complex social, environmental, and economic net zero carbon 2030 and 2050 climate change goals.

Another parameter of the design methodology was the equalising partnerships in terms of equity and enabling and empowering communities to participate as peers in city making. In this way, public engagement through ZC + considers deepening and widening democracy, a key underlying issue to the environmental crisis (Lele 2020) and building up compelling narratives to empower citizens and communities to take action (Carmichael 2019). Beyond attempting to promote small changes, the main aim of the urban game was to contribute to a huge shift in lifestyles and consumption patterns that are commensurate with the current climate crisis.

In its meaning and methods, ZC + embraces net zero strategies, for example, in adopting the idea of a carbon currency as a way to set the game mechanics, incorporating a timeline to resemble the pressing environmental situation being, literally, a race against time. However, many elements and the overall meaning of the game go beyond the narrow, quantifiable carbon counting. In this way, the environmental crisis and all actions and changes to be prompted at a neighbourhood level (enabled and facilitated by local authorities) are considered in an integral way, avoiding "tunnel" visions (Lele 2020, pp. 47–48). Hence, the meaning of the "plus" in ZeroCityPlus goes beyond net zero in a quantitative fashion and attempts to contribute holistically towards people and communities living and thriving in harmony with their environment (Drew et al. 2021).

The project directly relates to the two policymaking strategies by which, according to Carmichael (2019), societal and behavioural change towards net zero scenarios can be achieved. In its educational role, it aims to raise awareness and encourages community members to take specific actions towards significant emission reductions. As a tool for collective decision-making, it envisions to support and enable policies at a local level to help create social and political contexts that nurture public engagement in climate action, aligning with Demski's scheme of public engagement in action and decision and policymaking (Demski 2021).

One of the main objectives of ZC + was to unlock opportunities of public agency in policymaking when the participation process will happen at early stages (AHRC 2020). Normally done towards the end and almost as a box-ticking exercise, public consultation and participation implemented in advance are perceived as more meaningful. Furthermore, design prototyping through the scenarios created in the game gives chances to evaluate the viability and desirability of policy concepts. This project thus aligns with other designers trying to transform the dialogue between policymakers and citizens (AHRC 2020).

By moving across a timeline into the near future, fictional scenarios are created due to the effects of climate change, social and political contingency, and the agency of local people and authorities to face those challenges in the best possible way. ZC + becomes a speculative design tool that envisions a future where the accelerating negative effects of climate change become tangible and local, whilst communities and authorities work in a coordinated mode with the agency to respond through a number of lifestyle changing initiatives. This can encourage people to "provide critical feedback on ideas – such as policy options– in order to arrive at actions that are possible, plausible, probable and ultimately preferable." (AHRC 2020, pp.19–20).

Specifically, the first phase of ZC + as a design research project, in the context of the design partnership, had the following objectives:

- Understanding the state of the art at the intersection of the fields of public engagement, urban planning and design, environmental action and behavioural change, and games as a participatory tool.
- Creating, designing, and manufacturing the game to test it through a series of workshops to assess its functioning and impact in different contexts across England using feedback gathered from participants and the team's on-site observation.
- Evaluating the game's capacities and shortcomings in responding to the aims to support Urban Transcript's mission through a participatory tool.
- Setting out next steps of development for the scaling-up and refinement of ZC + prototypes, envisioning future evidence-based real-world implementation.

4.5 The Story Behind the Design Partnership of ZeroCityPlus (ZeroCity +)

Remix(c)ity is the direct precedent to ZC + and was developed by Urban Transcripts in partnership with Konzulat Studios for the MakeCity festival in Berlin (Urban Transcripts and Konzulat Studios 2018). It was conceived as a participatory city-making tool using a map of the city as a board, which approached the perception of the city through the emotions and experiences people have in their environments, framed as the "urban mix". It consisted of two rounds:

- diagnosis, where players allocated the urban ingredients and emotions that characterised the city and
- vision, where players redistributed and exchanged those elements to propose a new urban vision.

The game facilitated a participatory understanding of the city, about how its different elements could be intensified, spread, diversified—remixed, in other words—to produce a collective analysis and a shared idea of urban change.

Applying this initial framework to net zero strategies and the Sustainable Development Goals at a more local level, ZC + has better-defined aims and therefore is structured on how future scenarios are envisioned and informed by the players' actions and strategies. It also considers a different target audience, integrating communities with institutions (authorities, institutions, or others) involved in the improvement of the neighbourhoods' relationship with the environment.

The development of the project was based on an 8-month Design Exchange Partnership and an AHRC/UKRI funding received by the pilot round of the Future Observatory Research Programme, founded by the Design Museum in London. The partnership behind the game development benefitted by the experimental and innovation freedom created by the acknowledgement of the value in applying hybrid design research methodologies and collaborative active workshop frameworks.

4.6 Case Studies

Three case studies are presented in the paper, based on the three active community workshops that were organised to pilot the different prototypes/versions of ZeroCityPlus (ZC +). The selection of locations was based on decentralising the project and diversifying the types of communities we engaged with. Ethical considerations were taken into account, and the active workshops used methodologies of focus group discussions during the gaming sessions with members of the communities. Standard ethical processes of human-centred research were followed.

4.7 Prototype 1: Nottingham (the Meadows)

Following an extensive review of references, ZC + was conceived and its first iteration manufactured to be tested out in Workshop 1. It draws elements from its predecessor, working on the basis of a city plan. However, ZC + develops its own identity and distinctive game mechanics by integrating a timeline, cards, and player tokens whilst focusing on a specific area within a city, in this case The Meadows in Nottingham in February 2022. The venue was Gallery Zero at Nottingham Contemporary, a notable art gallery located at a walking distance from The Meadows, near the city centre.

In the spirit of a collaborative endeavour, the participants play like a team to defeat the area's carbon footprint—either all win or all lose. Using a specific place-based carbon calculator and governmental sources, the average carbon footprint per person per year of the target area is calculated and the players aim is to achieve net zero by the end of the timeline, set on a 10-year scale. This reflects the nature of the collective effort by communities and authorities coming together and organising themselves to tackle the environmental crisis. The players decide on a range of projects to be implemented in the area (including some of their own creation) and support them throughout the game with tokens, which have a value that goes against the figure corresponding to the area's carbon footprint.

From the players' feedback and the team's observations, the game presented a clear aim and mechanics and had visual and tactile qualities. The board was a successful tool of engagement in its own right. Nonetheless, the game resulted in being too complex and slow in its development and not fully accessible in its language. Players wanted to have more agency in the game, but randomness and luck were too predominant. The elements of collective deliberation and strategy needed more development, whilst other aspects of the game would have benefited from simplification. The venue was a comfortable, sophisticated, and well-provided place, but some participants suggested it could deter some low-income participants as they might feel intimidated.

4.8 Prototype 2: Reading (City Centre)

Based on the same principles and using many elements of the first iteration, ZC + improved in terms of player inclusivity, agility, and speed. Adjustments included the timeline reduced from months to seasons, the playing cards reduced from three to two piles, and the tokens were arranged in behaviours and resources, to match the new two different roles of players: Citizens and Institution Representatives, respectively. Introducing these changes brought the game closer to its original aims of focusing on the gap between policymakers and people. In this way, a greater balance was reached between reflecting the complexity of the variables involved in the environmental crisis whilst translating it into a game for people to understand its reach, effects, and how to possibly tackle it. Moments of deliberation were included for the playing of the cards, plus three reflective breaks to discuss game tactics and how it reflects the reality. The venue was the University of Reading's Urban Room, located in the Broad Street Mall, in the city centre and the workshop took place in March 2022. This venue was expected to draw on more participants and observers, since it is a commercial, popular setting in the city.

From the players' feedback and the team's observations, the game improved in speed and agility, although it proved challenging to be played in one go. It moved closer to its original aims and featured more deliberation opportunities and player agency, yet the creation of a collective strategy was still not taking place. The board was considered to be undeveloped in terms of its potential to engage with participants and the game itself, although there were clearer indicators of the educational value and level of engagement from the players. The game can still improve in terms of clarity, considering it is meant to play as a one-off for players to set the ground and give directions for actual collaboration and project developments for their neighbourhood.

4.9 Prototype 3: London (Hackney Central)

In its last stage of development, the game was refined in terms of tactile and physical attributes and its applicability in the context of real-world public participation practices. The board featured pictures and landmarks of the target area, also shifting its palette to brighter and more colourful tones, seeking to stimulate engagement. More structured stages of participation were set: before and after the game, to start exploring its framing into a larger process of strategy and action planning. The first was to convey the rules in a more effective and clear manner yet allowing participants to join in even when the game has started (reflecting the nature of a real-life community project). The venue, the Garden of Earthly Delights in Hackney Central, London, in June 2022, allowed for this approach as the rules of the game were placed along the way in through the garden, leading to the game location (see Fig. 4.1). A single but longer (lunch) break allowed for conversation and exchange of views between participants, and a final, more structured debriefing session was introduced to discuss how the events and actions in the game could take place in reality, discussing strategies for application of initiatives.

The venue chosen, the Garden of Earthly Delights (GED), is a community-led initiative for local gardening, crops growing and related activities—with the overall aim of bringing together people in a quiet natural environment, to educate and raise awareness on environmental action.

From the players' feedback and the team's observations, ZC + has become legible, featuring a clear challenge and mechanism to achieve a goal common to all players. The process invites thinking and raises awareness of the consequences of climate change and the overall environmental

Fig. 4.1 Still moment from the ZeroCity + community game session in Hackney central at the premises of the Hackney garden of earthly delights community. Photo by Carolina Vasilikou

crisis. Adjustments can still be done, like rule tweaks to improve dynamism, fine-tuning of the formula to define the value of tokens against the number of players, and include a tracking record for counting the tokens on the board.

As further considerations, the game had its maximum of players, so the inclusion of extra agents and events can be considered or splitting it into two sessions. The first to increase the game's credibility, the second to find a more substantiated way to define the carbon footprint tracking during the game. Overall, the players showed great interest in the game and provided rich, constructive feedback.

4.10 Conclusions: Outcomes of ZeroCityPlus (ZeroCity +)

The findings and critical analysis of the first phase of the ZC + project are presented below, with the lens on the development of the prototype into a functionable public participation process.

At this stage of development and having fulfilled the research objectives, we can consider ZC + as able to:

- engage participants in a collaborative game, which has elements of deliberation and fun; establishing a common ground where reality is reflected and upon which further discussion and exchange can evolve,
- raise awareness and engage players in understanding the potential effects of climate change at a local level, and how they can be addressed through a set of projects of their own election and creation,
- facilitate the co-creation of specific strategies and mutual commitments among players to tackle the environmental crisis from a bottom-up perspective, but it does not achieve by and within itself as a game,
- contribute to creating constituencies and capacities at a local level for collective action planning and agency in future change and resilience, given the previous achieving aims and understanding the game as part of a wider engagement process between authorities and policymakers, and communities.

One limitation in this evaluation is that the testing of the game in the community (workshops) included role-playing representatives or commitment from a local authority, next to the role of citizens. However, the game itself acquired more prominence in empathetic qualities, whilst the steps towards building strategies and commitments from the game need further development. These aims could be reassessed in the light of further more institutionally framed gaming sessions. As a tool, ZC + is still subject to both refinement and adaptation through further editing and scaled-up testing, depending on the framing of the game within a wider process of engagement to be discussed with clients (local governments, planning agencies, NGOs, organised communities).

4.11 Key Learnings and Recommendations

We are using the lens of design research in the prototyping of ZeroCity + and some key learnings included to adjust the game and explore its variations, whilst developing a framework for it as a tool for public engagement involving multiple stakeholders. The second phase of the ZC + project develops the (1) introduction of the consultancy to the stakeholders, (2) game induction, (3) gaming session structure, (4) reflection and set out ways forward, and (5) follow-up activities including mediation in meetings to develop and implement strategies or further gaming. The gaming session cannot be a single event of engagement, and Urban Transcripts must agree on ways to check on and inform further development, to ensure benefit for the involved communities and their living environments. Future development includes

- To keep the face-to-face group interaction around the gaming session for a more effective engagement as the core element of the consultancy, with other stages potentially to be held online, either synchronically or diachronically.

- To develop a database or archive of projects, initiatives, and behaviours at a local level that have a positive (and ideally measurable) impact in neighbourhoods and communities.
- To seek for a scientific basis to substantiate the value of tokens if measured in carbon emissions, when relating to specific actions and events, giving more consistency and credibility to the game, and to adjust its mechanics.
- To scale up and diversify the testing of ZeroCityPlus (and variations), involving more structured partnerships for participating in the game to reach the institutional level and testing it within the Urban Transcripts network as a means of internal dissemination, training, and engagement. The latter will be helpful for refining and testing tweaks, rule changes, new game elements, and so forth.

Considering ZeroCityPlus (ZeroCity +) as a game to be played just once or up to a few times by its players, it needs clarity, simplicity, a legible rationale, and straightforward communication to facilitate engagement and collaboration. On the other hand, the reality of the environmental crisis and climate change is incredibly complex—even at a local level. Similarly, to develop collective strategies and commitments in that context also proves to be complex—and strategy games normally involve a process of learning and trying for players to gradually master them. A fine balance between clarity and complexity has been and is likely to be the greatest challenge in future development of ZC+ .

Acknowledgements The authors would like to deeply thank the Future Observatory, a UK-wide research programme coordinated by the Design Museum in partnership with UK Research and Innovation and the Arts and Humanities Research Council for their generous funding and support during the 8 months of the project.

References

Abt C (1970) Serious games. Viking Press, New York

AHRC (2020) Design challenges for the future: public policy, pp 19–20

Anspach R (1973) Anti-Monopoly. University Games

Carmichael R (2019) Behaviour change, public engagement and net zero report. Imperial College London, pp 5–18

Castronova E, Knowles I (2015) Modeling board games into serious games: the case of climate policy. Int J Ser Games 2(3). https://doi.org/10.17083/ijsg.v2i3.77

Bristol Council (2021) Bristol one city plan. Retrieved 28 July 2022 from https://www.bristolonecity.com/

Demski C (2021) Net zero public engagement and participation. A research note. BEIS UK Government

Dodig M, Groat LN (eds) (2020) The Routledge Companion to Games in Architecture and Urban Planning. Tools for Design, Teaching and Research. Taylor and Francis, London

Drew C, Johnson J, Chadha S, Carlisle C, Burnett A (2021) Beyond Net Zero: a systemic design approach. Design Council, London, p 13

Duke R (1974) Gaming: the Future's Language. John Wiley, New York

Lele S (2020) Environment and Well-Being. A perspective from the global South. New Left Rev 123:41–63

Michael D, Chen S (2006) Serious games: games that educate, train and inform. Thomson Course Technology, Boston, MA

Play the city (2019) Games for cities. Retrieved 6 July 2022 from: http://www.gamesforcities.com/database/

Salen K, Zimmerman E (2003) Rules of play. MIT Press, The Game Design Fundamentals. Cambridge MA, pp 578–579

Sasse T, Allan S, Rutter J (2021) Public engagement and net zero: How government should involve citizens in climate policy making. Institute for Government, p 23

Serious Game Society (2014) Retrieved 18 Jan 2022 from: https://seriousgamessociety.org/

Van Slyke B, Stachiw A, McLeod M (2015) Co-opoly. TESA Collective

Sousa M (2020) A planning game over a map: playing cards and moving bits to collaboratively plan a City'. Front Comput Sci 2(37):1–12. https://doi.org/10.3389/fcomp.2020.00037

Urban Transcripts and Konzulat Studios (2018) Remix(c)ity: a participatory city-(re)making game. MakeCity Berlin. Retrieved 28 July 2022 from https://www.urbantranscripts.org/projects/

5 Partnership for Transforming Lives of the Urban Poor: Experiences of Unique Community-Led Housing Initiatives by Shelter Associates

Pratima Joshi and Geetanjali Deshmukh

Abstract

Architects at the India-based non-profit collective Shelter Associates (SA) are creating a paradigm shift in slum development by following a city-wide multi-stakeholder approach to urban development by recognizing that slum communities are an inevitable part of our cities. SA gives slum-dwellers a voice by arming them with data about their community, empowering them to take decisions making them partners in the developmental process. This data is the ultimate evidence of critical gaps in the service delivery system bringing the Urban Local Bodies (ULB), communities, and the civil society on a level-playing field. Understanding lived experiences, cultural practices, and preferences of slum-dwellers has helped SA architects break the chains of their academic discipline. Involving communities and ULBs in the process of designing is leading to innovations both in design and processes setting precedents for a multi-stakeholder approach to slum rehabilitation. Using data and technology, SA accurately targets the vulnerable communities and their needs leading to the efficient allocation of resources contributing to degrowth. As environmentalists, SA architects don't just consider land and other commons as resources but as life-sustaining forces and therefore nudge all stakeholders to jointly work towards the preservation of the commons. This article attempts to highlight the phenomenal success architects can achieve as facilitators and enablers owing to their technical abilities and the sensitivity and humane touch they bring to the process of social development. Their ability to take along diverse groups like ULBs, civil society, and slum communities for the transformation of the lives of the urban poor. Demonstrating the transformation that collectively designed built-in spaces bring about.

Keywords

Social housing · Multi-stakeholder approach · City-wide perspective · Urban development · Quality built-in environment · Spatial data

P. Joshi (✉)
Executive Director, Shelter Associates, Pune, India
e-mail: pratima@shelter-associates.org

G. Deshmukh
Manager (Strategic Communications), Shelter Associates, Pune, India
e-mail: geetanjali@shelter-associates.org

5.1 Introduction

The non-profit organization Shelter Associates (SA) was founded by three architects in 1994 with a passion to ensure equitable rights and

S. Hilal et al. (eds.), *Design for Partnerships for Change*, Sustainable Development Goals Series,
https://doi.org/10.1007/978-3-031-36993-3_5

access to essential services for the urban poor. Over time, professionals from diverse backgrounds including Engineers, Data Scientists, Geographical Information Systems (GIS) experts, and social workers joined SA to design and implement data-driven solutions in Sanitation and Housing.

SA is a pioneer in using spatial data for social impact in Maharashtra and its core sanitation project called the One Home One Toilet (OHOT) has facilitated individual toilets to 1.5 lakh individuals leading to vast improvements in their health and well-being.

SA has gained expertise in housing the poor through the implementation of three major resettlement projects in Pune and Sangli-Miraj, which are quite different from each other. These projects demonstrate a workable alternative to the typical slum rehabilitation projects that are prioritized by the government which fail to take cognizance of ground realities; do not view the issue of housing for the poor at a city-wide level; and are implemented in an intrusive and opaque manner.

The housing projects implemented by SA are: (1) based on accurate data which has been spatially organized; (2) generated from a city-wide approach that considers housing as an integral part of the city along with areas of employment, healthcare, education, and convenience; and (3) implemented in partnership with beneficiary communities.

Another path-breaking work SA undertook recently is the city-wide slum rehabilitation research project with ATE Chandra Foundation and Pramiti Foundation that advocates creating city-wide spatial data to have a clear understanding of the on-ground situation to identify vulnerable slums within the city and also proposes innovative yet simple solutions to design a resilient social housing for the poor.

Learnings from the city-wide research have led to a unique slum rehabilitation project at Bondre Nagar in Kolhapur which is currently being implemented under the "Beneficiary Led Construction" model under the Government of India's Pradhan Mantri Awas Yojana (PMAY) housing scheme.

5.2 A Time-Tested Approach to Social Housing

Three decades back when SA started studying the urban slum space, it observed that many social housing projects fail because they are designed without considering the city that it is situated in, its surrounding, economic activity, and practices of the communities. Lack of accurate on-ground data meant many missing pieces of information, critical to the acceptance of the very people it is intended for. This led to the development of SA's own approach to implementing housing solutions (Fig. 5.1).

(a) Granular Spatial Data: The settlement level and socio-economic data collected provide an accurate description of the lifestyle of the residents, their on-ground condition, and the issues they face while living in dire

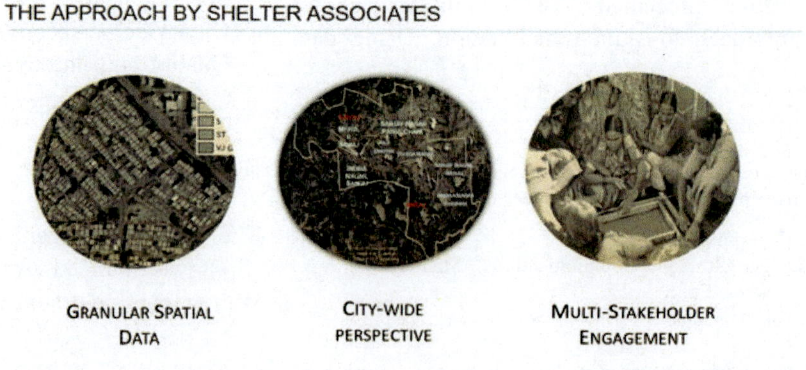

Fig. 5.1 SA's approach to social housing

conditions. The collected data is leveraged during multi-stakeholder interactions to ensure that extensive engagement with the community informs decisions in design development. It ensures (1) tenure security (2) well-designed internal spaces and (3) infrastructural amenities.
Furthermore, it also takes into consideration factors such as the location of (re)settlement, and the provision of amenities available to people within the neighbourhood. These factors are very important for holistic development but are often neglected during social housing.
The special aspect of our data-driven approach is the inclusion of community volunteers in creating data of their own slum. These volunteers, mostly young women are trained in conducting surveys, reading maps, and using digital survey tools. Data is the foundation of our work and co-creating data in this way leads to its acceptance by all involved.
The design developed with such an approach results in houses that are teeming with natural light and ventilation. They have unique areas such as (1) Spaces for keeping cattle, (2) Terraces to carry out traditional domestic chores such as making papads (thin crispy pancakes); exposing grain to the sunshine; cleaning beds. (3) Open spaces that are essential for forging community spirits and promote health and mental well-being.

(b) City-wide Perspective: Mapping and analysing the slum location along with their level of vulnerability help in prioritizing and planning the housing projects. It helps the ULBs in creating better city development plans, resource allocation, and planning logistics well in advance. A city-wide perspective on social housing leads to optimal utilization of land as a resource instead of the ad-hoc, unplanned social housing projects.

(c) Multi-stakeholder Engagement: SA advocates for involving both slum communities and ULBs in the planning, designing, and financial decision-making of the projects. All of SA's housing projects have been successful due to the active involvement of communities who have even contributed financially, taking microloans while the ULBs have supported by validating the designs and building common infrastructure. By bringing together multiple stakeholders on a common platform, SA creates a level-playing field that empowers the communities.

5.3 Previously Completed Housing Projects

(a) Dattawadi Housing project (1996–1998): Moving Mountains (Fig. 5.2)

Our first project was initiated in Pune, during the first rains of the monsoon, with the sudden demolition of an informal settlement known as Rajendra Nagar by the city administration. We worked with the community to ensure that 56 families were resettled into formal housing within a kilometre of the existing slum at Dattawadi with financial support from The Housing and Urban Development Corporation (HUDCO), a public sector undertaking under India's Ministry of Housing and Urban affairs.

This was a unique experiment in Pune city where the beneficiary families assisted with the design, monitoring, and construction of their own homes. Well-designed homes, customized sanitation facilities resulted in a healthier, hygienic lifestyle. The confidence they gained during their active involvement in the project led to a desire and concentrated efforts towards upgrading their lifestyle resulting in better education and employment options for the members.

(b) Kamgarputala Housing project (2003–2005): Disaster averted (Fig. 5.3)

In 1997, Pune faced one of its worst floods and the city administration invited us to carry out

Fig. 5.2 Community involved in constructing their own houses at Dattawadi

Fig. 5.3 Kamgarputala housing project still stands

detailed surveys of 6 informal communities along the river which flows through the city, the river Mutha. Kamgar Putala, one of the oldest informal settlements in Pune, was the worst affected by the floods due to its location at the confluence of the Mula and the Mutha Rivers. We worked with the 176 flood-affected families of Kamgar Putala to help them secure formal housing away from the flood-prone banks of the river.

The community accumulated the capital for contributing to the construction of new homes by forming micro-credit groups and building savings. Since flooding along the river Mutha was an annual occurrence and the community of Kamgar Putala was determined to find long-lasting solutions for the safety and future of their children.

Thus well-structured, liveable, and safe housing was designed. The floor area of each unit is 200 ft^2 divided into two bays. Each unit includes a toilet, a bathing area, a kitchen area, and a multipurpose living–dining–sleeping area in the front bay. The unique aspect was that the traditional way of living of the community was respected and accommodated. The community did not want a high-rise building, which was accepted. The community was used to cooking while sitting on the floor so low-raised cooking platforms were built. In spite of the small space, separate toilets and bathrooms were built to respect the community's wishes.

(c) IHSDP/City-Wide slum housing project, Sangli-Miraj-Kupwad (2009–2013): Multi-stakeholder success (Fig. 5.4)

SA was appointed as consultant for the Integrated Housing and Slum Development Programme (IHSDP) in Sangli-Miraj-Kupwad, a peri-urban area about 400 km south of Mumbai on the banks of the river Krishna, with a population close to 550,000. Our city-wide design proposal, informed by our poverty mapping and use of Geographic Information System (GIS) technology, optimized the use of the limited land available to meet the rehabilitation and relocation needs of the families living in 29 slums across the city.

The IHSDP had been introduced by the Government of India (GOI) for the improvement of slums in cities and towns under the Jawaharlal Nehru Urban Renewal Mission (JNNURM). The critical objective of the IHSDP was holistic slum development with a healthy, and enabling urban environment, by providing adequate shelter with security of tenure and basic infrastructure facilities.

The housing stock created would be sufficient to provide formal housing, with security of tenure and improved sanitation, to the families in 29 slums, almost half of the slum population of the peri-urban area.

A critical aspect of our approach was that it did not involve moving the beneficiary families to faraway sites on the edge of the peri-urban areas which lack services (transport, water, and drainage) and drives them deeper into poverty. Hence, relocation would not impact their access to places of employment, and amenities such as schools, health centres, emergency services, and markets.

The GOI approved this project for its innovative, holistic and inclusive approach, which could be emulated in other cities. This approach later became part of the government policy framed under Rajiv Awas Yojana (RAY), which requires cities to map its poor using GIS and remote sensing technologies.

Fig. 5.4 IHSDP Housing project—view of the society

5.4 City-wide Research Project

SA envisages an India where every citizen has access to basic infrastructure, secure tenure, and recognition of equal rights. To achieve this vision, SA undertakes slum rehabilitation projects using a process sensitive to the needs of all stakeholders, especially the community. Shelter Associates in collaboration with the Pramiti Foundation and funded by ATE CF had undertaken a research project to provide a framework for implementing social housing projects across 44 slums on government land in Kolhapur.

Through data collection, mapping, socio-economic surveys and discussions with the concerned stakeholders the research project provides a detailed on-ground scenario for affordable housing in the city of Kolhapur.

> Stage 01 | Gathering Data based on –
> (1)| Rapid Infrastructure Mapping (2)| Study of Planning rules (3)| Rapid Household Surveys
> This highlighted loss of judicious use of land, issues due to lack of holistic development and lack of inclusive planning.

Through the research project, SA has observed that the lack of comprehensive data available with the ULB makes it difficult to take a holistic view of the entire urban area. This can lead to piecemeal slum rehabilitation projects which represent a suboptimal use of the limited resources, such as tenable land and fail to leverage an economy of scale. It can also lead to hesitancy by the ULB in undertaking slum redevelopment projects due to complexities and conflicts that may arise during planning.

SA strongly advocated that the research follows the core principles of (1) a data-driven approach, (2) a holistic city-wide perspective to identify vulnerable settlements, and (3) including diverse multi-stakeholders in planning. It follows a process of federating the data collected on a unifying base map leading to a city-wide dataset that can be leveraged to make informed decisions in the planning and preparation of slum rehabilitation projects across the urban area (Fig. 5.5).

This ensures that an accurate profile of the surveyed area, whether a city, a neighbourhood, or an individual slum, is generated (Fig. 5.6).

SA advocates that a city-wide approach and identification of the most vulnerable slums in the

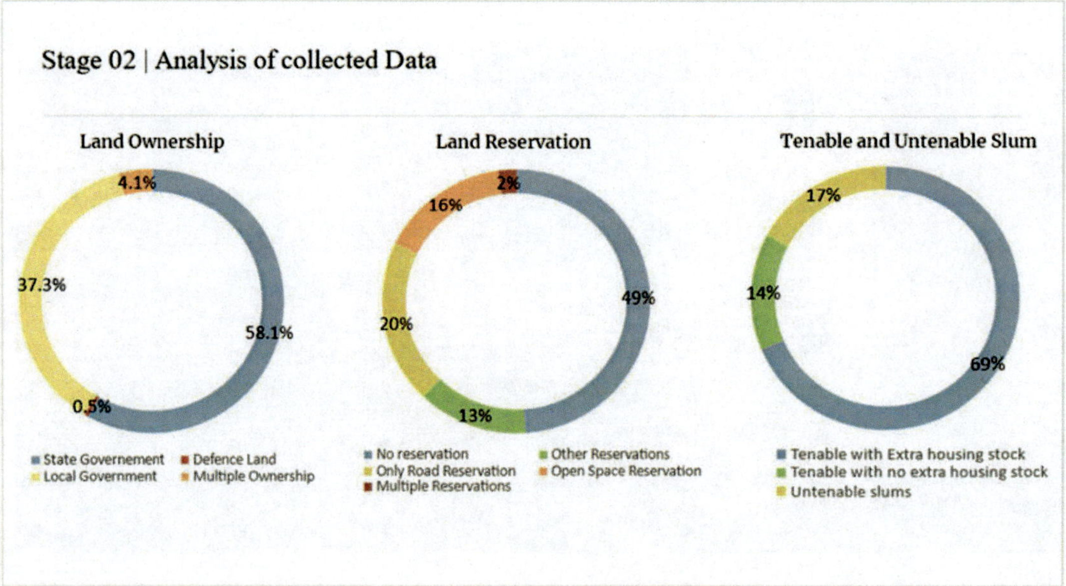

Fig. 5.5 Stage 02—analysis of collected data

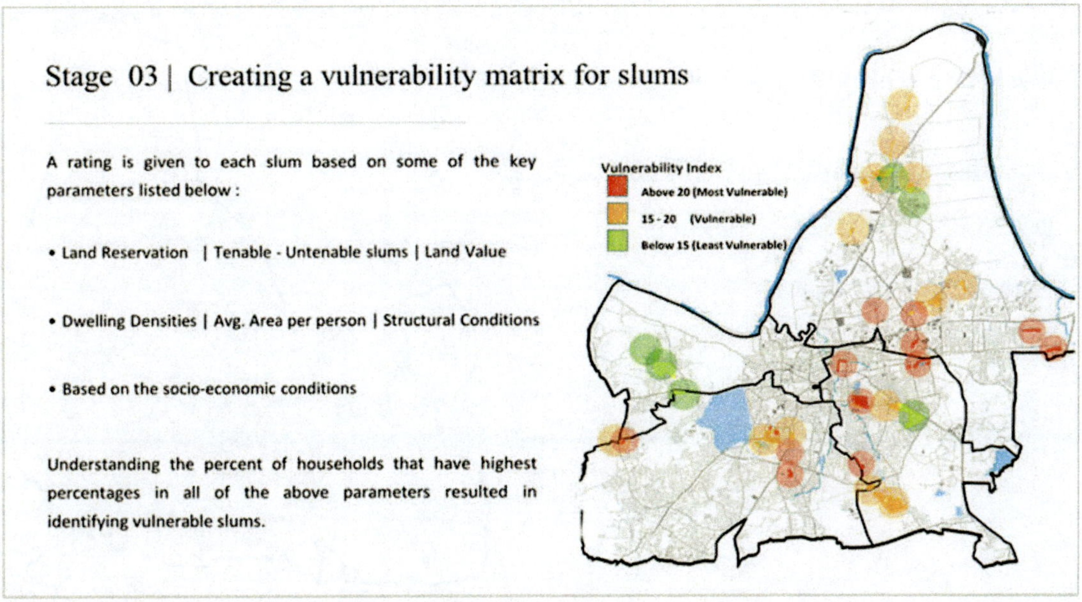

Fig. 5.6 Stage 03—creating a vulnerability matrix for slums

city will ensure optimum utilization of government land and an inclusive planning approach that will be beneficial for all concerned stakeholders (Fig. 5.7).

Solutions based on the vulnerability matrix (Fig. 5.8).

The finding, analysis, and solutions of the research project have been captured in detail in a toolkit. The proposed solutions in the toolkit can be easily accessed by like-minded organizations and government agencies as the solutions have the potential to be replicated in similar 2-tier and 3-tier cities willing to implement social housing programmes.

5.5 Live Project: Slum Rehabilitation at Bondre Nagar, Kolhapur

A few years back, the Urban Local Body (ULB) of Kolhapur City had proposed a slum rehabilitation scheme for the Bondre Nagar, a slum located at the edge of the city. The scheme proposed the generation of extra housing stock to become the saleable component in the design. This approach was questioned as there was no demand for extra housing in the area. Furthermore, the residents opposed the development of a multi-storeyed structure. Being farm labourers with lifestyles that were incompatible with a high-rise construction, the solution proposed by the PMAY department was unsustainable as it followed a top-down approach.

Voices from the community (Fig. 5.9).

Ram Dhere is one of the oldest residents of Bondre Nagar, and he used to work as a mason for almost 40 years, following which he quit his work in order to look after his ailing wife. "Every monsoon the gutter behind our house would overflow and flood all the houses in our area. This led to poor health and added on expenditure for us every year" says Rama Dhere. Meanwhile, his unmarried daughter Lakshmi used to earn income for the entire family through the ten cows they used to own. The family faced a series of unfortunate events, including the demise of Ramachandra's wife and Lakshmi's deteriorating health condition. "Not a single day passes by when I don't feel responsible for the hardships my family has been through. I wish I was capable enough of providing them a healthy and safe home", he adds (Fig. 5.10).

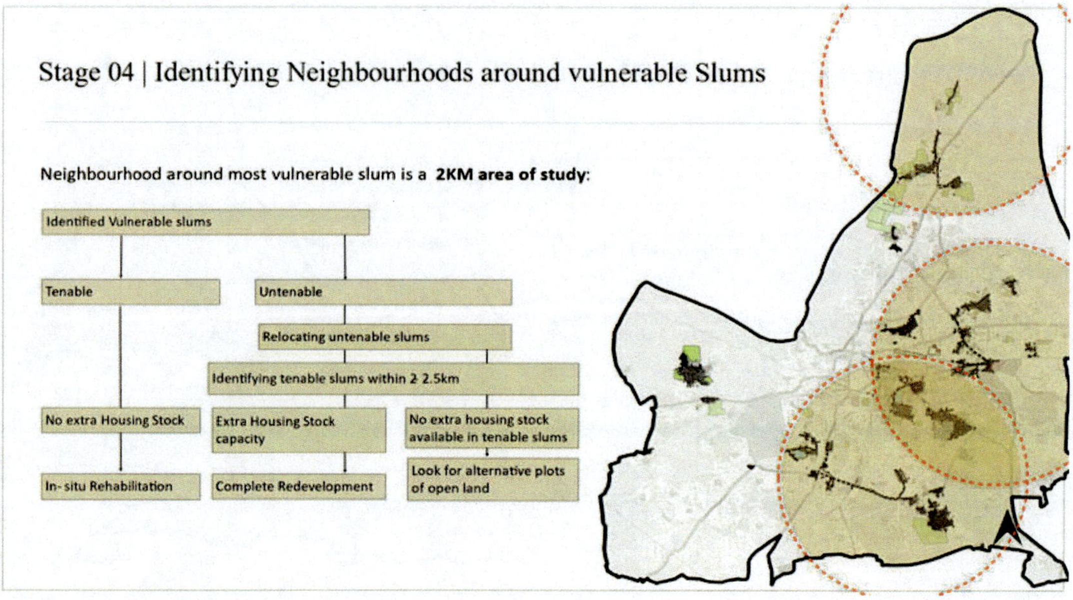

Fig. 5.7 Stage 04—identifying neighbourhoods around vulnerable slums

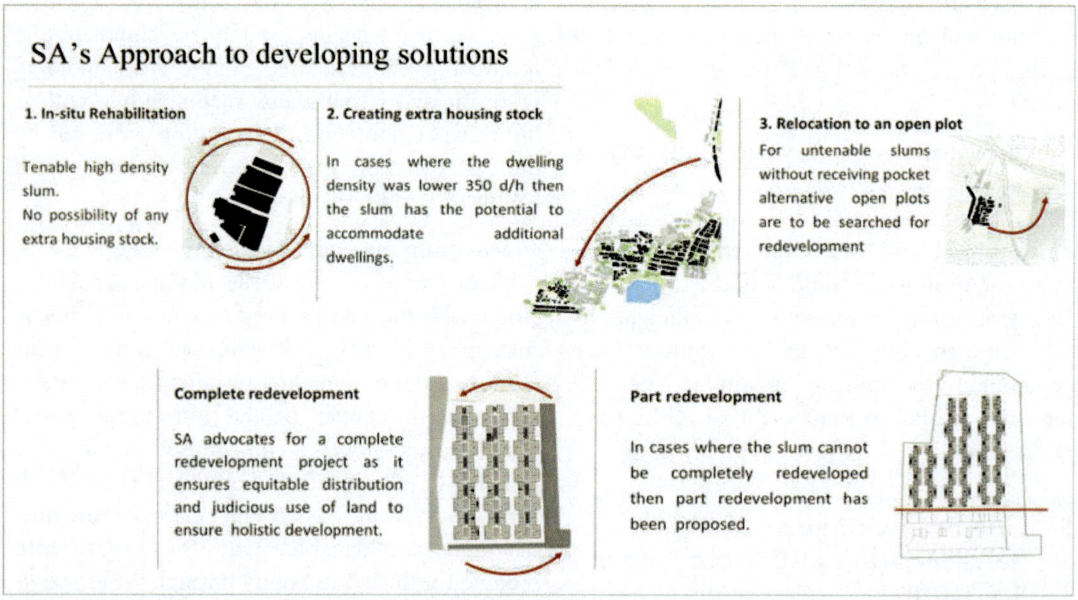

Fig. 5.8 Stage 05—SA's approach to developing solutions

There was a need for a holistic solution that had the approval of all concerned stakeholders with the interests of the community at the centre. SA at the behest of the community intervened with its multi-stakeholder approach and designed housing solutions based on the findings of the city-wide research project.

50-year-old Madhukar Gosavi lives alone in a shady house in Bondre Nagar, Kolhapur. During his youth, he contracted Chikungunya after

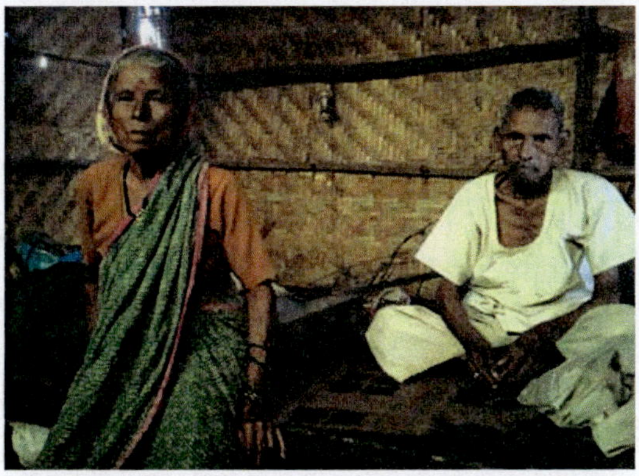

Fig. 5.9 Laxmi and Ram Dhere

which he lost the control over his limbs and thus his ability to walk on his own. But this was only the beginning of Madhukar's struggles. "Since I could not walk, my employer removed me from my job, and life came to a standstill. I had to be dependent on others for my survival. Earlier I had no toilet in my house. I had to make my way from the muddy road to the community toilet which was exactly at the other end of my house. Neighbours' glares and remarks only added to my discomfort and embarrassment. Shelter Associates intervened in our slum with a housing project with basic infrastructure and amenities and is supporting people like me to lead a life of

Fig. 5.10 Madhukar Gosavi

respect and dignity. Without their support, I would have continued to curse my fate and live in dependency with a poor quality of life" smiles Madhukar.

A well-designed redevelopment: For the rehabilitation of Bondre Nagar SA engaged the community in an inclusive process of designing. From taking door-to-door discussions to engaging the residents in a to-scale mock-up plan, SA has given a clear understanding of the space and its functionality. The developed design is a low-rise, high-density development consisting of G+1 tenements planned such that the adjacent tenements share a common wall and a rear open space that doubles up to form a courtyard. The tenements are designed to have adequate light and ventilation and the developed design has the approval of all concerned stakeholders. The open-to-sky spaces play a vital role in the Bondre Nagar redevelopment plan as it makes a decisive difference between liveable habitat and claustrophobia as a design upgrade for the community. Spillover spaces like courtyards and backyards not only cater to informal activities but also increase the indoor quality of living. Repetitive units make a cluster and such clusters create spatial richness by minimalistic means leading to a safer, healthier, and affordable living environment (Fig. 5.11).

Each house opens to an internal road leading to better accessibility. The arrangement of houses in a grid pattern ensures spatial requirements of the community. SA has followed all design by-laws, and the construction will be in RCC to provide permanent homes to the families. The settlement is well-organized such that road

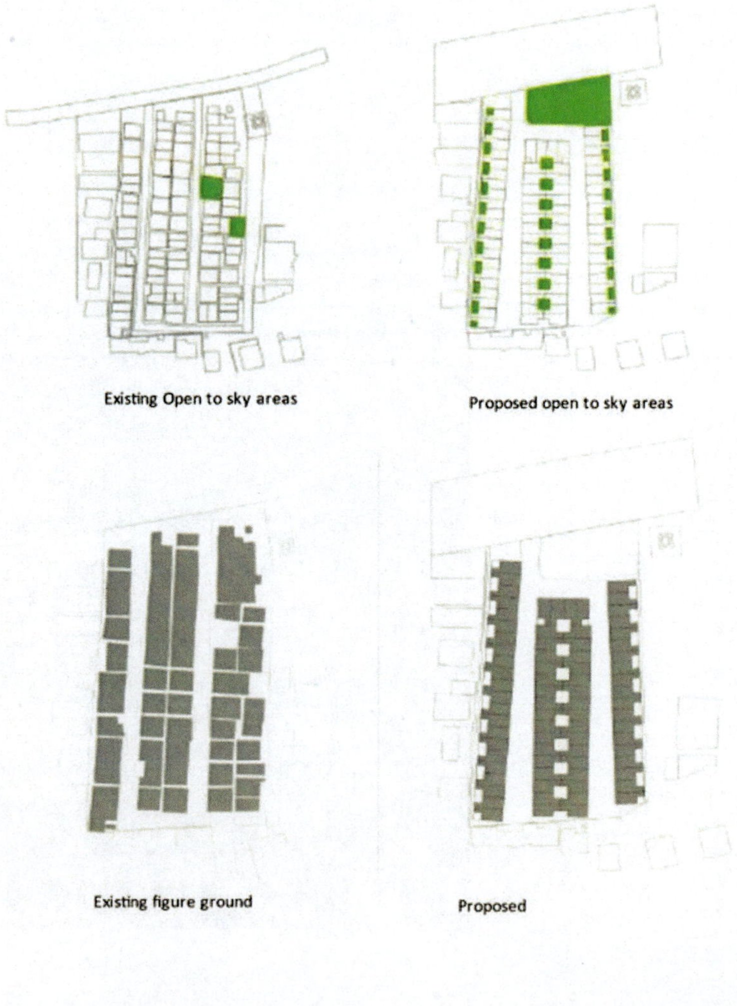

Fig. 5.11 Bondre Nagar redevelopment design for healthy living

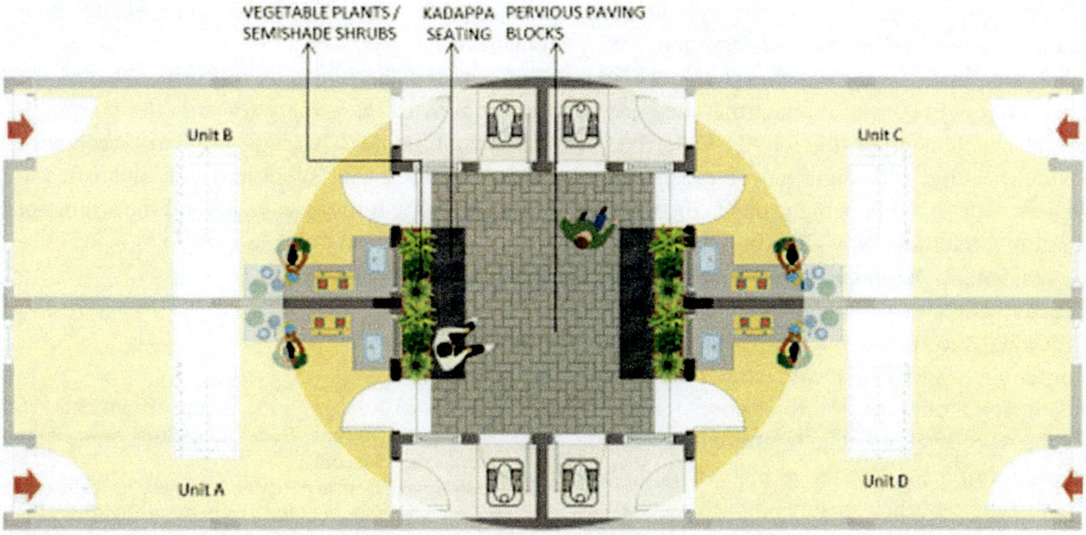

Fig. 5.12 Cluster plan showing open courtyards for informal activities

congestion will be avoided at all times enhancing firefighting capabilities (Fig. 5.12).

The Detailed Project Proposal for Bondre Nagar was submitted to the government under the Pradhan Mantri Awas Yojana's—Beneficiary Led Construction category and has received subsequent approvals. The land has been transferred in the name of the beneficiaries, thus providing tenure security. All approvals of various agencies include the town planning department. The residents have also formed a cooperative society and are applying for microloans. The Kolhapur Municipal Corporation is also actively involved in planning the common infrastructure. The site has been cleared, and the beneficiaries are now living in a transit camp nearby. We move towards the construction phase with great enthusiasm and hope for a dignified future.

Alongside Bondre Nagar, in the years 2022–24, SA would like to explore the translation of the research project into actionable on-ground impact for social housing projects that will have a city-wide impact. As SA strongly advocates the approach of focusing solutions for the most vulnerable slums within the city, planning future projects for them is of utmost importance. Projects that are planned around the vulnerable slums will ensure the optimum utilization of prime government land. Alongside monitoring the Bondre Nagar Project, SA would like to develop solutions for the vulnerable slums that are conducive to on-ground scenarios and in accordance with inclusive, multi-stakeholder involvement. SA would like to ensure that these settlements are undertaken for rehabilitation that benefits all concerned stakeholders.

5.6 Conclusion: Learnings from Diverse Experiences

In this almost three-decade journey, SA architects have faced many challenges like apathetic bureaucracy, unwanted political influence, and sometimes reluctance of the slum communities fearing change. We routinely face and shatter gender bias since most of our senior management and 60% of our field staff and community volunteers are women who have to face both the ULBs and the community.

We overcome these challenges with determination and passion to provide a secure tenure to the many urban poor in India so that our future generations can live and thrive in safe environments. "Shelter Associates strongly believes that everyone deserves to live in homes with adequate amenities, regardless of their backgrounds. Why do we frequently overlook the slums when it comes to growth despite the fact that they are an integral component of the city? We strive to provide healthy and dignified lives to slum-dwellers. Our social housing projects incorporate the culture, traditions as well as preferences of all the residents", expresses Ar. Pratima Joshi, Executive Director of Shelter Associates.

SA's holistic approach and involvement of the communities and ULBs have resulted in a robust housing for thousands of urban poor. Residents now have a dwelling that is structurally stable and well-designed. The sense of ownership and responsibility further ensure the maintenance, cleanliness, and hygiene of the space. Their lifestyle transforms into one that is in stark contrast to the vulnerable living conditions in the slums. It is also observed that they strive to maintain this transformed lifestyle by promoting the economic betterment of the family. They also have a sense of financial responsibility as a new lifestyle also comes with awareness of savings for repaying loans, maintenance of their apartments, etc. This results in upward economic mobility and the future generations greatly benefit from it.

Changing the built environment changes the perceptions of the society towards the marginalized communities. It leads to their acceptance and gradual assimilation into the mainstream. They can now forever leave behind the stigma of being slum-dwellers.

References

A case story on Kamgar putala. Retrieved on 25th Feb 2023 from D:\Tom Data July 2018\New Proje (shelter-associates.org)

City-wide Social Housing Research project in Kolhapur. Retrieved on 25th Feb 2023 from Research Project—Final Report .docx (shelter-associates.org)

Moving Mountains: A community's journey from a slum to a society. Retrieved on 25th Feb 2023 from Dattawadi_41.pdf (shelter-associates.org)

Relocation of Sanjay nagar slum to a transition camp. Retrieved on 25th Feb 2023 from Relocation of Sanjay Nagar slum to a transition camp (shelter-associates.org)

Shelter Associates website. Retrieved on 25th Feb 2023 from Shelter (shelter-associates.org)

Part II

Re-framing Commons

Check for updates

6 Rebirth of Commons with Collective Memory: A Study and Renovation Design of the Shanghai Confucian Temple

Yuming Hou

Abstract

The Confucian Temple Book Fair was once a famous and lively scenery in Shanghai. It plays an important role in the memory of local residents. However, in 2016, the Confucian Temple Book Fair began to migrate to official shopping mall, and the Shanghai Confucian Temple became a destination for a few tourists or students where students pray for their exams, which makes it much more desolate than before. While Shanghai Confucian Temple has a long history of being used as a public space or commons for local residents, which is inconsistent with its current management. There are three important transition points for the Shanghai Confucian Temple as a public space. It has been migrated many times and was built in 1855 on its present site. At the very beginning, the Confucian Temple was not only a temple to commemorate Confucius, who is one of the most important educators in China, but also a local agency for education. Then in 1927, government of Shanghai decided to reconstruct Shanghai Confucian Temple and make it a public park. Eventually in 1986, the Confucian Temple Book Fair was built with the fund of district government. Therefore, the author presents a renovation design proposal based on the study of the Shanghai Confucian Temple's history and its lack of public use at present, in the hope that this common with collective memory would reborn.

Keywords

Commons · Collective memory · The Shanghai Confucian Temple · The Confucian Temple Book Fair

6.1 Introduction

6.1.1 Background

Located in Huangpu District, the Shanghai Confucian Temple has a history of over 700 years. It was first built in 1292 and moved several times since then until it was rebuilt at its present location in 1855. At the beginning, this kind of temple was built to commemorate Confucius (B.C.551–B.C.479), who is the most important educator in China. Then they gradually became parts of schools in ancient China. After a time when schools were specially set in cities and divided from Confucian temples, these temples were used for different functions by different local governments. As for the Shanghai Confucian Temple, it was once a park in the 1930s and

Y. Hou (✉)
College of Architecture and Urban Planning, Tongji University, Shanghai, China
e-mail: 1037598932@qq.com

S. Hilal et al. (eds.), *Design for Partnerships for Change*, Sustainable Development Goals Series,
https://doi.org/10.1007/978-3-031-36993-3_6

then the book fair during the 1980–2000s. The partnership of this common was changing over time, from people worshipping Confucius, students, to citizens and intellectuals. Although it changes in use, it remains to be a common for local people. However, in 2016, the Confucian Temple Book Fair begins to migrate and Shanghai Confucian Temple totally turns into a destination which charges for tickets for a few tourists and is not a common any more. While not like many temples in China are located far from cities, the Shanghai Confucian Temple is located inside the city and surrounded by residential areas. As a result, future users for this common once upon a time need to be rethought, which would make better use of this place. This case can also be seen as a typical example for exploring new use for those historical buildings and commons inside cities which meets the SDGs.

6.1.2 Methodology

This article will first introduce not only the history of the Shanghai Confucian Temple as a specific buildings and a place for public events but also the history of the collective memory that it contains. Collective memory is another way of describing what the city actually is (Robin Monotti 2016). It refers to the shared pool of memories, knowledge and information of a social group that is significantly associated with the group's identity (Olick et al. 2011). Rossi writes in *the Architecture of the City*: One can say that the city itself is the collective memory of its people, and like memory it is associated with objects and places (Aldo 1984). Thus, the article treats the history of collective memory as important as the history of building and place. These studies of history are all from the perspective of common. Then, a renovation design proposal would be put forward based on this study of history. The article tries to use design as a method to further develop this study and explore the future use of the Shanghai Confucian Temple, which would also be part of the treatment for sustainable development goals.

The study of the building history and a place for public events would be based on secondary historical source which was developed according to the record about the Shanghai Confucian Temple in Shanghai Municipal Archives. As for the history of collective memory, the article will develop this part through secondary data from documents described the scenery of the Shanghai Confucian Temple in the old times and primary data from interviews with residents live nearby for more than ten years and used to go to the temple. Furthermore, the primary data also include visitors' opinions of the temple from Internet and social media, as well as current site observation.

6.1.3 Basic Information of the Shanghai Confucian Temple

The Shanghai Confucian Temple is part of the Shanghai Old Town Historic Landscape Protection Zone where the most traditional residential architecture and living pattern of Shanghai remains. However, the building density there used to be as high as 80% or more, with narrow street, old buildings in disrepair and little green land. At the same time, influenced by the Confucian Temple, there are education architecture around such as Huangpu Children's Palace, Huangpu Yangguan School and Shanghai Jingye Middle School (Fig. 6.1).

The temple includes the building groups of Wenmiao and Xuegong on two north–south axes, along with the garden in the east. Wenmiao group was built for ritual to commemorate Confucius with the main buildings including the Lingxing Gate, Dacheng Gate, Dacheng Hall and Chongsheng Ancestral Temple, while Xuegong group was built for teaching and learning with the Xue Gate, Yi Gate, Minglun Hall and Zunjing Pavilion. Additionally, in the east garden, there is a Kuixing Pagoda (Fig. 6.2).

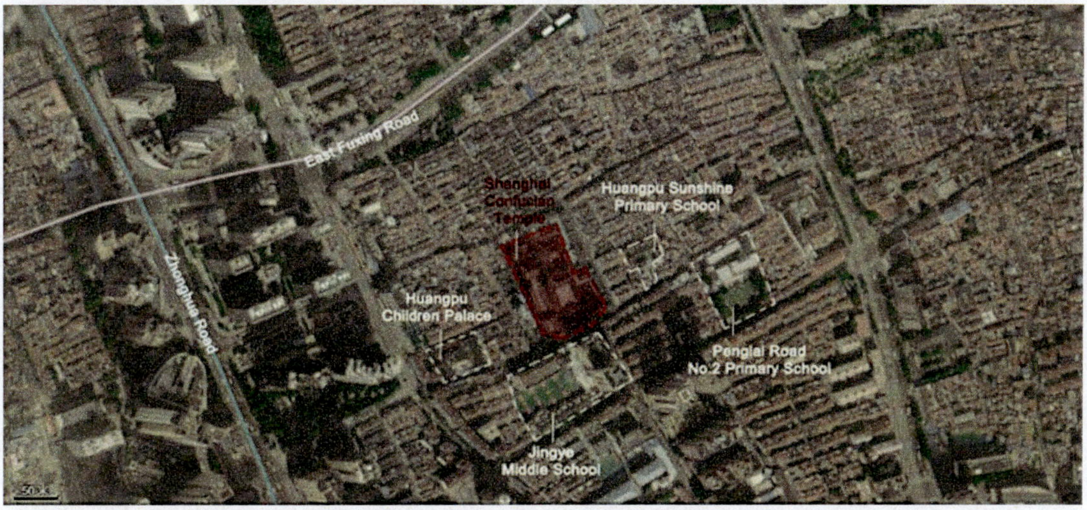

Fig. 6.1 Location of the Shanghai Confucian Temple. Drawn by the author based on Baidu Map

Fig. 6.2 Plan of the Shanghai Confucian Temple. Photo taken by the author

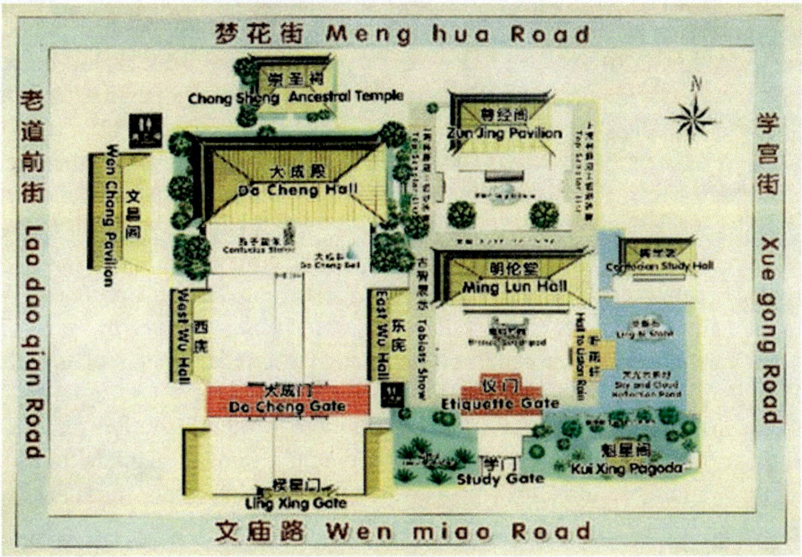

6.2 History of Place

6.2.1 Building History of the Shanghai Confucian Temple

The Confucian Temple is built to commemorate the famous Chinese educator Confucius. After Confucius died, the ruler at that time built a temple in Confucius' old living place in memory of him. This was the beginning of building the Confucian Temple. In the Tang Dynasty, worshipping Confucius became the highest honour for scholars, and the Confucian Temple gradually became an integral part of school at that time.

The Shanghai Confucian Temple was first built as the Wenchang Hall in 1292 by Tang Shicuo and his brother, which was also the Shanghai County School at that time. The buildings in the temple were added during the Ming and Qing dynasties but were destroyed during 1851–1864. After that, the temple moved to the present site. By 1927, the Shanghai Municipal Council proposed to convert the Confucian Temple into a park. In March 1931, the construction of the Confucian Temple Park began. However, after the completion of the first and second phases in October, the rest of the project was cancelled due to lack of funds. Then in May 1949, it was taken over by the Shanghai Municipal Military Administration and was transformed into the South Shanghai People's Cultural Hall. While in 1969, it was again destroyed by the Cultural Revolution. From 1997 to 2000, government allocated a large sum of money for renovation of it, and after three years, it was finally restored to its best state which is related to the one during the Qing Dynasty. As a result, the buildings in the Shanghai Confucian Temple are not very old. However, the traditions they represent date back to at least the seventh century (Сергій 2016).

6.2.2 History as a Place for Public Events

If taken as common, there are two important points in time of the Shanghai Confucian Temple.

The one is the reconstruction of the Shanghai Confucian Temple from 1927 to 1932. Before 1927, considering spreading Confucian culture to people, the Shanghai Confucian Temple Administration put forward a plan to transform the temple into a park to make it open to citizens. Then in 1927, the Shanghai Municipal Council officially proposed to convert the Shanghai Confucian Temple into a park. After the Kuomintang took over Shanghai, with the new national government's ambition to promote social education and modernization, the private associations and the municipal authorities worked together and finally succeeded in turning the Shanghai Confucian Temple into education hall and park (Guopeng 2020). In March 1931, the construction of the Confucian Temple Park began. After the completion of the first and second phases in October, the rest of the project was cancelled due to lack of funds. Subsequently, the management of the Shanghai Confucian Temple was transferred to the Shanghai Municipal Education Bureau. In 1932, it was again turned into the Shanghai Municipal Library, which is the first public library in Shanghai with more than 15,300 books. So far, the comparison of the Shanghai Confucian Temple before 1927 and after 1932 is shown in Table 6.1.

The other is the appearance of the Confucian Temple Book Fair. In early 1986, several students from Fudan University wrote a letter to the mayor of Shanghai, suggesting to set an old book fair at the Confucian Temple like the one by the Seine River in France. After the district cultural department received this proposal, they invited related associations such as publishers, the Administration for Industry and Commerce, and famous scholars to discuss the idea of setting the old book fair. The idea was unanimously

Table 6.1 Comparison of the buildings' functions of Shanghai Confucian Temple before and after renovation/Guopeng (2020)

Before renovation	After renovation
Dacheng gate	Current affairs exhibition, children reading room, entertainment room
Dacheng hall	Exhibition of sacrificial vessels for memorizing Confucius
Chengsheng temple	28, January Shanghai war exhibition, livelihood exhibition
Minglun hall	Lecture hall, classroom, reading room
Zunjing Pavilion	Library
Kuixing Pagoda	Reception room
Confucian study hall	Social education exhibition, northeast war exhibition, health education exhibition

The Transformation of Shanghai Confucian Temple in the Republic of China approved. After construction, the Confucian Temple Book Fair became the most important old books trading centre in Shanghai. It consists of two parts. The one is the Holiday Book Fair during holidays such as New Year's Day and National Day, in which discounted new books and magazines are sold. The other is the Weekend Book Fair, in which old books are sold and exchanged. As time passes, the Holiday Book Fair gradually disappears, whereas the Weekend Book Fair remains up to the 2010s (Zimin and Jian 1990). Since 1997, the district government has allocated funds for renovation of the Confucian Temple, and the district cultural department has also renovated the buildings used for old book fair. At the same time, in the northeast of the Confucian Temple, the Shanghai Confucian Temple Book Market was also developed. It was the largest neighbourhood-style book wholesale market in Shanghai.

6.3 History of Collective Memory

6.3.1 Study of Documents

To study the transformation of the Shanghai Confucian Temple from 1927 to 1932, I mainly use two articles. *The Transformation of Shanghai Confucian Temple in the Republic of China* points out the change especially in architecture' functions of Shanghai Confucian Temple before 1927 and after 1932. *The history and Current Situation of the Shanghai Confucian Temple* mentions the public events that took place in the Confucian Temple at that time. The public events can be classified into entertainment and social education.

At the beginning of the renovation in 1928, the government of Shanghai emphasized: "The temple need to be open to people rather than just closed without benefit. Minglun Hall can be used for lectures to spread Confucian culture. Buildings such as lecture hall, teaching room, reading room and teahouse are also need to be added". (Shanghai Municipal Archives 1928). It can be seen that the intention of the renovation is not only to provide a place to visit and rest but also for social education. The government temped to achieve democracy by teaching people and lead them to participate in the social and cultural activities (Shanghai Municipal People Education Institute 1933). However, at that time, the city suffered from the 9.18 Incident and the 1.28 Shanghai War, and the nation was in danger. Therefore, government promotes patriotism in the renovated Confucian Temple. There were exhibitions about the 1.28 Shanghai War, the revolution and current affairs. In addition, patriotism speech contests, activities of writing patriotic spring scrolls and publication of the New People's Journal were also held at the temple. On the other hand, in terms of social education, there were exhibitions about health education, public schools, libraries, lectures and citizen concerts. The temple also used for planting pox for children, publishing New People Picture Album and giving popular lectures.

The renovation of the Shanghai Confucian Temple has opened this sanctuary freely to all classes of people. It bridges the gap of gentry and normal people as well as the separation of the sacred and the secular. The opening of the Confucian Temple brought social equality to people in Shanghai, where there were many tenements and class separation (Guopeng 2020). This is written in A Night in the Temple Park, "It is absolutely equal for everyone here and this is the true publicness. No matter whether you are young or old, poor or rich, you can come here without a ticket". (Changgeng 1936) In addition, social education is included in the entertainment in the Confucian Temple Park. It is written in the Declaration of 1932: "In Confucian Temple Park, you can not only feel the fresh air, but also learn about flowers and trees from the billboard. You would know which categories and which families the plants belong to. In addition, in the 1.28 Shanghai War Memorial Hall, you can see the six hundred pounds bombs and all kinds of trophies that the loyal and brave soldiers took back. What is more, you can learn the use of sacrificial vessels for memorizing Confucius and some common sense about hygienics".

It is worth mentioning that the renovation plan was divided into two phases. The southern wall of the temple was completely removed in the first phase and replaced by stone pillars that are spaced approximately 1.65 m apart. Holly trees are planted between the piers. With this change, people can see the interior scene of the temple from the outside (Shanghai Municipal Archives 1931). Therefore, it is clear that the renovation paid much attention to the relationship between the people outside and the architecture. The Confucian Temple show welcome to passers-by at that time. On the contrary, nowadays, the high wall is rebuilt, the gates are closed and visitors will be charged.

6.3.2 Information from Internet and Social Media

Internet and social media also contain collective memory. Information from them may not be academic nor scholarly, but vivid. More importantly, this kind of information is the most accessible to the public, normal people who are not scholars. It means that this information is developed from collective memory and would create a new collective memory. Dianping (大众点评), the most popular application to make points of interest in China, was selected for this study. From Dianping, the comments and feelings of visitors to the temple could be seen. In addition, if searching for the Shanghai Confucian Temple through Baidu (百度), the most popular search engine in China, there are many blogs and websites with high hits that describe the Confucian Temple Book Fair and how the temple was in the past. Some of them are also selected to learn about collective memory. Above all, this part is written on the basis of information collected from online blogs, social media and websites. Thus, the credibility of this information is questionable, but it truly plays an important role in the study of collective memory. Through this information, the transformation of the Confucian Temple Book Fair in the past two decades can be partly depicted.

The book fair can be divided into two parts. The one outside the Confucian Temple focusses on the wholesale of various magazines and is open from Monday to Saturday, while the other was set in the plaza in front of Dacheng Hall inside the Temple. It focuses on the trade of old books and opens every Sunday. The book fair outside the Temple has been moved to the Shanghai Book Market since 2013 due to concern for fire disasters, disturbance on nearby residents and inconvenience in traffic. The Sunday Book Fair in the Temple is also proposed to be moved in 2020 for the same reasons, but this was earlier due to the epidemic. Among these two, the Sunday Book Fair is the more well-known and popular one. Therefore, the following discussion will mainly focus on the Sunday Book Fair, and the following Confucian Temple Book Fair refers to the Sunday Book Fair.

Many Shanghai locals say that the Confucian Temple Book Fair is an important part of their memories and is a paradise for book lovers. Also, many tourists comment in the Dianping that they went to the Confucian Temple Book Fair in purpose for its fame. It is clear that the Confucian Temple Book Fair is not only important to locals in Shanghai but is also a significant symbol of Shanghai for tourists.

However, the Confucian Temple Book Fair's prosperous period is gone. In 2001 and 2002, there would be 8000–9000 visitors every Sunday. At that time, 250 booths were set up, but that is still not enough to meet demand. The book venders had to queue overnight to get a booth. However, by 2013, there were only about 1300 visitors every Sunday, and the number of booths had been reduced to about 90. The later interview will also mention that the business of Confucian Temple Book Fair became worse and worse in the later years and was eventually moved away in 2020 (Fig. 6.3).

6.3.3 Interviews

There are five interviewees. All of them have lived nearby for more than ten years, which means they witness the prosperity of the past and the depression of the present Shanghai Confucian Temple. This makes them the proper interviewees to learn about the history of collective

Fig. 6.3 Notice of moving away the Confucian Temple Book Fair. Photo taken by the author

memory. Moreover, they are an old man who lives nearby and used to go to Confucian Temple, two residents in Menghua Lane, a shopkeeper on Wenmiao Road and another on Menghua Road. Through their different identities, different perspectives can be achieved. The interviews are mainly about the public events take place inside and around the Confucian Temple in the past and their feelings about it. Through their recollection of direct experience, the collective memory of the Shanghai Confucian Temple can be depicted. Furthermore, they are also asked about current nearby commons.

(1) **Reasons for Moving Away Confucian Temple Book Fair**

Through the information from internet, the two important reasons for moving away Confucian Temple Book Fair are fire disaster concern and disturbance on the nearby residents. However, in the interview, the two residents of Menghua Lane said they have become used to the "noise" and like this lively atmosphere. An old lady who lives there for more than 20 years said: "The old and traditional neighbourhood we live in is different from the new and modern one. We have a closer relationship and I enjoy it".

On the other hand, the shopkeeper on Menghua Road believes that the moving of the book fair is due to poor business. He said: "today, fewer and fewer people read books. So no one comes here for books anymore".

(2) **Public Events in the Past**

The old man who has lived in this neighbourhood for more than thirty years expresses his strong nostalgia for the lively atmosphere around the Confucian Temple in the past. He looked at the Lingxing Gate and said: "Twenty years ago, many activities took place here. There were cricket fights, booths and gathering people. The lively atmosphere at that time was especially good". At the same time, the old man also believes that the temple welcomed everyone in the past, which is different from how it states now. He said: "the administration hopes no one would come, so that it is convenient to manage".

The shopkeeper who has run a shop on Wenmiao Road for more than forty years also depicted the lively scene of the old Wenmiao Road: "When I was still at school, crickets, old coins and all kinds of gadgets were sold here. They are sold at booths by the road and there were fewer stores before". But now it is not allowed to set up booths by the Wenmiao Road and stores without business licence need to be closed. As a result, there are much fewer visitors than before. The shopkeeper also said: "It is for a bike to get through the Wenmiao Road before, but now even a car can get through it".

(3) **Current Nearby Commons for Public Events**

Residents say there are no commons like parks near this residential area and the nearest park is 1.9 km away, which takes more than 30 min to walk to. The old man who used to go to the Confucian Temple needs to ride a bicycle to get there, and it is inconvenient. Therefore, he seldom goes to the park and still takes a walk around the Confucian Temple although he cannot get into it.

6.3.4 Observation

When going to the Shanghai Confucian Temple at around 10a.m. on a Sunday morning, one would find that the plaza in front of Dacheng Hall looks empty and deserted after the book fair moved out (Fig. 6.4). There were no more than 15 visitors in total. Most of the visitors were students and their parents who came to pray for study (Fig. 6.5). In addition, there were four tourists who came to the Confucian Temple because they thought there might be too many visitors at another more famous temple in Shanghai. What is more, there were old women

Fig. 6.4 Empty and deserted plaza. Photo taken by the author

Fig. 6.5 Praying student. Photo taken by the author

Fig. 6.6 Elderly people exercising. Photo taken by the author

playing Tai Chi in the Confucian Temple (Fig. 6.6), which shows that residents may do exercise in the temple.

6.4 The Degradation of the Publicness of Shanghai Confucian Temple

On the basis of the study and field research above, the transformation of the Shanghai Confucian Temple as a common for public events can be concluded as follows.

In 1927, considering the social influence of the Confucian Temple, the government made the Confucian Temple open to every citizen and be responsible for social education. The Confucian Temple was turned into a park, and the fences were dismantled so that people could see the scenes and activities inside the temple. This was the highlight of publicness that the Confucian Temple could achieve in the next 100 years (Fig. 6.7).

The Confucian Temple Book Fair, which appeared in 1986 in the plaza in front of the Dacheng Hall, is also a significant part of the

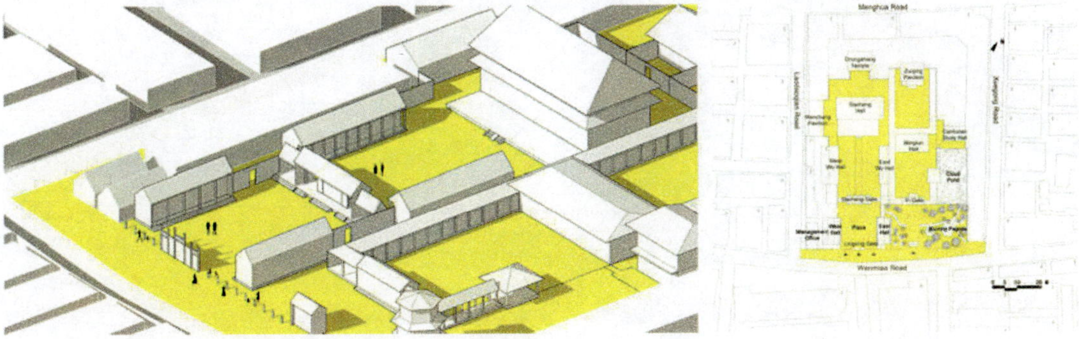

Fig. 6.7 Confucian Temple Park after 1932 (yellow = public space). Drawn by the author

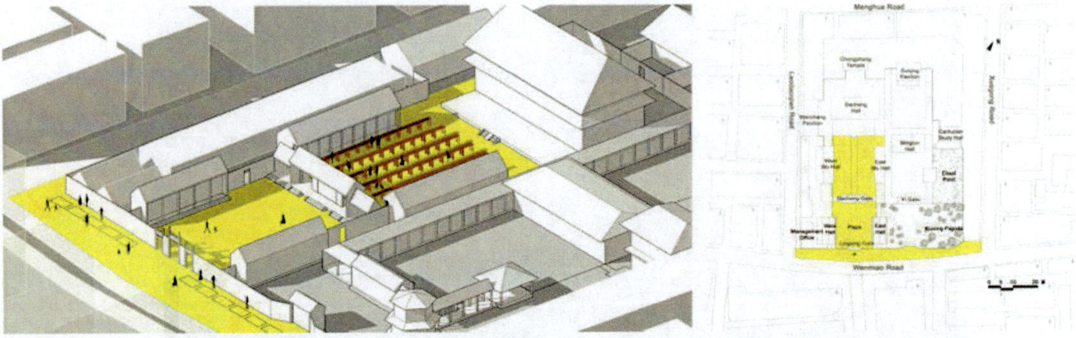

Fig. 6.8 Confucian Temple Book Fair in 1986 (yellow = public space). Drawn by the author

publicness. The book fair gradually became a landmark of Shanghai. Meanwhile, booths and trade by Wenmiao Road made it an important place for public events. During this time, although the temple has less space that is open to people compared to the previous park, public events still remained in the west of the temple due to the book fair (Fig. 6.8).

In early 2020, the Confucian Temple Book Fair was completely moved away. Only a few elderly people exercise and students pray for study in the temple. At the same time, booths on Wenmiao Road were banned. Public events could only take place on the sidewalk outside the temple. However, without booths and vendors, few people came here. The publicness of the temple dropped to a deep low point (Fig. 6.9).

Since the Shanghai Confucian Temple was turned into a park during 1927–1932, the area for public events of the temple has gradually shrunken. After the fences were built and the temple began to charge visitors, the book fair and booths on Wenmiao Road maintained a strong lively atmosphere. However, now the booths were banned, and the book fair was moved away, so area for public events was again shrunken and was very limited. In the later part of the article, a renovation design proposal would be put forward with the intention to make the vanished publicness of the temple properly reborn.

6.5 The Rebirth of Common: A Renovation Design of Shanghai Confucian Temple

When taking location into account, the Shanghai Confucian Temple is different from temples in the countryside, as it is surrounded by residential areas. Also, it is not like a religion hall because it is used to commemorate an educator and was once used as education hall. As a result, it is not

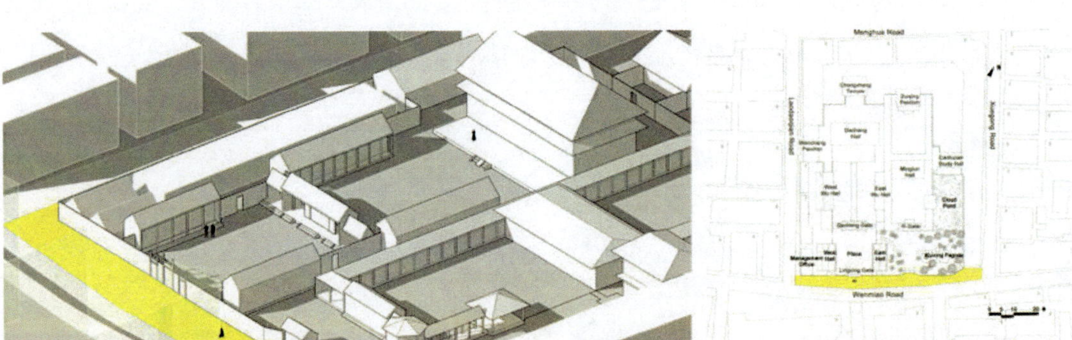

Fig. 6.9 Shanghai Confucian Temple in 2020 (yellow = public space) Drawn by the author

for a certain congregation, but a broader public, and will be expected for social education. Thus, related to its location and its relationship with education, the Shanghai Confucian Temple should be more open to the public compared to its current state.

Therefore, this proposal is put forward with the intention of improving the public appeal of the Shanghai Confucian Temple by renovating its entrance area (Fig. 6.10). The renovated area includes the management office, the garden in the west, the plaza in front of Dacheng Gate and Yi Gate. Firstly, the entrance gate at which visitors would be charged for tickets is moved from the Lingxing Gate and Xue Gate to the Dacheng Gate and Yi Gate. Then, the high fences at the Lingxing Gate are replaced with short stone piers, just like once it was in 1930s. Also, two entrances are added on Xue Gong Road and Wenmiao Road for the west garden. In this way, the plaza in front of Dacheng Gate and Yi Gate along with the west garden where the Kuixing Pagoda is will be open to the public. In addition, the management office area would be added with reading rooms, so that public events can take place here (Fig. 6.11).

After renovation, the entrance area consists of three parts. From west to east, they are the reading room, the plaza, the tea house and the pocket park (Fig. 6.12). The reading room can be used for social education related to the former Confucian Temple, as well as indoor public events. As for the plaza in front of Dacheng Gate, residents can do exercise and outdoor public events can take place there. The teahouse would be a connection between the plaza and the pocket park where people from both sides can rest. Lastly, the pocket park would make up for the lack of public green space in this residential area (Fig. 6.13).

Fig. 6.10 Current situation of the renovated area. Produced by the author

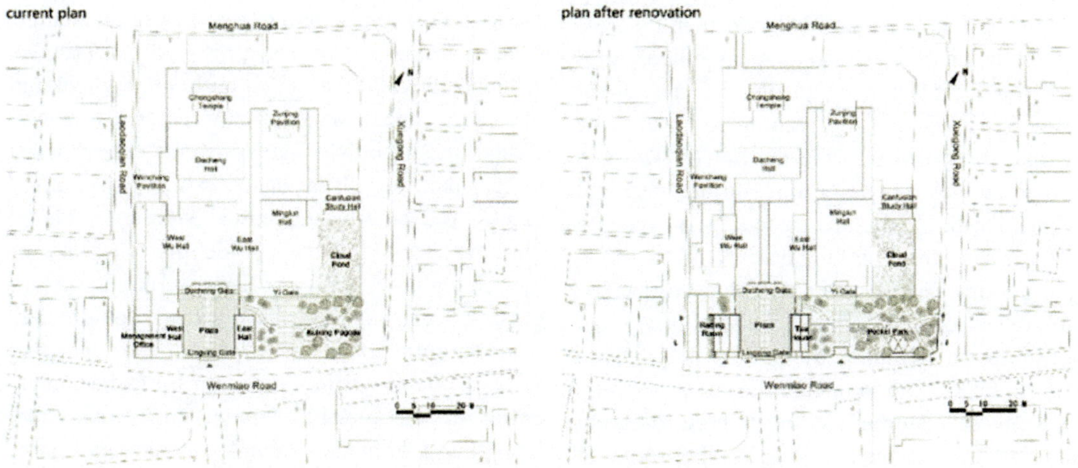

Fig. 6.11 Plan before and after renovation. Produced by the author

Fig. 6.12 Section. Produced by the author

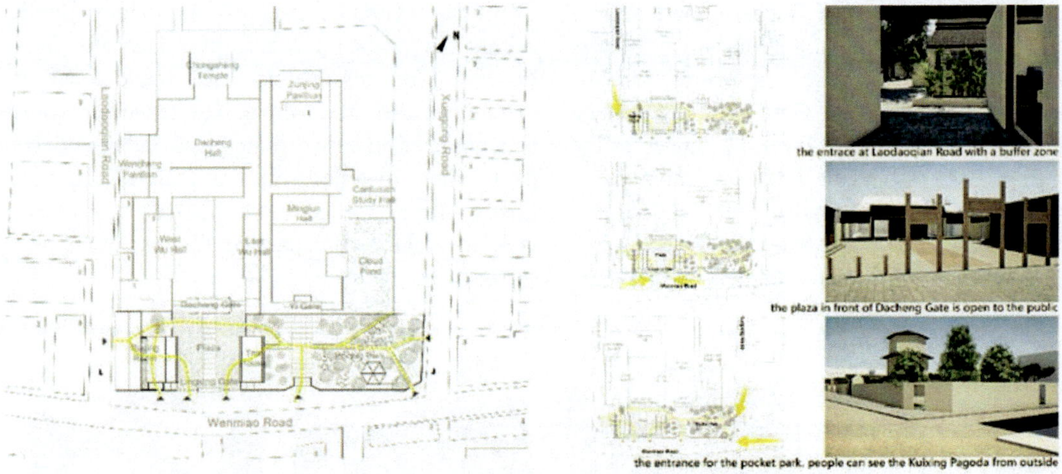

Fig. 6.13 Users' routes and different entrances. Produced by the author

Detail architecture designs for the reading room, plaza and tea house group are considered for the main entrance which is set in this group (Fig. 6.14). The West Hall is larger than the East Hall for it includes a management office. As a result, the West Hall and part of the management office are turned into the reading room and the East Hall is turned into the tea house. The gap between the new reading room and the west hall is planted with bamboo to provide good scenery for the users from both sides. Also, the two reading rooms can be connected by a trestle. The management office is moved to the northwest of this group. In addition, there is a buffer zone between the reading room and the street, so users would not be disturbed. What is more, different functions are arranged related to the route of users (Fig. 6.15).

6.6 Discussions and Conclusions

Although the renovation design is a small change if seen from a city scale, it may still make influence beyond the neighbourhood. By making the plaza open to the public again, it gives change to all kinds of public events taking place there, like the plaza dance which is popular among the elderly in China. Moreover, along with the special history and identity, the plaza of the Shanghai Confucian Temple could distinguish itself from other plazas, which could make some cultural events take place there, like the weekend book fair in the past or film screenings. Some of these public events may make the Shanghai Confucian Temple iconic and once again recognized beyond the neighbourhood.

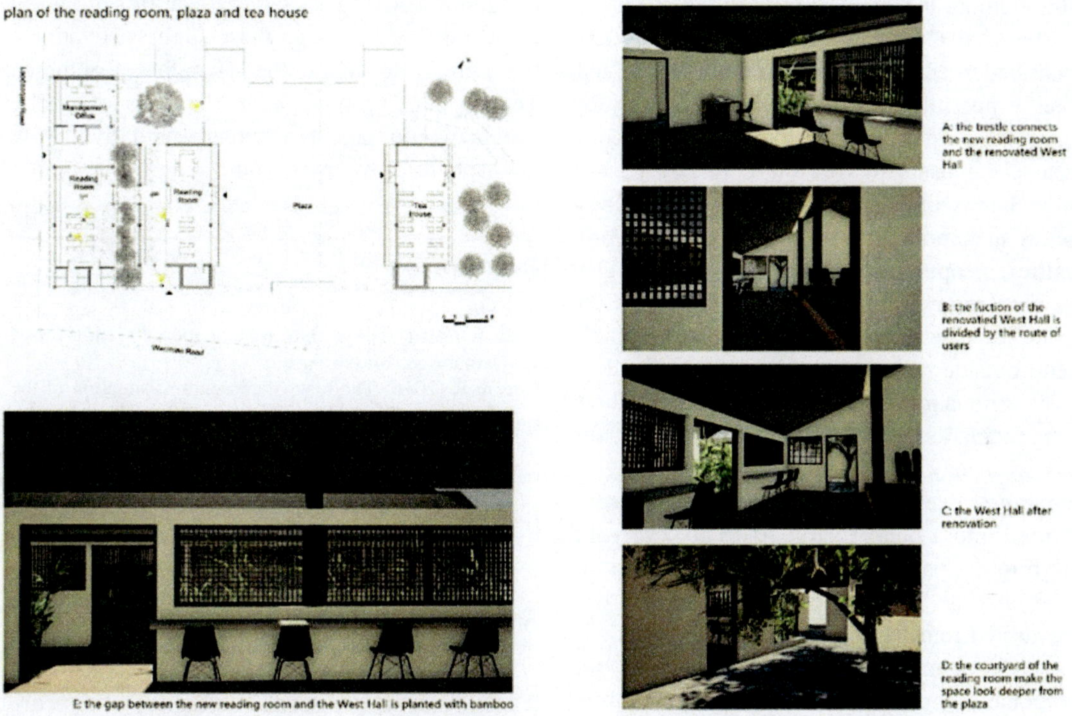

Fig. 6.14 Plan and perspectives for the reading room group. Produced by the author

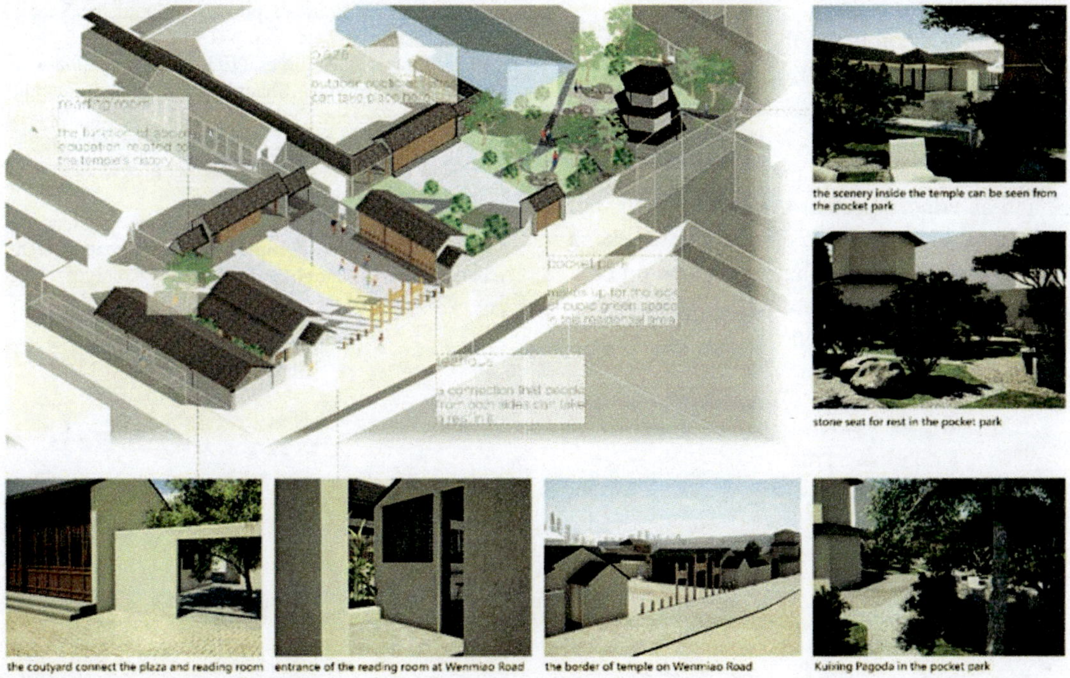

Fig. 6.15 Panorama of the renovation design proposal. Produced by the author

In addition, it not only reacts to the history of the Shanghai Confucian Temple, which is closely linked to education and cultural life, but also makes it possible to form a new community by turning the west hall group into a reading room. Because the users of this place probably have similar habits or hobbies, they can easily combine as a community. With this kind of communities, people from other parts of Shanghai may also come to this place, which makes the impact of the Shanghai Confucian Temple extend outside the neighbourhood.

The renovation proposal also considers about the participation of all people, especially the users who may not be interested in reading or any cultural events. As a result, the west garden is turned into a pocket park. After all, everyone can go to the park. This is also a reaction to the history that the Shanghai Confucian Temple was once turned into the Confucian Temple Park in 1930s. As mentioned in the former part, the Confucian Temple Park was absolutely equal for everyone here, and this is the true publicness. No matter whether you are young or old, poor or rich, you can come here without a ticket. (Changgeng 1936).

Ultimately, according to the documents, social media and field research, the publicness of Shanghai Confucian Temple has declined since its renovation during 1927–1932 and people show nostalgia to this common. Firstly, it was a park and an education hall, then a book fair, but now it is a tourist attraction and more like a temple in the countryside than in the central city. However, the Shanghai Confucian Temple has a history as a common as well as a location to be a common. Thus, this study-base design proposal is put forward. Through these slight renovations, the aim is to make the Shanghai Confucian Temple a common again. All kinds of public events would take place there, and the lively atmosphere of the past would reborn.

References

Aldo R (1984) The architecture of the city. MIT Press, Cambridge, Massachusetts

Сергій K (2016) The history and current situation of the Shanghai Confucian Temple. Ukrainian Studies 00:195–204

Changgeng (1936) A night in Confucian Temple Park. In: Women's resonance, vol 9

Guopeng Z (2020) The transformation of Shanghai Confucian Temple in the Republic of China. Chin Cultur Stud 2:27–39. https://doi.org/10.15990/j.cnki.cn11-3306/g2.2020.02.004

Olick JK, Vinitzky-Seroussi V, Levy D (2011) The collective memory reader. Oxford University Press. ISBN 9780195337419

Robin Monotti G (2016) Aldo Rossi's city of collective memory. Retrieved on 21 Dec 2022, from https://nulluslocussinegenio.com/2016/10/17/aldo-rossis-city-of-collective-memory/

Shanghai Municipal Archives (1928) Shanghai Municipality works committee's public works of the Shanghai Confucian Temple. File No. Q215-18090

Shanghai Municipal Archives (1931) The first phase of Construction of the Shanghai Confucian Temple renovation, File No. Q215-18090

Shanghai Municipal People Education Institute (1933) Overview of the Shanghai Municipal people education hall 18

Zimin Z, Jian L (1990) Strengthen management at the Shanghai Confucian Temple book fair (general technical report). Shanghai Municipal Archives, File No. Q215-18090

7 Investing in the Rejuvenation of Urban Water Commons—Vayalagam Experiences of Madurai

S. Elamuhil, M. P. Vasimalai, and Arsh Saluja

Abstract

The landscape of Madurai urban, the temple city of Tamil Nadu, India, has been hydrologically embedded with hundreds of traditional surface water bodies well-connected as cascades with the natural drainage network as well as the rivers. The design of these water bodies has adopted a style quite fitting to the gradual fall of the contours and the system of decentralised village administration. Villages in Madurai are named after the water bodies signifying their reliance on these surface water bodies. The city of Pandiyas expanded five times in just four decades into a hub of service sector. It opened up greater and wider opportunities, pulling non-natives, and accelerating land-use conversions, converting water bodies and their agricultural land into development zones. These conversions have resulted in the deterioration of urban water bodies. Despite the urban pressures, the city has 55 irrigation tanks, 44 ponds, and 12 temple tanks, still serving the urban. Increased water stress, fall of groundwater reserve, and floods in the last decade pushed the urbanites towards the conservation of the water bodies. DHAN Foundation in collaboration with key stakeholders, enabled the urban communities in rejuvenating 16 urban water bodies, by introducing functional components such as recreation, rejuvenating some of the traditional practices, and evolving combating mechanisms against urban pressures. This paper highlights various expressions of the community rejuvenation processes that would help in better investment in urban water commons for marching towards 'sustainable cities and communities' (SDG-11).

Keywords

Surface water commons · Cultural commons · Sustainable cities · Rejuvenation · Community investment

7.1 Waterbodies and Communities in Past—Madurai as an Agrarian Society

The landscape of Madurai, the temple city of Tamil Nadu, has been hydrologically embedded with hundreds of traditional surface water bodies well-connected as cascades with the existing drainage network as well as the rivers. The meteorological characteristics of Indian Monsoon and geological characteristics of Deccan

S. Elamuhil (✉) · A. Saluja
Dhan Foundation, Centre for Urban Water Resources, Madurai, India
e-mail: elamuhil@dhan.org

M. P. Vasimalai
Dhan Foundation, Madurai, India

S. Hilal et al. (eds.), *Design for Partnerships for Change*, Sustainable Development Goals Series,
https://doi.org/10.1007/978-3-031-36993-3_7

Plateau necessitated the creation of these surface water bodies to store rainwater during the monsoon and utilise the same at a later date (Ratnavel 2006). The layout, structure, and construction of these small water bodies bring out the ingenuity of the community that has adopted a style quite fitting to the gradual fall of the contours and the system of decentralised village administration (Shanmugham 2005). The 'self-revenue generative' nature and direct dependency of rural living on the revenue were the key factors behind the existence of these water bodies over the millennium. Each village in this region has three distinct water harvesting and storage structures—the Irrigation tanks known as *'Eri'*, *'Kanmoi'*, or *'Kulam'*, the drinking water storages known as *'ooranis'* and the 'holy' temple tanks.

The villages of Madurai take their names after the waterbody that sustained their living and livelihood. *Kodikulam, Thattankulam, Kananendhal, Kadachanendhal, Thennaneri, Anthaneri, Koothiyarkundu, Kosavankundu, Karupayoorani, and Sekkanoorani* are some of the villages that are named after the water bodies the villages were dependent on. The suffixes—*kulam, endhal, eri, kundu, oorani*—refer to water bodies. One of the ancient Tamil literatures refers Madurai city as *'Maadakula Keezh Madurai'*, meaning Madurai city was established downstream of *Maadakulam* Irrigation Tank, indicating the communities' significant reliance on the traditional surface water system (Narasimman 1936).

Long ago, these water bodies were accessed for the water stored, the clay and silt deposited, the soil moisture in the tank bed, the floral and aquatic species. The ayacut farmers, landless agriculture labourers, livestock and duck rearers, lotus cultivators, nomadic shepherds, fishers, washers, and potters centred their life on these water bodies (Balasubramanian 2003). Even though these tanks and channels were built by individuals, villagers, kings, queens, traders, and others, they were operated and maintained by the local 'people institutions'. The people's organisations had well laid out rules and regulations imbedded as social norms to manage water effectively (Palanisami 2006).

Historically, these water bodies have thrived as rules were conceptualised at two levels: operational and collective choice. Operational rules were to regulate the participants while collective choice stipulated to the conditions for adopting, enforcing, and modifying operational rules (Visvanathan 2021). Notable feature in the early Tamil kingdoms was that there was little or no private property in land. The tanks and their *ayacut* (tank command area) were common property, the land was redistributed every year among the cultivators to keep equality among them as the yield from each land parcel could vary according to soil fertility and availability of water (Sakthivadivel 2004). The decline of land holding as 'common property' started from the fifth century AD by means of donations or gifts to chiefs by kings. Kings, noblemen, or the villagers made endowments for tank management (Gurukkal 2007). After colonisation, the waterbodies were taken over by the British Indian Government and later by the Free-Indian Government (Mosse 1999). Despite several structural, functional, governance, social, and cultural changes, these waterbodies survived because they were not just waterbodies but were the 'common pool resources', the water 'commons'.

7.2 Commons as Poromboke in Urban Fabrics

The city of Pandiyas, post-Independence expanded as the *'Thoonganagaram'* (Sleepless city), a hub of administration, tourism, heritage, agro-market, and centre for the service sector. It opened up greater and wider opportunities, pulling non-natives, and accelerating land-use conversions, converting water bodies and their agricultural land into development zones. In the process of urbanisation, the *ayacut* farmers converted their farmland as build-ups, and so the dependent landless 'agri-labourers' and tank-dependent marginals moved towards 'non-agrarian sector'. The fragile functional relationship with 'water commons' led to weathering of the socio-cultural expressions of the communities embedded with these water bodies. Without sufficient willingness to create, maintain, protect, and improvise, these commons have been at risk of conversion, deterioration, or encroachment. It is not just the 'livelihoods' but also the 'living' that was also not centring these water bodies. The urban communities now meet their 'domestic water' needs from 'public' water systems, 'private' bore wells or both. The process of urbanisation convoluted the idea of waterbodies from 'commons' to 'poromboke' (wasteland), then began the tragedy of commons (Bakker 2007).

The ownership of these water bodies is with the state departments and municipal corporations. The state departments treated these 'resources' just as *'Neerpidi Poromboke'* (water-spread wasteland) (Aparna Watve 2021). To reduce the land pressure of urban sprawl, irrigation tanks without cultivable command area were targeted for the establishment of state infrastructure such as administrative buildings, housing complexes and courts; the *ooranis* (ponds) were converted into urban infrastructures such as bus stations, schools, and healthcare centres. The built-up area of Madurai City has increased five times (from 13.5 to 67.5 km^2.) in just four decades. These conversions have resulted in the deterioration of urban water bodies engulfing 30% of the total water-spread area. But all these conversions were undertaken with the consent of the majority of local population.

It's not just the conversion of waterbodies but also the conversion of waterways that have affected the traditional water system, which is cascading in nature. The conversion of waterways into lined channels, to avoid encroachments, has ignored or blocked some of the subdivision channels which fed the irrigation tanks. While conversion of the tank command area into the residential plots, the field channels were also annexed. This has led to cutting-off of feed to some of the minor water bodies such as *'Yendhal[1]' and oorani.* The ponds that were solely dependent on the subsurface water, became barren as the cultivable lands turned into impervious build-ups.

7.3 'Barren Waterbodies'—A Free Ride Commodity

Both the urban individuals and the state treated the 'barren waterbodies' as a 'free ride commodity'. It has been taken for granted and misused. As the boundaries of water bodies are not clearly defined and imposed, it was encroached initially by dumping construction debris. Communities dispose their solid and liquid waste in the waterways and waterbodies, 'wetting' the wasteland with untreated wastes. Lack of relevant waste management infrastructures in the urban systems dilutes the stringent actions taken by 'regulatory bodies' against the disposers (Kessides 2005). Plastics and alcohol glass bottles form the bed of the waterbodies in the age of Anthropocene.

7.4 Encroachments—Individual, Collective and Marginals

Encroachment of waterways by residents is more common than encroachment of waterbodies. When the 'influential' individuals encroach the waterbodies/waterways, the encroachment characteristic is 'permanent'. The build-ups constructed are often within the waterbody and 'expensive'. The dependency of the local or marginal communities on the 'influential' individuals to meet their intricate social, political, economic, and water demands, turns as 'the social license to encroach'.

Establishment of deity or small religious structure in waterbody is adapted as a strategy by 'individuals' or a 'group' to inculcate 'permanency' towards their encroachments. The religious connectedness, festivals, and rituals in 'worship', a form of cultural commons turn as the defence mechanism against the encroachment eviction. When such encroachments are socially accepted by the local community, eviction becomes nearly impossible.

[1] Yendhal is a minor irrigation tank located in the command of the major irrigation tank. It is fed either by one of the sluices of the major irrigation tank or by the sub-surface flow that seeps from the flood-irrigated *ayacut.*

When the waterways and waterbodies are encroached by the 'non-influential individuals', the encroachments are generally in the form of 'add-on' structures such as a compound wall, toilet, or a septic tank, that shall be evicted whenever the regulations are imposed. Similar to the urban sprawl, these encroachments also have the sprawling effect (Brueckner 2001). Encroachment by an old individual resident, builds confidence of the later residents to encroach, ultimately turning into collective encroachment. Though the sense of 'collectiveness' among the individuals gives strength to the violators, 'the fear' of eviction is expressed as temporary structures constructed in the fringe of the waterbodies/waterways.

Cultivating short-term crops such as small millets, vegetable, greens or fodder in the tank bed, well utilising the soil moisture is one of the traditional practices in the irrigation tank-fed rural areas. It has been the 'usufruct' right of landless agricultural labourers to practice tank-bed cultivation. As the tanks are seasonal and so the dependent cultivable command area. The tank-bed cultivation practice builds resilience of the landless farm labourers against seasonal migration, supplementing the food and nutritional security. To ensure equity and avoid encroachment, construction of wells and fencing the 'usufruct' zone are restricted socially by the villagers, thereby the norms are never violated by the marginal individuals. The government departments in urban pose this socially regulated 'usufruct' right as an 'encroachment' and the economically weaker landless tank-bed cultivators as 'encroachers'. Consequently, the 'untapped' soil moisture of the tank-bed favours invasion of *Prosopis juliflora* and *Ipomoea carnea.*

Lack of space management in the sprawling urban pushes economically poor local immigrants towards the 'ignored' waterbodies, as first point of free 'access' to establish their settlements. A patch of such individual settlements in the bunds of the waterbodies is often constructed out of mud, tin sheets, thatches that are low cost and dismantlable in nature, characterising the encroachment nature as 'temporary'. Wherever the regulatory system remains insensitive towards such encroachments for a long period, it graduates from 'temporary' to 'semi-permanent structure'.

7.5 Community Towards Urban Waterbodies

Despite the urban pressures, the city has 55 irrigation tanks, 34 ooranis, and 12 temple tanks. Water insecurity arose in the twenty-first century, with prolonged failure of monsoons, fall in groundwater table and inundations. The fall in the land and rental value due to water issues pushed certain residential associations towards the conservation of waterbodies in their vicinity. Livelihood insecurity became another triggering factor behind the landless agrarian urban communities to involve themselves in the restoration of waterbodies. Though several organised groups, and individuals pressurised the government to invest in the restoration of waterbodies, the allocation of investment towards the respective waterbodies relied heavily on the 'purchasing capacity' of the population. Such tapped investments were predominantly 'piece-meal', direct, and less-democratic.

Emerging social concern over water resources, catalysed by the success of community-led movement such as '*Tarun Bharat Sangh*', pulled youth volunteers to engage in such restoration activities. Though the communities as individuals or collectively initiated investing in the waterbodies, it was accelerated by the corporate investments as part of their social responsibility. Despite the investments, the restoration of waterbodies was less effective. It was realised that the local communities are devoid of the sense of 'commons' or the 'trusteeship' which degenerated the restored waterbodies and turned the investment less beneficial.

This sense of commons and trusteeship was the basis for *'Kudimaramathu'*, the traditional socio-cultural system that restored waterbodies every year, preventing from deterioration. *Kudimaramathu* is a citizen engaged process of restoring water structures in which water users volunteer labour on a household and/or acreage basis. *Kudimaramathu* was done once a year, generally in summer or before the onset of the monsoon. If a household was unable to contribute labour, it contributed in kind (grain or money). This system was strengthened by legends *(Thiruvilaiyaadal puranam)* nurturing the system as a 'cultural commons'. Post-colonisation, such traditional practices degenerated with the state wanting to control common pool resources (Manisha Shah 2018). Though the state government revived the *Kudimaramathu* concept in 2017, it was never the same with the 'social capital' that involved in restoration. Implementation of the scheme fell short because of the void in the 'capacities' of the 'social capital' built around the waterbodies (Rajendran 2018).

7.6 DHAN's Approach in Waterbody Restoration/Rejuvenation

Sustainability of the traditional water system relies on the reconceptualisation of these 'waterbodies' as 'water commons'. The sense of water commons comes not only with the community but also along with their socio-cultural norms and relationship with the natural elements and resources (Thompson 1993). DHAN Foundation, a not-for-profit development organisation, enables the people institutions to re-invent their waterbodies as 'commons'. From 25 years of DHAN Foundation's experience in water sector, it has been realised that the sense of 'commons' is vital in post-rejuvenation investment. When the rejuvenation is driven by the demand stream, at the grass root levels, i.e. local communities, starting from planning, the sense of 'common property resource' is naturally nurtured among the communities. With the community-invested restoration model, DHAN Foundation has enabled local communities to restore more than 4,000 water commons in rural India by promoting 'water commons association' called *'Vayalagam'*. Thereby, DHAN Foundation catalyses the demand stream for benefitting from the supply stream (Water commons rejuvenation investments) in a better way. Since 2017, the 'well-proven' community model was tested in the urban fabrics of Madurai. So far DHAN Foundation has promoted 16 *Vayalagams* which rejuvenated urban water bodies, by introducing functional components such as recreation, rejuvenating some of the traditional practices, and evolving combating mechanisms against urban pressures. In this model, every rejuvenation process starts with community.

7.7 Commons: In Search of Community

A major challenge with managing urban commons stems from the existence of multiple uses and users that compete for the resources (Vinay Gidwani 2011). Because of shared rights of usage, unrestricted access, and unregulated use, the accountability for maintaining commons is largely not established and users often resort to blame-game on the face of dwindling resources. A major hurdle in making commons management work sustainably is to shift from competition to cooperation, which is heavily dependent on existing social capital. The societal nature; dependency on waterbodies; and cultural maturity of the communities in the vicinity of the waterbodies and their complex interactions determine the possibility of retrieving the sense of 'commons' in the rejuvenation process.

Though some of the waterbodies fall under urban fabric, the dependent communities still retain some of the socio-cultural practices from their traditional 'rural' past. Restoration/rejuvenation serves as a process of retrieving socio-cultural practices. The process of formation of conservation associations becomes simple as they still have the direct functional dependency and acquired cultural commons over the waterbody. As they lead the restoration process, the collective investment during and post-rejuvenation is high.

In the new urbanising zone, the interactions and negotiations among the 'new residents' and landless natives are vital in the successful rejuvenation of the water commons. The friction that exists among them, adversely affects in sharing the common pool resources. As the new economically sound residents conceive the water commons as a recreational and groundwater recharge space, they tend to perceive the economically weaker native landless who use the pond for domestic purposes, as 'polluters'. The social injustice is expressed even in the conceptualisation of rejuvenation of waterbody by the addition of fence, removal of access Ghats and further amplified post-rejuvenation by restricting the entry to specific timings. Though such rejuvenated waterbodies remain undeterred, it loses the characteristics of commons. Creating platform of interactions and common ground during conceptualisation of rejuvenation, slowing down the rejuvenation implementation process by phase-wise implementation, facilitating collective investment and action to acquire the sense of 'commons' are key in transforming such 'waterbodies' as 'water commons'.

The very early stage of 'social capital' development for the urban water commons is that their 'heterogeneous individual-inhabitants' begin to see themselves as a group. Sharing expectations in their relations of living together; and the preservation of the socio-cultural elements that make the group, as a 'group' are vital elements (Bilgrami 2021). Very old urban establishments have developed their sense of 'group' by fulfilling their residential needs collectively. As they express good signs of cultural norms among them, the waterbody rejuvenation is led by the community with few initial triggers. Starting from the rejuvenation planning process, the level of investment is high throughout the process and post-rejuvenation. The rejuvenation process serves as a platform for the communities to practice the acquired sense of 'commons'.

In 1974, at Harveypatti, Madurai's first township established by the Scottish Harvey brothers (Muthukrishnan 2014)*, a group of youngsters established 'Atlas Gym'. The 'Atlas Gym' served as a space for youngsters to nurture themselves as 'weight lifters' and 'wrestlers'. The gym produced a lot of policemen and soldiers from the locality. The Atlas gym weathered after two decades but not the acquired cultural commons such as 'equity, self-regulatory, shared stake, democracy, serving the needy, collaboration, and oneness'. After 34 years, the same 'boys group' had another platform to re-express their sense of 'commons'. They promoted the 'Sooravalaimedu tank conservation Vayalagam' association in 2018, for rejuvenation and conservation of the waterbody. The committee was democratic, with the best mix of experienced women, men, and youth of diverse socio-political-economic backgrounds. The association completely redesigned the rejuvenation plan that was suggested by the City Corporation. They were highly consultative with the Corporation officials and other stakeholders during the co-planning process. They introduced the 'walking track in the tank bund and a children's park', the 'women and children' element in the rejuvenation plan compromising a few engineering structures, to ensure social justice. They expressed strong community negotiation in the eviction of encroachments. Their accountability was expressed during the implementation by analysing the effectiveness of the implementation on daily basis and maintaining transparency in the accounts handling size of 20 lakhs. The association has invested about 2 lakhs in the rejuvenation and also maintained the same on their own on a rotational basis. They are also in the process of transferring this 'selfless collective trusteeship' to the next generation to continue the legacy* (Figs. 7.1 and 7.2).

It's neither the unresolved litigations nor the technical infeasibility, the community itself has hindered the rejuvenation processes. There were few cases where the heterogeneous communities with an adverse functional relationship with the waterbodies such as collective encroachment, collective disposal of solid waste, and sewage have shunned the rejuvenation process. Despite having the scope of investment, even after several triggers, the rejuvenation process has been dropped due to the lack of community interest.

Fig. 7.1 Arial view of filthy ignored Sooravalimedu Tank in 2019

Fig. 7.2 Arial view of Sooravalimedu Tank in 2021—post-community-led rejuvenation

7.8 Way Forward

Zero encroachment, zero disposal, zero wastewater discharge, zero invasion of exotic species are some of the long-term visions which demand strong 'socio-political-ecological' conviction, continuous research and sensible investment. Furthermore, reconceptualising the urban water commons for future functionalities to connect the local communities with the water commons, stakeholders' acceptance for the redesign without affecting the hydrological balance are the prerequisites. But the real challenge lies in facilitating development of the cultural commons and nurturing the sense of 'commons' towards these water structures in the collective conscious of the urban communities.

Enriching the 'collective sense of conservation and its vernacular openness' of the rejuvenated urban waterbodies is a greater challenge as most of the rejuvenation pushes the marginalised communities away from the rejuvenation process and so the rejuvenated water commons. As the transformation of a 'public space' (*Poramboke*) into 'community commons' is a long-term process, rejuvenation investments through short-term (3–6 months) projects shall not be helpful. Long-term partnerships (3–15 years and beyond) among the stakeholders with shared vision and community driven, community-invested phase-wise implementation will consolidate the sense of 'commons' among the communities. Developing opportunities and platforms to build the sense of 'commons' as the 'capacity of the population' (UNU-INWEH 2013) is the core for cities like Madurai where the relationship with water and water commons is 'cultural'. The spirit of the water commons is not the 'exclusion or exploitation' but the sense of 'collective eco-sensitive practices' along with a new framework of trusteeship which makes cities and human settlements inclusive, safe, resilient, and sustainable (SDG-11).

Acknowledgements DHAN Foundation, thanks all the stakeholders, Madurai Corporation, TNPWD-WRD, HCL Foundation, Hi-tech Arai. Pvt. Ltd. in specific for investing in the rejuvenation of urban water commons. It is because of their investment and support; we are building the notion that 'water commons are not just a land use or a land cover but direct testimonials of our living standards and the social temperament'.

References

Aparna Watve VAIM (2021) The need to overhaul wasteland classification systems in India. Econ Pol Wkly 56(40):36–40

Bakker K (2007) The 'commons' versus the 'commodity': alter—globalisation, anti-privatisation, and the human right to water in the global south. Antipode 39(3):430–455

Balasubramanian RK (2003) Poverty, private property and common pool resource management: the case of irrigation tanks in South India, Kathmandu: South Asian network for development and environmental economics (SANDEE)

Bilgrami A (2021) The cultural commons - a philosophical analysis—Part I. Econ Pol Wkly 56(26–27):45–50

Brueckner JKEM, MK (2001) Urban sprawl: lessons from urban economics. In: Brookings-wharton papers on urban affairs, pp 65–97

Gurukkal R (2007) Shift of trust from words to deeds: implications of the proliferation of epigraphs in the Tamil South. Indian Hist Rev 34(2):16–35

Kessides IN (2005) Infrastructure privatization and regulation: promises and perils. World Bank Res Obser 20(1):81–108

Manisha Shah RS (2018) Will Kudimaramathu make communities "think tanks" again? Int J Eng Technol 7 (4):6878–6883

Mosse D (1999) Colonial and contemporary ideologies of 'community management': the case of tank irrigation development in South India. Mod Asian Stud 33 (2):303–338

Muthukrishnan A (2014) Vikatan. [Online] Available at: https://www.vikatan.com/government-and-politics/literature/madurai-history-the-arrival-of-madura-mills-and-its-effect-on-the-society. Accessed 11 Sep 2022

Narasimman (1936) Research on Thiruvilaiyadal Puranam. In: Umamaheswaram TV (ed) Tamizh pozhil. Karandhai Tamil Sangam, Tanjore, pp 193–200

Palanisami K (2006) Sustainable management of tank irrigation systems in India. J Dev Sustain Agric I (1):34–40

Rajendran S (2018) Tamil Nadu revives Kudimaramathu: ancient wisdom of water management. Econ Pol Wkly 53(6):18–20

Ratnavel SM (2006) In search of ancient wisdom - irrigation tanks, 1st edn. DHAN Foundation, Madurai

Sakthivadivel RPG, TS (2004) Rejuvenating irrigation tanks through local institutions. Econ Pol Wkly 39 (31):3521–3526

Shanmugham CRJ (2005) Technology of tanks—the traditional waterbodies of rural India, 1st edn. DHAN Foundation, Madurai
Thompson E (1993) Custom, law, and common right. In: Customs in common: studies in traditional popular culture. The New Press, New York, pp 97–184
UNU-INWEH (2013) Water security and the global water agenda. United Nations University, Ontario
Vinay Gidwani AB (2011) Urban commons. Econ Pol Wkly 46(50):42–43
Visvanathan S (2021) Reinventing the commons. Econ Pol Wkly 56(8):37–43

8

Contemporary Retrospection of Local Tradition—The Design of Cultural Service Center and Local Workshops of Ruyi Village, Shuikou Town, Yao Autonomous County of Jianghua

Xu Haohao, Wei Chunyu, and Ouyang Sheng

Abstract

Based on a villager activity center in the southern minority area of Hunan Province designed by the author, this paper explores an architectural design strategy that adapts to the characteristics of climate, production and lifestyle, and the characteristics of minority communities. Through interviews with villagers and field research in the early stage, the regional climate characteristics and spatial location characteristics of the town are studied and applied to the location of buildings and the design of external space. Then we combined the typological research method to analyze the space characteristics required by the daily life, social activities, and etiquette activities of the ethnic minority community where the project is located, namely "Gaoshan Yao", to obtain the space life prototypes of local people, such as the hall and patio, eaves gallery, stairs and sun terrace, and to explore the relationship between these "prototype" spaces and people's daily social life. In the result part of the design, this paper mainly responds to the two main problems of "Community culture construction" and "low-cost construction". The first problem is "the construction of public activity space in ethnic minority communities". We propose a cross-cultural design strategy based on architectural typology, that is, how to adapt the characteristics of ethnic minority cultural life etiquette activities in the design of architectural space, and present a certain degree of "heterogeneity" and "modernity". The second question is "how to deal with the cost limit and improve the construction quality as much as possible during the construction process". In the process of architectural design and construction, due to the limited budget and the low level of construction technology in the region, the architect team decided to form a construction team with local villagers to carry out the construction. The architect designed some special masonry structures composed of bricks and concrete to provide a good space experience while meeting the budget. In short, this paper aims to explore how to integrate the needs of a minority immigrant community into the design of a public building and form a new activity center and spiritual center of the villagers' "community" through a design research example.

Keywords

Community culture building · Space prototype · Architectural typology · Minority architecture · Low-cost construction

X. Haohao (✉) · W. Chunyu · O. Sheng
School of Architecture and Planning,
Changsha, China
e-mail: xuhaohao@hnu.edu.cn

S. Hilal et al. (eds.), *Design for Partnerships for Change*, Sustainable Development Goals Series,
https://doi.org/10.1007/978-3-031-36993-3_8

8.1 Introduction

An important symbol of community architecture in the early twenty-first century is the conflict and game between a large number of replicated designs of globalization and industrialization and traditional local architectural culture. This trend is intensified with the spread of international style in the second half of the twentieth century and with the spread of globalization as the dominant world outlook. How architects express cultural identity in contemporary vernacular architecture design, build "local identity" and respond to people's spiritual needs and space needs in the community has become an important proposition for in-depth thinking. (Salman 2018). From a global perspective, Hassan Fathy and Balkrishna Doshi responded to the proposition of this era through some strategies in their designed buildings (Mito 1992). The African architect Francis Kéré has completed very high-quality architectural works in some areas of Africa, such as Ganduo Primary School (Wei 2022). However, unlike their practices in Africa or India, architects in China may face some special problems in rural areas. For example, many rural industries in China are dying, and a large number of young people have left the countryside, forming a hollow village dominated by 60-year-olds. In addition, China is also a multi-ethnic country with 56 ethnic groups, and each ethnic group has its own unique culture. Under such conditions, it is also very necessary to study carefully how architects get involved in such communities.

Based on a village activity center building designed by the author in the southern minority areas of Hunan Province, this paper discusses an architectural design strategy that adapts to the climate characteristics, production and lifestyle, and the characteristics of minority communities. This paper focuses on the analysis of architects and the strategies, mechanisms, and tools they use in their work to respond to the two main problems of "shaping local community culture" and "low-cost construction". In short, this paper aims to explore how to integrate the needs of a minority immigrant community into the design of a public building, and form a new activity center and spiritual center of the villagers' "community".

8.1.1 Background

The project is located in Shuikou Town, Jianghua Yao Autonomous County, Yongzhou City, Hunan Province. The location is a new town formed due to the relocation of immigrants. The original Shuikou Town was submerged due to the construction of a reservoir. Today's Xinshuikou Town is located in a relatively flat area surrounded by mountains. It is a new town formed by the overall relocation and expansion of the old town on the basis of the tree-like distribution of the original natural villages. Therefore, it is a mixed new community formed by the relocated residents and the original residents. The government expects to build a small rural cultural building complex for residents in the local area that integrates village affairs, villagers' activities, book reading, literature and art performances, local culture and tourism promotion, and display and sales of characteristic products.

8.1.2 To Rebuild the "Local Identity" of the Community, How Can Architects Intervene in a Minority Community?

The project is located in Shuikou Town, Jianghua Yao Autonomous County, Yongzhou City, Hunan Province. The location is a new town formed due to the relocation of immigrants. The original Shuikou Town was submerged due to the construction of a reservoir. Today's Xinshuikou Town is located in a relatively flat area surrounded by mountains. It is a new town formed by the overall relocation and expansion of the old town on the basis of the tree-like distribution of the original natural villages. Therefore, it is a mixed new community formed by the relocated residents and the original residents. The government expects to build a small rural cultural building complex for residents in the local area that integrates village affairs, villagers' activities, book reading, literature and art performances, local culture and tourism promotion, and display and sales of characteristic products.

For many years, the architect team in which the author works has been committed to the study of architectural regional typology (Chunyu 2009). In recent years, it has also been studying the relationship between tradition and contemporary in architecture, trying to explore the combination of "community and local tradition" in design. Design cases such as Tianhan Cultural Park in 2018, Wantouqiao Town Center, and even the early Chinese Academy Museum (Chunyu and Erxi 2020), and this project is another design case based on many previous projects.

8.1.3 Considering the Local Climate and Economic Conditions, How Do Architects Deal with the "Sustainability" of Buildings?

Jianghua County where the project is located belongs to subtropical monsoon climate. This climate is characterized by high temperature and rain in summer, cold and dry in winter, and more precipitation throughout the year. There is basically no snow all year round. In terms of lighting, this is one of the most sunny areas in Hunan Province except the Dongting Lake area. The average annual lighting time is 1422 h. Therefore, sunlight is the object that needs to be considered in the architectural design. From the local traditional houses, we can see that the windows are relatively small and adapt to the strategy of strong sunlight, high-temperature sun protection in summer and wind protection in winter. And because there is a lot of local precipitation, with annual average precipitation of 1240–1490 mm and high air humidity, in this architectural design, if we want to have better "sustainability" and energy-saving effect, we can learn from some local traditional architectural forms, because we can find some space forms that adapt to the climate in many local dwellings.

8.1.4 Methods

Through interviews with villagers and fieldwork, we first studied the regional climate characteristics and spatial location characteristics of the town and interviewed many local famous Yao craftsmen to conduct in-depth investigation and analysis of the construction process, basic form, spatial configuration and spatial use relationship of Jianghua Yao traditional houses, and further explore the principles and order of its spatial organization structure. It is also applied to the site selection and external space design of buildings. Then the designer combined the typological research method to analyze the space characteristics required by the daily life, social activities, and etiquette activities of the ethnic minority community where the project is located, namely "Gaoshan Yao", and obtained four types of "prototypes" of local people's space life, such as the hall and patio, eaves gallery, stairs, and balcony, and discussed the relationship between these "prototypes" space and social life in the community. Finally, the "prototypes" of these spaces are combined and applied to the architectural design.

8.2 Design Process and Strategy

The former Shuikou old town is located in the flooded area of the expansion project of Qiantianhe Reservoir, and now the new Shuikou town is located in a relatively flat area surrounded by vertical mountains. It is a new town that was relocated and expanded from the old town on the basis of the tree-like distribution of the original natural villages. Overlooking the whole town, the houses designed and built by the government, such as resettlement houses, new schools, and office buildings, present a regular array, which is in sharp contrast with the original natural villages derived from the bottom-up in the region. Shuikou town as a whole presents an interesting overall spatial form of tree-like distribution of "old" natural villages and planned construction of "new" resettlement buildings (Figs. 8.1 and 8.2).

Fig. 8.1 Aerial view of Old Shuikou Town and New Shuikou Town (from GoogleEarth)

Fig. 8.2 Juxtaposition of natural villages and planned settlement areas

8.2.1 Four Types of Local Buildings

The whole town also presents the appearance of mixed houses built in different periods. After field investigation, we found that it can be divided into four types according to the age (Table 8.1; Fig. 8.3).

8.2.2 The Most Important Thing in Minority Community Activities is "Ceremony"

"There are no mountains and no Yao in Nanling". Jianghua Yao Autonomous County is the county with the most concentrated Yao population in the

Table 8.1 Analysis of building types in different periods in the village

	Type01	Type02	Type03	Type04
Image of buildings				
Construction time	Before 1990	Before 1990	1990 ~ now	2015 ~ now
Material	Wood and stone	Clay brick, wood, stone	Red brick, concrete red brick	Red brick, concrete, and coating
Designer	Designed by villagers themselves	Designed by villagers themselves	Designed by villagers themselves	Designed by some architects
Constructor	Built by villagers themselves	Built by villagers themselves	Built by villagers themselves	Built by government
Architecture plane layout	Free	Free	Free	Fixed
Building plane bay	3	3	3	3
Typical plan	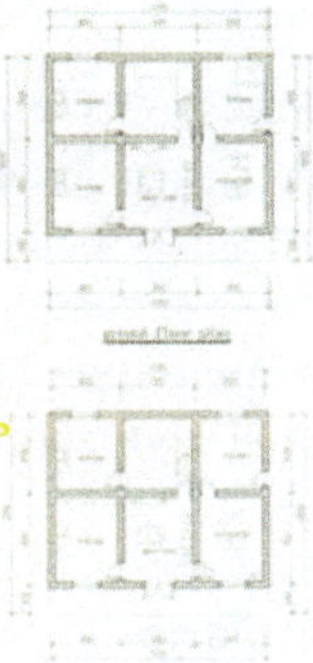	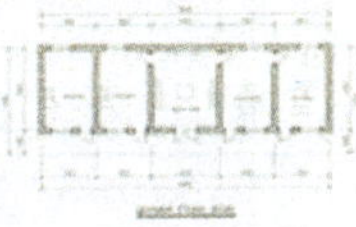	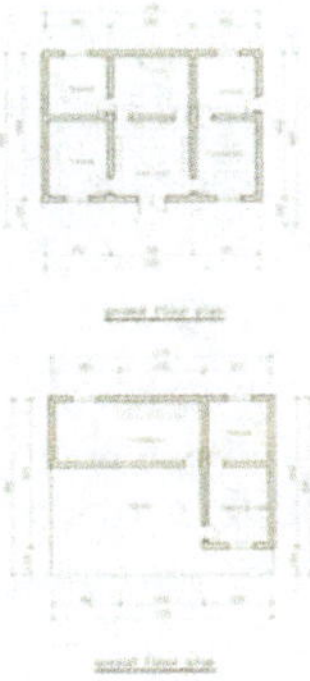	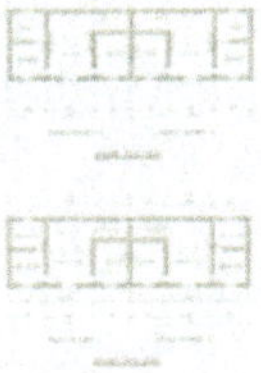

hinterland of Nanling, China, and is known as the "Yaodu of Shenzhou". The Yao villagers in Shuikou Town used to live in the mountains and forests, and were called "Gaoshan Yao". They believe in Wang Pan as their ancestor, and every year on October 16th of the lunar calendar, grand and solemn sacrificial activities are held in all parts of the county, which are called "Wang Pan Festival" and "the second Spring Festival" of Yao people. The typical folk activity is Yao's Long Drum Dance, which belongs to folk festival dance, usually accompanied by suona and gongs and drums, and can vividly reflect the personality and feelings of Yao's family.

The custom in rural daily events is actually a kind of folk belief, which is the heartfelt awe of the villagers. Many custom activities are inseparable from the ceremony space. Traditional folk buildings such as ancestral hall, patio, courtyard, hall, and studio are all important spatial nodes and places for folk ceremonies. The building carries daily activities such as villagers' entertainment and office, as well as local ritual activities such as sacrifice and long drum dance, which urges us to deliberately construct a space place with both material and spiritual significance in the design.

After migration, Yao people often return to their original place of residence. Such cultural and emotional support is their strong family-oriented concept. "Home first" and "fire pond" are two core aspects of Yao family concept,

Fig. 8.3 Four building types in different periods in the village

which play a vital role in immigrant life. For example, several important festivals in a year, such as "Celebrating the New Year before the Chinese New Year", "Celebrating the Qingming Family" at the beginning of the year, are all held in their original places of residence, and immigrants will return to their original places of residence to participate in such activities, which becomes their emotional bond with their original places of residence. It can be seen that the Yao people attach great importance to the communication with their family ancestors, and these religious ceremonies, which play the role of communication, have promoted the emotional, cultural, and communication ties between the relocated Yao people and their original places of residence, and made the relocated Yao people swing back and forth between their original places of residence and the immigrant places (Fig. 8.4).

8.2.3 "Prototypes" of Traditional Buildings

After careful investigation of the houses in Shuikou town and its surrounding villages, we summarized four kinds of space prototypes in local dwellings, such as hall and patio, eaves, stairs and paths, and sunny balcony (Table 8.2). Then we will use these spatial prototypes in our new design, trying to respond to the cultural elements in traditional community life from the space.

8.2.3.1 Hall and Patio

The hall is the most important space in the local old and new dwellings. It not only undertakes the daily life of the Yao people, but also undertakes special ceremonial functions such as marriage, funeral, remembering ancestors. Through the contemporary translation of the hall and the patio, the gallery, the atrium, the patio and the stage construct the spatial sequence of "platoon gate-courtyard patio hall". The stage in the atrium is the core space where rituals and rituals coexist.

To the village public relations, the ceremony can produce the collective emotion, and it is the foundation of organization belief, thought, moral standard, and culture. In the village system, the villagers make use of the emotion produced by the ceremony to trigger the ethnic interaction. In the design, the atrium stage is

Fig. 8.4 Typical "ceremony" space of Yao villagers' activities

constructed into a space with a sense of ritual and a sense of "heaven and earth", which is expected to activate villagers' interaction and enhance villagers' cohesion and cultural identity (Fig. 8.5).

8.2.3.2 Eaves

Eaves is another typical space element in the local traditional houses. Yao people like to enjoy the cool, chat, drink tea, and even eat meals under the eaves on the first floor after they come home from work. Through the continuation and remodeling of the gallery space, the building adapts to the hot and rainy climate of the local summer, while enhancing the openness and public participation. In order to break up the monotonous column net array of eaves corridor space, the design translates the row doors of the traditional residential buildings, implants the frame column into the sheet wall, and adds the architect's "perceptual disorder" in the "rational order" of the structure by rotating 45 or 30°, as if the sheet wall is free from the external wall of the building, which increases the level and interest of the eaves corridor space (Figs. 8.6, 8.7, and 8.8).

8.2.3.3 Stairs and Paths

The organization of the stairs and traffic routes of the local villagers' self-constructed houses is characterized by the premise of function and intensiveness, which is straightforward and clear. Therefore, indoor and outdoor stairs usually appear together in self-constructed houses, connecting the vertical and indoor and outdoor spaces in series. Inspired by this, the staircase not only carries the vertical traffic function but also constructs the traffic system in the building as a spatial element together with the gallery platform.

West of the site is a big tree, the main elevation eaves at the end of the lounge near the big tree location designed a hanging closed run stairs. The staircase is deliberately closed, and when pedestrians walk through a dark staircase to the three-story outdoor corridor, they see the tree within reach, trying to create a subtle dialogue between the staircase and the tree (Figs. 8.9, 8.10, and 8.11).

8.2.3.4 A Sunny Balcony

The roofs of local villagers' self-built houses are usually flat roofs that can be accessed by people

Table 8.2 Prototype in Yao Local architecture and the translation in the new building

Prototype	Hall and patio	Eaves	Stairs and paths	Sunny balcony
Images of buildings				
Prototype abstraction			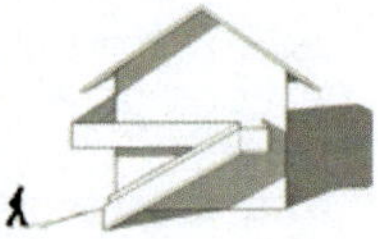	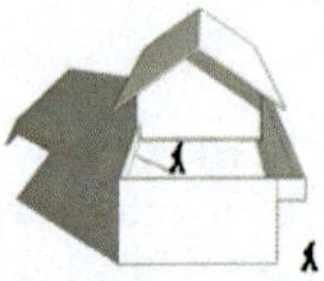
Design translation				

and often have recess relationships to serve as balconies for their daily lives. According to the actual needs, the roof of the special workshop and cultural service center is also designed as a flat roof. As an outdoor extension space of the building, there are open stairs.

As the top space of the building, the roof has the natural property of opening to the mountains and the sky. The flat roof in the building is not only a public platform for villagers, but also a public activity space and observation platform (Fig. 8.12).

8.3 Design Results

8.3.1 Site Plan

The ideal form of a new building is not to leave the old village, but to "grow naturally" from the

Fig. 8.5 A few villagers chatting in the atrium on a heavy rainy day, this is the same as our space experience in traditional architecture

Fig. 8.6 Gables and steps

Fig. 8.7 Main entrance facade real view

Fig. 8.8 Composition of the eaves porch

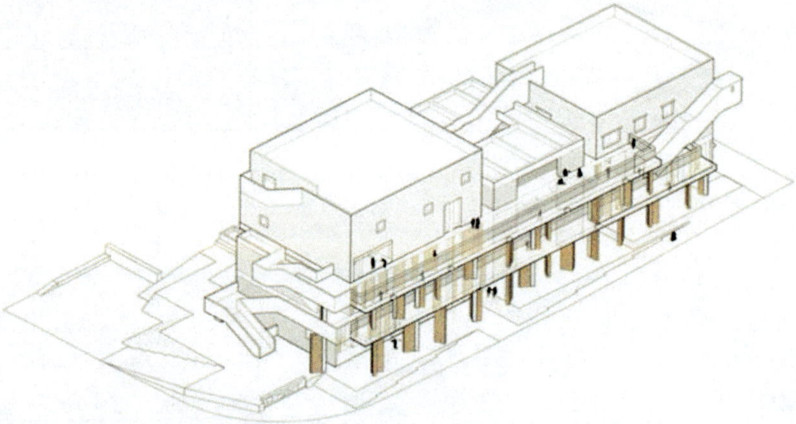

old village. The project base is located between two hills on the northern edge of Shuikou Town, and there are several ordinary red brick villagers' self-built houses scattered on its south side. We think that the building should continue the texture and form of villagers' self-built houses in an appropriate way as far as possible and become one of many villagers' self-built houses (Fig. 8.13).

8.3.2 Organization of Space

In Shuikou Town, Jianghua, the new and old dwellings of Gaoshanyao are all laid out with the hall as the center, and the most common way is the three-span type; that is, the hall is in the center and the houses are arranged on both sides. At the same time, the overall layout is simple, regular, economical, and applicable. The project base has a height difference of about two meters from west to east. In the design, the site is divided into three terraces along the height difference, with the atrium stage as the center, the cultural service center, and the characteristic workshop separated on both sides, and they are connected in series through the corridor, which is independent and interrelated (Figs. 8.14, 8.15, 8.16, 8.17, and 8.18).

Fig. 8.9 View from the stairs of the side court of the cultural service center to the atrium stage

Fig. 8.10 Staircase and the side court over the theater

Fig. 8.11 Staircase from exterior

8.3.3 Community Building and Local Community Participation in Construction

This project is a government poverty alleviation project with limited project budget. At the same time, Shuikou Town of Jianghua Yao Autonomous County is located in a remote area, and the local construction technology and conditions are relatively backward compared with modern cities, and the construction period of the project is also strict. According to careful research, we choose common materials that can be bought locally, such as red bricks, cement bricks, and concrete, which are consistent with the villagers' self-built houses, and adopt a frame structure system with regular column spans, giving full consideration to site contingency and process fault tolerance, hoping to achieve "finding a balance between economic conditions, technical level, and architectural art". We invited local villagers to participate in the construction of the

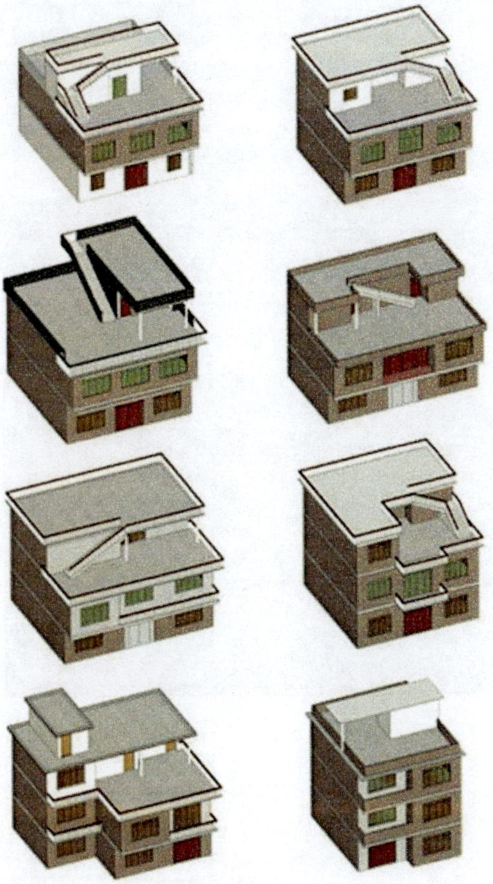

Fig. 8.12 Common types of self-built houses for local villagers, every building has a sunny balcony

house, and we also arranged for an architect to live in the local area, eat and live with the local villagers, explain to the villagers why the design should be like this, and solve some specific problems in the construction process (Fig. 8.19).

8.3.4 Detail Design

In our design, some spaces with public attributes are constructed with fair-faced brick walls both indoors and outdoors. On the basis of comprehensive consideration of structural load and construction cost, the masonry technology of building external walls has been appropriately improved. The sandwich insulation system of 120 red brick + 80 insulation layer + 120 red brick is adopted, and the ring beam and constructional column are hidden in the wall on the premise of ensuring structural safety. At the same time, in order to further strengthen the stability of the wall, we asked the bricklayer to join the ding brick method according to the construction experience to increase the Rachel effect of the wall (Figs. 8.20 and 8.21).

In order to increase the richness of the building, we have made some micro-expression designs on the texture of the external wall of the building. The first floor of the cultural service center and the characteristic workshop adopts the masonry method of picking out 20 mm one brick and putting it in, and the second floor adopts the method of picking out 20 mm one brick and putting it in two bricks. The third floor of the cultural service center adopts one skin brick to pick out 20 mm, and three skin bricks are collected. Because of the large volume of the third floor of the characteristic workshop, the ordinary flush masonry method is used to create some slight differences as blank space, and the whole building conveys the sense of craftsmanship and modernity of the building by laying subtle texture details.

8.4 Result and Discussion

Through this project, we get some conclusions:

1. The creation of community atmosphere in architectural design and the extraction and application of local cultural elements are very important. Architectural typology is an effective design method, which we have used extensively in this case.
2. About the "modernity" thinking in the creation of modern architecture, as a designer, we should use modern design methods, but more importantly, we should connect with

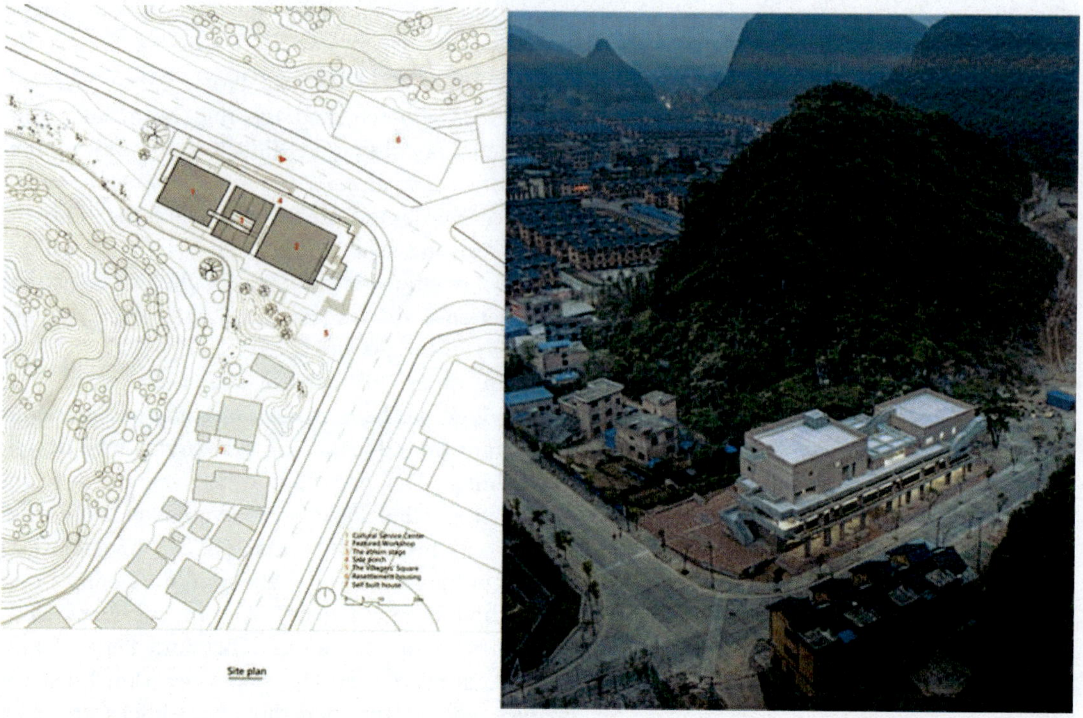

Fig. 8.13 Site plan and north-facing panoramic aerial view

Fig. 8.14 Yao children holding activities in the plaza and atrium stage (left), villagers are rehearsing the dance of the Yao (right)

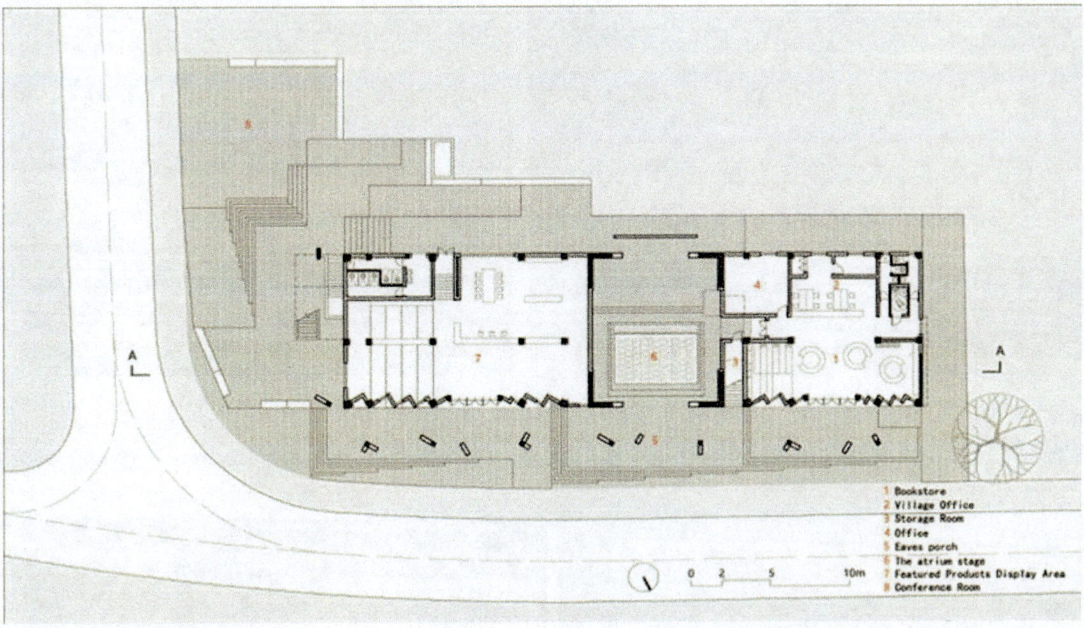

Fig. 8.15 Ground floor plan

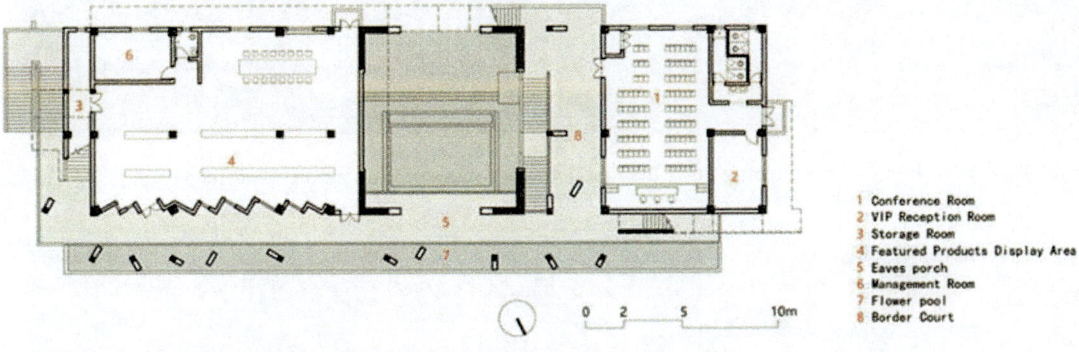

Fig. 8.16 Second floor plan

local communities and think about culture. We need to re-examine the construction logic behind some rural buildings, rather than the design logic of modern industrial products.

3. At the early stage of the design, the architect investigated the needs of local villagers and responded in the architectural scheme. After the building was completed, many local villagers expressed their appreciation.
4. Integrating designers into the local community and working with local people is a way to save construction costs, but it requires more input from designers and communication with people in the community. In this project, we specially arranged for an architect to live and work with local people all the time, spend a lot of time explaining this project, asking people questions, and answering their questions.

Fig. 8.17 Yao children holding activities in the plaza and atrium stage(left), villagers are rehearsing the dance of the Yao (right)

Fig. 8.18 A mini library in the building (left), the first-floor showroom of the specialty workshop (right)

Fig. 8.19 Local red bricks and hollow concrete blocks, we also continue this combination of materials in the new building

Fig. 8.20 Local red bricks and hollow concrete blocks

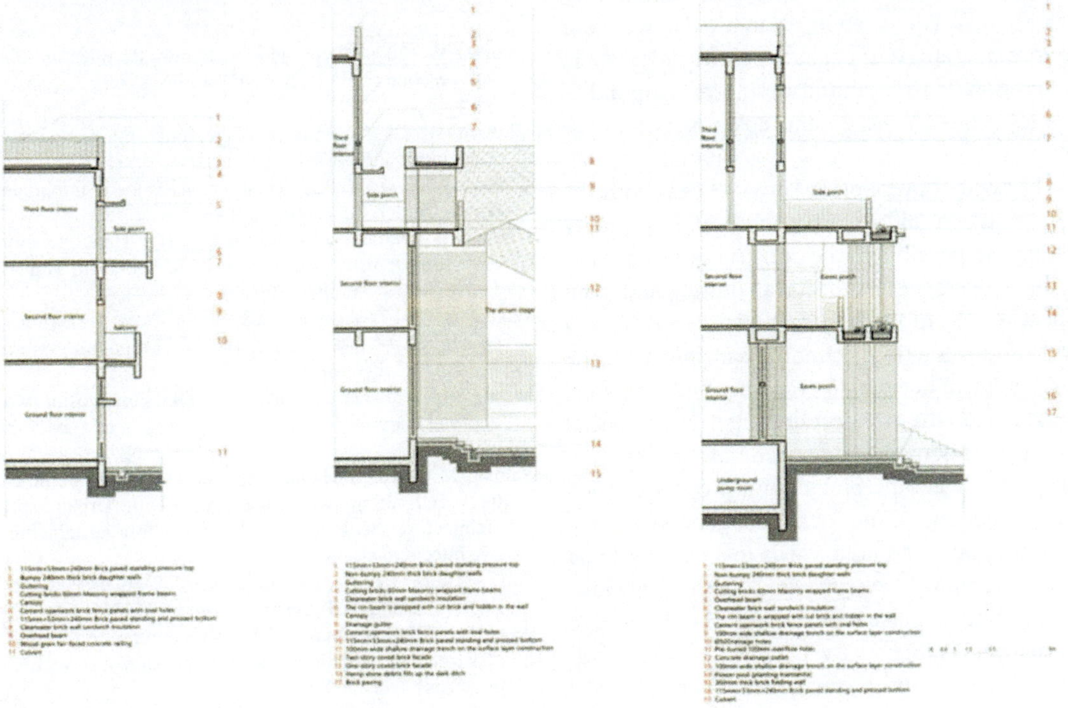

Fig. 8.21 Wall detail drawing

8.5 Conclusions

Based on this design case, this paper studies the design and solution strategies of "building community culture" and "building with low cost." The first problem is "the construction of public activity space in minority communities." We put forward a cross-cultural design strategy based on architectural typology; that is, the design of architectural space adapts to the characteristics of etiquette activities of minority cultural life, and presents certain heterogeneity and modernity. We found that the villagers' self-built houses in villages also have its very important characteristics. As a famous Chinese Architect Jiakun (2015) said, "Although the ordinary villagers' self-built houses in natural villages are not very old or special, they are the most common part, but they are widespread and close to daily life, and they still have folk spirit". In this design, we didn't use the symbolic expression of Yao characteristics, but based on the derivation of "in the ground", we chose to make our designed buildings consistent with the current villagers' self-built houses. At the same time, we tried a different way of space construction by extracting some "prototypes" from local traditional houses, trying to make local villagers feel a certain "familiar" atmosphere in a modern building. In today's international style, as architects, we try to explore a strategy to express cultural identity in contemporary architectural design, to respond to people's spiritual needs and space needs in the community.

The second question is "how to deal with the cost limitation and improve the building quality as much as possible in the construction process". In the process of architectural design and construction, due to the limited budget and the low level of construction technology in this area, the architect team decided to construct with the local villagers, and the architect designed some special masonry structures composed of bricks and concrete to meet the budget and provide a good space experience. The local villagers' red brick self-built houses are like weeds that take root and sprout naturally everywhere, with tenacious vitality. They represent the universal existence of the present, reality, daily life, and natural ecology, which we think can be understood as the spirit of "on the ground". As Shu (Shu and Luo 2016), the less material affectation, the closer it is to the original basic skills, and the closer it is to ordinary daily affairs, the more spiritual and transcendental it is.

Acknowledgements This paper is supported by the following fund projects, the National Natural Science Foundation project (52178013), the value mining and communication strategy research of Huxiang red buildings (heritage) from the perspective of media, the subject of Hunan Provincial Social Science Achievement Review Committee, (XSP22YBC153), Research on the construction of design teaching resources based on air-ground collaborative 3D reconstruction and virtual reality (the Ministry of Education's collaborative teaching content and curriculum system reform project 202102185009)

References

Chunyu W (2009) Type and interface - design thinking and practice of Wei Chunyu's studio. World Architect 3:94–95

Chunyu W (2010) Practice of regional interface type. J Architect 2:62–67

Chunyu W, Erxi L (2020) Tianbian Jichang - Wantouqiao Town Center. J Archit 12:50–56

Chunyu W, Haohao X, Jiansong L (2012) Heterogeneity and isomorphism — from Yuelu Academy to Hunan University. J Architect 3:6–12

Jiakun L (2015) Another talk about "low technology strategy" — take Xicun Besen compound as an example. Collection 8

Mito AS (1992) Hassan Fathy and Balkrishna Doshi: two regional architects in the context of modern architecture

Salman M (2018) Sustainability and vernacular architecture: rethinking what identity is. In: Urban and architectural heritage conservation within sustainability. IntechOpen.

Shu W, Luo Q (2016). Renovation of Wencun Village, Fuyang, Zhejiang Province, Nashan Nashui 11:86–91

Wei He (2022) Francis Kéré and contemporary rural construction in China: contrast, critique and revelation. Architect 2:49–54

9 Conto Até Mil. Visual Activism Narrative

Elena Parnisari

Abstract

Today's urban environment is defined by social, political and architectural fragmentation. Groups of different ages and ethnicity occupy the space in diverse patterns and structures. Thus, they take part in its evolution with contrasting involvement and connections. Public spaces have become fragile environments where socio-spatial inequalities affect children and their caregivers most. However, proximity and co-responsibility can still be found inside the neighbourhood unit, made visible through the diverse communities' actions. Albeit, what is public in the space? Commonly what belongs to everyone, no one takes care of, at least in the south of Europe and almost everywhere outside this continent. Through the implementation of urban diagnostic workshops and the making of a documentary with Roma children and their caregivers, developed at Contumil social housing neighbourhood in Porto, Portugal, the academic research aims to show that neighbours are full participants and learners through action and reflection on the urban environment as a community. Over interviews and footage of daily life in the neighbourhood's public spaces, we listen and understand the space through the eyes of children and mothers to build knowledge together. We show children's socialisation and cultures: the relationship between children's everyday life and the neighbourhood scale. We understand neighbourliness as a collective means of the public space. Thus, we aim at re-framing the *communing* in the neighbourhood unit through children's participation in their evolution. This considers the contrasting involvement of community experience and connection in *righting* the city.

Keywords

Social housing neighbourhood · Children · Roma community · Interviews · Urban diagnostic workshops · Storytelling · Portugal

9.1 Introduction

This visual essay aims to re-frame the urban environment as the place where learning happens through imagination and personal interpretation in a fragmented and changing context.

The urban design of places explicitly impacts children's growth and daily life. Diverse communities are still unaware of their right to participate in projects actively.

E. Parnisari (✉)
FAUP—Faculty of Architecture of the University of Porto, CEAU—Center for Studies in Architecture and Urbanism, Porto, Portugal
e-mail: elena.parnisari@live.com

S. Hilal et al. (eds.), *Design for Partnerships for Change*, Sustainable Development Goals Series,
https://doi.org/10.1007/978-3-031-36993-3_9

How do we turn children's voices into their urban rights?

9.1.1 Motivation

Children's rights are becoming benchmarks for national, regional and local policies following the 1989 UN Convention on the Rights of the Child.

Though, whose right, what right and to what city? (Marcuse 2009). The urban space became specialised, forgetting the needs of younger generations, women, the elderly, disabled people, the poor and immigrants (Tonucci 1996).

If we contemplate children's perspectives, a city designed for them would be thriving for everyone (ARUP 2017).

Portugal is one of the most unequal countries in Europe (UNICEF 2013), with a child poverty rate of 33.9% (INE 2019). It is essential to break the cycle of inequality by considering urban public policies and design measures for children (Sarmento 2018).

An urban laboratory in Porto social housing neighbourhoods is set to record the possible feasibility of urban design interventions targeted to children.

9.1.2 The Project

The 'Conto até Mil' documentary project, presented through interviews extracts and photographs in this visual essay, complements the case study of the participatory action research 'Searching for equitable, inclusive and caring urban neighbourhoods: measuring children's right to the city through urban design. An urban laboratory in Contumil and Lagarteiro social housing'.

This Ph.D. research faces the question of inclusiveness through the indicator of children in social housing neighbourhoods.

Children, considered inclusive beings, do not perpetuate ethnic segregation. If they were considered indicators of urban inequality and determinants of inclusive urban design, they could become public connectors and produce guidance, exploring their co-responsibility in urban communities.

Urban diagnostic workshops and interviews took place with 15 children from the Roma community in Contumil, Porto, Portugal, between January and September 2022.

These activities aimed to affirm theory and practice with direct participation and cooperation of the community together with the association Equipa de Rua Oriental, the only organisation in the neighbourhood that carries out activities with children (Figs. 9.1, 9.2, 9.3, 9.4, 9.5, 9.6, 9.7, 9.8, 9.9, 9.10, 9.11, 9.12, 9.13, 9.14, 9.15, 9.16, 9.17, 9.18, 9.19 and 9.20).

> "It's a good neighbourhood, where we get along well with everyone, and our children feel safe here. There are no problems, it's quiet, it's clean. I feel really at home here. I was born here, I got married and stayed here. Even if people who live here are not from our ethnic group, they know us, and they know my girls."

> "Sometimes we do little parties or lunch inside our houses. What we celebrate is São João.[1] We set up something beautiful. Then we all eat together and grill the sardines together. We put little rugs on the ground, something pretty."

> "When I leave school I go rollerblading with my friends, and go to the sports field to play hide and seek."
> "When I leave school, I always go rollerblading with Dejanira and Afonsinha when it's good weather."

> "I don't feel comfortable in the city centre. The city centre is where everything goes and where people from outside go. And I wouldn't feel comfortable in that, but here I do. For example, here at my door, there aren't even cars passing by, there are only cars on the other side of the neighbourhood. So even my little boy I leave him here at the door and my older one looks at him because there are no cars. In the city centre, you can't do that!"

> "Sometimes I have a snack here at the door with my neighbours. Sometimes I'm over here hanging out the laundry, I talk to those neighbours over there. Sometimes I talk to the upstairs neighbours. I know almost everyone here. If I go up there with my daughters to the sports field, I talk to the neighbours from that block or the other blocks."

[1] The festival of St. John is celebrated all over the city of Porto, Portugal, on the night of 23 June.

Fig. 9.1 Entrance to one of the blocks in Contumil

"Sometimes we're all here at my door, one from one building and the other from another. Even if they are not of my ethnicity, we talk a lot. And we gather there in the street and chat. It's normal."
"I spend a lot of time outside here with the neighbours. We get along very well with everyone from our ethnic group, and without being from our ethnic group."
"The next-door neighbour is our friend. These people here are all friends."

"When you live in a neighbourhood you live more with the neighbours. The children have more freedom and play more, and if you live in bigger central areas, it is no longer like that."
"I often use my smartphone to play on TikTok. But It's always better to play in the street with friends."
"When I leave school I stay with my brother and I go with Zezinho, my cousin. He lives up there and I live right here in block 10."
"My girls have many friends, even though they don't belong to our ethnic group. They get along very well at school and here in the neighbourhood. They all know each other and play together."

Fig. 9.2 Building where the Equipa de Rua Oriental association is located

"I always play with the other kids and with my cousin Tayris. My mother let me play with her because with my aunt and cousin I am safe."

"Here in the neighbourhood, there isn't that thing of saying 'oh, we're not going to play with them because they are Romani'. And the other way around.

In the street, when they play, they are all equal. In our neighbourhood in Contumil, I don't think there's this prejudice. Our neighbourhood is not racist."

"Here we are all equal, so we get along very well. There was a time when I needed something, and I didn't ask anyone for anything. They came knocking at my door and gave me bags of things and I didn't ask anyone for anything, and they weren't Romani."

"They lost their joy when they lost the freedom to go to the street."[2]

9.2 Conclusion

Diverse ethnic groups participate in the evolution of public space with contrasting involvement and connections of community experience and living.

Re-framing is recognising the specificity of cultures to define the meaning of community through them.

Re-framing is creating a meeting space of relationship with neighbours, towards a dialectical and unitary understanding, to establish visual responses to enhance the community life of cultures.

Communing is fighting segregation, isolation and stigma through space for sharing knowledge at the service of communities and starting from them.

[2] She refers to the months of lockdown during the COVID-19 pandemic.

Fig. 9.3 Contumil sports field

Fig. 9.4 Introduction to neighbourhood architecture and urban dimension

Fig. 9.5 Children leaving the neighbourhood to go on a school trip

Fig. 9.6 Children waiting for the metro

Fig. 9.7 Residents of the neighbourhood gave new use to the concrete structures in the public space

Fig. 9.8 People do not have enough space to store their belongings inside their homes. So, they take over the public space of the neighbourhood

Fig. 9.9 Residents control the public space from their homes

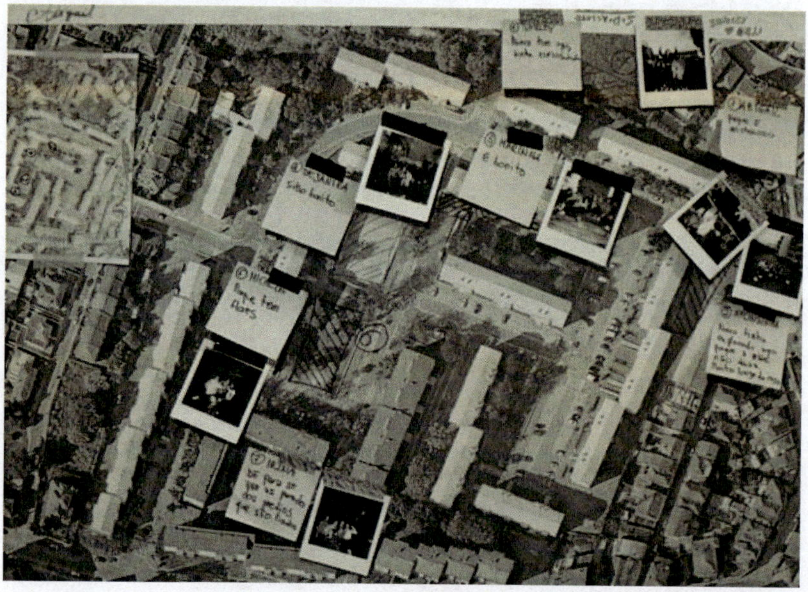

Fig. 9.10 Map made with the children during an exploratory walk of the neighbourhood, as part of the urban workshops

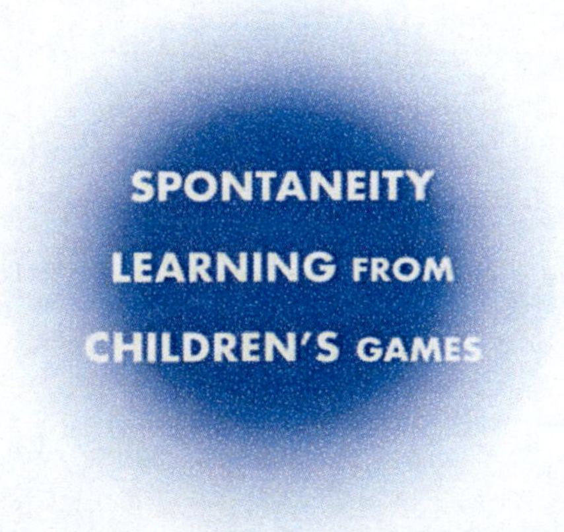

Fig. 9.11 Introduction to the neighbourhood sociality

Fig. 9.12 Children sitting outside the Equipa de Rua Oriental association

Fig. 9.13 Public space is almost all occupied by residents' parked cars

Fig. 9.14 Children of different ages share games and spaces

Fig. 9.15 Children skipping rope during the activities of the association Equipa de Rua Oriental

Fig. 9.16 Children dancing in the public space

Fig. 9.17 Introduction to the neighbourhood participation

Fig. 9.18 Portrait of the family of Maria da Graça during the interview

Fig. 9.19 Children during the filming of the documentary Conto Até Mil

Fig. 9.20 Party to celebrate the 20 years of the association Equipa de Rua Oriental

Acknowledgements This work was supported by the Portuguese Foundation for Science and Technology (FCT) through the individual research grant 2021.06908. BD. The interviews were conducted with mothers and their children from Contumil social housing neighbourhood in Porto, Portugal between 18th April and 20th May 2022: Argentina, Dejanira, Lisandra, Maria da Graça, Maria José, Marinha, Miguel, Shirley. All photographs and images are by Elena Parnisari, taken between May 2021 and September 2022

References

Arup (2017) Cities alive. Designing for urban childhoods. Retrieved February 18, 2021, from: https://www.arup.com/perspectives/publications/research/section/cities-alive-designing-for-urban-%20childhoods

INE (2019) Instituto Nacional de Estatística - Anuário Estatístico de Portugal : 2019. Lisboa. https://www.ine.pt/xurl/pub/444301590. ISSN 0871-8741. ISBN 978-989-25-0525-1

Marcuse P (2009) From critical urban theory to the right to the city. City 13(2–3):185–197. https://doi.org/10.1080/13604810902982177

Sarmento M (2018) A Sociologia da Infância portuguesa e o seu contributo para o campo dos estudos sociais da infância. Contemporânea 8(2):385–405. https://doi.org/10.4322/2316-1329.065

Tonucci F (1996) *La città dei bambini*. Laterza, Bari

UNICEF Comité Português (2013) As crianças e a crise em Portugal. Vozes de crianças, políticas públicas e indicadores sociais. UNICEF, Lisboa

10 Commonalities Past and Present: Rethinking the Role of History in Designing for the Local Neighbourhoods

Yifan Wang

Abstract

This visual essay explicates the 'commonalities' of historic neighbourhoods in Quanzhou, Southeast China. It focuses on place-making of Quanzhou's socio-spatial system 'pujing', where each locality had its own sacred territory of a 'genius loci' and a temple for local deity. It adopts the method of architectural ethnography, with it involving architectural drawings, temporal-spatial analyses, and in situ records of daily practices. Through the consistency and continuity revealed in the neighbourhoods' pattern of life and daily rhythm, the fieldwork research takes seriously the need to investigate the hidden logic in everyday life—secular and sacred, past and present. The intention is to inspire a critical attitude when designing for people who live with their common tacit knowledge, aiming to promote dialogue across a vast multitude of actors.

Keywords

Commonalities · Place-making · Historic neighbourhoods · Les Espaces Vécus · Architectural ethnography

Image Credits All figures are originally produced by the Yifan Wang, Yingjun Li, Sixiang Peng, Xiaolei Ma and Wentai Cui.

Y. Wang (✉)
School of Architecture, Southeast University, Nanjing, China
e-mail: yifan.wang@seu.edu.cn

10.1 Introduction

The concept of 'genius loci' in contemporary architecture usually refers to a location's distinctive atmosphere or a 'spirit of place'. In Roman mythology, however, the genius loci was the guardian spirit of a place (Leatherbarrow 1993). Similarly in Chinese popular religion, the god of each of its smallest territorial divisions was simply known as the 'Locality God'. These cults of locality in popular religion were conceived of as belonging to the 'people' as against the prevailing orthodoxies, or to the locality as against the outside world and its regulatory powers (Wilson 1983).

The sacred atmosphere of localities once characterized the physical appearance and the community life of Chinese cities, especially those in the south (Yang 1961). Typical of such cities is Quanzhou, a port city in Southern China better known by its Arabic name Zayton, whose unique atmosphere has been enjoyed by travellers and literati from both ancient and modern times. In this city, one can find thousands of temples of various folk beliefs, often small but

S. Hilal et al. (eds.), *Design for Partnerships for Change*, Sustainable Development Goals Series,
https://doi.org/10.1007/978-3-031-36993-3_10

central, symbolizing families, neighbourhoods, guilds or other societies, and therefore always carefully designed and built.

Throughout China's late imperial period, an integrated system of urban socio-spatial divisions called 'pujing' (literally meaning 'wards and boundaries') was practised in the city of Quanzhou. Pujing was a spatial organization of both territorial cults and local administration. Each locality had its own sacred territory of a 'genius loci' on which a temple was sited and divinity was manifested. Despite being abolished in the past century, pujing survives and has enjoyed a partial revival in recent decades. It is the temples scattered throughout the city, either maintained or rebuilt, that have shaped the city's most distinctive characters.

For long, the perception of traditional Chinese 'space' and 'city' has been biased towards palaces, official temples, fortifications and the Emperor's presence in city-building, which represent elite culture. Hence, small places in Quanzhou, which are located in a city far from the centre of the empire and in a context that is relatively insignificant compared to the palaces and monuments, are often easily overlooked by official historical writing. While the outside world is aware of the outstanding value presented by the magnificent sites in Quanzhou for UNESCO's recognition of this world heritage city (Rowlands 2020), little is known about the rest and the majority of the city, that is likewise the setting to those magnificent monuments—as Fernand Braudel put it (Braudel 1973): 'Unfortunately, we know more about these great palace scenes than about the fish market…' It is in the formation of pujing neighbourhoods and its 'spirit of place' that the indispensable folk life of the old city is vividly and concretely laid out, and the heritage city of Quanzhou today can be seen as the continuation of 'les espaces vécus', or the 'lived spaces'.

This paper is rooted in John Friedmann's concept of place in the cities of China—places, or small spaces in the city, are 'spaces of encounter where the little histories of the city are played out'. They are shaped, on the one hand, by the people who live there and, on the other hand, by the state through planning, regulation and ordinances (Friedmann 2007). The complexity of the small spaces represented by pujing system is thus not simply determined by the populace and the cultural roots of the localities but is formed in the social process where interaction between the official and the folk, the outside and the local is constantly going on. It reflects the overlap between top-down and bottom-up systems, which act subtly differently in the hierarchizing of urban space, the operation of communities, the management of local authority, and the narrative of local history. In the process of interaction between the two systems, places are also sites of incorporation and contestation. The tension between them evidently makes pujing an urban system where heterogeneous cultures and diverse communities coexist. Hence, this paper proposes an architectural ethnographic perspective on place-making, focusing on the heterogeneous scenes of the pujing neighbourhoods undergoing a modernization process in urbanism.

10.2 The Duality of Place-Making Exemplified by Pujing

In his essay 'Reflections on Place and Place-making in the Cities of China', John Friedmann illustrates the meaning of place with reference to a Folk Religious Temple Tsu-Sze, noting that the small temple and its adjacent precinct represent a place. Drawing on Henri Lefebvre's combined use of 'everyday life' and 'les espaces vécus' (Lefebvre 2004), he emphasizes two aspects of place: the physical space that is being lived in, and the patterns and rhythms of life that are inherent in it. Furthermore, he argues that the study of urban China needs to be informed by a sense of its own past (Friedmann 2006).

Historically and endogenously, place-making in urban China results from the intersection of the state with the everyday life of the people—an intersection that, not to be overlooked, also alludes to the duality embodied in Chinese religion, namely the dualistic relationship between the official cults of the seigneurial courts and the folk rites of the unadorned wilderness, which has long been regarded by sinologists as a key clue to understanding Chinese religion and, in essence, Chinese society (De Groot 1892; Freedman 1974; Granet 1975). In this paper, we have then a highly typical case in which courtly and popular practices have both worked in a reflective way. The literal meaning of 'pujing' indicates an overlap between the two systems, as the author will further illustrate in this article, the Chinese character 'pu' represents a top-down administrative system of urban governance, while the character 'jing' represents a more spontaneous system in Chinese religious life.

The essence of the pujing system has been disputable among scholars. The description of pujing in ancient documents and relics generally refers to locations and built-up areas, and there are references to the 'lords of pu' and 'lords of jing', which are names of the deities by local inhabitants. In the official gazetteer, each unit of pu are recorded, whereas the traces of jing are missing, indicating that there seems to be a distinction between the two. These materials, together with ancient maps from Southeastern China, provide clues from which we may be able to trace the root and differentiation of the system. In ancient times, pu originally means a span of distance equivalent to ten Chinese miles and was used as an official unit for distance measurement. It could be associated with the 'pu-yi' system of imperial postal stations where official documents were conveyed, as well as the imperial baojia system—the most enduring administrative apparatus in China. The term pu was also adopted as an organization for the militia units in garrison towns of Quanzhou's jurisdiction.

In the walled city of Quanzhou, wards or precincts known as pu comprised a chartered area, and each ward was composed of one or two or more neighbourhoods referred to as jing. The term jing, originally meaning 'boundary' or 'territory', was commonly a territorial division for popular ceremonial culture and, in particular, territorial cult festivals and processions.

Investigations of pujing compiled by local historians and folklorists since the 1920s have confirmed that pujing played an important role in the social life of local inhabitants as a tradition. In the 1980s, a folklore investigation inclusive of all the pujing temples and precincts laid the foundations for subsequent studies, among which the most influential was 'Empire and Local Worlds' by Mingming Wang.

While Western Sinology inquires into popular religion through the 'great tradition and little tradition', the distant origins of grassroot cults and official cults, and the social roots of the Chinese cosmological system of classification (Ahern et al. 1981; Feuchtwang 1992; Meulenbeld 2015; Sangren 1987; Wolf 1974), local scholars undoubtedly prefer to use a particular case to reveal the projection of such conceptual framework in specific localities. Mingming Wang's work reflects on the sinological theories with reference to Quanzhou, examining in depth the relationship between localities and the Empire during the Ming and Qing dynasties, in the sense of a case illustration of a more general point—the interactions between administrative space and ritual folk appropriations—and then asks, 'Are these small places part of the state geo-political structures, or are they autonomous enough to comprise their own socio-cultural universe?' As suggested by the case of pujing, locally there is a centrifugal manner which does not comply with administrative intervention of integrating socio-cultural diversities into a state structure but instead facilitates the creation of grassroots ceremonial culture and local socio-economic activities, transforming the top-down administrative spatial system into an 'alternative' spatial system (Wang 1995).

10.3 An Architectural Ethnography of Pujing

In this light, the place-making is an active process of the contest between different social forces —the family, the local community, the state and the wider world, not merely a hegemonic construction of space. With all the contestation, a spatial system utilized to govern society is altered in terms of its function and meaning. Pujing system is ritual (faith-based) as well as administrative (order-based) and is constantly in flux, negotiation, transaction, and implementation, presenting a mixture of institutional and customary, of orthodox and heterogeneous.

Hence, when confronted with the diversity of behaviours and spaces, a certain level of redundancy must be kept in the depiction of pujing. Not only is it necessary to make in-depth field observation and in situ ethnography of participants and their behaviours in space, but it is also necessary to add a temporal dimension to the analysis and representation of the small places. Therefore, a research-by-drawing perspective that brings both people and space into focus—architectural ethnography—is introduced in this paper as an important method for dissecting the formation and operation of spatial and social culture.

This paper delves into the pujing neighbourhoods in the historic core of Quanzhou, drawing on field research and workshops between 2015 and 2020, to observe and map not only the spaces at various scales, but also the daily practices of local communities through participatory observation and numerous interviews. In response to Friedman's argument for understanding place in terms of the historical continuity of the patterns and rhythms of Chinese urban life, this paper argues that the commonalities of pujing can be summarized in three propositions:

1. The secular activities interwoven with the sacred ones.
2. The daily rituals intertwisted with ceremonial festivals.
3. The actual function of commemorating in the temples alternated with memories of the places or localities.

Inspired by the concept of commonality, the author augues that in Quanzhou, the historical settings should not be solely observed through the lens of archaeology. Rather, the city is replete with 'small places' that actively engage with life and the present moment. The pervasive presence of antiquities actually creates a 'bricolage', where the past and the present coexist.

10.3.1 Pujing as 'Les Espaces Vécus'

See Fig. 10.1.

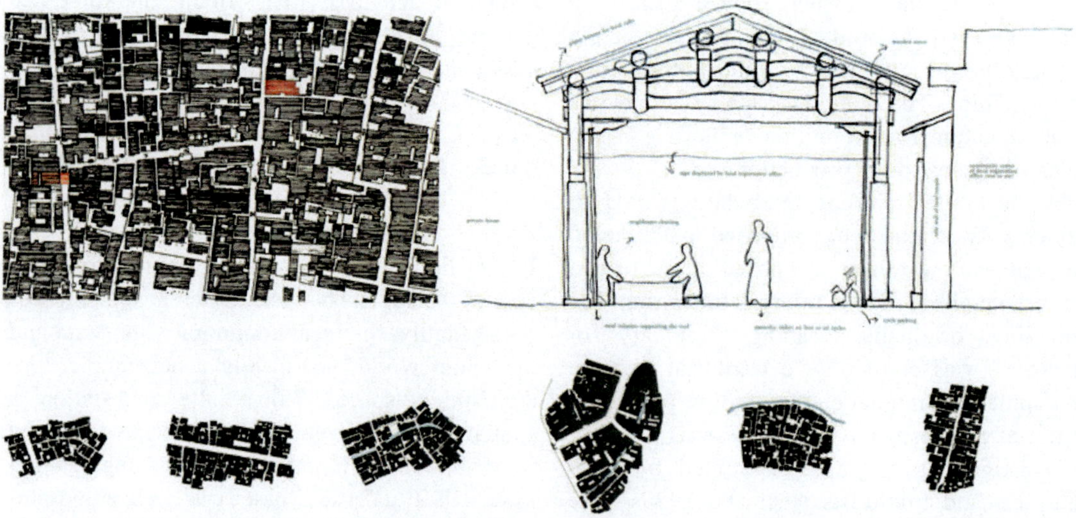

Fig. 10.1 Pujing temples sited in the maze-like streets and alleys of Quanzhou West Street District (upper left); a cross-section showing the daily communication and activity in the neighbourhoods (upper right); the streets and alleys in the urban fabrics of Quanzhou (lower)

10.3.2 Pujing Temples

See Figs. 10.2, 10.3, 10.4, 10.5 and 10.6.

10.3.3 Typology of Pujing

See Figs. 10.7, 10.8, 10.9, 10.10, 10.11, 10.12, 10.13, 10.14, 10.15, 10.16, 10.17, 10.18 and 10.19.

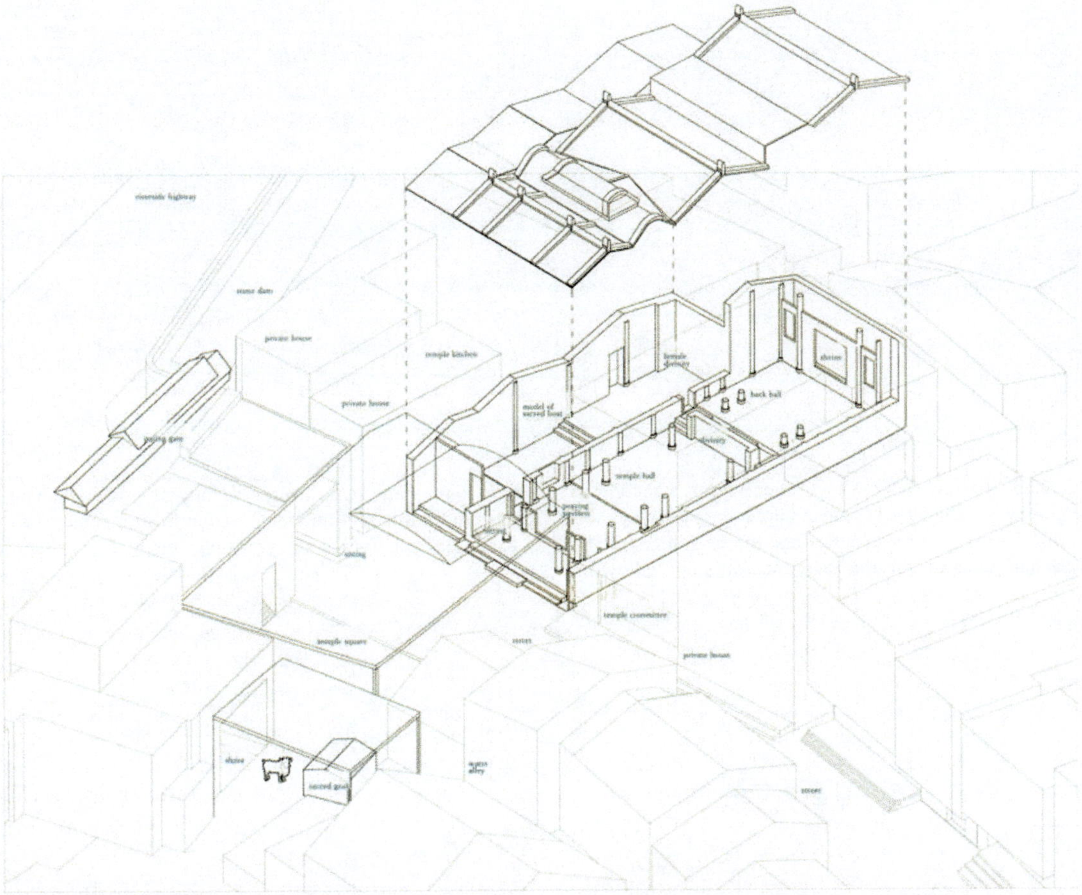

Fig. 10.2 Temple of Fu-Mei, with the small square in front of it surrounded by residences of the neighbourhood and affiliated ritual buildings, including a hut for the sacred goat

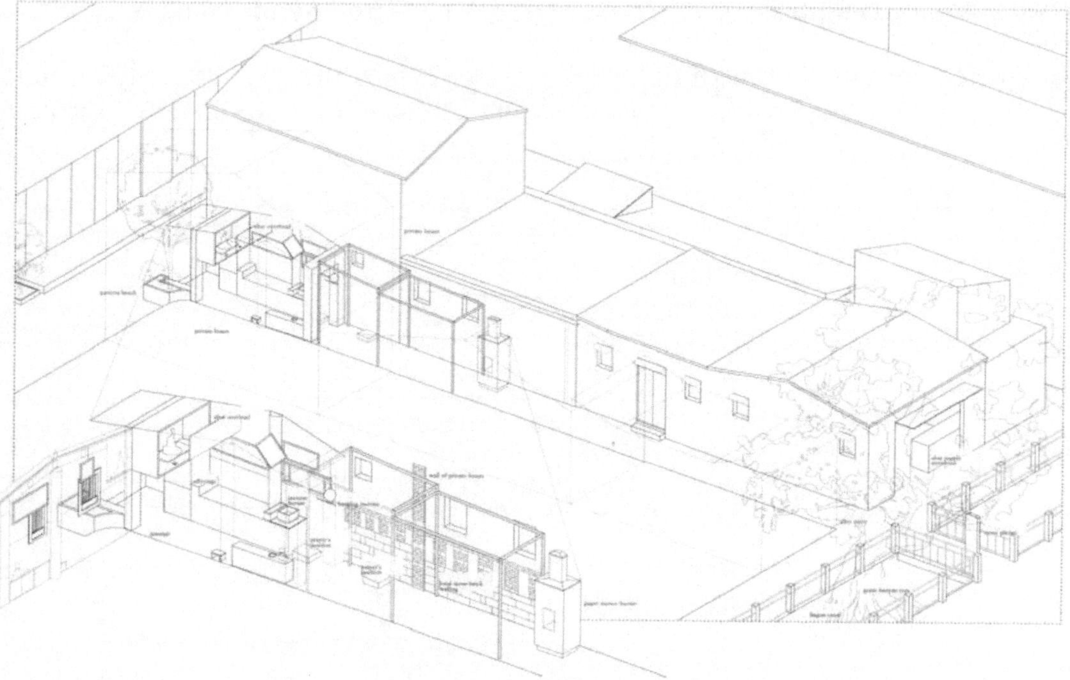

Fig. 10.3 Temple of Zu-Shi, nestled within a section of the alleyway—the altar is placed above the alleyway, half of which is used for the secular function of traffic, and the other half is used for the religious function, allowing people to pray and burn incense as they pass through. The levels within this section create a unique arrangement for both the daily activities and spiritual rituals

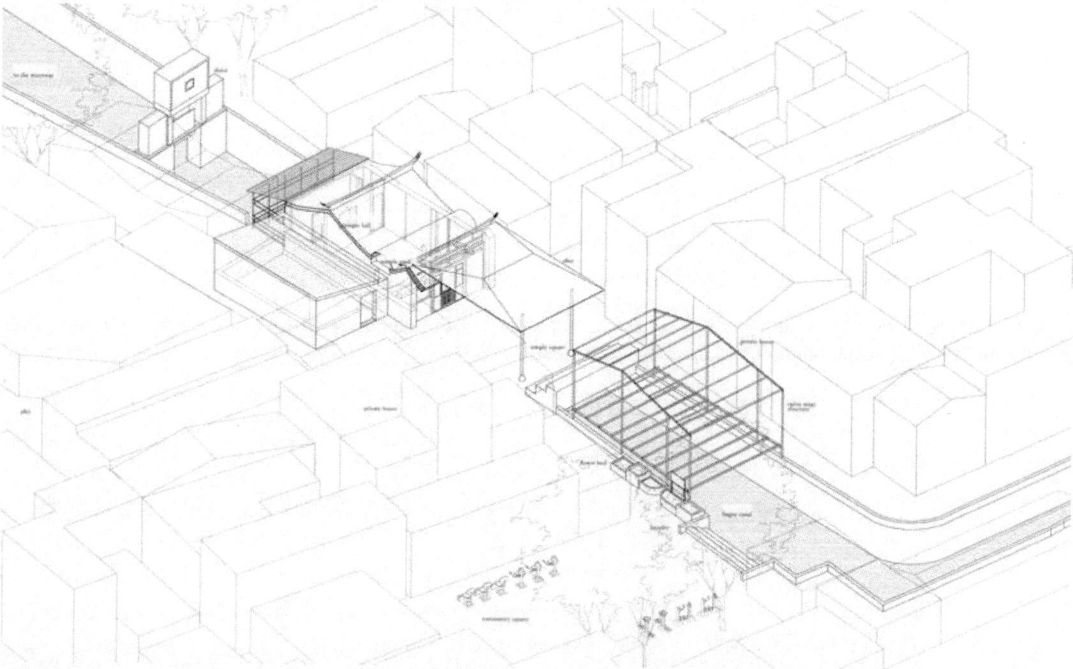

Fig. 10.4 Temple of Tong-Jin, located on the Eight Trigrams Canal (Ba Gua Gou). The stage of the temple was built atop the canal. The stage frame made of steel is supported on both banks of the canal. When the local deities are offered with operas during festivals, the stage is covered with wooden boards that were usually kept in the temple. People gather in the small square in front of the temple to share the operas

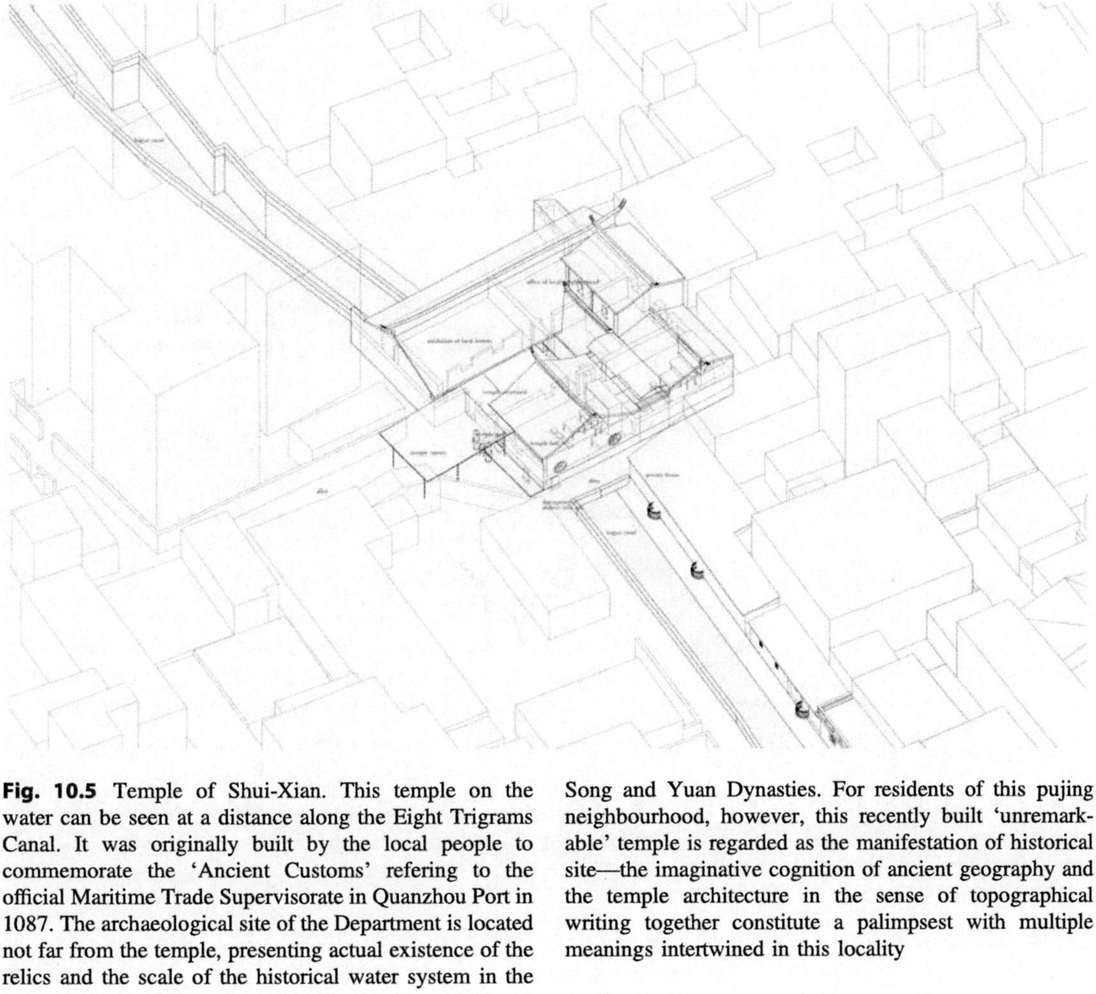

Fig. 10.5 Temple of Shui-Xian. This temple on the water can be seen at a distance along the Eight Trigrams Canal. It was originally built by the local people to commemorate the 'Ancient Customs' refering to the official Maritime Trade Supervisorate in Quanzhou Port in 1087. The archaeological site of the Department is located not far from the temple, presenting actual existence of the relics and the scale of the historical water system in the Song and Yuan Dynasties. For residents of this pujing neighbourhood, however, this recently built 'unremarkable' temple is regarded as the manifestation of historical site—the imaginative cognition of ancient geography and the temple architecture in the sense of topographical writing together constitute a palimpsest with multiple meanings intertwined in this locality

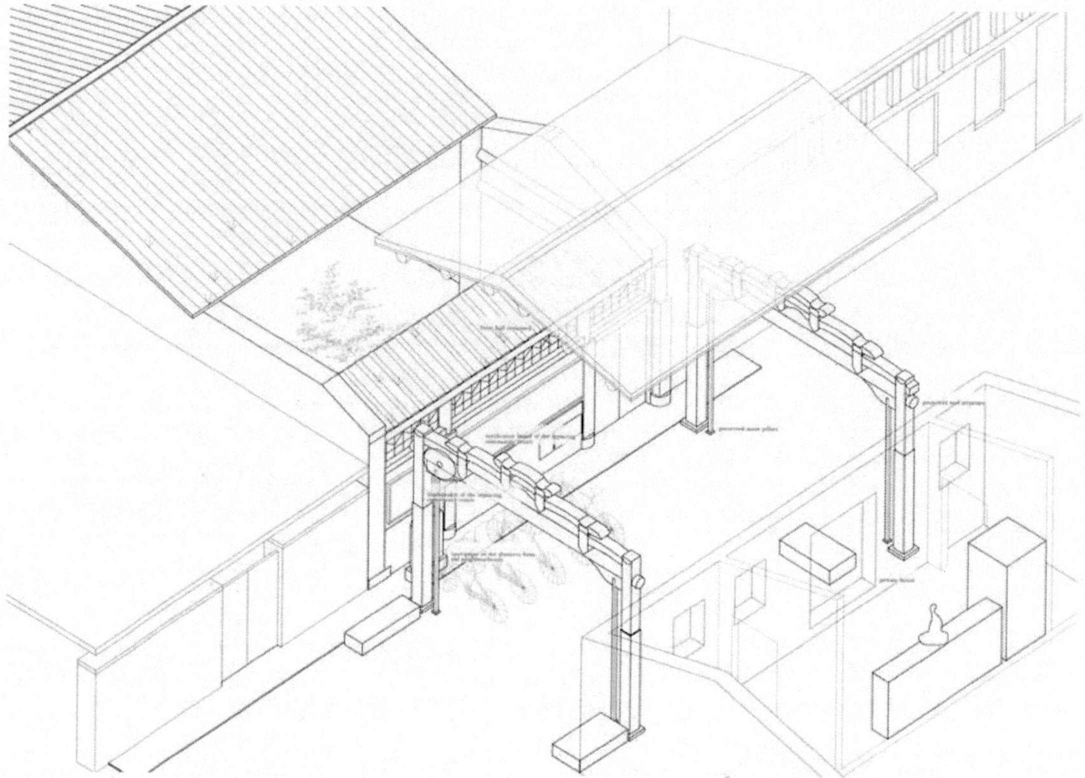

Fig. 10.6 Temple of Qi-Shi, which was known for the goddess patron of motherhood. The temple building remains from the Qing Dynasty to the present day bear witness to the dramatic changes that the pujing neighbourhood has undergone over the past century

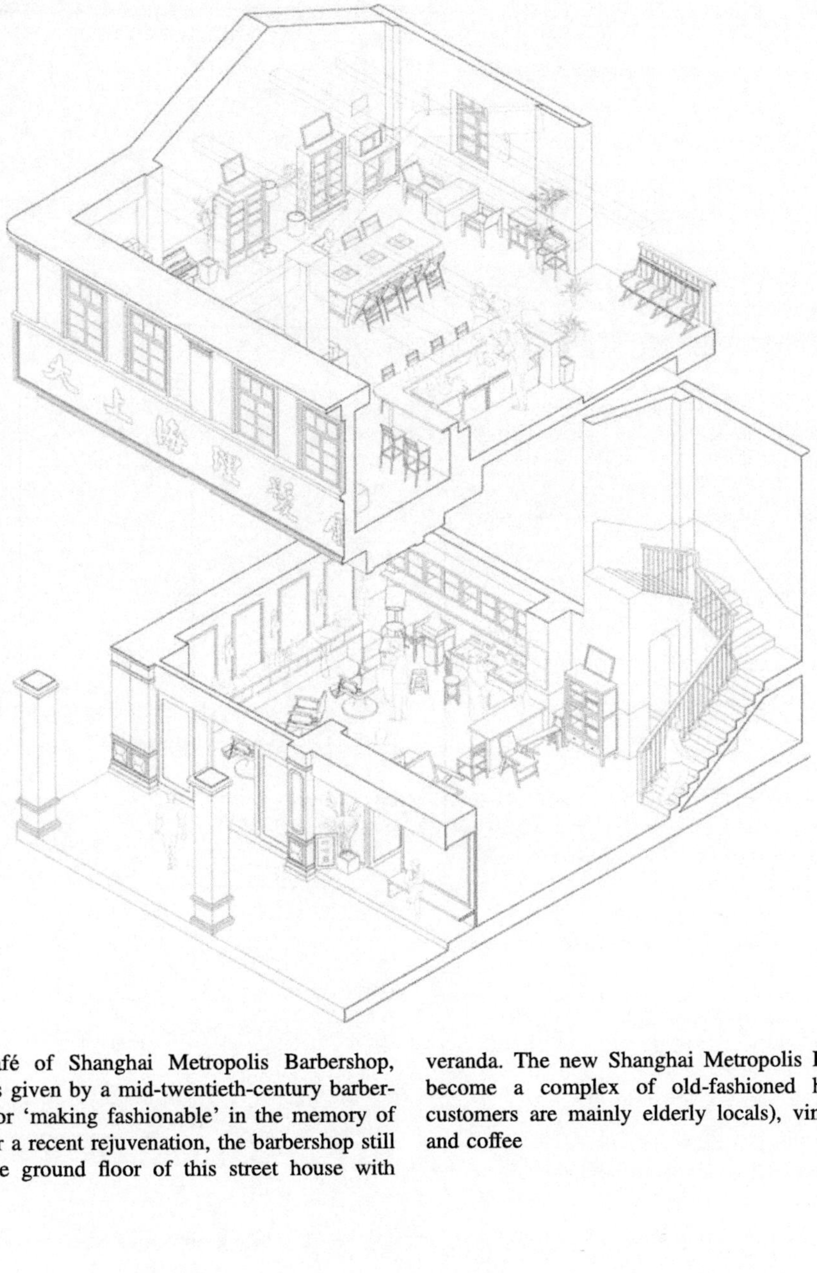

Fig. 10.7 Café of Shanghai Metropolis Barbershop, whose name is given by a mid-twentieth-century barbershop known for 'making fashionable' in the memory of the local. After a recent rejuvenation, the barbershop still operates in the ground floor of this street house with veranda. The new Shanghai Metropolis Barbershop has become a complex of old-fashioned haircut (whose customers are mainly elderly locals), vintage, antiques, and coffee

Fig. 10.8 Nanyin Opera street house

Fig. 10.9 Florist's in a stone house

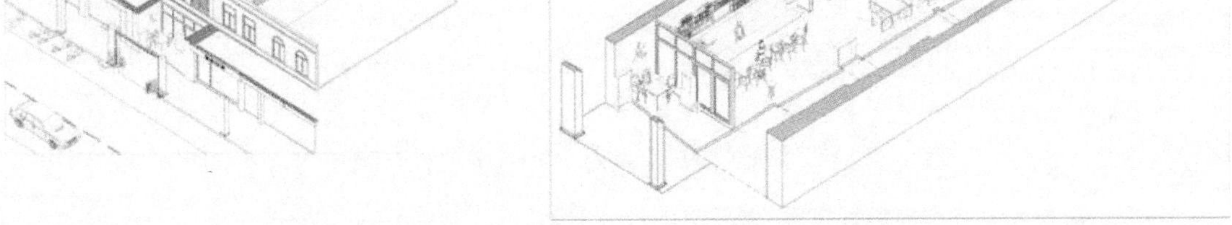

Fig. 10.10 Street house with veranda of Qi-Qi Store

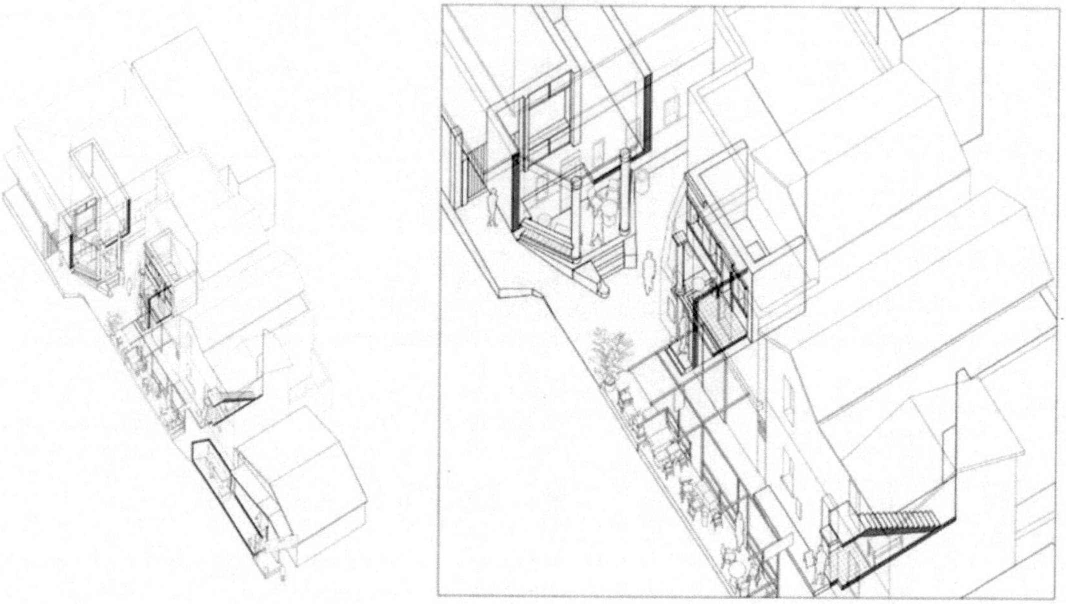

Fig. 10.11 Courtyard-lane named 'Distant'

Fig. 10.12 Street market near the temple of Huang-Di

Fig. 10.13 Street market near the temple of Shui-Men

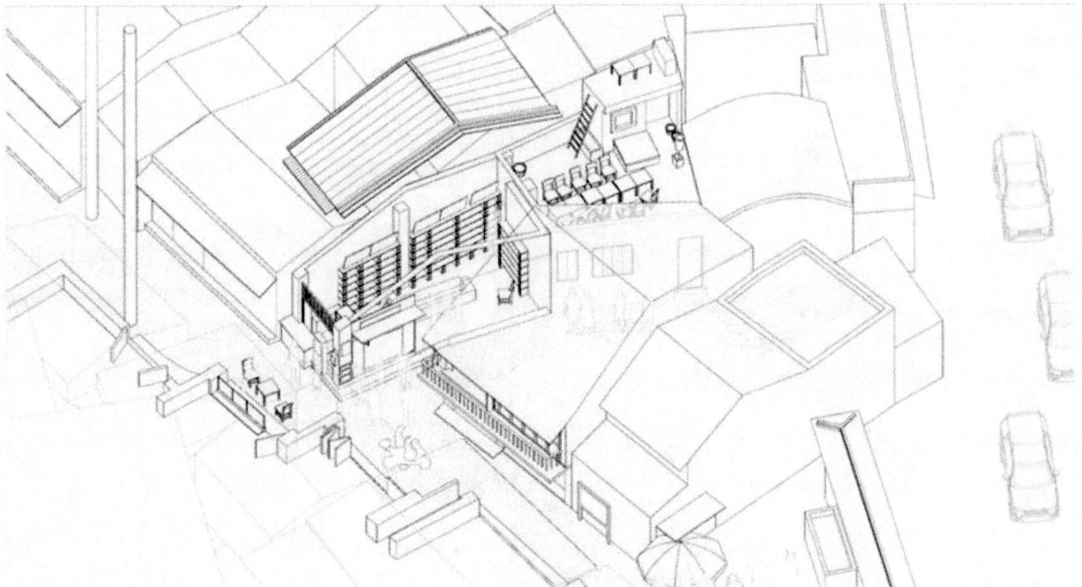

Fig. 10.14 Jie-Zi bookstore

Fig. 10.15 Hou-Cheng market

Fig. 10.16 Gardener's in a dilapidated house

Fig. 10.17 Square in front of the temple of Pu-Dong

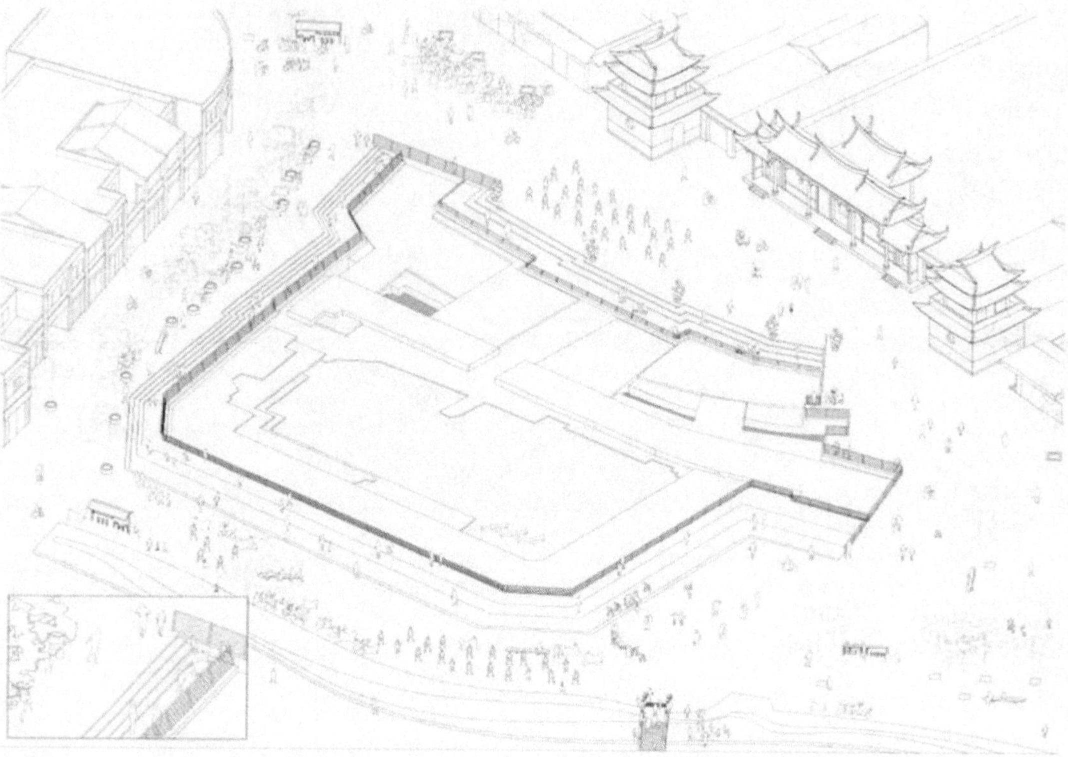

Fig. 10.18 Public space at the ancient gate of De-Ji

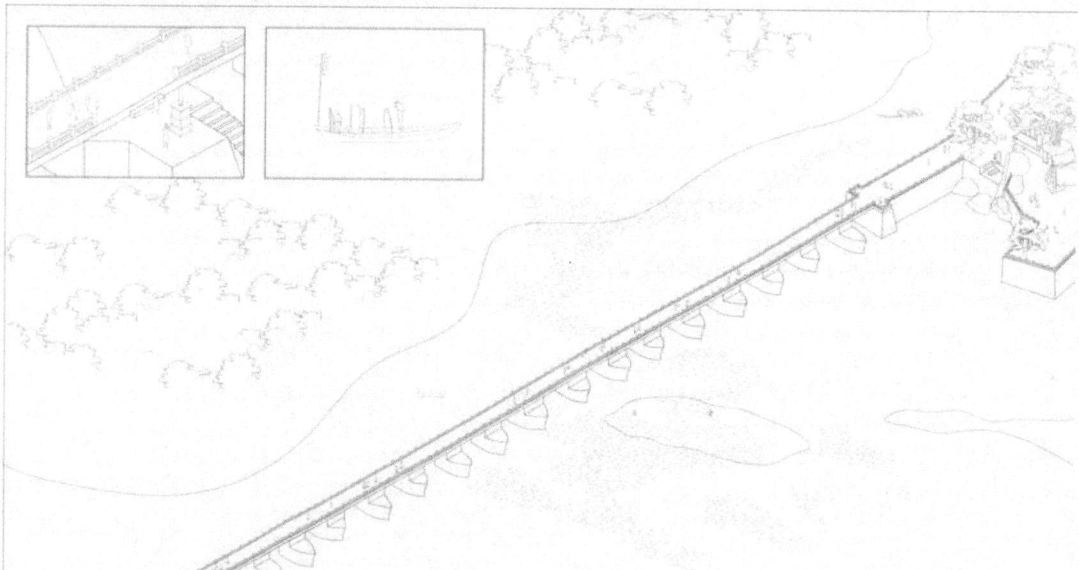

Fig. 10.19 Public space at the ancient stone bridge of Luo-Yang

10.4 Conclusion

In concluding the article 'Reflections on Place and Place-making in the Cities of China', Friedman proposes to understand place-making as a contested process. On the one hand, the place resists the vicissitudes of life with its humanistic binds and ordinary scenery, inspiring people to sublimate their feelings for the place and believe in the spirit of the place. Through the case of Pujing, this paper discusses the tangibility of the spirit of place, whose influence continues to this day. On the other hand, what the small places confront is the overwhelming power in rewriting local spaces and modes of life. Capital as a collective actor has the power to erase an existing community and instead build new housing anywhere. In the face of this crisis, this article regards pujing as a daily place-making practice and expresses the defence of small places.

The city of Quanzhou is not isolated from the outside world, where time is stagnant. A large number of new buildings and businesses in the ancient city indicate that it is also included in the global production and consumption system like all the other cities. However, in the Pujing neighbourhood where the traditional city is little known to outsiders, it seems that an ancient city invisible from the outside can be found in the architectural ethnography.

References

Ahern EM, Lamley HJ, Gates H (1981) The anthropology of Taiwanese society

Braudel F (1973) Capitalism and material life, 1400–1800. HarperCollins 1973:430

Feuchtwang S (1992) The imperial metaphor: popular religion in China. Routledge

Freedman M (1974) On the sociological study of Chinese religion. Religion and Ritual in Chinese Society, 19–41

Friedmann J (2006) Four theses in the study of China's urbanization. Int J Urban Reg Res 30(2):440–451

Friedmann J (2007) Reflections on place and place-making in the cities of China. Int J Urban Reg Res 31 (2):257–279

Granet M (1975) The religion of the Chinese people. Blackwell

De Groot JJM (1892) The religious system of china: its ancient forms, evolution, history and present aspects, manners, custom and social institutions connected therewith. Brill Archive

Leatherbarrow D (1993) The roots of architectural invention: site, enclosure, materials Cambridge University Press Cambridge

Lefebvre H (2004) Rhythmanalysis: space, time, and everyday life (Trans. S. Elden and G. Moore). Continuum, London and New York

Meulenbeld MRE (2015) Demonic warfare: Daoism, territorial networks, and the history of a Ming novel. University of Hawai'i Press

Rowlands M (2020) Areas in flux: temples, cults and the organising of the maritime silk road. 建筑创作 (Archit Creation) 2020(02):52–61

Wilson S (ed) (1983) Saints and their cults; studies in religious sociology, folklore and history. Cambridge University Press, Cambridge, 40

Sangren P (1987) History and magical power in a Chinese community. Stanford University Press

Wang M (1995) Place, administration, and territorial cults in late imperial China: a case study from south Fujian. Late Imperial China 16(1):33–78

Wolf AP (ed) (1974) Religion and ritual in Chinese society. Stanford University Press

Yang CK (1961) Religion in Chinese society: a study of contemporary social functions of religion and some of their historical factors. University of California Press

11 Land Dignity—Learning from Squatting Movement

Husam Abusalem

Abstract

Drawing on the spatial and discursive practices of Ciumara Ranni in squatting and activating abandoned land, this paper borrows knowledge from the ecovillage experience as an alternative way to land reviving and building restoration beyond state monopoly and tax reliance. It also questions how we can utilise the ontological understanding of the concept of dignity in the process of inhabiting the land and the common places beyond modernism and into active participation. By putting the embodiment of dignity and its performativity at the centre of the human-land relation, this paper aims to articulate and propose a definition of the term land dignity and its meaningful role in spatial studies, restoration, and social housing policies.

Keywords

Dignity · Commons · Ciumara Ranni · Ecovillage · Squatting

11.1 Introduction

Critical approaches to architecture, spatial design, and city planning have been centred around safeguarding, achieving, and employing social concepts such as justice, well-being, equality, sustainability, and inclusivity. Yet, dignity is not a concept that is often seen in the process of architecture making or spatial design despite its wide use in other fields. This paper aims to bring the concept of dignity to the centre of the discussion and its meaningful role in spatial studies, restoration, and social housing policies. It does so by taking a look at the performativity of dignity and its embodiment in the making of *Ciumara Ranni*, an ecovillage in the inland of Sicily. The aim of exploring this case study is to learn from their experiences of land revival and community-making to find alternative ways of enriching our understanding of the concept of dignity beyond its humanistic monotony as well as finding an alternative way to engage with the issue of being unhoused. But first, let us dig into a theorisation of dignity.

The concept of dignity in contemporary writings is often celebrated as a moral concept that signifies inherited respect for all beings as equals. The first article of the Universal Declaration of Human Rights states: 'All human beings are born free and equal in dignity and rights' (1948). A similar understanding of the concept is also found within Judeo-Christian

H. Abusalem (✉)
Department of Culture, Religion and Social Studies, University of South-Eastern Norway, Drammen, Norway
e-mail: husam.abusalem@usn.no

S. Hilal et al. (eds.), *Design for Partnerships for Change*, Sustainable Development Goals Series,
https://doi.org/10.1007/978-3-031-36993-3_11

beliefs where the dignified value of human beings is stemming from the fundamental idea of humankind's creation in the image of God. A similar approach could be observed in the Islamic reading of the word, though dignity is not an inherited character from human to human, it is a gift from God to all humans, and it is indeed seen as an admirable gift that one shall actively thrive to safeguard and celebrate through being a good human and follow God—it is important to keep in mind that safeguarding dignity also necessitates the possibility of losing it (Bani-Sadr and Schroeder 2017). A monolinguistic approach to the notion of dignity links it to its Roman juridico-political root '*Dignitas*', meaning a person of rank or status. Though '*Dignitas*' was solely for the *res publica*, the rest of the society and the ordinary citizens do not have '*Dignitas*' (Douzinas 2019). A thicker description of the concept could be captured by looking into the Kantian understanding of dignity as 'an unconditional and incomparable worth'. He considered that only humans are worth dignity while every other object in this world has a price. Moreover, Kant argues that all humans—with no regard for their social, racial, political, class, or any other modern categorisation—has an equal intrinsic dignity that is their worth. This human dignity is embedded in their existence as persons and cannot be earned or deprived of. Kant continues his argument that people shall endeavour to make choices and take actions that suit such high status of dignity and its intrinsic value which placed humans above all (Hill 2014).

Though the various aforementioned conceptions of human dignity define the notion from different perspectives in response to various contexts, they still seem to neither act as an inclusive of the unhoused nor account for their values and ways of living a dignified life. The concept has been narrowed into a virtue of how to treat one another, yet it still fails to include or provide respect for the dignity of the marginalised communities and those the society left behind. The majority of our societies around the world are still judgmental of the unhoused communities. The unhoused communities are still treated as undignified beings who failed to follow societal expectations and dogmas and therefore deserve no respect. Why is that? Why do societies treat the unhoused as undignified? And why does the unhoused communities' way of practising dignity seem so distant from its theoretical description?

Following the previous brief introduction on the conceptualisation of the idea of dignity and its contemporary use, let's take a look at its manifestations in the day-to-day life of the inhabitants of a squatted land who are legally registered as unhoused or homeless persons and how the concept of dignity is present in their relation to each other and the land. These manifestations render strong evidence of the failure of theorising dignity as an intrinsic value. The aim of exploring this case study is to learn from their experiences of land revival and community-making to find alternative ways of enriching our understanding of the concept of dignity beyond its humanistic monotony as well as utilising the concept in urban studies and social housing projects.

11.2 Methodology

Homelessness is often seen as undignified and a manifestation of individual failure, and it may seem like a counterintuitive methodology to learn from homelessness and illegal squatting about the meaning of dignity as they are certainly stigmatised and do not conform to society's expectations and rules. However, I believe that the strength and the persistence of the case study at hand are in the essence of the embodiment of dignity and its performativity not only to humanity but extended to nature. And in that embodiment, better ways to approach social issues related to architecture, social housing, commons, and homelessness can be found.

To better understand the position of this paper, it is rather beneficial to keep in mind the multi-layered nature of homelessness. Sommerville (2013) explains homelessness by shedding light on more of its aspects besides the lack of adequate shelter. He argues that homelessness entails the lack of five different layers that are necessary for a dignified living. The first is the physiological layer which is concerned with the lack of shelter and body comfort. The second is the territorial layer and its connection to the lack of privacy, ownership, and feeling of home. Those two layers are the most visible ones. The third, in turn, is the emotional layer and the lack of love, whether it is self-love or external love, lack of joy, lack of intimacy, and closeness. The fourth is the ontological layer which is an implication of the lack of rootedness, lack of community or sense of belonging. The last is the spiritual layer which is the lack of hope, purpose, and belief in oneself (Somerville 2013). Often, the spiritual layer and its struggle get intensified as a consequence of the other layers. With that in mind, the remedy to homelessness is not a mere temporary shelter as seen in many homeless shelters or even social housing projects where homeless people are treated as undignified, irresponsible, and passive. It is only remediable through tackling all the layers of homelessness. That can be done by learning from success stories of self-organised initiatives where unhoused people performed their dignity by resisting societal pressure and forming alternative communities. The paper at hand chooses to use the term 'unhoused' instead of 'homeless' whenever possible. That is to shift the frame from blaming the victims of homelessness to the lack of societal support. The term unhoused also articulates the lack of shelter while leaving the idea of making a home outside the societal normative ways of doing so possible.

In the subsequent section, I will provide the reader with a thick description of the case study Ciumara Ranni which is an ecovillage located in Sicily. The description is based on fieldwork, conversations with the community members and other activists and scholars who attended the Decolonizing Architecture Summer School of 2022, a tour and round discussion by the community members, oral history, and written resources published online and in newspapers. The methodological approach in this article follows Gary Thomas's definition of case study research. The article is based on a holistic analysis of a case study for the purpose of theory seeking that emerges from local knowledge and practices (Thomas 2011). It aims to illuminate an alternative framework of living through the acknowledgement of the unhoused people's dignity and autonomy. The description in the following section is not just factual narration, but it is also a discursive storytelling text of the embodiment of dignity, with attention to the embodiment of the idea rather than the term as it has been used in a limited way by the community. Yet the embodiment of the idea of *the dignity of the land* is very present in their everyday life. The discussion section of this paper is spread throughout the case study text to better connect the practice and the theory.

By putting this embodiment of dignity at the centre of the thinking process, I aim to articulate and propose a definition of the term *land dignity* and its meaningful role in spatial studies, restoration, and social housing policies.

11.3 Ciumara Ranni

The case discussed in this article is based on a community of people who formed an alternative ecovillage named Ciumara Ranni. The ecovillage lies alongside the banks of the great Carrubba River, located in Carlentini, Sicily. The ecovillage put to use a series of abandoned windmills, wineries, residential buildings, animal farms, storages, and water canals (*Saqiya*), most of which are partially destroyed. The ecovillage was founded by Roberto Barbagallo with the idea of reviving what others have left behind in an attempt to create a space with food sovereignty and energy independence through practising the methods of permaculture. Even though the odds were against the sustainability of such an initiative, Ciumara Ranni is now a decade old and inhabited by at least 15 permanent residents whose legal residential status is registered as homeless persons. I visited the ecovillage as part of a research visit to Sicily in relation to the difficult heritage summer school in 2022 which focused on finding alternative ways to inhabit a separate neighbouring space called borgo ~~Rizza~~ (we strike-through the name Rizza as it stands to commemorate a fascist militant. (DAAR 2020)). The visit to Ciumara Ranni was organised by Francesca Gattello and Steffie De Gaetano as part of their event 'ready landscapes'. The text at hand lacks any photographs or digital visuals as per the community's request not to use any digital equipment in the ecovillage. The ecovillage is off-the-grid and free of electronics or radio waves. The exact location of the ecovillage is also not disclosed as per the community's request. If one is interested in visiting the ecovillage, it is recommended to physically go to Sortino and ask for directions. The ecovillage is welcoming to anyone who respects their community and ways of living.

11.4 The Collective Memory of the Place

The area, adjacent to where the great Carrubba River flows, was seasonally inhabited by several families of farmers. Surrounding the river, there is a series of abandoned and partially damaged mills, storages, summer houses, wineries, and water canals which are referred to in Sicily as (Saqiya\ساقية) which is the Arabic word for an agricultural water system. The majority of the man-made structures are in a state of ruins except for very few relatively recent and illegally built summer houses. According to the local oral history and the inhabitants of Ciumara Ranni (2022), most of the structures that were built prior to World War II had been destroyed by the fascist regime following the establishment of '*Ente per la Colonizzazione del Latifondo Siciliano*' (ECLS), or the Entity of Colonisation of the Sicilian Latifundia. The entity was established by the fascist regime in 1940 as part of a large movement to colonise and modernise the Sicilian Latifundia which included the creation of urban and agricultural projects such as borgo. It also had the responsibility of controlling, restructuring, and enforcing authority over the agricultural lands on the island. Therefore, in order to take control over the local agricultural production, such as wheat and other grains, the fascist regime banned the use of the local mills surrounding borgo ~~Rizza~~ and destroyed the mills and the winery, enforcing the local farmers to either be dependent on borgo ~~Rizza~~ and the entity of colonisation or be displaced. Yet, the local farmers disobeyed this form of dictatorship and continued to use the ruins of the mills at night to make flour as some of the wheels were still functional. Milling and social gatherings often occurred when the moon was full, providing them with enough light to see the way without the need for man-made lighting. The buildings as well as the agricultural lands in this area were then abandoned for at least two decades prior to the creation of Ciumara Ranni (Ciumara Ranni Inhabitants 2022). Further research is needed in this regard as it is based on oral history and speculations. That being said, Distretti and Petti (2019) further describe the internal colonisation in Italy in their paper '*The Afterlife of Fascist Colonial Architecture—A Critical Manifesto*'. They underline the fascist regime approach of weaponizing architecture and urbanism as tools for controlling the lands and the inhabitants of Sicily as well as creating 'unity' between the 'metropole' and its 'peripheral' worlds. The researchers continue to highlight the vast similarities between the colonisation of Libya and Eritrea and that of Sicily as they were all part of the fascist modernist propaganda for new urbanity in the colonised lands. In the south of Italy, local populations and their methods of sovereignty and land use were considered uncivilised and backward by the fascist regime (Distretti and Petti 2019). Accordingly, the agricultural lands then were claimed by the fascist regime and the local ways of farming and living were uprooted under the name of planning, development, modernism, and industrialisation. Hence, that led to the creation of a series of monuments, towns, various buildings, infrastructures, and agricultural villages such as borgo ~~Rizza~~ on the one hand, and to the explosion of the local farmers, the destruction of their structures, and the eradication of their techniques and ways of living on the other (Distretti and Petti 2019).

11.5 Inhabiting the Ruins

According to personal communications with the community inhabiting Ciumara Ranni at the moment (2022), the space grew out of Roberto's inspiration to create a self-sufficient, vegan-oriented community detached from the modern reliance on mass-produced food and polluting energy. Hence, free access to the land and drinkable water is essential to sustain the independence of the ecovillage and grow local crops. Ciumara Ranni's journey started on social media, over ten years ago, with Roberto looking into 'wwoofing Italia' for a neglected land with access to drinkable water and the possibility of a rent-free agreement with the owners. Several options emerged through the exploration and after several attempts, the location of Ciumara Ranni was picked to be alongside the banks of the great Carrubba River. The ecovillage was named after the name of the river in the local dialect out of the belief in the importance of the very precious drinkable water of the river which does not dry out even in the Sicilian hot summers. The running water gives the space its unique character and prospect for the possibility of actualising a sustainable community (Ciumara Ranni Inhabitants 2022). In a newspaper published in 2012, Federica Motta tells us that the project took off as a squatting action in an abandoned and illegally build summer house found beside the river and by planting seeds in its surrounding. Then, she mentioned that after long conversations with the owners of the abandoned lands, a contract between Roberto and the owners was sorted which entailed a rent-free loan of 16 hectares of agricultural land that the river flows in and which hosts 12 types of buildings and other historical architectural elements mostly in a state of ruins. The contract was agreed to be extended over a ten-year period under the condition of maintaining the land and making improvements on its situation at the time of signing. By the end of the contract period, if the conditions of land improvement were met, the two parties agreed that part of the land will be sold to Ciumara Ranni. The ecovillage started with Roberto as the only permanent resident of the ecovillage inhabiting the illegally built summerhouse which was the only structure in an inhabitable condition. Roberto then fixed it and transformed it to become the communal house of the ecovillage accommodating up to five guests. Several people were also spending the daytime in the area helping in its restoration through permaculture and natural architecture practices. A kitchenette made out of the mud was installed in the communal house. The community made use of the existing water system (Saqiya) to get running water to the kitchenette (Motta 2012).

11.6 Community-Making

At the time of writing this text, Roberto was no longer part of the community, and the ecovillage is back as a squatting space. Yet, the community inhabiting Ciumara Ranni at the moment (2022) informed us that the decade-old ecovillage grew to become a community of around 15 permanent residents (with two kids that were born and live in the community) as well as approximately 20 temporary residents inhabiting the majority of the ruins and establishing two new structures out of alternative materials. The exact number of inhabitants on a given day would be known by counting the people around the dinner table, which is an evident practice of hospitality and openness. The playing factor in limiting the number of inhabitants is the capacity of the communal house, especially in the winter. For the reason that it functions as a shelter from the constant rain. While in the summer, the ecovillage can accommodate a larger community as people are able to sleep outside in nature, in tents, or in the rest of the ruins that are not inhabitable.

When it comes to the legalities of their residential status at the local authorities, many of the inhabitants are registered as homeless persons, and two mentioned that they registered Ciumara Ranni's location as their home address and were accepted by the authorities. It is worth mentioning that a member of the community referred to the community as homeless with air quotation marks then continued to agree with the term and took pride in that status and the fact that they are inhabiting nature (Ciumara Ranni Inhabitants 2022). The fact that Ciumara Ranni inhabitants drew attention to their status as homeless persons, not only shamelessly but with pride, is by itself an epiphany of a dignified person with autonomy. The success of Ciumara Ranni challenges the common idea that homelessness is a predicament that only includes those who failed society. In the contrast, homelessness here is a dignified concept that embodies living in harmony and as equals with one's surrounding nature without dominating it. Perhaps one would not imagine a dignified space when looking at ruins of abandoned structures, but Ciumara Ranni managed to challenge the modernist imagination of what home should look like bringing back nature and community at the centre of a dignified living.

11.7 Beyond Authenticity

As water could be one of the largest challenges when it comes to inhabiting and making use of abandoned land, especially with the extended droughts the world is going through due to climate change, the inhabitants of Ciumara Ranni make use of the drinkable water of the great Carrubba River. The community (2022) stressed the importance and sacristy of the river water as the central element of Ciumara Ranni. They believe that by drinking this water and irrigating the plants, they become part of the nature they inhabit and build a spiritual bond with it. They utilised the river water by renovating, so far, around 400 m of the said to be 1500 m of old Saqiya water system. The use of the existing water canals provided Ciumara Ranni with running drinkable water throughout the inhabited area all year long. Except for the communal house, as it is located higher than the restored water canals. For it, they installed a mechanical water pump which works on solar panels. Lighting in inhabited structures is also dependent on solar panels. The renovation was done following local knowledge and materials through communicating with the older generations in the surrounding towns. A drip system has also been installed to irrigate the plants without the need for electrical pumps as the system is dependent on gravity as well. They also built a functioning compost toilet with access to running water for hygiene. The compost is then used for growing plants.

It is essential to keep in mind that none of the community members inhabiting Ciumara Ranni at the moment had any professional experiences in neither building and restoring architecture nor other related skills such as plumbing or electric wiring. Their approach is largely based on ambitions, experimentation, and lots of trial and error. They believe that the best way to know people is to communicate while doing activities together, which could be experimenting with water systems, electricity, restoration, or food making. With time, loads of experimentations, trials and errors, the space has become inhabitable and wheelchair accessible while keeping the natural atmosphere of the area. The success in inhabiting the abandoned area questions the inflexibility of authenticity in restoration projects which often becomes an obstacle in sustaining life in a newly restored place. It also challenges the human-centric approach to restoration where architecture is separated from its surrounding nature. In Ciumara Ranni, nature is not seen as a threat to the ruins they inhabit but as an inseparable part of their environment.

11.8 The Ethics of Food

The community of Ciumara Ranni's food resources is either self-sufficient or bought directly from local farmers. For example, they get quinoa, chickpea, and other beans from a neighbouring farm, and rice and other grains from the north of Italy. In the winter, they are largely dependent on eating from the vast varieties of wild food that can be found in the area. They mentioned that the land could provide at least 17 different wild food in their surroundings during the winter. In the summertime, their dependence shifts to farming the land and conserving the excess vegetables to be used for the rest of the year. The community also utilises a large number of wild trees in the area by grafting them with a fruit-producing variety of trees; around 50 new grafts are made every winter. Excess food that cannot be conserved is often traded with other farmers for crops that they cannot or did not plant in the area. In so doing, the community splits away from capitalism and the dependency on cash.

11.9 Sustaining a Dignifying Community

Currently, the ecovillage inhabitants aim to sustain the space as a free, democratic, and dignified space where living in harmony with the land is the norm. They endeavour to keep Ciumara Ranni alive as a parallel system disconnected from modernity and its dependence on consumerism, hierarchies, and power structures. The decision-making process in the ecovillage usually takes place in the weekly discussion circle. The circle provides a flat structure where everyone present is equal and all decisions have to be owned and decided by the whole group. They have a decision-making system based on the concept of a consensus where not everyone has to be on board with every decision, but everyone has to have a common ground, and no one has the need to block the proposed idea even if it does not fit their agenda fully. If an agreement is not achieved after the discussion, the group respect the strong position of the person and the proposed idea would not be passed. Most often proposed ideas are discussed and proposed spontaneously prior to the circle and a general agreement is achieved. This approach is working in the ecovillage because of three reasons. First, the group is relatively small. Second, there is common trust in the group. Third, the group shares a common point of view on where the ecovillage is heading.

A similar approach is also implemented when welcoming newcomers. The inhabitants of the ecovillage would not treat newcomers as guests and play the host role. Instead, once newcomers are inside the group, they get introduced to the commonalities of the community and if the newcomers feel like a good fit, they are part of the community and they consequently start experimenting, working, and participating in the weekly circles alongside the rest of the inhabitants. When new member joins the community, they get to choose where they see their skill set or potential collides with the group and they join the work, or they just spontaneously get involved and grow interested in being part of the living community. This approach builds upon previous knowledge and exchanges skills between the inhabitants. There are no written rules in the ecovillage, it is based on the principles of mutual respect, love, and active participation in upkeeping and reviving nature. In so doing, Ciumara Ranni shapes a sense of self-worth and dignity for its members. However, this openness and hospitality could sometimes be taken advantage of; in such situations, the community utilises the weekly circle to communicate any misunderstandings and share worries and discomforts. If a newcomer does not respect the dynamics of the group and does not show signs of improvement and assimilation, they will be expelled.

11.10 We Owe the Land

As the community of Ciumara Ranni (2022) mentioned in our meeting, land use in Sicily is not solely based on owning the land through paperwork but also by inhabiting the land and being present. In other words, the land belongs to the ones who grew the seeds and cultivated them. They continue to argue that the idea of landownership in itself is problematic, as the land belongs to the plants, and people owe the land and the plants the care they need in return for food. And the larger the community is, the more land they need, and the more land they occupy and therefore protect. Parallels could be made to the concept of masha' (مشاع) in Arabic (roughly translated into the commons) where public lands are common lands, and they are accessible to those who continuously and actively inhabit it through living and aggregating the land. Hence, one can only use what they are able to activate and no more. Once the land is no longer activated, it goes back to the public realm, opening a chance for others to use the land (Siraj et al. 2011). Several similar land commons existed in Sicily prior to the fascist regime restructuring of land ownership.

However, this non-capitalistic approach is not the general term of land use in Italy leading to crashes with landowners, neighbours, and surrounding communities. Therefore, difficulties with landowners happen frequently; the community even mentioned that they had been threatened to be killed and their ecovillage burned at one point. That being said, the Ciumara Ranni community mentioned that they are in a rolling basis conversation with most of the landowners. These conversations are often tricky and unexpected as anything is possible to trigger and ignite conflicts. After countless conversations, the community in Ciumara Ranni convinced some of the owners that they are there to be part of the land and take care of abandoned areas only to eat. They are not there to take over the legal ownership of the land or make a profit out of it. Some consensus has been made with the owners of the land and the surrounding community. Yet, the members of Ciumara Ranni never all leave the area together; a group must always be present in the land at all times in fear of wildfire, or man-made fire that is claimed to be wild, the herds of cows that belong to the neighbour, or any other threats. Moreover, stereotypes of the new age communities and their alternative approach to life play a role in resurfacing some of the crashes. Nevertheless, the community keeps good relations with the surrounding towns and volunteers in cleaning the surrounding nature and the streets as gestures of kindness, well intention, and hospitality. In so doing, Ciumara Ranni can be framed as a living example of detaching from the modernist assumptions of public spaces and the need for urbanity. It is an extraterritorial space isolated from any modernist design practice but rooted in commonality as a discursive participatory community. Having said that, ethical dilemmas can be present concerning imposing this alternative social life modality on children. However, in a discursive manner, how come we do not question the ethical dilemmas concerning imposing the hegemonic social life modality on children? This suppression of alternatives is in itself part of the hegemonic project of modernity.

11.11 Towards Demodernisation

With the act of 'define and order' being the essence of modernity and rationalism, concepts, or parts of them that do not fit within the grid of modernity and its dogmas, will be suppressed in an attempt to eliminate vagueness as well as ambivalence, and seek precision and concision. This modern need to 'define, categorise, and order' has, to an extent, stigmatised outer conceptions of an idea beyond its hegemonic understanding. And to break free of stigmatisation, one needs to assimilate through an active acceptance of a non-stigmatised form of life and patterns of knowing (Bauman 1990). Therefore, effectively alienating what is outside the grid of modern cultural conformity and subordinating them as chaos, as an accidental surplus of knowledge. What, however, happens if one departs from the so-called chaos and embraces its ambivalence? What happens when we learn dignity from the unhoused and squatted lands? Perhaps dignity is not meant to have a unified definition, but rather an embodied, performative one as evident in the case of Ciumara Ranni.

Having said that, the modernist project and its suppression of alternatives do not only exclude communities such as Ciumara Ranni from being integrated into society but also create a reaction from these communities against the world as we know it. This anti-modernist reaction leads to a refusal of modern-day tools, electricity, equipment, knowledge, and a detachment of social welfare. Therefore, those who refused to assimilate ended up segregated and left behind. Though in the specific case of Ciumara Ranni, there is not a total anti-modernist reaction from the community nor total suppression from the larger society to the Ciumara Ranni. There is some kind of mutual agreement and tolerance between all the stakeholders within the larger context which allows alternatives to emerge. In this mutual tolerance dwells the seeds of demodernisation, one that not only opposes the suppression of alternatives but also supports them and calls for connectedness between the alternatives. By no means Ciumara Ranni is a demodernisation project nor can it survive for a decade without the tolerance and support of the surrounding community. The ecovillage can also not be foreseen as a large-scale solution for similar challenges. But it is indeed a sign of a possible liberation of praxis beyond modernity. Demodernisation then means having a diverse, ever-growing set of contextual solutions that emerge from praxis and in response to the specificities of each case, instead of an imposed set of criteria or guidelines.

11.12 To Conclude

The decade-old ecovillage taught us a lot and effectively challenged the common understandings and established conceptions of homelessness, home, and dignity. First of all, the ecovillage's success in keeping its sustainability and autonomy outside the typical societal norms and distanced from capitalism and consumerism is a reminder of the possibility of alternative forms of living. Second, the fact that many of the inhabitants are also registered as homeless persons acknowledges that unhoused people are dignified people who are capable of forming a supporting community, unlike the societal stigmas. It is also a reminder that homelessness is not just a lack of shelter, but a lack of community. Therefore, it is not a problem for the individual who fell into it, but a sign of societal failure and lack of proper support beyond mere shelter. This is also a reminder of the importance of shifting the terminology from the 'homeless' to the 'unhoused'. That is to shift the focus to the lack of shelter due to societal failure instead of an individual one. Third, Ciumara Ranni challenged the modernist imagination of what home looks like and brought back the integration with nature and community at the centre of the process of homing. They also challenged the modernist understanding of public space by turning the whole ecovillage into an inhabited common. Forth, the community taught us that not only do we find dignity within us and in the place we call home, but we also embody and perform the idea of dignity to externalise our senses beyond its status or passive intrinsic understanding. Therefore, land dignity is the mutual dignity of all the elements of a given space. It dwells in the relations between the elements and their treatment of each other in our shared habitat. This includes the human-to-land relationship as vividly evident in the case study, to treat the land with dignity, is to not own it but owe it. Land provides life, and we owe our lives to the land by taking care of nature. This is a lesson that human rights conventions, urban studies, and architectural restoration approaches could benefit immensely from.

References

Austin JL (1962) How to do things with words. Harvard University Press, Cambridge

Bani-Sadr A-H, Schroeder D (2017) Dignity in the middle east. In: Dignity in the 21st Century. Springer, pp 65–88

Bauman Z (1990) Modernity and ambivalence. Theory Cult Soc 7(2–3):143–169. https://doi.org/10.1177/026327690007002010

Butler J 1956(2004) Undoing gender. Routledge, New York

Ciumara Ranni Inhabitants (2022) Personal communication

DAAR (2020) Towards an entity of decolonization. http://www.decolonizing.ps/site/palazzo-delle-esposizioni-roma/

Distretti, Petti (2019) The Afterlife of Fascist colonial architecture: a critical Manifesto. Future Anterior J Hist Preserv Hist Theory Crit 16:47. https://doi.org/10.5749/futuante.16.2.0047

Petti (2022) Personal communication

Douzinas C (2019) The story of dignitas. In: The radical philosophy of rights. Routledge, pp 22–31

Hill TE (2014) Kantian perspectives on the rational basis of human dignity. In: The Cambridge handbook of human dignity. Cambridge University Press, pp 215–221

Motta F (2012) Ciumara Ranni, villaggio vegano a Sortino L'idea green di un ex informatico catanese. https://meridionews.it/ciumara-ranni-villaggio-vegano-a-sortino-lidea-green-di-un-ex-informatico-catanese/?refresh_ce. Accessed 10.9.22

Siraj S, Peters B, Åsa J, Wagner R (2011) Islamic principles and land opportunities for engagement. United Nations Human Settlements Programme (UN-HABITAT)

Somerville P (2013) Understanding homelessness. Hous Theory Soc 30:384–415. https://doi.org/10.1080/14036096.2012.756096

Thomas G (2011) A typology for the case study in social science following a review of definition, discourse, and structure. Qual Inq 17:511–521. https://doi.org/10.1177/1077800411409884

United Nations (1948) Universal declaration of human rights

12 Contextual Approach of Tactical Urbanism as a Tool to Mitigate Social Segregation

S M Kaikobad

Abstract

Population surge and fast-paced urbanization are continuously changing the urban landscape of the 400-year-old city, Dhaka, the capital of Bangladesh. While people of all backgrounds experience the consequences of urban planning, it is the underrepresented community that suffers the most due to the lack of socially just public space and inclusiveness of place. This paper aims at illustrating how the combination of urban informalities, contextual reuses of public infrastructures, and user-oriented planning approach of transcending urbanism can turn into an effective chair to mitigate social segregation through an evocative, interactive, and integrated space-making enhancing accessibility for diverse community. The paper focuses on an evidence-based context study, Gulshan, planned as a high-class residential area, established in 1961, now turning into a mixed-use Central Business District (CBD), welcoming more people from all backgrounds. With the setbacks of conventional place-making approaches, the true potential of social spaces is evidently underutilized with a lack of memory or cultural value where segregation is alarming. Socially just urban civic space is needed to foster the culture of community building in busting and overpopulated cities like Dhaka. Tactical urban intervention by making proper use of demand-based multilayered "space" with efficient use of resources is essential for marginalized populations to enjoy a sense of belongingness and ownership. This paper discusses the constellation of urban elements to become interactive means for civic functionality to mitigating social segregation with cohesive connectivity between people, nature, and the built form.

Keywords

Social justice · Urban place-making · Contextual · Urban utilization of resources · Inclusive design

12.1 Introduction

12.1.1 Background

A city is a place where people get fascinated by its availability of resources and different opportunities for earning a livelihood. The healthy development of physical, intellectual, emotional, and social aspects depends on the hierarchy of open space contributing to the quality of life (Mowla 2005, Farida 2010). For a densified city like Dhaka, where every year more than 300,000

S M Kaikobad (✉)
Department of Architecture, Brac University, Dhaka, Bangladesh
e-mail: saad.m.kaikobad@gmail.com

S. Hilal et al. (eds.), *Design for Partnerships for Change*, Sustainable Development Goals Series,
https://doi.org/10.1007/978-3-031-36993-3_12

additional inhabitants have to be accommodated, it is hard-hitting to meet demand. (Hackenbroch 2013). The living condition in Dhaka is getting compromised making no sense of space, which is resulting in a lack of social interaction, environmental dereliction, and the cities into a concrete jungle (Hackenbroch 2013). In the path of involuntary life, public spaces play a central role in the creation of inclusive communities with enriched cultural diversity (Low et al. 2005). Unfortunately, for developing countries, with the name of security and maintenance, access to public space has been personalized ignoring issues of equity and justice (Fitzgibbons and Mitchell 2019).

By the year 2050, metropolises will be accommodating more than 66% of the world population (UN 2014), considering such circumstances, city governance for mega cities like Dhaka will be more challenging. Without emphasizing social conditions harmonizing the city's living standard, productivity will decline in the multilevel gated community (Islam 2015). Lack of interaction makes people impassive, and scarcity of land is overtaking the open spaces; thus, through reviewing primary and secondary sources, the paper analyzes several scholarly arguments on the term social inclusion and interaction and the impact of public space; articulates several interventions on policy planning implementation utilizing existing resources.

12.1.2 Aim and Objectives

This paper focuses on the urban-built environment and the importance of socially just civic space and interaction between its user groups. It aims to create a framework for the efficient use of underutilized space to be included at the policy level stating the significance of social inclusion on the socio-economic development of the city. This will also reflect how public–private partnerships can work on bringing out more accessible spaces for diverse community through small interventions and tactical urbanism with an incremental impact.

12.1.3 Methodology

The paper reviews scholarly secondary data to establish the relationship between informal urbanism, interactive space-making, and their impact on micro- to macro-scale. With primary data from a questionnaire survey, site study, several mapping with operationalizing Jane Jacobs's urban analysis method, and interviews with expert stakeholders, the research was conducted in Gulshan, a residential area turned into commercial business district of capital Dhaka to identify potential roles of interactive spaces within the built fabric to mitigate social segregation.

12.2 Literature Review

12.2.1 Tactical Urbanism and Inclusive City

Tactical urbanism is defined by quick, often low-cost, and creative community-driven initiatives by activists, planners, and policymakers seeking to drive toward sustainable cities as a powerful tool for urban transformation (Lydon and Garcia 2015). The notion of tactical urbanism as a bottom-up process starts with the involvement of multilevel stakeholders and public participants, overpowering the bureaucratic top-down process. (Alisdairi 2014). With rapid urbanization, this has become an effective tool for all inhabitants to participate in the use and shaping of all urban space creating a sense of ownership among them. (Purcell 2008).

Dhaka, the capital city of Bangladesh, is currently in the 11th position and will reach 6th by 2030 (UN 2014) in terms of the gradual increase of city population having 44,100 people per square kilometer. In attempts to make cities more livable, it is important to empower citizens as change agents to trigger solution-driven, collective action (Hou 2010). Through the planning-by-doing approach, citizen interaction, attention to perceived shortcomings, widen citizen engagement, and develop a deeper understanding of people's priorities can be achieved. Such interventions are typically low-cost and can contribute to civic pride, belongingness, and stewardship (Elrahman 2016).

With the rapid growth of the economy and perception of social status through the social hierarchy, the system of exclusion and gap between citizens is increasing as a product of modern capitalist society (Low and Smith 2006). In this divided city, minorities and unempowered groups have no place in participating or upholding human and citizen rights and liberties. The impact of social isolation on the tangible and the intangible aspects aggravates inequalities and discrimination and paves the way for the stratification of the population. (UN-HABITAT 2008). Through tactical urbanism of citizen participation, the right to the city can be established for deeper restructuring of social relations and missing interaction. (Harvey 2012; Hintjens and Kurian 2019). The urban theorist Henri Lefebvre conceptualized the right to the city as the right to access, use, and enjoy the city and fully participate in the production of urban space (Levebvre 1996, 2003). In response to diversity and the importance of accepting and respecting it, the concept of "inclusion" has been developed. (El-Din and El-Zafarany 2018).

According to the UN-HABITAT in their report on the state of the world's cities 2012/2013, equity and social inclusion integrated with quality of life enhance environmental sustainability, productivity, and infrastructure toward city prosperity (UN-HABITAT 2012). The Open Working Group's 2030 Sustainable Development Goals 8, 10, 11, and 16 all refer to inclusion in the creation of a sustainable inclusive urban future (UN 2014).

12.2.2 Social Interaction over Isolation

With the rapid increase in global urban population, cities have been increasing in size and density and more than half the world's population is living in urban environments (UN-HABITAT 2012). Within the rigid structures of a consumerist society, due to its scale and excess of the city, the significance of personal interactions between the inhabitants is becoming more elusive. (El-Din and El-Zafarany 2018). Socially excluded individuals are more susceptible to various types of mental health problems and depression is at the top (Vichealth Mental Health and Wellbeing Unit 2005). A sense of belongingness and inclusive and diversified supportive relationships are important factors of well-being and healthy behavior patterns (Wilkinson 2003).

Though until the 1970s Marxist and critical approaches to social inequities had mostly ignored the role of space, social interactions and pursuits have been established as an integral part of the community's well-being (El-Din and El-Zafarany 2018). The spatial aspect of the public realm offers opportunities for a wide range of interactions with nature as well as others, limiting social segregation and creating opportunities for self and community identity expression (Region 2010) where the society reinvents itself. Such open spaces create a scope of interaction within the city as they have a great influence on both physical and psychological health, social development, and community ownership to make a stronger bond with the city and people leading to better living conditions (Shaftoe 2008).

12.2.3 Diversity and Homogeneity of Public Realm

The diversity of users of the public space determines its contribution to society which is considered along many aspects such as gender and sexuality, ethnicity, culture and faith, socio-economic class, age groups, capability, and political expression. In 2011, UN-HABITAT and Project for Public Spaces (PPS) signed a cooperative agreement to harness the ability of public space for transforming cities for common good. A healthy public realm promotes democracy and diverse community cohesiveness creating opportunities where people meet and the city takes shape (El-Din and El-Zafarany 2018). Public places, as participatory landscapes, are considered to be one of the most effective means to perform particular social interactions (Barbui ca 2012; Sennette 1974). The aspects chosen by Anholt fall within the spectrum of the public realm and are considered both the reflection and the driver of happiness, inclusion, diversity, and well-being in cities. (The 10 Happiest Cities in The World 2013). Communal diversity is directly related to the interaction between different types of people (Stutter 2017).

The spontaneous development of spatial strategies in the current status of society compelled to ensure that people meet only the desired group members of their same class, domain, or background. In the composition of interrelated heterogeneous networks, this homogeneity is hindering cities' inclusive growth and specifically in the public realm. (Grimaldi and Sulis 2009). The presence of security, both active and passive, is an overt feature of public spaces dominated by the user group, and it has been reported repeatedly that one of the major reasons for crime in Dhaka city is the lack of proper and adequate recreational facilities (Siddique 1991: 315) where social inclusion, as a promotional element, is considered to be essential to reduce poverty and improve well-being (Oxoby 2009).

12.2.4 Transformation of Public Places of Dhaka: Past to Present

Urban civic space plays the primary role of "urban collector" in the current process of re-imagining urbanity in the global arena (Ravazzoli and Torricelli 2017). Dhaka city has gone through different sociocultural changes which can be traced through its settlement pattern over time. Along with this physical and morphological conversion of Dhaka City through different ages, characteristics and patterns of public places also have undergone major changes (Nilufar 1997). Under three different periods of time indigenous, colonial and post-colonial times, most of the changes were significant (Figs. 12.1 and 12.2).

Indigenous Public Place

Old Dhaka from the pre-Mughal period had been developed through an intuitive approach where people of Dhaka inherited the habit of socialization at outdoor places from the very beginning. This led to the formation of traditional places such as—gali (streets), morh (node), mahalla (neighborhood), and chowk (market square) (Mowla 2003). According to Mawla, the important characteristic of the indigenous pattern of urban fabric was winding the irregular and intricate street networks directed toward the market square. Streets were not mere sidewalks or just a place to pass through, but a space for interaction and accommodating the extension of family activities. To some extent, their significance was more than as a civic space rather than as a path. This indigenous urban typology, as an interactive thoroughfare, brought a massive change in social, political, religious, and cultural demands and became the regional identity and matter of pride for the users. (Mashrur Rahman Mishu 2014).

Colonial Public Place

Since 1757, British rule forced a great impact on the settlement pattern, whereas the previous traditional urban spaces had a strong sense of enclosure, roads, and plots divided the land during the new colonial layouts (Siddiqua 2011). New houses were being built within the plots in a bungalow pattern with individual private spaces as administrative wards for the authority only. The indigenous pattern was covered with more functions, and users, on the other hand, colonial typology features a formal street pattern, segregation of functional areas, and the repetition of similar features (Mowla 2003).

Present public space

In the last few decades, the social groups on the higher power spectrum are creating protected enclaves, in housing and work, as well as leisure spaces due to the susceptible feeling of fear (El-din and El-Zafarany 2018). People at present are more home-bound and, thereby, has changed into

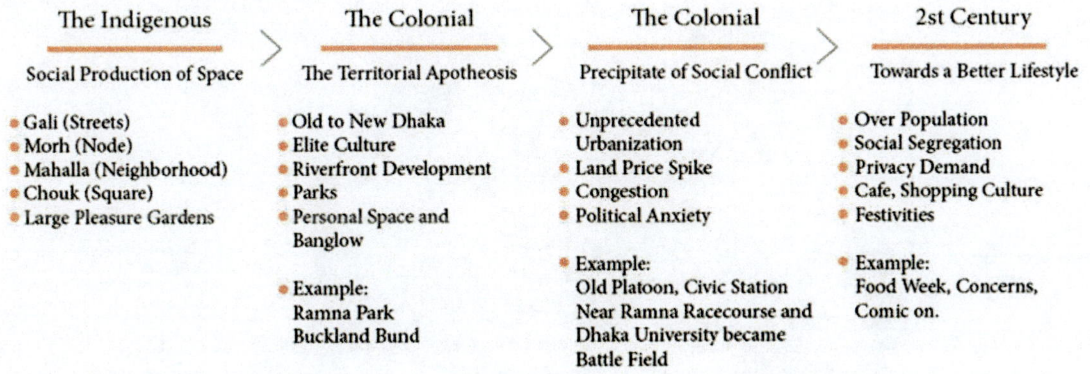

Fig. 12.1 Change of social interactive spaces throughout the time 2005 (*Source* Habib, "The post-Colonial Public spaces and Cultural Diversity")

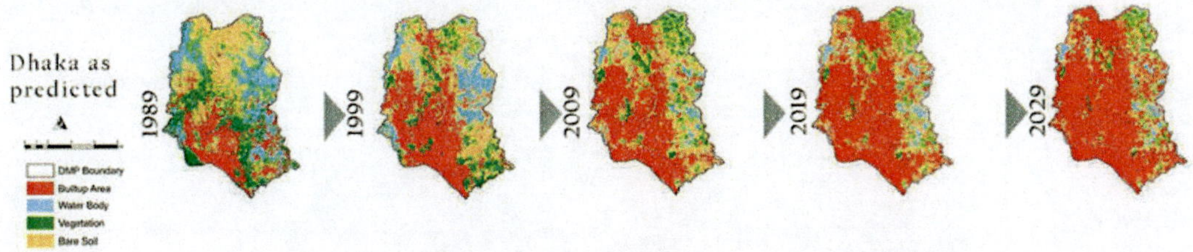

In the old part of city there is only 5% open space while in New Dhaka 12 % of land is green and open. The total amount of open spaces in greater Dhaka is about 17% to 18% and the total stock of public open spaces is hardly over 5000 acres (Mowla, 2011)

Land Cover Type	1999		2009		Change of Area (km²) 1999-2009
	Area (km²)	%	Area (km²)	%	
Built-up Area	108.564	24.35426	166.885	37.438	**+58.321**
Water Body	51.078	11.45838	60.337	13.536	**+09.259**
Vegetation	130.115	29.18882	98.518	22.100	**-31.597**
Low Land	62.926	14.11625	40.716	9.134	**--22.210**
Fallow Land	93.087	20.88229	79.314	17.792	**-13.773**
Total	**445.770**	**100**	**445.770**	**100**	**0**

Fig. 12.2 Built area coverage (*Source* Detailed Area Planning, DAP)

a more individuated, self-centered, and mechanical society in turn (Nilufar 1997). Higher demand for land, overpopulation, and managing the increase of vehicular traffic with failed policies has changed the city formation to satisfy the needs of a motorized population and capitalist group. Ignoring the significance of public places and their essence toward interaction, diversity, and exchange, streets have become "spaces for cars" and urban spaces "spaces for parking" (Davis 1990; Newman and Kenworthy 1999; Mitchell 1995; Sennett 1992).

In the old part of the city toward the south, there is only 5% open space while in new development toward the north Dhaka has 12% of open green. The total amount of open spaces in greater Dhaka is about 17–18%, and the total stock of public open spaces is hardly over 5000 acres (Mowla 2011).

Plot-based urban development and government's focus on the infrastructure development only overseeing the current demand of civic quality of life is creating a sharp difference between social livability and the built environment. Restaurants are being only source of social civic offering, and a large group of community is missing out the quality-of-life lacking publicness of space (Fig. 12.3).

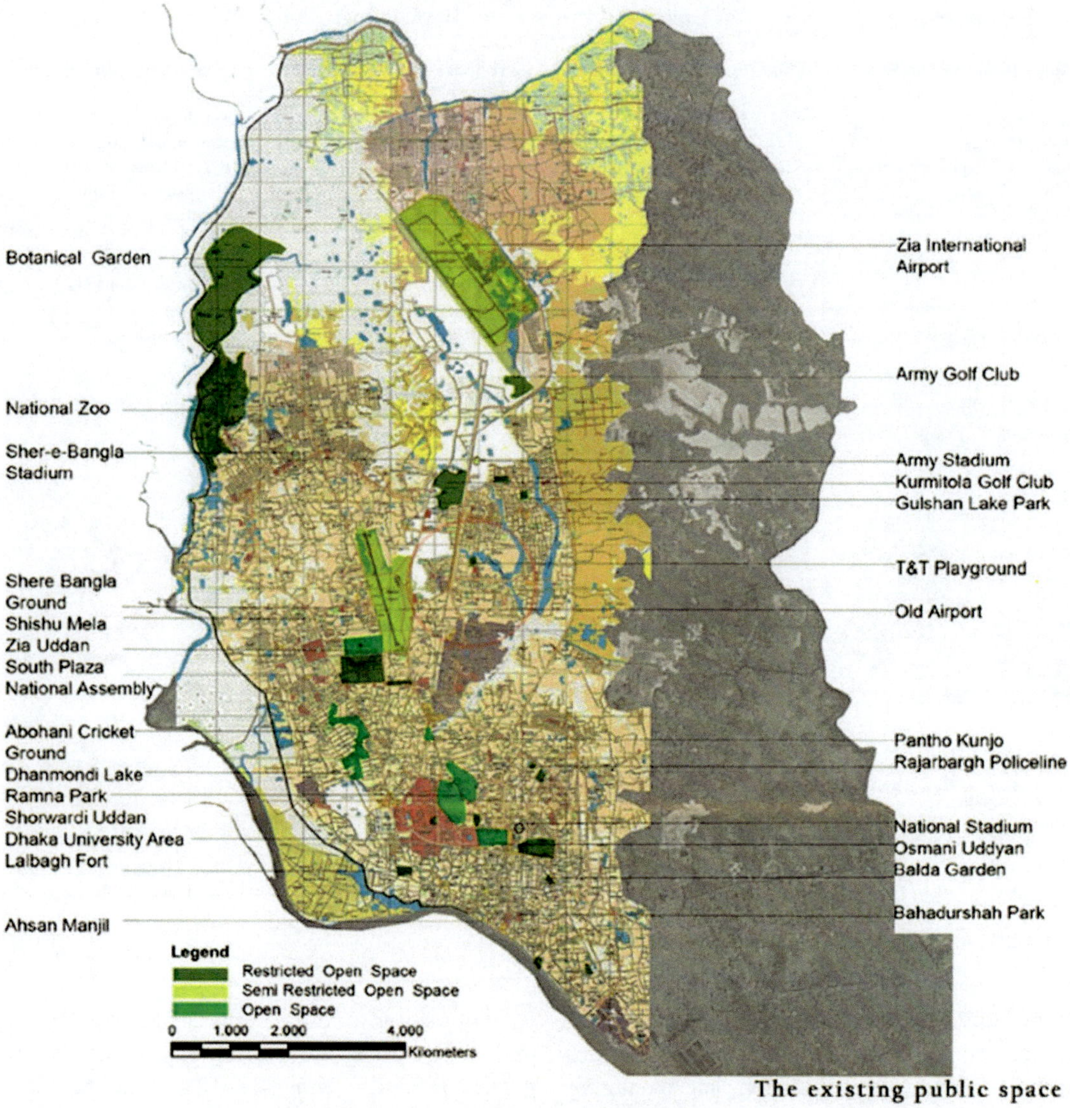

Fig. 12.3 Major existing open spaces of Dhaka city (*Source* (Habib 2010))

12.3 Study on The Turning CBD, Gulshan

12.3.1 Survey- Documentation Method and Approach

Space and place are intrinsic parts of our being in the world having healing capacity and the creation of memory (Athill 2010). These open interactive public spaces around the world have four key qualities: accessible; people are engaged in activities there; the space is comfortable and has a good image; and finally, it is a sociable place: where people meet each other and take people when they come to visit (Roushan 2013). As a study, the Gulshan area was selected having multiple user groups for mixed-use purposes, which converted into an important commercial business district along with the fusion of residential units. The primary focus of the study was to identify potential spaces for tactical urban space-making and creating interactive space. The study indicated that this district consists of 50% residential, 20% commercial, and 12% diplomatic area coverage. The rest 18% consists of open spaces like parks, lakes, and slums which have limited accessibility.

The conventional and unconventional survey helps to understand the city fabrics, accessibility, co-dependent districts, building use, and urban voids. A questionnaire survey brought out the user group of the area with income range, age, thoughts, and expectations of the user group. Counting onsite hard mobility (motorized vehicle) and soft mobility (pedestrian and bicycle) helps to determine the user flow pattern, access, and hierarchy of publicness of different spaces (Figs. 12.4 and 12.5).

Three districts share a symbiotic relationship where culinary culture is strongly established as a means to the social gathering. Parks are mostly used by a specific user of the adjacent homogenous residents of the neighborhood. Gulshan is rapidly turning into CBD due to its geographic location on the new Dhaka, social and economic status, and the increasing demand for commercial space, where entertainment is perceived on occasional visits to numerous restaurants, cafe, and hotels, facilitating one class of people (Figs. 12.6 and 12.7).

Despite having multiple entertainment facilities like park, shopping center, and rich culinary, a certain class of people access these facilities. The parks have time-based public accessibility but have restrictions and specific dress code. Moreover, the formal approach of design to these facilities creates agitation to middle and lower income group of people questing their belongingness and identity to the certain built environment.

12.3.2 Focus of the Study

Social interaction depends (Mawla 2008) on accessibility where a good range of mobility foster the creation of more sustainable public spaces

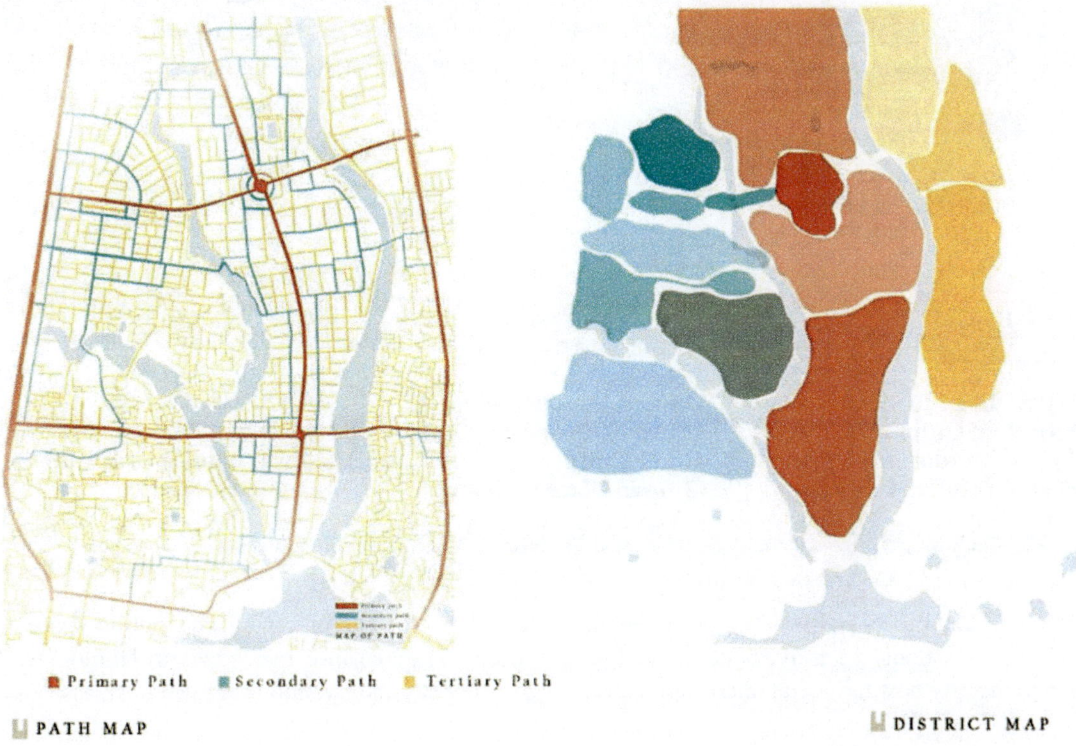

Fig. 12.4 Study map, Gulshan (*Source* Author)

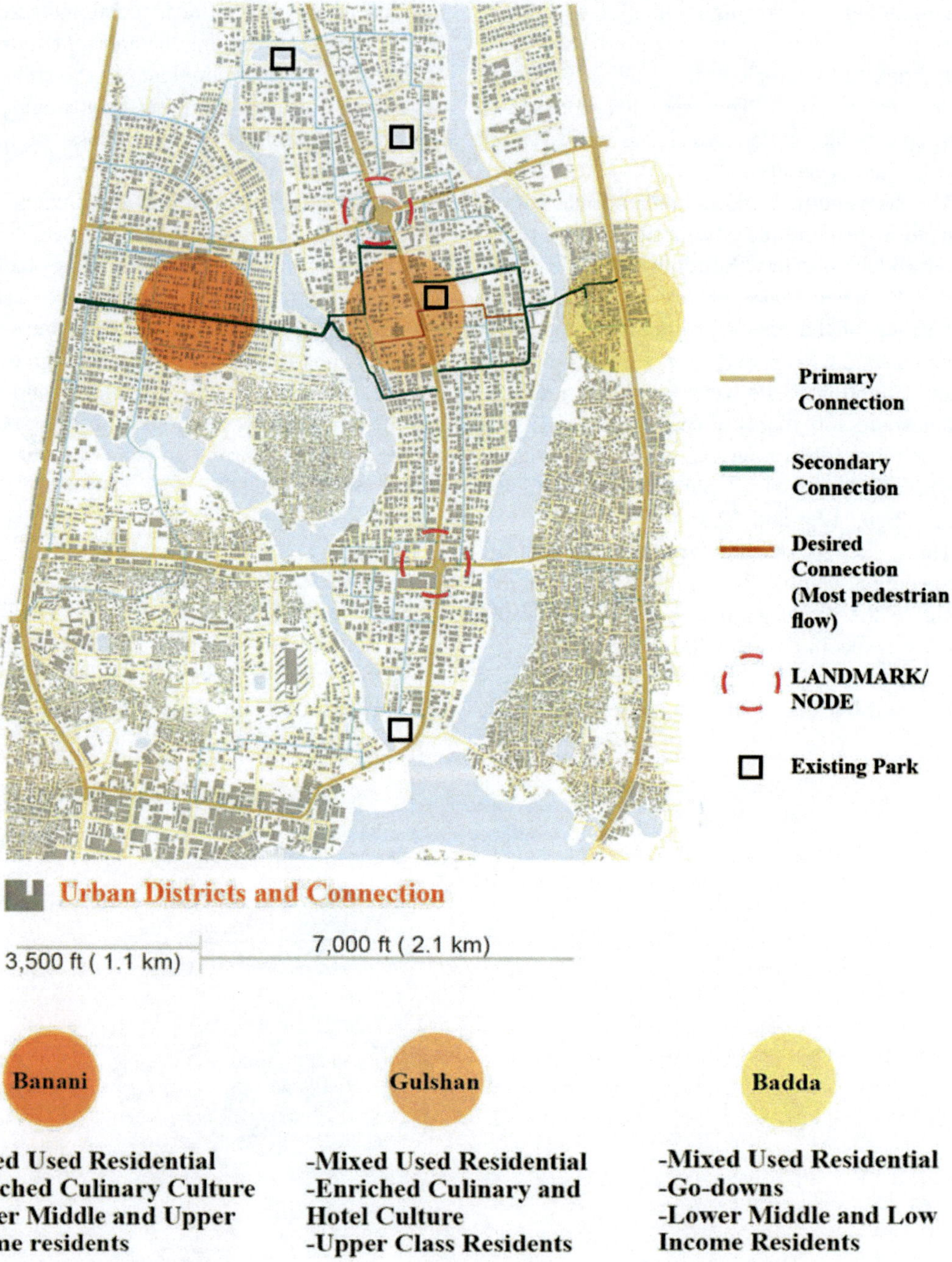

Fig. 12.5 Three districts, Banani, Gulshan, Badda and Established Public spaces at Gulshan (*Source* Author)

(Gehl et al. 2006; PPS 2014b). The primary data were collected along the movement of the user group to identify how the social interaction works within the context of established public space, pedestrians, parks, in-between spaces of buildings, government property, and streets (Fig. 12.8).

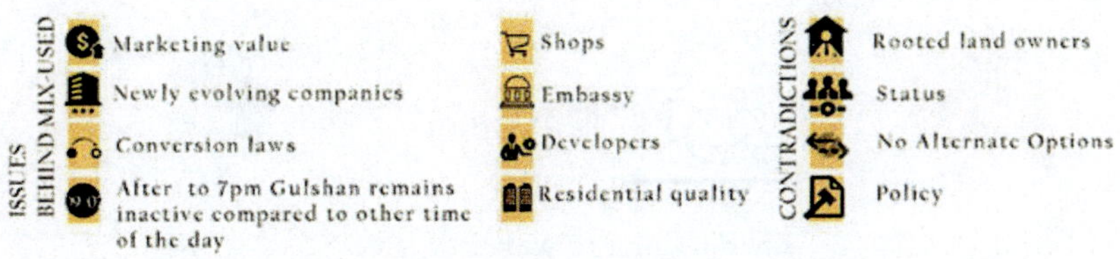

Fig. 12.6 Three reasons behind CBD formation (*Source* Author)

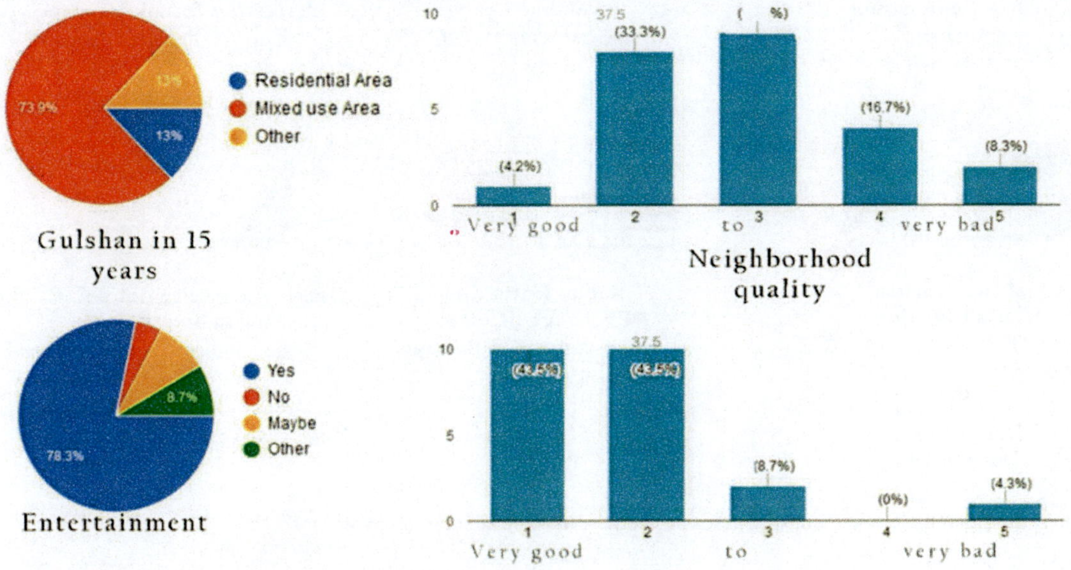

Fig. 12.7 Survey data on neighborhood quality and opportunities (*Source* Author)

12.3.3 Primary Finding

Working on some key points of physical and virtual space can bring back the identity of the interactive spaces. The "shared space" is a concept that was devised to attempt to include the different groups of pedestrians in the vehicular flow (Joyce 2012; Hamilton-Baillie 2008). Different districts of the city have a connection through Gulshan, where most of the districts serve Gulshan being the commercial business district. Although the study initially focused on the connecting path between two important landmarks and nodes, Gulshan 1 and 2, the study indicated a more promising and strong existing connection between the secondary nodes/edges of the three districts (Fig. 12.9).

The pedestrian flow through this secondary connection are four times higher than the established connection between primary landmarks (Fig. 12.11). Diverse people use this walkway as an interchange between districts and interestingly, followed by an adjacent walkway of four open spaces with civic character (Fig. 12.13). The study directed the project into the negotiation and policymaking on the success of this transitional path by rejuvenating the urban civics spaces, the engaging connectors between its inhabitants creating scope for an interactive and shared society.

The pedestrian linkage between the three interdependent districts has been ignored due to the informal urban social character of the connecting nodes. The human flow follows these public or semipublic spaces which offers a larger scope of interaction.

Policies need to be revisited to accommodate the need of user group and to incorporate these

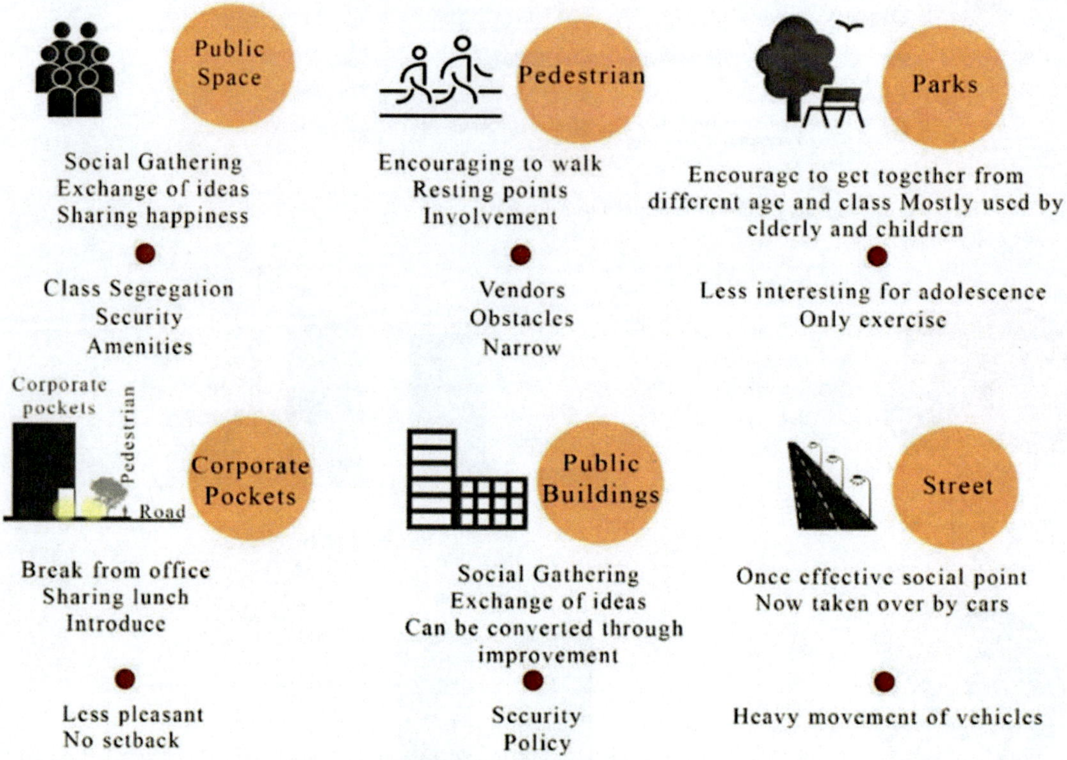

Fig. 12.8 Opportunities and constrains within the potential social space (*Source* Author)

underutilized spaces into tactical urban innovations to offer more public spaces (Fig. 12.10).

The connecting path adjacent to the open spaces are social resources that can be use in urban transformation. People tend to follow open space rather than walk side by a wall, and this established pathway can be used as assets to fill in the urban void to interactive spaces through tactical urbanism (Fig. 12.12).

The districts are connected through vehicular roads separated by two lakes and lack proper pedestrian connections where the movement of people is four times higher through this secondary connection than the primary road. Pedestrian connection should be established to enrich socio-economical values and to create ownership with improved neighborhood quality, social amenities, and activity space.

Five urban-built elements (Fig. 12.13) were discovered in the transition having potential civic quality and to work as an anchor to enhance community identity and social interactive space (Figs. 12.14, 12.15 and 12.16).

12.3.4 Constellation of Urban Pockets and Their Accessibility

The extensive study and gathered data on user activity patterns, communication routes, and social behavior and widen citizen engagement have directed the project to the realization of a deeper understanding of people's priorities with a much-needed connection between three major interdependent urban districts, Banani, Gulshan, and Badda. With a priority of none vehicular movement and commute pattern of mass people, five underutilized yet high-potential spaces as the urban connector was identified to initiate the negotiation on the inter- and intra-relation of the elements of the civic, urban-, and landscape-built environment.

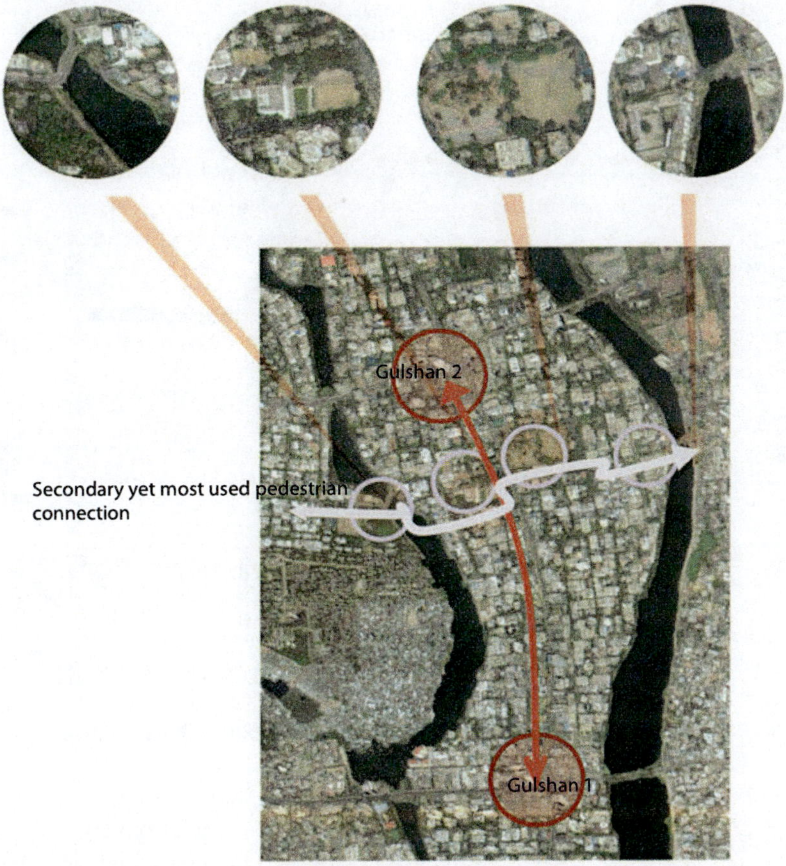

Fig. 12.9 User enforced alternative pedestrian connection between districts (*Source* Author)

Banani 11 no bridge

Banani 11 no bridge is the secondary connection linking Gulshan and Banani itself. One side of the bridge is covered with trees on the government land, which leads to the most active culinary street of the capital being the primary social hub, and another side is the residence and mixed-use building backing the lake with negative spaces in the setback. The bridge offers civic quality space for the adjacent neighborhood having different vending and entertainment (Fig. 12.17).

Gulshan Central Mosque

The mosque is maintained under Gulshan society having a large setback from the adjacent vehicular road. With a grilled boundary barrier, the setback offers a lot of green for the passerby. People mostly use the mosque premises during five times payer, and the front field is used for sports by the students in the evenings.

Rajuk Park and Youth Club

Rajuk park previously designed as an amusement park and recently converted into a society park offers a space for multilevel users. Kids from adjacent low-income community play different sports, while the children from upper and upper middle class use the club for recreation and sports.

Badda Jhilpar Bridge

Badda Jhilpar is most used secondary entrance with pedestrian linkage between North Badda and Gulshan. Illegal informal urban vendors occupy the bridge congesting it, as well as serve on a large-scale vegetable and food supply satisfying daily need of the adjacent neighborhood (Fig. 12.18).

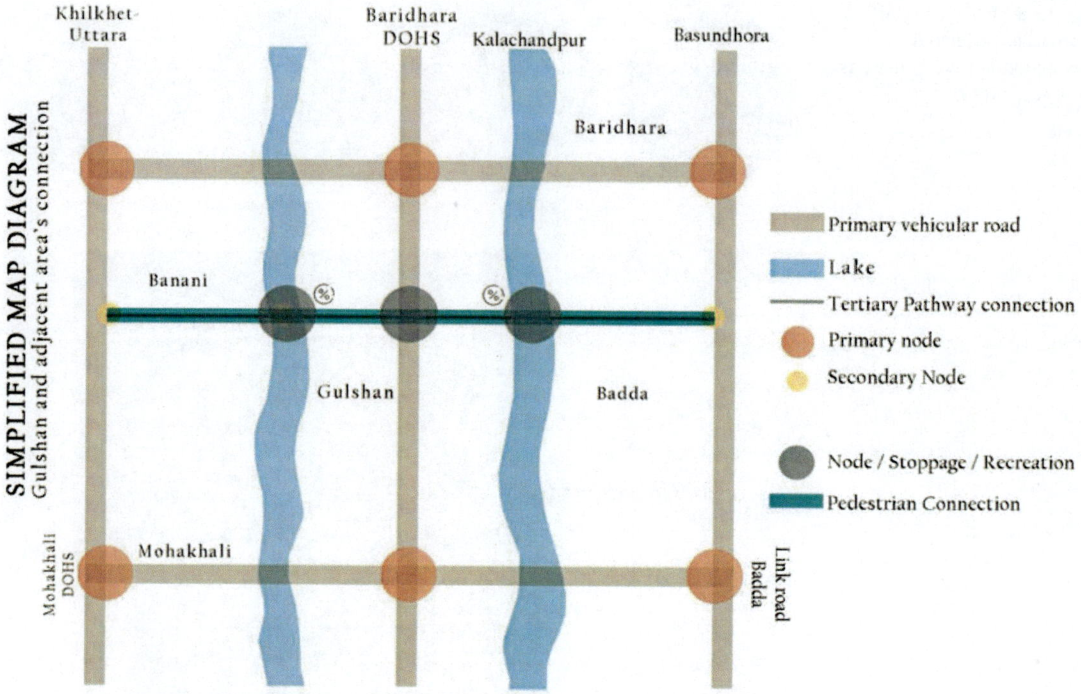

Fig. 12.10 Simplified diagram of connections and accessibility (*Source* Author)

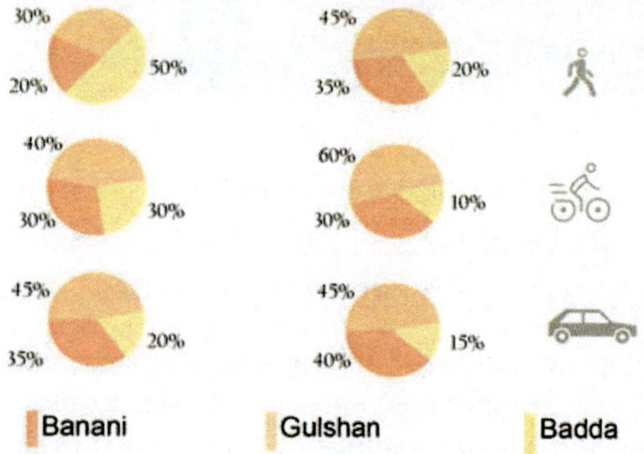

Fig. 12.11 Accessibility between interdependent districts (*Source* Author)

12.4 Idea Development and Conceptualization

12.4.1 Urban Voids and the Negotiation of Threshold

The integration of underutilized urban voids and public spaces, social programmers and economic activities can increase the use of the new public spaces, making the programmers and activities inclusive and accessible. In this sense, public spaces are intended to accommodate a wide range of social functions. Interconnected road networks are also considered important to integrate informal urban elements with their surrounding areas (Yanliu 2016). Civic facilities are public parks and plazas; sidewalks and main streets; community centers and public buildings; commercial establishments such as coffee shops, pubs, and diners; private establishments with public functions such as airports and malls; cyber-civic space; households, temples, and schools; and insurgent spaces—spaces of protest such as government buildings, corporate headquarters, and public squares (Douglass 2002: 353). Negotiating the threshold of urban voids and including them in policy and governance can create a scope for tactical urban intervention where citizens can actively participate in the development process for a developing country like Dhaka, where space is considered a luxury (Figs. 12.19 and 12.20).

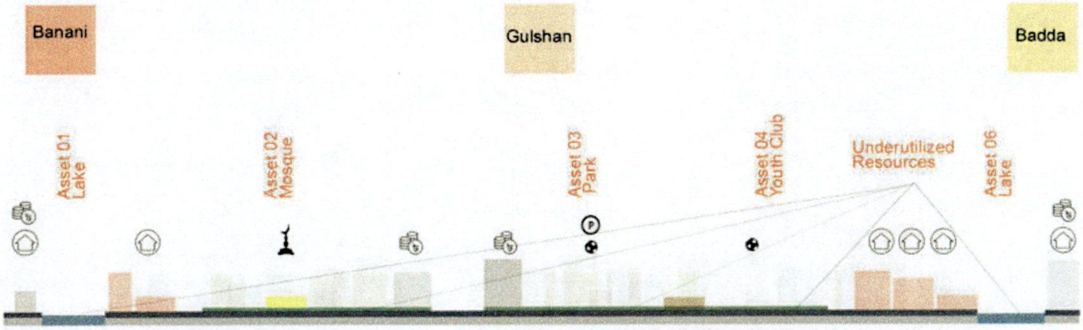

Fig. 12.12 Simplified sectional diagram of connections and accessibility (*Source* Author)

Fig. 12.13 Opportunities with five underutilized/self-indulge urban resources (*Source* Author)

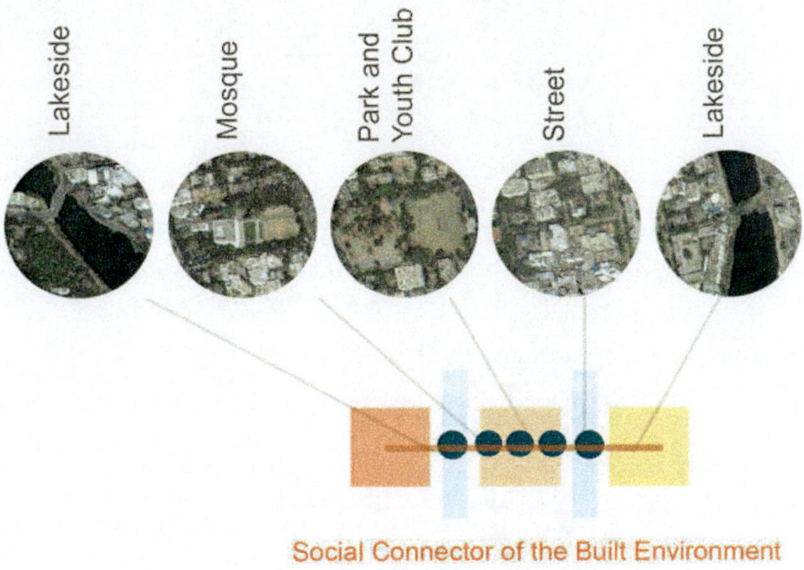

12.4.2 Right to Access and Informality

The paper analyzes urban informality as an "organizing logic" results in a specific mode of the production of space and thus in continuously changing spatiality of power. Based on the analysis of negotiations of access arrangements to public space at one specific locality, informality can be conceptualized as part of a triad of interwoven "spaces," where filters like time, use and access can be a tool of assessment (Kaikobad 2017). This analytical framework is intended to contribute to an understanding of how accessibility and use rights of public space are established in the megacity of Dhaka (Fig. 12.21).

12.4.3 Social Activators of Tactical Urbanism

Pathway

Streets and pathways are considered the backbone and main reservoir of public interactive space and social exchange. (Lydon and Garcia 2015). Tactical urbanism includes various approaches, which point out interim activation initiatives of different property types, developed and undeveloped spaces with micro-scale approaches with retail, arts, and recreation (Elrahman 2016).

Diverse Social Performance

Creating the scope of festivals is one of the established ways of creating relations between

Fig. 12.14 Use map of the existing fabric (*Source* Author)

Fig. 12.15 Activity map (*Source* Author)

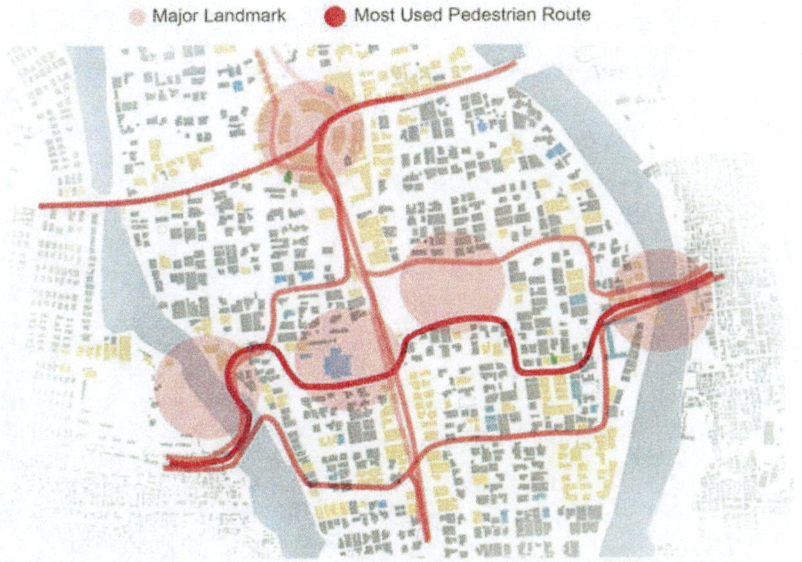

Fig. 12.16 Pedestrian route (*Source* Author)

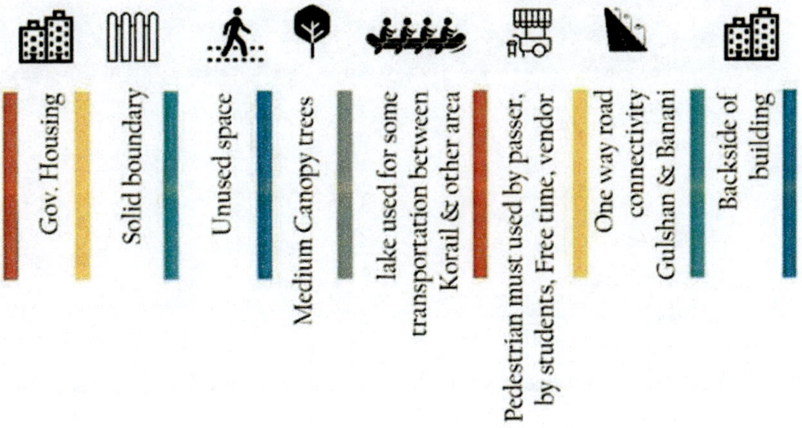

Fig. 12.17 Layers at the node, Banani 11 no bridge (*Source* Author)

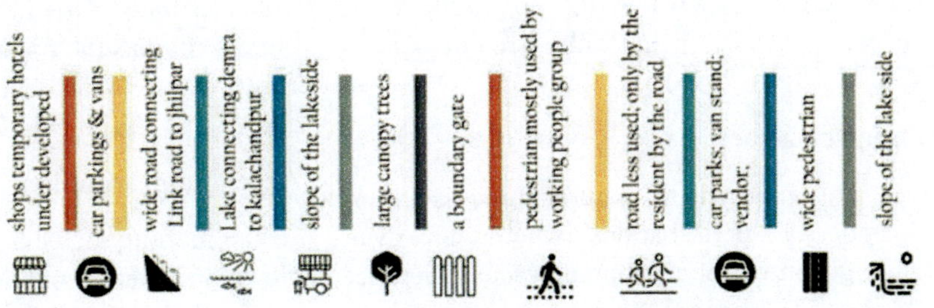

Fig. 12.18 Layers of the node at Badda Jhilpar (*Source* Author)

users from different domains, classes, cultures, and ethnicities (Pineda 2022). Public spaces play a central role in the creation of inclusive communities and more specifically, in the formation of public culture and in enriching cultural diversity (Ravazzoli and Torricelli 2017). An important aspect of creating an inclusive city that accommodates and respects diversity is the

Fig. 12.19 Identify and increase social activities incorporating urban resources (*Source* Author)

public policies focused on that aspect and the official bodies responsible for upholding those policies, especially in the public realm (El-Din and El-Zafarany 2018).

Urban Pop Space

Cities are exposed to natural processes and anthropogenic perturbations where the ever-changing notions often contain vacant, derelict, and negative spaces. This is considered an inevitable byproduct of urban development processes in the context of political, economic, and industrial circumstances. (Kim and Kim 2012; Németh and Langhorst 2014). Such spaces are most likely to occur adjacent to transitional areas between different morphological patterns of urban districts. This integral part of the contemporary city has the potential short- and long-term contribution to tactical urban innovation (Narandžić and Ljubojević 2022).

Access and Mobility

Urban mobility has the potential to create livable cities and better public space (Ravazzoli and Torricelli 2017). According to a report published by the EU, "a sustainable city must have attractive open public spaces and promote sustainable, inclusive and healthy mobility [..]" (European Union Regional Policy 2011). Rebalancing the relationship between public spaces, streets, and

Fig. 12.20 Impact on the wider linkage through activity and inclusion (*Source* Author)

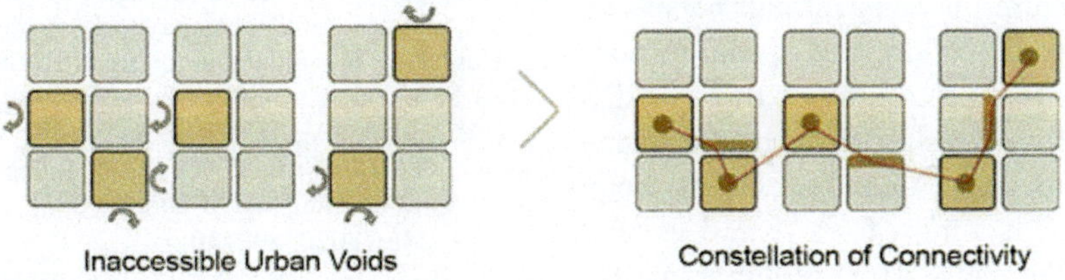

Fig. 12.21 Connectivity and access to resources (*Source* Author)

urban mobility through the integrated use of spaces and multiple means of movement, the reestablishment of social civil quality can be achieved (Ravazzoli, and Torricelli 2017).

Context and Champions

The description of Sigrid Reiter (2004) on three dimensions of public spaces such as coherence, co-existence, and contextuality refers to the idea of a collaborative bottom-up approach of urban space-making. Coherence refers to the degree to which public spaces become places enriched with a collective identification. Co-existence denotes the ability of public places to favor the gathering of differences; and contextuality refers to the integration of public spaces into the local context.

Potential champions and allies' initiatives including academia, professional organizations, and government agencies need to be established in the policy, design, and implementation. Assessment of transformative activities through a collaborative process will identify features through prompt community initiative and temporary installations to establish social interaction and identify problems. These tactical citizen-driven interventions can also test-drive ideas for developing localized spaces/activities through short-term action but with implications in the long term.

12.5 Conclusion

At the livability of Dhaka city, it has become imperative to develop an adequate amount of quality public places, especially at the time when public realm is under challenge. Social interaction is the foundation of social networks, and all kinds of social relationships and community engagements are essential to achieve social sustainability (Bramley 2009). Interactive and inclusive civic spaces trigger a sense of responsibility among citizens with higher social capital. Addressing informal urbanism and increased accessibility of shared resources in policy and planning creates an opportunity to enhance communal and social bonding in the emerging built environment (Kaikobad 2018). The quality open spaces in the city are the crucial driving force for culture and creativity with a sustainable social connection. Through tactical and context-specific urban intervention, minor changes and minimal design with inclusive programming can stimulate momentum for constant and persistent development of space-making and spatial justice.

Heterogeneity is the main theme of a healthy society, addressing diversity in the public realm is a right to all members of the society without discrimination and is a must for a shared society (Ravazzoli and Torricelli 2017). Policy must be revisited to address the multilayered use of potential urban spaces with an incentive-based public and private relationship to ensure inclusivity of space-making. The user-oriented decision-making process must include marginalized groups of society. Planners of the cities need to execute extensive research across the vast spectrum of diversity in society before laying plans and designs. While design is a nonlinear process to address contextual contemporary needs, the top-down conventional urban planning is influencing social segregation. This research will further investigate the symbiotic relationship of contextual urban-built environment with the adjacent community, and its influence on the behavioral pattern of its users for urban rejuvenation through contextual approach.

Acknowledgements The author acknowledges the support of the city people and core user group who generously took part in the survey sharing their valuable information. The author is grateful to Professor Zainab F Ali and Fuad H Mallick, who was inspiration throughout the time. I am grateful to Professor Adnan Morshed to me push boundaries and devote into scholarly researches. The intellectual input of Associate Professor Mohammad Habib Reza realized the research work effectively and efficiently. In the end, I am thankful to all the researchers, who are continuously shaping and enhancing the knowledge procreation process.The research paper is a shortened and adjusted version of S.M. Kaikobad's thesis submitted for Bachelor of Architecture, "Breathing Life in Space: Interaction over isolation" (2017). http://dspace.bracu.ac.bd/xmlui/handle/10361/10263.

References

Alisdairi LK (2014) A cry and a demand: tactical urbanism and the right to the city. M.Sc. University of Washington

Athill D (2010) Beauty and a love of life. People and Places, Commission for Architecture and the Built Environment

Bramley G (2009) Urban form and social sustainability: the role of density and housing type. Environ Plann B Plann Des 36(1):30–48. https://doi.org/10.1068/b33129

Davis M (1990) City of quartz. Excavating the future in Los Angeles. Vintage, London

Douglass M, Daniere A (2008) The politics of civic space in Asia: building urban communities. https://doi.org/10.4324/9780203892770

El-Din AS, El-Zafarany NA (2018) Diversity and inclusion in the public space as aspects of happiness and wellbeing. J Urban Res, vol 28. https://doi.org/10.21608/jur.2018.88384

Elrahman AA (2016) Tactical Urbanism "A pop-up local change for Cairo's built environment". Procedia—Soc Behav Sci, vol 216. https://doi.org/10.1016/j.sbspro.2015.12.032

European Union Regional Policy (2011) City of tomorrow. Challenges visions, Ways foreword

Fitzgibbons J, Mitchell CL (2019) Just urban futures? Exploring equity in "100 Resilient Cities." World Dev 122:648–659. https://doi.org/10.1016/j.worlddev.2019.06.021

Gehl J et al (2006) Close encounters with buildings. Urban Des Int 11:29–47

Grimaldi V, Sulis P (2009) In-between spaces for social interaction—new public realm and the network society. The new urban question—urbanism beyond neo-liberalism. In: The 4th international conference of the international forum on urbanism (IFoU), Amsterdam/Delft

Habib K (2005) The post-colonial public spaces and its cultural diversity: the case of national-cultural representative public spaces of Dhaka

Hackenbroch K (2011) Urban informality and negotiated space—negotiations of access to public space in Dhaka, Bangladesh. https://doi.org/10.1080/02513625.2011.10654019

Hackenbroch K (2013) The spatiality of livelihoods: negotiations of access to public space in Dhaka, Bangladesh. Megacities and Global Change/Megastädte und globaler Wandel—Band 7. https://doi.org/10.25162/9783515103404

Hamilton-Baillie B (2008) Shared space: reconciling people, places and traffic. Built Environ 34(2):161–181

Harvey D (2012) Rebel cities: from the right to the city to the urban revolution. Verso, London

Hintjens H, Kurian R (2019) Enacting citizenship and the right to the city: towards inclusion through deepening democracy? Soc Incl 7:71–78. https://doi.org/10.17645/siv7i4.2654

Hou J (2010) Insurgent public space: guerrilla urbanism and the remaking of contemporary cities. Routledge, New York

Ijla AM (2012) Does public space create social capital? Int J Sociol Anthropol 4:48–53. Retrieved from http://www.academicjournals.org/IJSA

Islam T (2015) Dhaka, the city of rich and poor, an over view of spatial justice: 'whitening' and 'blackening' of spaces. In: Conference paper. World urbanization prospects, s.l.: United Nations

Islam MS, Barua S, Ibrahim MA, Shuvra DF (2015) "From Urban Space to Urban Place": the case of Gulshan South Park, Dhaka, Bangladesh. AIUB J Sci Eng (AJSE) 14(1):163

Joyce M (2012) Shared space in urban environments: guidance note. Flow Transportation Specialists LTD, Auckland

Kaikobad SM (2017) Breathing life in space—interaction over isolation. Submitted as A Partial Requirement for The Degree of Bachelor of Architecture, BRAC University. http://dspace.bracu.ac.bd/xmlui/handle/10361/10263

Kaikobad SM (2018) Street food vending: exploring social inclusion and interaction in informal urbanism. In: Great Asian streets symposium/pacific rim community design network/structures for inclusion

Levebvre H (2003) The urban revolution. University of Minnesota Press, Minneapolis

Lin Y (2016) The strategic urban project approach for informal settlement upgrading in Brazil, Colombia and Indonesia: vision, action and partnership. Chin J Urban Environ Stud 4(2):1650017. https://doi.org/10.1142/S2345748116500172

Low S, Taplin D, Scheld S (2005) Rethinking urban parks: public space and cultural diversity. University of Texas Press, Austin

Lydon M (2012) Tactical urbanism 2: short-term action, long-term change. http://issuu.com/streetplanscollaborative/docs/tactical_urbanism_vol_2_final?e=4528751/2585800

Lydon M, Garcia A (2015) Tactical urbanism: short-term action for long-term change. The Streets Plans Collaborative, Inc. Pages 26, 36, 90, 171

Mitchell D (1995) The end of public space? People's Park, definitions of the public, and democracy. Ann Assoc Am Geogr 85(1):108–133

Narandžić T, Ljubojević M (2022) Urban space awakening—identification and potential uses of urban pockets. Urban Ecosyst 25(1). https://doi.org/10.1007/s11252-022-01219-6

Newman P, Kenworthy J (1999) Sustainability and cities. Overcoming automobile dependence. Island Press, Washington D.C.

Oxoby R (2009) Understanding social inclusion, social cohesion, and social capital. Int J Soc Econ 36(12):1133–1152

Pineda VS (2022) What is inclusive and accessible public space? J Public Space 7(2):5–8. https://doi.org/10.32891/jps.v7i2.1607

Purcell M (2008) Recapturing democracy: neo liberalization and the struggle for alternative. Ur-ban Futures. Routledge, New York

Ravazzoli E, Torricelli GP (2017) Urban mobility and public space. A challenge for the sustainable livable city of the future. J Public Space 2(2):37. https://doi.org/10.5204/jps.v2i2.91

Sennett R (1992) The fall of public Man. W.W. Norton & Co Inc., New York

Shaftoe H (2008) Convivial urban spaces: creating effective public places. Earthscan, London, Sterling

Short JR (2021) Social inclusion in cities. Front Sustain Cities Specialty Grand Challenge. https://doi.org/10.3389/frsc.2021.684572

Siddiqua A (2011) Emergence of open-spaces for dense Dhaka: searching for solutions from traditional settings. Thesis submitted for the degree of bachelor of architecture at the Ahsanullah University of Science and Technology, Bangladesh

Stutter N (2017) The social life of street food: exploring the social sustainability of street food in Hanoi, Vietnam. Thesis is submitted in fulfillment of the requirements for the degree of Doctor of Philosophy School of Geography and Planning Cardiff University

The 10 Happiest Cities in The World (2013). Retrieved from Fast Coexist: http://www.fastcoexist.com/1681359/the-10-happiest-cities-in-the-world

Vic Health Mental Health & Wellbeing Unit (2005) Social inclusion as a determinant of mental health and wellbeing. VicHealth, Sydney

UN-HABITAT (2008) State of the world cities 2010/2011. Nairobi: Earthscan on behalf of United Nations Human Settlements Programme (UN-HABITAT)

Wilkinson M (2003) Social determinants of health: the solid facts. World Health Organization, Geneva

Part III
Re-framing Design

13 Imagining Urban Village Design Grammar: Unravel Pattern Language to Form an Alternative Approach to Desakota's Urbanism

Satria Agung Permana,
Syarifah Ismailiyah Alathas,
and Ilya Fadjar Maharika

Abstract

The discourse on urbanism in Indonesia is still in its infancy. Existing development disparities and difficulties between urban and rural areas motivate people to migrate to large cities in search of opportunities. As a result, urban growth has been dispersed, making settlement supply challenging and forcing people to reside in *desakota*s, which are urban residual spaces (urban villages). Regardless of the subjective quality of environmental well-being, the design of these spaces revealed an exploratory spatial quality. If these inventive uses of space disappear, it will be a pity that municipalities have such extensive and practical plans to convert these areas into new ones. In search of an alternative strategy for future development, this essay examined a critical record and analysis of spatial quality using *desakota* in Jogoyudan, Yogyakarta, as an illustration. This essay argues for inventive vocabularies to advocate *desakota* urbanism development in Indonesia by applying Christopher Alexander's Pattern Language lenses to decode and improve the design grammar.

Keywords

Desakota · Urbanism · Pattern language · Design grammar · Yogyakarta

13.1 Introduction: The Urgencies of the Desakota Design Framework

Typically, Indonesian cities consist of formal and informal areas known as *desakota* or *kampung*, an urban village. Rapid urbanisation made it possible for villagers to gradually migrate to larger cities in search of opportunities and a better quality of life. In contrast, surviving in a city is difficult; they also lack insurance and social security. After arriving without a place to live, they built a shelter in urban, informal open areas, gathered more people, and gradually became communal settlements (Dobbins 2009). They were eventually contained in illegally constructed squatter settlements that the government tolerated. As a result, the city became fragmented and dispersed, leading to complex social and environmental issues.

In the latter days, there were two major streams of discourse: pragmatic top-down participatory planning and romantic bottom-up participatory

S. A. Permana (✉)
Department of Architecture, Delft University of Technology, Delft, Netherlands
e-mail: permanasatriaagung@gmail.com

S. A. Permana · S. I. Alathas · I. F. Maharika
Department of Architecture, Universitas Islam Indonesia, Yogyakarta, Indonesia

S. Hilal et al. (eds.), *Design for Partnerships for Change*, Sustainable Development Goals Series,
https://doi.org/10.1007/978-3-031-36993-3_13

planning. In Jakarta, for instance, two contradictory projects are currently underway: the forced eviction of former Kalijodo residents to make way for public space and the revitalisation of Kampung Akuarium, which depicts a more modest mixed-community apartment. In a separate case in Yogyakarta, Ayodiya (2014) found that the majority of Code riverside kampong residents opted to remain rather than relocate. The government has also considered implementing a tourism strategy in kampong, albeit with only temporary support. In addition, the majority of flat house programmes designed to revitalise the kampong area into a single large structure are occupied by newcomers rather than the original residents.

In the past, YB Mangunwijaya took an innovative approach to riverside kampong preservation, assisting residents in rebuilding the settlements. This project was awarded the Aga Khan Award in the 1990 cycle for its modest social and cultural improvement that thwarted the government's eviction plan. In her book *Tempat Terbaik di Dunia*, Roanne van Voorst describes her other harrowing experiences while living in a poor neighbourhood on the riverbank of Jakarta. She explains the vital role they play in the community ecosystem due to the strong sense of communal ownership, togetherness, hospitality, and helping one another. Future alternative design lenses in architecture and the urban realm could be influenced by the collective sense and how we perceive design and space. However, the optimal strategy for executing proper planning is still being sought. This paper speculates the use of a pattern language to envision grammar as a medium or strategy for comprehending micro formal structures that play as the "cultural DNA" and source of inspiration for future urbanism development.

13.2 Method and Materials: Design Grammar as a Medium

Design grammar is frequently argued as a computational or algorithmic design process in architectural discourse (Pauwels et al. 2015) As seen in Fig. 13.1, design grammar modifies specific design rules to achieve a desirable or effective design outcome. In further discussion, Pauwels elaborates on George Stiny's early use of grammar as analytic rules to generate additional shapes. To employ shape grammar, the designer must possess basic cultural DNA, which will be referred to as grammar later in this paper, as a starting point for exploring the potential design within the discussion. However, designing in the *desakota* field requires user and designer dialogue and synthesis. In this instance, design grammar would closely describe the cultural DNA of each design language observed in

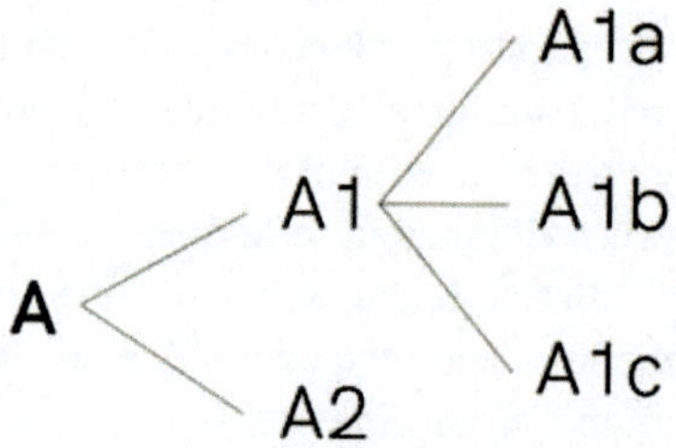

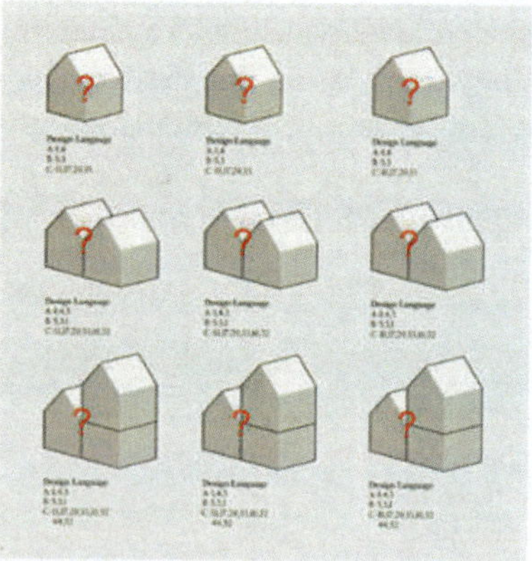

Fig. 13.1 Grammar and design permutation. *Design permutation deliver options during design process as a result of collection of options from one DNA/grammar. The grammar, hence, harmonised with another grammar to process a new design form* (*Source* authors' drawing)

the site's daily life. Or perhaps harmony between grammars can produce a distinctive design without compromising the context's identity.

An architect or designer is not an artist who can solve a complex design problem (Lawson 2005). Lawson mapped design issues that involved the designer, user, client, and legislator to ease the tension. A branch of possibilities results from this top-down relationship. In this situation, the designer may take on the planner's role or choose to collaborate with the community. Multi-discipline integration is necessary when working with another stakeholder, and communicating a design mind is difficult as well, especially with Indonesian literacy levels still below average (Kemenko 2021). As a result, the designer should indeed work hard to promote the plan. We were surprised to discover a variety of creative space production techniques in the wild, demonstrating that their capacity to produce organic design is also creative in terms of adaptability, such as making the congested street in between a vibrant public area.

In top-down planning, residents are compelled to accept and live in a generic community; it is a form of instant architecture, contextless design desired by certain communities due to rapidly changing pop culture (Maharika 2018), which is frequently found in private–public development. In Indonesian communities, the eastern tradition of viewing one's neighbours as members of one's family is still very much alive. This type of typology is, therefore, inappropriate for them. Under these conditions, bottom-up planning is currently the most effective method of controlling this development, with the designer acting as an activist or mediator to translate and provide them with extensive support. Unlike the preceding, this requires a substantial amount of time and resources. What if, then, the designer and the residents could discuss the design using the same grammar? Then, we could incorporate the designer's forward-thinking ideas into a design that is familiar to locals.

This question prompted the discussion to delve into Christopher Alexander's Pattern Language in order to comprehend community space production as a bridge between stakeholders in the design process. Alexander never advocated design as a collage; he viewed patterns as a complex interaction between humans and their environment. To decipher the genetic code of the built environment, Alexander examined the entire pattern, from the regional scale to the architecture. Alexander simplifies narrative pattern components into language that the majority of individuals can comprehend. The pattern, rather than the function, addresses the specific activity and options in order to facilitate the formation of space relationships by accommodating those options. Due to the fact that literacy in Indonesia is still on the rise, a visual image is preferred to complete the narrative, which is subsequently referred to as design grammar and language. Language exists because the level of identities and differences serves as a basis for further development by means of continuity, resemblance, repetition, and adaptation (Foucault 1966). Each design of tectonics should therefore have its own grammar (Fig. 13.2).

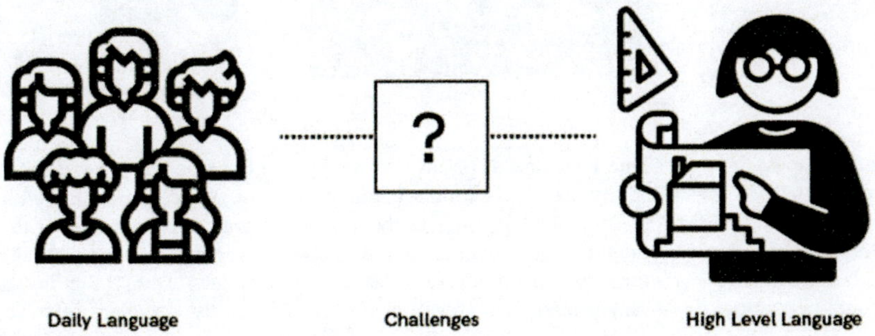

Fig. 13.2 Challenges between two worlds. *What the resident say? What the architect say? What the government say? What the other actor say? This questions are often found in the discussion. How to come up with same language become the challenge* (*Source* authors' drawing)

Design grammar is perhaps part of the more general concept of pattern in which type is then developed for architectural design. As Plowright has elucidated, architectural patterns incorporate social information into the formal composition of buildings. A space's shape, volume, adjacency, characteristics (light, sound, textures, mood), and distribution reveal how a person uses it (where elements are found within the space) (Plowright 2015). In this sense, following Durand, design grammar perhaps is also a tool for communicating people (and their culture) and architects (with their preconception of type, of geometry of space and mass, etc.).

13.3 Exploration of Desakota Design Grammar in Jogoyudan

Consider the work of Balkrishna Doshi at Aranya Community Housing, he designed each home based on the evolving and adaptable preferences of each resident. Half-to-Complete Housing, designed by Alejandro Aravena, allows the user to complete the design. The grammar of these designs allows people to romantically incorporate what they know into the built environment, which may create a loop that makes it difficult to adapt to future uncertainty.

This persistent problem's solution is a language property with specific recurrence, an archetype; it is discovered, not created. The pattern eventually evolved into a synthesis. Although it is too early to declare it an approach, it may serve as an alternative source of inspiration for enhancing the design process through a better understanding of each perspective. As a case study, we will examine the examples of each pattern at three different Jogoyudan scales such as village area, neighbourhood, and alleyways architecture.

As shown in Table 13.1, these innovative and novel patterns would not exist without their organic adaptability to manage spaces over years of survival in a small space. We could document

Table 13.1 Three example patterns found in Jogoyudan

Pattern Communal Public Street *Scale* Village Area	*Pattern* Everywhere is Playground *Scale* Neighbourhood	*Pattern* The Back Door *Scale* Architecture

Because of the limited space in Jogoyudan, the residents made the most of the street's potential. With fewer vehicles, the street is not only a commuting space but also an interactive communal network	Due to a lack of public spaces in Jogoyudan, the children made use of every available space in the village, including the street. In some cases, they even use the rooftop as a playground	Because the organic street network resembled a building with no right front door, a backdoor that also functioned as the front door was created. During the day, the door is mostly left open. However, only certain people, such as family members or relatives, have access to it

Source field observations, 2018

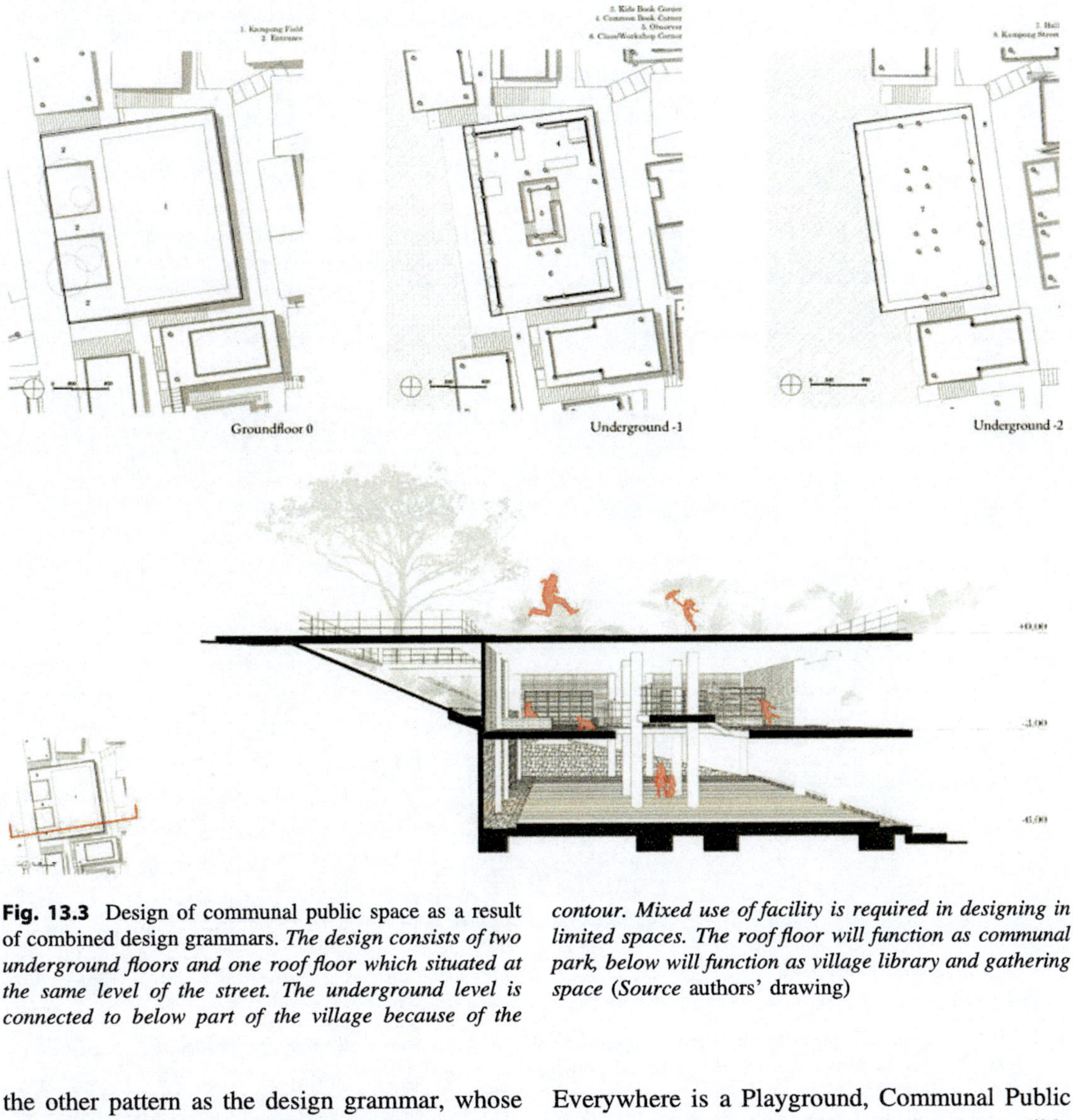

Fig. 13.3 Design of communal public space as a result of combined design grammars. *The design consists of two underground floors and one roof floor which situated at the same level of the street. The underground level is connected to below part of the village because of the contour. Mixed use of facility is required in designing in limited spaces. The roof floor will function as communal park, below will function as village library and gathering space* (*Source* authors' drawing)

the other pattern as the design grammar, whose numbers are still evolving due to the large number of distinct spaces created. The focus then shifts to grammar usage, which requires additional research.

As an illustration of application, this essay presents a strategy for redesigning certain areas of Jogoyudan. It was necessary to address the lack of public space. Due to space constraints, we must maintain the limit in a way that allows for a variety of activities. In this instance, we multiplied and combined one function of the area with the field and village centre. Using the grammar of Everywhere is a Playground, Communal Public Street produced a mixed-use design compatible with the existing contour (Figs. 13.3, 13.4 and 13.5).

13.4 Conclusion

Reading *desakota* or heterotopia space necessitates a particular technique in order to obtain the necessary DNAs. Instead of a conclusion, we conclude this essay with a reflection and a provocation to form concerns about these

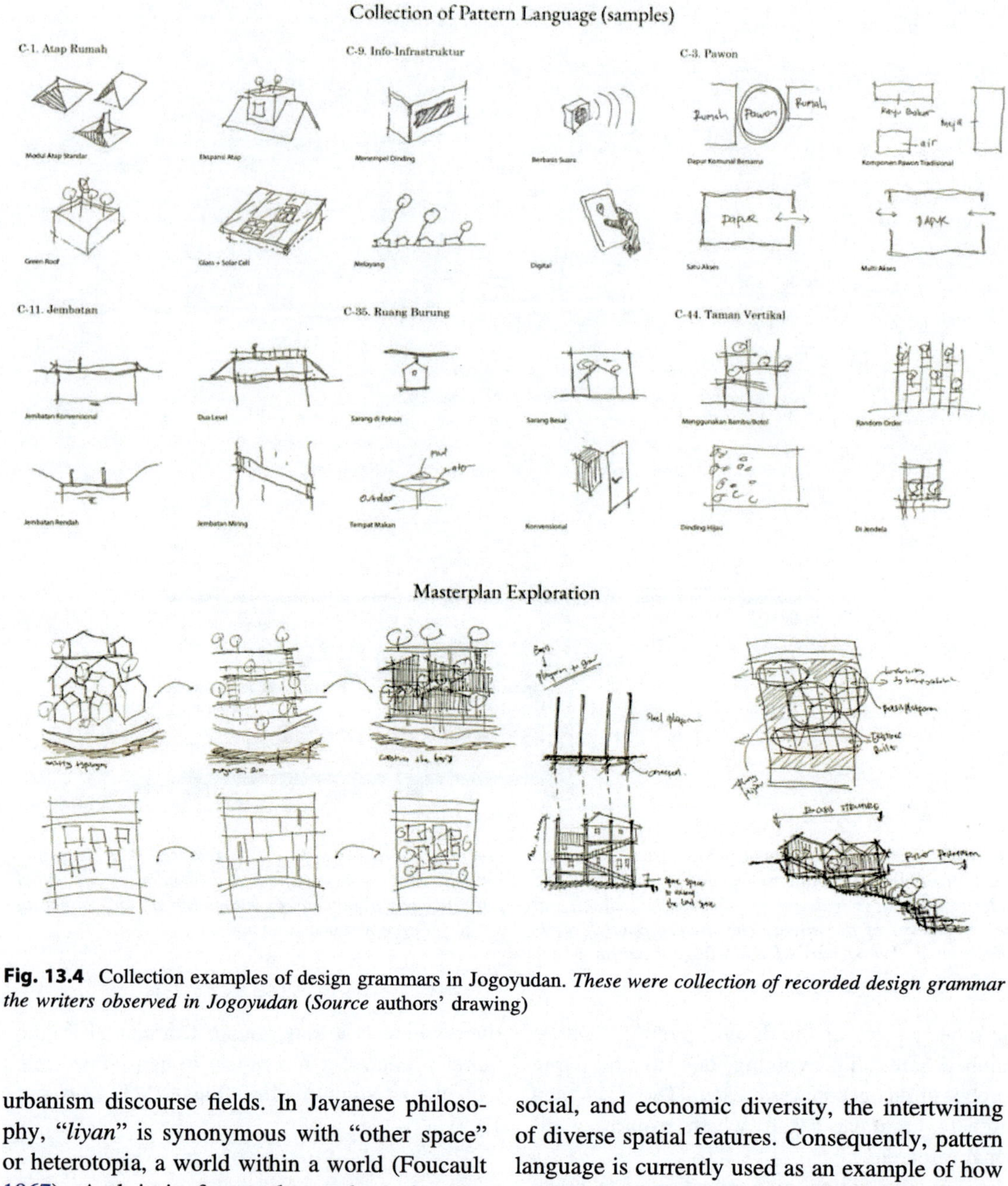

Fig. 13.4 Collection examples of design grammars in Jogoyudan. *These were collection of recorded design grammar the writers observed in Jogoyudan* (*Source* authors' drawing)

urbanism discourse fields. In Javanese philosophy, "*liyan*" is synonymous with "other space" or heterotopia, a world within a world (Foucault 1967). And it is frequently used to describe informal settings. Culturally, *liyan* is embraced on multiple levels, as a unified entity that celebrates the life of difference. It is about spatial relationships, not visuals. As a result of cultural, social, and economic diversity, the intertwining of diverse spatial features. Consequently, pattern language is currently used as an example of how to read the spatial conclusion of what the space is intended to be, necessitating a grammar that is as simple as possible to comprehend. However, there are alternative ways to advance urbanism discourse in Indonesia.

Fig. 13.5 Collection of design permutation. *From those grammars, new design was developed as a product of harmonisation between the languages. The redesigned parts were some buildings in Jogoyudan area (Source* authors' drawing)

Acknowledgements This essay is a reflection of the authors' undergraduate thesis project, which was supervised by Ilya F. Maharika and Syarifah Ismailiyah Alathas. Their constructive guidance throughout the works is greatly appreciated. Not forgetting the reviewers, Yu Sing and Wiryono Raharjo, who also expand comprehension. In addition, Gregory Bracken who guided the supporting writing and theory for this reflective writing. And finally, the LPDP Scholarship from Indonesian Government that made this study journey possible.

References

Ayodiya NRP (2014) Policy model for north code village settlement on the bank of code river. J Reg Urban Dev 10(1):22–23

Alexander C, Ishikawa S, Silverstein M (1977) A pattern language: towns, buildings, construction. Oxford University Press, New York

Dobbins M (2009) Urban design and people. Wiley, New Jersey

Foucault M (1966) The order of things: an archaeology of the human sciences. Les mots et les choses: Une archéologie des sciences humaines (1966). Translated from the French by Jay Miskowiec. MIT Press

Foucault M (1967) Of other spaces. "Des Espace Autres," March (1967). Translated from the French by Tavistock/Routledge. Routledge

Kemenko PMK (2021) Indonesia's literacy level is concerning, coordinating ministry for human development and culture prepares a national literacy cultivation roadmap. https://www.kemenkopmk.go.id/tingkat-literasi-indonesia-memprihatinkan-kemenko-pmk-siapkan-peta-jalan-pembudayaan-literasi

Lawson B (2005) How designers think. Biddles, Ltd., Oxford, United Kingdom

Maharika IF (2018) Umranisme: exploration of architectural intentions to build Adab. Universitas Islam Indonesia, Yogyakarta

Pauwels P, Strobbe T, Eloy S, Meyer RD (2015) Shape grammars for architectural design: the need for reframing. Shape grammars for architectural design: the need for reframing | SpringerLink. Retrieved 14 Jan 2023, from https://doi.org/10.1007/978-3-662-47386-3_28

Plowright PD (2014) Revealing architectural design methods, frameworks and tools. Routledge

Design for Fragility—13 Stories of Humanitarian Architects

14

Esther Charlesworth and John Fien

Abstract

This paper explores the themes of fragility, design, disaster and architecture as explored by the authors new book: *Design for Fragility: Thirteen Stories of Humanitarian Architecture*. The book profiles thirteen built projects that have transformed the social and economic fabric of the communities' lives for whom the projects were built and sdesigned with. *Design for Fragility* sought to go beyond just detailing the architects' motivations or final design/built form of the project. The aim was to investigate these thirteen diverse projects and innovative design practices to understand what implications they might have for architectural practice at large, through analysing:

1. The experiences and perceptions of geopolitical fragility—or precarity—that directed the particular spatial response by the architects.
2. The specific typology of the project, whether that be a housing, health, children's, or a First Nations project.
3. The personal values that influenced the architects to work on humanitarian/community projects and how consultation occurred with often contested project stakeholders.
4. The experiences of the design team as well as project managers, occupants, and donors of the built project, exploring what they deemed successful about the project, and what, if any, were its limitations.

Design for Fragility builds on Charlesworth's *Humanitarian Architecture: 15 Stories of Architects Working After Disasters* (Routledge 2014), which explored the role for architects in exercising 'spatial agency' while designing shelter and settlement projects for communities after disasters. Since that time, the humanitarian architecture movement has expanded globally with the prominence of design agencies including the MASS. Design Group and Architecture Sans Frontières (ASF) International. Supporting the rise of humanitarian architecture has been the emergence of dedicated master's degrees in the last decade that are training the next generation of design and transdisciplinary professionals to work in the disaster and development sectors. These include the RMIT University Masters of Design, Disaster and Development [MoDDD] in Melbourne, Australia, and the UIC Master of International Cooperation Sustainable Emergency Architecture in Barcelona, Spain.

E. Charlesworth (✉) · J. Fien
Department of Architecture, Rmit University, Melbourne, Australia
e-mail: esther.charlesworth@rmit.edu.au

S. Hilal et al. (eds.), *Design for Partnerships for Change*, Sustainable Development Goals Series,
https://doi.org/10.1007/978-3-031-36993-3_14

Keywords

Disaster architecture · Design for Fragility · Humanitarian architecture

14.1 Purpose

This paper explores the themes of fragility, design, disaster and architecture as explored by the authors new book; *Design for Fragility: Thirteen Stories of Humanitarian Architecture*. The book profiles 13 built projects which have transformed the social and economic fabric of the diverse communities for whom the projects were built and designed with. They include:

1. The SOS CHILDREN'S VILLAGE by Urko Sanchez Architects in Djibouti
2. The HABITAT FOR ORPHAN GIRLS by Zav Architects in Iran
3. BHOLU 16 by The Anganwadi Project in India
4. The MATERNITY WAITING VILLAGE by The Mass Design Group in Malawi
5. The ANANDALOY CENTRE by Studio Anna Heringer in Bangladesh
6. The SYNAPSE SUPPORTED ACCOMMODATION INNOVATION FACILITY by People Oriented Design [POD], Australia
7. THE SANTA PROJECT by Breathe architects in Australia
8. The GREEN SHELTER PROJECT by Yasmeen Lari in Pakistan
9. THE PILOT SHELTER PROGRAM by UNHCR+CRS in Uganda
10. THE GAHINGA BATWA VILLAGE by Localworks in Uganda
11. 'BILYA KOORT BOODJA' [Centre for Nyoongar Culture and Environmental Knowledge] by Iredale Pedersen Hook in Australia
12. The CAKAUDROVE WOMEN'S RESOURCE CENTRE by Architects Without Frontiers in Fiji
13. 'The 20 K PROJECT' or 'DAVE'S HOUSE' by Rural Studio in the USA

Our book *Design for Fragility* sought to go beyond just detailing the architects' motivations or discussions of the design final built form of the project, through exploring:

1. The experiences and perceptions of social/political/environmental fragility—or precarity—that provided a design challenge and directed the particular spatial response.
2. The specific typology of the project, whether that be a housing, health, children's, or a First Nations project.
3. The personal values that influenced the architects to work on humanitarian/community projects and how consultation occurred with diverse and often contested project stakeholders.
4. The experiences of the design team as well as project managers, occupants, and donors of the built project, exploring what they deemed successful about the project, and what, if any, were its limitations.

This book follows Charlesworth's *Humanitarian Architecture: 15 Stories of Architects Working After Disasters* (Routledge 2014), which analysed the role for architects in exercising 'spatial agency' while designing shelter and settlement projects for communities after disasters. Since that time, the humanitarian architecture movement has expanded globally with the prominence of design agencies including the MASS Design Group and Architecture Sans Frontières (ASF) International. Supporting the rise of humanitarian architecture as both a design movement and practice has been the emergence of dedicated master's degrees that are training the next generation of design and transdisciplinary professionals to work in the disaster and development sectors. These include the RMIT University Masters of Design, Disaster and Development [MoDDD] in Melbourne, Australia, and the UIC Master of International Cooperation Sustainable Emergency Architecture in Barcelona, Spain.

14.2 Fragile States

More than two decades ago, Esther began her research and design practice on cities destroyed by civil conflict, disasters, and social marginalisation while living in Beirut, 2000–2002. Since that time, the themes of war, terrorism, trauma and rebuilding after floods, fires and tsunamis have grown exponentially in the public consciousness. This has been driven by the compounding impacts of social inequality caused by globalisation, the enduring impacts of colonialism, worldwide increases in inter-ethnic conflict, and the devastating impacts of climate change.

Twenty years ago, we saw the destruction of the World Trade Centre and the Bamiyan Buddhas (Afghanistan). The following years witnessed wars in Iraq, Afghanistan, Sudan, Syria and Yemen and, since the Indian Ocean Tsunami of 2004, we have experienced the Global Financial Crisis of 2008 and devastating earthquakes, droughts, hurricanes, floods and tsunamis from Haiti to the Philippines. It seemed almost inconceivable that yet another 'un-natural disaster', the 2020–2002 COVID-19 pandemic, could further deepen already existing social, economic, and environmental fragilities, globally.

As we completed this paper in December 2022, Ukraine is still under siege from Russian troops with over 13 million Ukrainians internally displaced or fleeing to neighbouring countries in search of shelter, peace, and a semblance of safety and stability. In the three-month period from February to May 2022, floods engulfed our own 'backyard' in Eastern Australia with many residents, especially the less privileged, living on the cheaper land of floodplains, now displaced and calling themselves 'climate refugees'. At the same time, the small island states of the Pacific and the Caribbean including Tuvalu and the Bahamas are preparing for permanent climate-induced displacement. While each conflict, famine and disaster is radically different in its causes, scale and wider impacts, feelings of helplessness seem to pervade most of us watching the news from afar.

Yet, there are signs of hope that offer us light amongst the sheer human bleakness of war, disasters, and pandemics—and architecture is emerging as one of the elements in that 'light' through its capacity for spatial and social transformation.

This might seem a grand claim about the capabilities and potential of the architecture profession. Indeed, only eight years ago, when working in the Philippines after Typhoon Haiyan for World Vision, architect Brett Moore lamented: '[T]he number of architects, planners and landscape architects equipped to work with disaster and development professionals in rebuilding in the aftermath of conflict, political and social instability, floods, fires, earthquakes, typhoons and tsunamis is still chronically low'.[1] Yet Brett Moore acknowledges the light that architecture is offering to deal with these 'wicked problems', through the emergence of new models of innovative design practice education. Building on these complex geopolitical and spatial challenges, *Design for Fragility: 13 Stories of Humanitarian Architecture* explored thirteen design projects from Rwanda and Iran to Bangladesh, Fiji, and Australia as evidence of the transformational changes underway in architecture as the profession begins to respond to the complex challenges of our fragile planet.

14.3 Why Fragility?

'Fragility' might appear to be a non-academic term to use in this paper and as the focus of future priorities for architectural practice and education, especially when words such as 'marginalisation', 'hazard', 'risk', and 'vulnerability' are *de rigeur.* We chose to use 'fragility' deliberately to develop a spatial perspective on the concept of the 'fragile state' and to demonstrate in a practical way how architecture can be used as both a political and spatial tool for building community resilience, before during and after disasters.

The OECD defines 'fragile states' as situations in which increasing exposure to serious

[1] Charlesworth (2014).

risks combined with insufficient coping capacity prevent governments and/or communities from managing or mitigating those risks.[2] In this definition, fragility is both caused by and, in turn, intensifies crises related to poverty, conflict, and displacement, the breakdown of civil institutions, displacement, and extreme environmental hazards. This reflects a 'systems thinking' approach to the concept of fragility in terms of its multi-dimensional properties across economic, environmental, political, cultural, and social causes and impacts.

These systemic properties of fragility are found in the diverse geographic and cultural contexts of the thirteen case studies in 'Design for Fragility'. However, all of the projects profiled demonstrate how appropriate architectural interventions at key crisis points, can redirect a community's recovery from a state of fragility to a state of stability and resilience. As we see in the interviews with architects, this involves identifying the complex drivers of risks in the social and ecological setting they are working within, and then developing design projects to address the specific impacts of those risks on people's lives. This can also be achieved through developing design and infrastructure projects that mitigate exposure to future ecological or social threats in such a way that the adaptive coping capacities of communities and families are strengthened.

To take but one example of the thirteen case studies: the Gahinga Batwa Village designed by Felix Holland and LocalWorks. Designed as a settlement for eighteen Batwa families dispossessed from their culture in the mountain forests of Rwanda to make way for a national park and gorilla tourism, the village and houses not only replicate a culturally appropriate typology for the Batwa, but also provide the land tenure and economic security for them to help build their own homes and develop skills in a sedentary agricultural economy. This is what humanitarian architecture has the capacity to do for communities and families impacted by disaster and poverty.

Traditionally, interventions in fragile states have been of the 'macro-kind'—building integrity and transparency in government, managing the causes of violence through mechanisms for conflict resolution, and building bridges, roads, and other hard infrastructure. A humanitarian approach to architecture is more modest in scale but is the same process. Importantly, this involves viewing the communities with whom we work as partners not 'clients' and collaborating with them from the initial schematic design phase of any project and then long after our architectural work is actually completed.

14.4 Why Design Matters

The case studies in *Design for Fragility* range from the Gahinga Batwa Village in Rwanda to the design of a maternity village in Malawi and then to a design/build programme for poor rural communities in the rural south of the USA. Together, the thirteen stories exemplify the growing cadre of architects using design to facilitate the transition from fragility to resilience. The narratives and interviews also represent diverse geographic and business models of architecture, ranging from commercial architectural firms (e.g. Iredale Pedersen Hook [IPH], Breathe, LocalWorks, and People Oriented Design [POD]) to architects who have largely shifted their projects into the international development and humanitarian fields (e.g. the MASS Design Group, Yasmeen Lari, and Phoebe Goodwin). What unites the selected architects is a desire or 'spatial agency'—as Jeremy Till[3] frames it—to work with and for communities impacted by war, poverty, fires, floods and dispossession in order to rebuild their livelihoods through modest scale housing and civic infrastructure projects.

These are not 'one-off' or token projects by the architects that we have selected. Rather, they represent critical choices of design collaborations that have created long-term social, environmental and economic impacts for the disaster or poverty-impacted communities, into which they were invited to work.

[2] OECD (2016).

[3] Schneider and Till (2009).

Design for Fragility is a companion to *Humanitarian Architecture: 15 Stories of Architects Working After Disaster*, which analysed the origins of humanitarian architecture and the expanding role for architects in designing projects for communities after disasters. Since then, the humanitarian architecture movement has expanded across design education and practice with the prominence of firms such as the MASS Design Group (named the 2020 Architecture Innovator of the Year by *WSJ Magazine* and the 2021 Collaborative Achievement Award by the American Institute of Architects) and the rise of the national organisations associated with *Architecture Sans Frontières International.*

Humanitarian Architecture is thus now recognised as pioneering a new era in design and spatial agency. Supporting this rise in social activism in design is the body of theory developed since the Indian Ocean Tsunami in 2004 by architects such as Jeremy Till, Marie Aquilino, Cynthia Smith, Cameron Sinclair and Eric Cesal.[4] Their books and projects have variously identified the relationship between design and emerging theoretical constructs including spatial agency, design responsibility and urban resilience. Linked to this rise in the discourse of social agency and architecture have been the significant global policy platforms of 'The New Urban Agenda' and development of the 17 Sustainable Development Goals' (SDGs). Both policies recognise that ending poverty is a complex task and must go hand-in-hand with strategies that improve health and education, reduce inequality, and encourage economic growth—while simultaneously tackling climate change and protecting our increasingly fragile world.

14.5 Through Post-Colonial Eyes

As we completed *Design for Fragility*, British architect, Lord Norman Foster had visited Kharkiv in Ukraine mid-2022 to discuss the city's rebuilding through a master plan developed by 'the best minds with the best planning, architectural, design, and engineering skills in the world'. Yet, with most of that city's residents displaced by the Russian invasion, exactly who the future city is for, and what housing, health, education, business, and civic infrastructure they will want or need, are far less clear? Can design be effective on an almost *tabula rasa* site such as Kharkiv? Apparently so because Lord Norman is but one of a long list of architects signing up for potential recovery projects in Ukraine, just as they were in Sri Lanka after the Indian Ocean Tsunami in 2004 and in Haiti after the 2010 earthquake.

This 'trauma-glam' syndrome[5] is a strong reminder of the failings of the Marshall Plan developed to rebuild Europe after World War II and the zeal to rebuild cities like Warsaw, even when the bombs were still falling before that war had actually ended. While it is deeply admirable, and even sometimes necessary, to create a spatial footprint of a city's future beyond war or disaster, the architects in this book would no doubt stress a need for caution and humility in the ways we work with vulnerable communities, and understand Bruce Nussbaum's question 'Is humanitarian design the new imperialism?'.

Are designers the new anthropologists or missionaries, come to poke into village life, 'understand' it and make it better, their 'modern' way? Is the new humanitarian design coming out of the USA and Europe being perceived through post-colonial eyes as colonialism? Are the American and European designers presuming too much in their attempt to do good?

14.6 How do we know what we know and according to whom?

Acknowledging the diverse impacts of the case studies is the central way in which this book seeks to make its mark. However, architecture is not usually regarded as an evidence-based profession. As Ariadne Labs commented in an

[4] Awan et al. (2011), Aquilino (2012), Smith (2007), Sinclair (2006), and Cesal (2010).

[5] See https://www.dezeen.com/2022/04/21/norman-foster-reconstruction-kharkiv-ukraine-war/.

evaluation study commissioned by the MASS Design Group, health care and other designs 'are rarely informed by empirical evidence of what does and does not work. The consequences of a limited evidence base include the propagation of designs that are low in value, expensive and even potentially harmful to patients'. And that was a report primarily on designs for health and medical facilities and, if design for that sector is deficient in its approach to evaluating impact, little attention beyond the occasional post-occupancy evaluation, can be expected of design for other sectors.

Similarly, Brett Moore (Chief of Shelter, UNHCR) has often lamented to us that the lack of evaluation data on the impacts of alternative designs for certain refugee shelters compared with the extensive data sets on the impacts of health, education, social protection and legal support for refugees. This situation makes it comparatively difficult for the shelter sector to argue for additional resources despite the fact that the health, education, safety and family stability of refugees depend upon adequate and appropriate housing. Related to this, the late International Federation of the Red Cross (IFRC) Chief of Shelter, architect Graham Saunders, always emphasised the critical need to evaluate shelter programmes as a basis for them to be 'scaled up', so that the design of a refugee settlement, a typhoon-resilient house, or a shelter for dispossessed children could be recontextualised and adapted for potential use elsewhere.

The issue of the need for evidence-based design in architecture is a complex one, unlike in medicine, law, or engineering where it is non-negotiable in developing future health, legal or structural policies and solutions. Our effort to analyse impact assessment reports on the thirteen case study projects was frustrated by the lack of any systematic evaluations done on the projects except for the Maternal Waiting Village in Malawi. Strong anecdotal evidence was provided for the other twelve projects, which was rich and authentic and reflected the lived experience of project benefits. However, it was not comprehensive or independent which meant that there is often little evidence to guide architects in the design of future projects or reporting to donors in ways that might actually encourage future architectural commissions. In an era of massive cost-cutting in the built environment and design sectors, there is often little budget or will to undertake independent evaluations of the impact of projects. Nevertheless, the COVID-19 pandemic has taught us that numbers do count, such as in determining what is the best vaccine or policy directive to deal with the consequences of the disruption it caused to both rich and poor, globally. However, the wisdom of transferring the concept of 'vaccine treatments' to architectural practice is a problematic analogy and indicates that newer, qualitative, and community design-based approaches to evaluation are critically needed.

14.7 Poverty does not Exclude Aesthetics

'You are not a 'real architect'! We have often heard this throw-away line from colleagues in the design profession regarding why one would choose to work in the humanitarian or international development sectors. They see working for vulnerable communities, whether paid or pro bono, as something to do after hours or as marginal, at best, to the greater mission of corporate architecture. This perception is thankfully slowly changing through the rise of extraordinary design practices such as the Mass Design Group in the USA and Breathe in Australia; firms that deliberately undertake pro bono design projects as part of their organisational structure.

In all thirteen practices and projects explored, innovative design and beauty have been key drivers of the design of spatial solutions for families and communities impacted by war, poverty, or disasters. They are not projects that result in beautiful yellow trace sketches but no actual building. Rather, many of the projects and architects profiled in this book, including Anna Herringer, Urko Sanchez, Mass Design, Breathe and Iredale Pedersen and Hook, have all won esteemed design awards for their built works, including the including Aga Khan Award for

Architecture and the American and Australian Institutes of Architecture awards.

However, in addition to such accolades, the sheer beauty in the execution of the designs of these thirteen projects has significantly transformed the lives and livelihoods in the communities where they were built. Indeed, if the aesthetics of architecture can do so much culturally, mentally, and spiritually for the one or two per cent of the general public who are generally the only ones able to afford architects, then it is easy to imagine the significant additional impacts that architectural beauty can bring to the lives of the poor and marginalised. Without such visceral and spatial delight embedded in our housing, health, and civic buildings, they could be seen to become structures to exist in, rather than buildings that exalt us.

References

Aquilino M (2012) Beyond shelter: architecture and human dignity. Metropolis Books, New York

Awan N, Schneider T, Till J (2011) Spatial agency: other ways of doing architecture. Routledge, London

Cesal E (2010) Down detour road: an architect in search of practice. MIT Press, Boston

Charlesworth E (2014) Humanitarian architecture: 15 stories of architects working after disasters. Routledge, London

OECD (2016) States of fragility 2016: understanding violence. OECD Publishing, Paris

Schneider T, Till J (2009) Beyond discourse: notes on spatial agency, in Special issue on 'Agency in architecture: reframing criticality in theory and practice'. Footprint: Delft Archit Theory J, Issue 4, pp 97–112

Sinclair C (2006) Design like you give a damn. Metropolitan Books, New York

Smith C (2007) Design for the other 90%. Cooper Hewitt, New York

Steps to an Ecology of Design

15

Angelo Bucci

Abstract

The environment in which we live is almost all designed. Everything around us comes from a design process that, as defined by Thackara in his book "*In the bubble*," determines 80% of the environmental impact of a product or service. Why design process generates this major environmental impact? Probably the main mistake made is to consider the object of design as an object by itself, without considering all those dynamics, relationships and interactions that a project generates and the influence that the designed buildings, objects and services, have on society, environment and economy. Burkhardt, in his book "*Design is Invisible*," points out to us that no project is neutral, all projects point society in a direction, and this fulfills designers with responsibility, who, in every project, should make everything fit into a system of cultural, political, economic and environmental sustainability that should be intelligible in the designed product or service and that, through their presence in the world, are capable of directing change toward a better future. This paper aims to analyze what might be the tools of a new design paradigm that will enable future designers to engage with the complex dynamic system in which we are involved, being aware of the fact that each individual project must play a role in a broader vision and preserve a flexibility necessary to accommodate the diversity and changes that will arise. Interdisciplinarity and foreshadowing of future scenarios will be the main tools of a necessary meta-design upstream of any project.

Keywords

Paradigm shift · Foreshadowing future scenarios · Partnership · Interdisciplinarity · Speculation · Creative approach · Complexity

15.1 Steps to an Ecology of Design

Most of what surrounds us are the result of design: from objects to spaces to intangible elements such as services, travel, and networks. The world we live in, our environment, derives from design choices, some of them more conscious and some others less, which have shaped our existence and will continue to do so, determining an ecology (understood as the relationship between human beings and the environment, which is increasingly the built environment), made up of constructs and beliefs that are not always sustainable.

A. Bucci (✉)
Università Europea del Design Pescara, Montesilvano (PE), Italy
e-mail: a.bucci@dezignstudio.it

S. Hilal et al. (eds.), *Design for Partnerships for Change*, Sustainable Development Goals Series,
https://doi.org/10.1007/978-3-031-36993-3_15

Often it is the design process that makes this system unsustainable since it is this phase that, as defined by Thackara (2005) in his book "*In the bubble,*" determines 80% of the environmental impact of a product or service. But why does the design process generate this large environmental impact?

In the field of architecture and design, a belief that needs to be reconsidered is that the object is designed as an element in its own right, simply fulfilling a function and falling into the aesthetic category of "beautiful." In fact, this purely modernist view does not allow us to think about the actions and interactions that the product generates around it, or how we can structure a new culture of beauty that, as Tamborrini (2009) suggests in his book "*Design sostenibile—oggetti, sistemi e comportamenti,*" makes us appreciate more what is sustainable, in a vision that is more concerned with society, the environment, and people.

The main theme of design, in short, should be the development of processes rather than products: processes of awareness toward environmental protection, social and cultural processes, processes of sustainable development that start from the grassroots and from people, because, as Bateson (1972) says in his book "*Steps to an Ecology of Mind*": "…ours is not the only way of being men (and women, ed.): it is conceivable that we can change it."

This hoped-for and necessary change toward a new paradigm capable of holding together environment, society, and economy must be stimulated through abstract ideas capable of maintaining that necessary flexibility which, again to quote Bateson, "harmonizes with that of the environment to give rise to a complex dynamic system, open to gradual changes in even fundamental characteristics."

Again, Bateson, during a lecture in 1970 (New York, 1970, "Restructuring the Ecology of a Big City"), gives a definition of "High Civilization," in which he explains that civilizations change and evolve and that it is unwise to think of going back to living like the Australian Aborigines because it would take us back to the same point we are today, but we must somehow maintain the level of satisfaction of our needs and what he calls the "wisdom" acquired in the course of evolution. For this reason, we cannot think of stopping designing; on the contrary, we must design with an ever-increasing attention to diversity, which, again according to Bateson, creates "the flexibility and 'pre-adaptation' necessary to deal with unpredictable change." Resource use, respect for the global complex system, and sustainability, in its broadest sense, should, therefore, be the design beacons of this hoped-for civilization.

Unlike all those designers of the Modern Movement who mistakenly believed that the perfect design would solve all the world's problems and issues, it is flexibility, the ability to adapt, to change, that becomes pivotal.

In the field of Product Design, as early as 1972, in the MoMA exhibition curated by Ambasz "*Italy: the new domestic landscape,*" those designers whom Ambasz called protesters and nonconformists—motivated by a deep concern for the role of designers in a consumerist society—pushed their research and proposals toward the theme of flexibility. This found fulfillment in objects capable of relating to environments, of transforming themselves according to the needs of a precise moment, to the point of prefiguring a new nomadism of young people, ready to be rootless and willing to change.

But, although very interesting and perfectly consistent with the times, those design researches found great difficulty as the attention of companies and markets continued to focus on the aesthetics and primary function of the object, on that "good form" that has polarized the development of the discipline of Design and Architecture, diverting everyone's attention away from the experiments of the late 1960s and early 1970s, which had much to do with what we would now call resilience.

The "antagonistic" approach pursued by those young designers and architects had a political role in society and sought to question what was the kind of future that has to be designed, before designing the objects. The importance of this approach can be found in a research by Carl Disalvo (2012), of Massachusetts Institute of

Technology (MIT), entitled "Adversarial Design," in which the author makes us think about the importance of the antagonistic approach in democracies and how this approach (of course non-violent but critical-constructive) can be implemented through design. In his research, Disalvo shows us how there can be a Politic Design (which he differentiates from Design for Politics) and how this tool is capable of bringing about discussions and consciousness-raising. He draws our attention to the political role of design, in short, a role that was clear to architects and designers in the early twentieth century but seems to have disappeared from our horizon today, barring sporadic instances.

What has happened to those antagonistic visions? Why does design have such a low political value?

Today we live in a product society, everything is designed but everything is thought out considered only by focusing on the object of design, without questioning what this creative act generates in the complex system. The result is overproduction, of products and services, which often fail to focus the frames that enclose the problem they are meant to solve, and so generating more problems, in terms of pollution, cultural depletion, resource exploitation, and so on. Although the objects of design have to fulfill functions, we can no longer overlook the importance that these products and services have for society in terms of raising awareness, developing a path, and communicating; we cannot pretend that the product of design does not change the world, that it is neutral, as Burkhardt said (Burkhardt, "Design is invisible," 2012).

We must resume considering design as a tool for reflection, rather than as the solution to something, leaving each design with its own degree of flexibility that will allow it to adapt to the peculiar situations of that historical period, to technical and technological possibilities, to environmental and societal issues, in short: being flexible in design means not being self-referential but opening up to the world and leaving useful room for maneuver so that the product of design is not yet another object thrown into the world, but interacts on multiple levels with society, becoming a stimulus for reasoning, discussion, and evolution.

All this is made possible when, before the design of the product itself, a meta-design is developed involving many more figures, from different disciplines, capable of developing design scenarios toward which the design will be directed, both in the case of products and architectures and, of course, services.

These scenarios, therefore, cannot be developed by architects or designers or engineers alone; rather, there is a need for other figures to join these designers so that the scenario reaches the right degree of viability and is culturally, politically, economically, and environmentally sustainable. They assume great importance, for example, in scenario prefiguration: sociologists, anthropologists, engineers and energy experts, economists, agronomists, geologists, and the list could go on and on.

We must, as Bateson says, accumulate an "uncommitted potential for change," and that is precisely the purpose of having a well-structured future scenario in mind. Moreover, in order for this potential to always be at its highest level, scenarios need to be reviewed and critiqued continuously, so that they are updated periodically, again ensuring the necessary flexibility.

The problem of constructive criticism of products, architecture and design in general is, today, another important issue. At a time in history where political correctness has become the dominant culture, we need to pay more attention on what is a critique, able to rise a discussion and new points of views, and what an offense. As I can see, today there is a fear of posing criticisms about what is produced by design, reducing the possibilities for reflection on the existence of other approaches or other solutions. This is because critique is increasingly viewed in its negative sense rather than as a time for reflection and eventual evolution.

The exercise of prefiguring futuristic scenarios is never an easy exercise, but it always leads, at the very least, to reflections, which, unfortunately, we are not addressing.

Trapped in a need for immediacy of results and certainty of goals, we have forgotten the importance of long-term planning. In fact, also even though foreshadowing future scenarios has always been difficult, given the ease with which they can change, having a long-term goal allows, at the very least, to direct common efforts toward a hopefully better future than the one we are living. This is planning: projecting forward into a vision and beginning the journey to reach it.

When we go walking or climbing up in the mountains, we do not just look at our feet for fear of stumbling, but we must, often, look up to see if the direction we are taking will lead us to the right path. Here, this is the same purpose of scenario prefiguration, we do not have to forget to look down, so we deal with the day-to-day issues, but without losing sight of the ultimate goal and also changing our steps, our decisions, to reach it.

Design, in most cases, has lost this capacity for vision and has been subjected to market needs that have caged it in the very short term, in producing and selling to go right back to designing, in a vicious circle that does not allow for in-depth research, limiting possible action to solving technical issues more than conceptual ones, let alone having the time for scenario prefiguration and for comparison with other professionalism!

This, I believe, has been the greatest loss of design in recent decades, and this is what prompted me to write about the importance of Design in society. I am very far, therefore, from those who say that one should no longer design; on the contrary, one should do it much more, but with consciousness and awareness, comparing with other professions on possible futures or, simply, on how to stimulate more and more people to have a purposeful approach to the future.

I have always liked to define the designer as an accumulator of information to be returned to society through projects, and this definition traces, in part, the definition of "bond-formers" given by Thackara in his book "*In the bubble*." But to form bonds or to give back information, we must be aware of our role and we can't approach design out of that glossy vision someone has of this profession (and which, by the way, in 90 percent of cases isn't true at all).

Regaining consciousness of one's role means, first and foremost, regaining possession of the founding values of design, of the cultural and social aspect of this discipline, of its close relationship with the community, a community that must be listened to and not used as a purchasing basin, a community of propositional consumers (prosumers) who, however, must be brought out of the shadow cone in which they find themselves, and sensitized to the fact that they do not have to be necessarily passive market players, because they are the market.

To this end, the objects and services designed must become moments of reflection, first of all for the designers in the moment of design and, secondly, for the people who will use them. Thus, the designers to come will have to learn how to get out of the trap of "good form" as an end in itself and begin to reflect on good form that is also a stimulus for change, which is capable of becoming an action and of having a political role in education, stimulated by the designs themselves. That it is an incentive to improve interactions between people, in society. In short, we will have to pay much more attention to those invisible elements of design that direct us to new behaviors, new habits, and new paradigms to be structured.

Interdisciplinarity becomes increasingly important for this very reason: we cannot know everything but we need to be together and design together to achieve such goals. Sharing ideas and comparing different visions allow everyone to be an active part in building the reference scenario, participants in the change, each with his or her own knowledge and contribution to the structuring of a path.

Of course, we cannot expect to arrive at such a level and planning capacity immediately.

Arriving at such a prefigurative and design level, in fact, involves a path of "deutero-learning" (a process through which one "learns to learn") that involves trial and error which must be taken into account.

If this path had started in 1972 (the date of publication of "*The limits to growth*," the research paper that, first, tried to raise awareness of the concept of sustainability) we probably would have had the time needed to absorb the mistakes, due to the deutero-learning process, and focus everything on raising people's awareness, without the need for obligation or coercion.

But this, unfortunately, has not happened, and because of the time lost and, to prevent the time needed for change and the inertia of the system from taking us over the edge, legislative actions are needed to interrupt certain behaviors and help designers find a field of action ready to accommodate a different approach to design.

However, even if political support is really important today, it is still necessary a real education about behaviors and encouragement of certain choices so that design becomes more than just drawing objects, buildings, or systems.

The tools and places to stimulate these processes exist, even if they are underexploited in this sense, and they are all those places of meeting and confrontation, from universities to trade fairs.

Unfortunately, the situation of research, not only in Italy, is not the best because it is subject to market rules as everything else. The most funded fields and research are those that have the prospect of generating a fairly quite large market in a fairly quite short period, and therefore, we fall back on the usual short-term view.

The path seems even more impassable if we look at it from the perspective of the private individuals, companies, or trade shows. No matter how many meetings, seminars, workshops are organized, the goal is always just to publicize some product or some brands, taking away as much space as possible for discussion.

But past experiences have always taught us that great ideas and perspectives have always come from dialectical clashes and even the most heated confrontations. In the Deutscher Werkbund, for example, the dialectical battle over "moral" or "immoral" ornamentation was not an easy one to deal with, and it was never completely resolved. Anyway, all the discussions and contrasts developed in the Deutscher Werkbund were a stimulus to the Modern Movement, the Bauhaus, and all the movements that followed.

When Ambasz came to Italy for his project, he was faced with a real conflict, with very sharp positions of some designers, and hard contrasts, very often, in the discussions that arose around Product Design and consumer society. Out of those contrasts came some of the most important figures in the well-known Made in Italy Design.

Why, today, do we have so much difficulty in confronting each other and exposing opinions and reflections, even personal ones or simply in contrast to mainstream thinking?

What seems to us to be a comfort zone, comfortable and without too much thought, is perhaps annihilating us and, accustomed to living this way, we are now addicted to this condition in which it seems more comfortable to do as we have always done, because changing things seems impossible.

During a lecture, D. L. Meadows (who was head of the group of researchers who published the book "*The Limits to Growth*") gave a very simple example of what should be done to safeguard our planet. Meadows asked the audience to cross their arms. Done the first time, he asked them to do it again, and after that had happened, he put the focus on the fact that, sure enough, the position of the arms was the same as the first time. This, he explained, happens because it is a habit. Then he challenged the audience to cross their arms the opposite from the way they had already done it and to repeat this experience three or four times. In conclusion, he emphasized how this simple example manifests a profound difficulty for humans to change their habits, when asked, however, we do it with difficulty the first few times but, as we do it again, a new habit will begin to structure itself in our brains, a new pattern, which, eventually, will become the norm.

What if we try crossing our arms in a different way?

It would not take that long for a new paradigm to become a habit as well.

At the end of the day, the first step needed is willingness.

This willingness for change must take shape through the structuring of working groups that have the aim of defining future scenarios, through speculative design, and begin to define strategic planning of the steps to take to direct us toward a possible and preferable future.

As Peccei (founder of the Club of Rome and funder of the research "The limits to growth") said, either we wake up and start directing our efforts in a direction or we will have to submit to the transformations that will take place (and some are already taking place) hoping, simply, that everything will be okay.

I firmly believe that anyone who calls himself or herself a creative or a designer feels, now, the blood boiling in his or her veins, watching things from a perspective that makes them feel useless. The designer has and must have, always, a momentum to do something to improve. Well, now it is the time.

Now it is the time for creative people, of all fields (because creative people are not only artists or designers, but they are also those who, in their field, are able to go beyond the obstacle, harnessing intuition, reasoning, intra-logic, experience, creativity in short), to begin to determine their role in society and make it clear how important their contribution can be in the future perspective, how design culture, understood as not limiting oneself to today but thinking about the future, can help this and future generations so that they can continue to design in turn.

These are the aims of this design partnership: to have an explorative approach able to define how to sensitize people and influence the complex system we live in with small and big changes, and to do so, a new sensibility needs to be instilled.

If we want to break down those erroneous assumptions that Bateson posits as the "roots" of the ecological crisis (ranging from: "Man vs. environment"; "Us vs. other men"; "It is the individual that matters"; "We can unilaterally control the environment"; "We can expand indefinitely"; etc.) we must begin to form a new aesthetic and social sensibility toward the world (natural and built) and truly perceive ourselves as part of this world. It is precisely this concept of consciousness formation that so many important authors refer to, and it is the key element in enacting real change to move us toward sustainable development.

We, therefore, need scenarios, general ideas, before we need products and services, and these scenarios will have to be flexible and revisable when necessary, so that they do not lock up the practical activities, the strategic planning resulting from them, in a structure that is too rigid and, consequently, too ill-suited to the fluidity of complex dynamic systems. We must indulge, as planners and as people, in this flexibility and make ourselves advocates of the necessary changes and any mistakes we might encounter, and stop thinking that we have the solutions for everything, that's such an egocentric vision. Let us begin this process of training from ourselves and not make excuses when we are called upon to develop a new project, because, as Bateson says, "ecological processes cannot be mocked."

Let us form prosumers through our projects, let's educate them about acceptance, conscious use of resources, social ethics, sustainability, and, above all, let us train new designers to have different, divergent, better visions of the future and give them the tools to implement them.

This, I think, is the job that is incumbent on designers, even when they are designing the simplest object, and that is to make it fit every time into a system of values of cultural, political, economic, and environmental sustainability that are intelligible in the designed product or service and that, through their presence in the world, are capable of directing change toward a better future.

Everything we do is part of a complex system with counterintuitive behaviors, maybe it will take little or maybe it will take a lot of commitment and sacrifice, but now it is the time for a change to take place and it is good for design and architecture to be an active part of the change.

Don't you think so?

Bibliography

Ambasz E (1972) Italy: the new domestic landscape—achievements and problems of Italian design. New York Graphic Society

Argan GC (1951) Walter Gropius e la Bauhaus. Einaudi

Bateson G (1972) Steps to an ecology of mind. Chandler Publishing Company

Burkhardt L (2012) Design is invisible. Blumenthal S, Schmitz M (eds)

De Fusco R (2007) Made in Italy—storia del design italiano. Edizioni Laterza
Disalvo C (2012) Adversarial Design. The MIT Press
Florida R (2011) The rise of the creative class. Basic Books
Maldonado T (2008) Design Industriale, un riesame. Feltrinelli
Meadows DH, Meadows DL, Randers J, Behrens WW (1972) The limits to growth. Potomac Associates
Tamborrini P (2009) Design sostenibile—oggetti, sistemi e comportamenti. Electa
Thackara J (2005) In the bubble—designing in a complex world. MIT Press

Check for updates

Regional Criticism: A Passive Resistance to Planetary Urbanization

16

Chensi Shen

Abstract

While the theory of Planetary Urbanization is valuable in understanding the realities of widespread urbanization and capital expansion, it also constructs an over-linked spatial approach, ignoring non-urban processes and the subjective efforts of architects. Emphasizing the specific regional character and taking the rural imagination to a higher level, Regional Criticism is a factual resistance to global urbanization. However, this resistance often becomes powerless to declare in a socio-economic context with its substance of compromising globalization. The isolated and scattered resistance based on architectural forms does not shake the social foundation and is difficult to have a voice. Although the resistance of Regional Criticism successfully portrayed non-urban living landscapes but did not create spaces outside the urban, the practice ultimately confirmed the validity of the Planetary Urbanization theory. This article uses Wang Shu and Lu Wenyu's rural reform project "Wandering Wen Village" as an example to illustrate how a typical Regional Criticism practice encountered difficulties in reality and ultimately absorbed them as part of the urbanization process. By analyzing how this resistance interacts with local communities, capital and political stakeholders, the article argues to rethink sustainable design ideologies in a changing social context. Rethinking the role that architects should play in sustainable transformation can be aided by an interdisciplinary approach to analysis, which can help architectural discussions go beyond the confines of the manifestation practice and attempt to understand and affect the socio-economic-political processes of an entire territory.

Keywords

Planetary urbanization · Regional criticism · Community participation · Interdisciplinary approach · Sustainable design

16.1 Introduction

Tariq Jazeel advises caution concerning the Planetary Urbanization theory, arguing that the urban theory "without an outside" brings a research methodology that can be overly urban-centrism that is limited within the urban studies knowledge production (Jazeel 2017). Although it contributes to understanding the globalized urban process, it simplifies the complicated, broad and

C. Shen (✉)
Tianjin University, Paris 1 Panthéon-Sorbonne University, Tianjin, China
e-mail: chensi.shen.tju@gmail.com

S. Hilal et al. (eds.), *Design for Partnerships for Change*, Sustainable Development Goals Series,
https://doi.org/10.1007/978-3-031-36993-3_16

multiple-dimensional social-spatial analysis in the reality of multiple modernity. Using the example of Geoffrey Bawa's architectural practice of Regional Criticism in Colombo, which goes against the colonial nature of the city and towards a pure imagination of the countryside, Jazeel states the practice of Regional Criticism provides an outside "supplement" to the Planetary Urbanization theory. By presenting an attempt to create an urban "outside", like Regional Criticism in architectural discourse, he asks for a multidisciplinary conversation to also create an "outside" for urban studies so that the city and urbanization are not the only lenses through which spatial studies are conducted.

On the other hand, Regional Criticism and the vernacular approach to architecture that it encourages are frequently seen as environmentally friendly sustainable strategies. Such an idea originates from the natural adaptation of its forms and materials to the environment and society, the non-industrial mode of building and material production, the sensitivity to the environment, the ancient energy wisdom, and the aesthetic character of the cultural rebuttal to modernization, globalization, and colonial influences. It also represents an approach to design that respects the community by adapting materials and spatial features familiar to residents. This article aims to demonstrate, however, that Regional Criticism has not been successful in establishing what Jazeel refers to as "the city outside", even though it has been conceptualized as a kind of resistance to widespread urbanization. On a practical level, it only appears in a rebellious look but proved to be assimilated by globalized urbanization eventually. Its sustainability has also been called into question, particularly when architects have gone to greater measures to replicate traditional craftsmanship and believe that the residents will continue to live a particular way in pursuit of a certain ideal. However, significant logic is revealed from its specific social and economic process and the obstacles encountered by architects. To realize the critical-historical perspective of "architecture as an institution", as Tafuri (1977) puts it, the architectural critique can also benefit from urban studies setting that offers a more comprehensive political-economic analysis, such as Planetary Urbanization.

Taking Wang Shu and Lu Wenyu's rural reconstruction project Wandering Wen Village as an example (Xu 2019), this article tries to identify the issues emerging in this resistant process, to argue that the existence of Regional Criticism as an ideology attests to the challenge that the subjective efforts of spatial producers pose to Planetary Urbanization theory as a methodology, even though that challenge does not succeed in its purpose in reality. The causes for this failure must be examined in light of the complexities of the local community and the complex process of realizing the architectural design. For star architects in particular, the fame label has become a type of capital in the process of architectural realization. The architect's strong personal position, the community's ineffective social participation, and the simplification of the local developer's vision of a better future living all have impacts on the project. It requires not only diversity in design and participation but also an interdisciplinary manner of thinking in architectural critique to comprehend this power structure and effectively develop sustainable architectural models.

16.2 An Ideological Resistance

Regional Criticism, as a theoretical school, began in the 1960s. It appeared as the resistance to global production and indiscriminate landscapes (Zhu 2013), trying to create a contemporary regionalism that awakens territorial characteristics. On the other hand, Planetary Urbanization is a theory that emerged after the 2000s, proposing that in a universally connected globalized capital economy, urbanization has become the most important social space process on the planet's surface and has encompassed all other natural, geographical, and economic processes. The uneven urbanization spatial development cannot be captured through discrete regional models and the idea of dual urban-rural separation is no longer applicable. A broadly connected perspective should be used to look at cities and the process of urbanization (Brenner and Schmid 2015).

Although the former is not a direct resistance to the latter, the Regional Criticism, which aims to nail its practice with the territory to oppose mobilization, naturally resists the connection tendency of Planetary Urbanization. The over-linking implication of Planetary Urbanization theory leads to the disintegration and demise of meaning. Planetary Urbanization explains the reality from an over-constructed and over-processed perspective, paying too much attention to connections, forgetting about entities and obscuring self-logic within the territory. The excessive construction of the political economy ignores the social and cultural level, which creates forces of resistance at the local level. Although it needs to be acknowledged that, a wide range of geographical methodologies has practical significance to help us to understand the reality of the process of capital expansion through cities as spatial tools, the systematic efforts of Planetary Urbanization theory have obliterated the "non-systematic" elements and the possibility of other non-urban spatial processes.

Therefore, it's precisely because people are trying to find self-worth and identity in an increasingly homogeneous and all-connected global system that Regional Criticism has a presence as existence. This desire is reflected in the resistance to the endless reproduction of globalized images in space. However, it is not so much a de facto resistance to globalized urbanization as it is an ideological one for four reasons.

First, the premise of Regional Criticism is "region". This demarcation of boundaries is a direct negation of a no internal–external methodology.

> For Heidegger, the boundary is not the line at which something stops, but rather the contour within which something begins its 'presenting'. (Frampton 1983)

The boundary of the region is where the regional culture begins to "present", and the differentiation that Regional Criticism attempts to evoke can only be manifested through isolation. However, borders also limit diffusion effects and drive isolated resistance to become a weak declaration. Architects are limited to the areas they are rooted in, need to work in a familiar context and devote strong energy and emotional connection to territory. This limitation is reinforced by the requirement to use local architectural language and materials, which makes it challenging to be copied in a circulating commodity market. Therefore, "anti-association" Regional Criticism cannot be produced worldwide quickly, for instance, the work of a global architect like Zaha Hadid easily gains the right to speak.

Secondly, Regional Criticism itself has the nature of a global sense. Since appeared, it has attempted to create a coordinated regionalism while approving world values with the essence of compromise. It hopes to reconcile the contradictions between global and national identity. It couldn't resist becoming part of universal value, especially when indigenousness becomes one of the means to showcase the cosmopolitan. For instance, to establish a "world city", Doha invested in several museums exhibiting indigenousness, trying to create an image of indigenous Islam but also welcoming to the world (Levitt 2015). Furthermore, Massey proposes that extensive spatial and time connections have been "compressed" in the process of "place" generation. Global production and consumption are involved in the construction of a sense of place (Massey 1991). At this level, every territory has mobility and connectivity and pure unpolluted indigenousness doesn't exist.

Thirdly, "place" and indigenousness are difficult to exist as fixed entities. The definition of a place may require the exact point in the space–time coordinate system and extraction from a certain period in history. This extraction is based on cultural aesthetics and represents a discipline of lifestyle, a template for architects to create an ideal life through spatial guidance. It requires delicate sensibility and has an elitist tendency, which can easily become the arrogance of the middle class, causing separation from the territory itself.

Indigenousness could be encouraged by the political institution, for it symbolizes important regional identity in the development process. However. it can't be much too fragmented, for the state wants to create a simplified and unified identity. Media dissemination also causes distortion. If the regional architectural practice has become a destination of online celebrities and global tourists, then should we consider it a success in highlighting regional characteristics or a failure to retain indigenous life?

Moreover, the resistance to Regional Criticism is prone to encounter many obstacles in practice. Due to the interference of politics, economy, social relations, media, capital, institutions, etc., it is very easy to shake its original intention and be assimilated into part of urbanization itself. Its consistency often dies during the production and consumption of space, leaving floating symbolic resistance to imagination and compromise. Regional Criticism has a nature against the globalized economic system and marketization. However, architecture is the co-product of the socio-economic process and if in practice, regionalism is a critique that only appears in the form, it will likely be absorbed by the globalized environment and can only be a so-declared resistance.

Therefore, the Regional Criticism is more of ideological resistance, but not meaningless. It shows us the difficulty of resisting the disorderly spread of urbanization, especially when this tendency has been deeply embedded in the socio-economic soil of the post-neoliberal era. Regional Criticism proposes a right beyond cities and urbanization, a right to villages and territories, and a right to identities and memories.

At a time when the fruits of modernism are spreading throughout fast urbanizing areas and beginning to provoke criticism on many sustainability issues, Regional Criticism architects are being embraced by the architectural community and their practise and ideas are being encouraged by architectural schools. It is considered not only environmentally sustainable but also a prescription for reviving cultural values and community relations. Although it employs vernacular language, it varies from authentic vernacular architecture in that it is still created by trained professional city architects, goes through a whole industrial process, and then culminates in the presentation of a circulating photographic product. As it intervenes in the vernacular society with a rebellious position towards urbanization, how does this approach work in reality, especially viewed and participated by the local communities? What exactly is the outcome of the typical Regional Criticism practice's attempt to create "urban outside"? How should the sustainable value of the process be assessed? It is also interesting to discuss the word resistance itself: resistance must be a rebellion against some fait accompli, so it first acknowledges the victory of its opponent. Therefore, ideological resistance is hard to realize its claim perfectly in reality. By analyzing the economic and social processes in the rural practice of Wang Shu, I will try to clarify the success and failure that occurred with the "resistance" of Regional Criticism in the real world.

16.3 The Rebellious Architects with a Rebellious Project

Since his student days, Wang Shu has appeared as a resister, using radical language to criticize the discipline of architecture in China, "the negation and refusal attitude has always been consistent" (Li 2012). As representative architects of Regional Criticism, Wang and Lu's architectural aesthetics have always been rooted in the territorial logic of China. They even use this language in urban public buildings, creating a heterogeneous spatial experience of "rural painting" to express the anti-regulation of urbanization. In 2012, they began their first rural reconstruction practice "Wandering Wencun Village" in Fuyang, Hangzhou. In 2019, Wang was interviewed for this project, with the report titled "The Pritzker Prize Winner Wang Shu's Rural Salvation Warfare: The Core of Chinese Culture is in the Country, Not in the City" (Xu 2019).

The architect expressed his attitude directly against urbanization and its dominance. However, the project started ironically because of the city itself. As soon as Wang won the prize, Fuyang District went to him with a museum project in the suburb, for the memory of "old mountain life", as part of the new city plan. He agreed but proposed to let him reform a village in exchange. The government of Zhejiang province approved that, for they hoped to have a successful template, as "a reproducible model of a beautiful village in Zhejiang region". This even sparked a competition between local governments as they competed to invite the most renowned architect. Wang chose Wencun, a very remote, inconvenient village, with agricultural culture and genuine traditional texture, to intentionally "avoid excessive commercialization, urbanization and tourism" (Xu 2019). A such choice had opposing voices within the government because they wanted to invest in public projects with more development values, but Wang's reputation made them compromised.

Wang Shu's research aims to construct a sustainable mentality of learning from the past. This research has been done for 10 years in the villages in Zhejiang Province by architects and their students. "Learn from the ancient village and know how to conserve land so that everyone can develop an intensive, ethical, and self-disciplined way of life", says the architect. "Make the new house feel like it has grown spontaneously out of the existing structure." (Xu 2019) Wang tried to rebuild rural civilization in Wencun. To encourage villagers to join, the local government invested in buying original homesteads and subsidized families who chose to participate in the project. The team tried to encourage the participation of the local community from the very beginning. A team of students stayed in the village for several days, living and eating with them together to observe and interview their life routines. At the sketch stage, the architects have been visited villagers and ask about their opinions, then change their designs accordingly, even if it is just a roof or a wall.

However, Wang Shu's dream of restoring a traditional, culturally informed, and sustainable life is difficult, especially now that contemporary life has become the dominating voice. Although Wang Shu finds that the villagers living in the new patios consist a beautiful scenes as ancient paintings, it was difficult to convince villagers to accept the old civilization when they yearned for a "modern and wealthy" life. "We don't want old wet patios. We want to live in modern European houses." (Xu 2019) The villagers believed the courtyard design, which was translated from traditional architecture, became a waste of homestead. To persuade villagers to accept, the local government made concessions, changing their policy to ensure the yard within 10 m^2 would not be calculated in the homestead area, so they can enlarge their living spaces. There were also problems in rural social relations. The designers' 1.8-m canopy was just built, and a neighbour knocked it off with a knife and refused to compromise because he thought it was an invasion of his private space. "The villagers would like larger bedrooms or living rooms, but Wang Shu doesn't agree. He thinks there should be life separated in different functional spaces." (Xu 2019) Only half villagers accepted to buy the house after its accomplishment, as they think the designed spaces with the idea of the architects are not easy for their life. The villagers who bought the house also changed the original space: the second and a half storey, originally designed for drying and storing rice, was sealed up by the villagers and transformed into a room with a terrace; a glass roof was also installed above the open patio (Xu 2019). Finally, Wang Shu and Lu Wenyu accepted the engineer's proposal: the facade was not moved, and the internal space could be changed according to structural requirements.

After the project was completed, the designers' purpose was only partially achieved. Under the door bucket, villagers gathered to peel edamame. The rural landscape Wang had envisioned was back. The most important change is the function of the project, which runs counter to the architects' original vision to avoid commercialization. While the old generation is not all accepting Wang's design, their children, who work in the cities, understand the commercial value of the famous architect's design at once and persuaded their parents to buy the houses. Soon, residents saw the opportunity to bring tourism, for "rural houses" built by a famous architect are the best experiential "B&B". The local Government has designated Wencun as a "B&B demonstration village", it went popular on the Internet and some houses were rented out to commercial companies.

The village has gained popularity and jobs, and more and more young people began to return from the city to the village to join the development of tourism. But local authorities found it difficult to promote this example, even though the second-phase project has not yet been realized. In 2016, a real estate company listed in Hong Kong signed a contract with Wencun, claiming to invest 3.05 billion yuan furthermore, but the funds haven't yet been implemented (Xu 2019). At present, the local government is still working to attract investment and tourism, they hope the mechanism can be more flexible to allow capital involving residents together.

However, the development project is also encountering difficulties. First of all, the techniques that were applied were unique, which brought uncontrollable costs, for the materials are hard to find and fabric from a mature factory. Tracing the building materials once used in the villages is also a difficult situation. The architects' design also confused structural and cost engineers. It's not a reproducible assembly-line product that they were familiar with, while architects insisted on using local craftsmanship and materials, whose prices were difficult to evaluate. The project's engineers expressed trepidation: "It is difficult to recreate Mr Wang Shu's beginning point and vision due to the cost, complexity, and uniqueness of the process. Such an architect is better suited to public projects." (Xu 2019) Although some architects stated online that Wang Shu's Wencun project was only a manifestation of his personality, according to local managers, the more practical issues with the capital return plan prevented the project's second phase from starting, not because of aesthetic concerns. "If the village first establishes a tourism development firm, the villagers will share in the rights to the leasing and sale of the historic homes before approaching outside businesses and investors about joining forces on the project. Such development creation is sustainable when the village serves as the major body and the government above it provides policies and advice." (Xu 2019) It is obvious that the true sustainability dispute surrounding this project goes beyond the debate around the architect's choice of architectural language and instead focuses more on development strategies. The young people who returned also have concerns about the sustainability of tourism's future, with the current monotony of the B&Bs does not keep visitors for a long time. But without more investment from the top-down approach, it seems difficult to continue the project.

Wang Shu believes that resistance is the way to sustainability. " For people to discover an appropriate value framework, whether you're talking about ecology or sustainability, there must be a need for architects to have a critical perspective, then identify the best architecture inside that value framework."(Xu 2019) Although the original starting point of creating a place not polluted by urbanisation, in the end, the architect's fame brought urbanisation and tourism to this remote village. The architect feels "exhausted with an impossible mission", while the shift of tourism eventually comforts him with the "future potential of the countryside".

16.4 Conclusion

While this Wang and Lu's Regional Criticism practice was a precious effort to create an "outsider" of the urban process. Architects worked hard to create a "non-urban" and "non-connected" living place and a micro-social system, elevating the rural image to a higher level of an ideal. By using local architectural language and getting close to nature, architects tried to protect regional characteristics and social relations, which are independent of the overall relations of large-scale urbanization and industrial landscape. In this typical case, architects, residents, government, constructors, and capital parties are all part of the realization process of an ideological design product. While the complex conflicts between these different actors are fascinating, the most striking outcome demonstrated is the resistance and compromise of the regional practice with urbanization, and the result of it finally contained by urbanization in the capital process.

The communication between inter-disciplined methodologies can be very useful in light of analyzing this process. Planetary Urbanization theory is valuable in breaking cities from isolated concepts and understanding the capital process as a whole, but overemphasizing the dominant role of urbanization ignores the existence of the rural landscape and the subjective initiative of architects as space producers. Yet, the theory still stands well the practice of Regional Criticism was indeed incorporated into the globalized urban process. The villagers dreamed of modern urban life, and people in the city wished to return to the territorial sense, while the capital captured this opportunity to develop tourism. Eventually, it has become another global trend, even an "elite trend", bringing both urbanization and gentrification. The template-type propaganda symbol was a top-down process with the driving force of government so the policy provided support. The characteristics of regional architecture relied on specific craftsmanship and materials, requiring so much effort that the architect could only insist on form in the end. It can be inferred that the practice of Regional Criticism is not an "ordinary product" but a "luxury product", its point-like influence is limited and difficult to become widespread. As a result, we can still find this resistance passive. On the other hand, the existence of architectural practice is also an important factor to be considered in regional urban theory to get a holistic understanding of urban phenomena. It requires more dialogue and awareness between architecture and human geography, the two disciplines of spatial studies, to reach a more completed urban theory and a more successful architectural practice.

It is necessary to rethink the sustainable design discourse with the indigenous imagination, especially facing a changing community with different generations and beliefs. Although the architects tried to repaint the traditional rural landscape by inviting the residents into the discussion, this participation may proofed inefficient and still architect-centred. The lack of reliable financial management to generate collaboration between villages and local authorities obscured the potentiality. It failed to generate a genuine linked community but still become a manifestation of the architect. It calls us to rethink these regional practices as popular sustainable design methodologies, taking the transformation of partnerships and communities into real account.

Resistance must acknowledge reality. As the socio-economic and cultural manifestation, architecture cannot win victory alone with monomer forms and cultural criticism. If architect-led resistance to the present is a way towards sustainability, more needs to be done after the declaration. Architects must understand and adapt to social and economic environments, and use politics and institutions as weapons, to develop more powerful methodologies that could affect the entire territorial scale, while bringing changes into society. This implies not only the expansion of architectural theory and disciplinary boundaries into political economy and cultural sociology but also the importance of the scalability of sustainable architectural practices confronting a broader reality of alienation and simplification.

16.5 Images

See Figs. 16.1 and 16.2.

Fig. 16.1 Wanderring Wencun Project with a grandma welcoming tourists, Retrived From Nanfang People Journal, 2023.1.14; https://nfpeople.infzm.com/article/9009

Fig. 16.2 Interior photo of Wencun B&B; Retrived From Nanfang People Journal, 2023.1.14; https://nfpeople.infzm.com/article/9009

References

Brenner N, Schmid C (2015) Towards a new epistemology of the urban? City 19(2–3):151–182. Retrieved 14 Jan 2023, from https://doi.org/10.1080/13604813.2015.1014712

Frampton K (1983) Prospects for a critical regionalism. Perspecta 20:147. Retrieved 14 Jan 2023, from https://doi.org/10.2307/1567071

Jazeel T (2017) Urban theory with an outside. Environ Plan D: Soc Space 36(3):405–419. Retrieved 14 Jan 2023, from https://doi.org/10.1177/0263775817707968

Levitt P (2015) Artifacts and allegiances: how museums put the nation and the world on display. University of California Press, Oakland, CA

Li X, 李翔宁 (2012) (Architecture as resistance-Wang Shu and his architecture). 作为抵抗的建筑学——王澍和他的建筑. (World Architecture) 世界建筑5:30–31

Massey D (1991) Global sense of place. In: Space, place, and gender. Routledge, London, pp 146–156

Tafuri M (1977) Architecture and utopia design and capitalist development. Mass. u.a. MIT Press, Cambridge

Xu L, 徐琳玲 (2019) (Pritzker prize winner Wang Shu's rural rescue battle: the core of Chinese culture is in the countryside, not in the city) 普利兹克奖得主王澍的乡村拯救战: 中国文化核心在乡村, 不在城市. (Southern people weekly) 南方人物周刊 582. Retrieved 14 Jan 2023, from https://nfpeople.infzm.com/article/9009

Zhu L, 朱亦民 (2013) (Modernity and regionalism——Interpretation of "critical regionalism-six points to resistance architecture") 现代性与地域主义——解读《走向批判的地域主义——抵抗建筑学的六要点》. (New architecture) 新建筑 3:30–36

17 Sustainability in the Pluriverse: Learning from Global Futures

Joel Peter Weber Letkemann

Abstract

If the project of making architecture is a project in building the future, it is worth asking what imaginations inspire such futuring. This paper argues that discourses of architectural futurity can be augmented and inspired by discourses in global futurism. While bearing similarities to science fiction, these global futurisms are an impetus to re-frame the discourse of architectural sustainability in radical ways. The essay brings together scholarship in global futurisms from Bodhisattva Chattopadhyay with post-colonial anthropologist Arturo Escobar's concept of the "pluriverse." As the ambitions represented by the UN sustainable development goals enter into dialog with speculative storytelling traditions from across the world, such dialog reveals that there is not a singular or universal practice of sustainability which the design disciplines can easily adopt. The stark contrasts between western science fiction and global futurism are a reminder that sustainable architectural futures can be imagined as technological fixes to an existing global order, or within very different social, economic, and political frameworks. As an example of one such fiction, this essay discusses "Reunion," a novella by New Delhi born Vandana Singh, for its inspirations for architectural practice. In particular the novella describes an experimental settlement called Ashapur, as well as an infrastructure which combines technologically-oriented futures with one which privileges community with both human and non-human relations. Next, this essay makes some tentative associations between certain architectural practices and discourses in global futurism in order to argue for the many possible futures already lying latent within global architectural practice.

Keywords

Global futurism · Science fiction · Sustainable architecture · Vandana singh · Reunion

17.1 From the Future to Cofutures

When we talk about the future, what future are we talking about? The big-F future is an image that congeals from combined discourses of entertainment, advertising, financial speculation, government policy, and any number of other discourses. If we take these discourses as evidence, *the future* has occupied our imaginations for some time, forming a collective if not stable image; we might expect technological or esthetic

J. P. W. Letkemann (✉)
Department of Architecture, Design and Media Technology, Aalborg University, Aalborg, Denmark
e-mail: jpwl@create.aau.dk

S. Hilal et al. (eds.), *Design for Partnerships for Change*, Sustainable Development Goals Series,
https://doi.org/10.1007/978-3-031-36993-3_17

innovation while existing sociopolitical or economic structures remain intact, becoming a future very like the present. Literary theorist Fredric Jameson warns us that the project of imagining the future is always at risk of "coloniz [ing] the future" according to the dominant perspectives and privileges of the time in which such a future is described (Jameson 2007, p. 228). However, given the specter of climate change, growing inequalities, and a precarious public realm, the need to imagine new sustainable and equitable ways of living has never seemed more urgent.

While the western worldview still dominates the big-F future, existing storytelling traditions from around the world open up a pluriverse of possibility for sustainable and equitable futures beyond western forms of knowledge. The field of architecture is also not monolithic, and many architects from many subject positions continue to seek out new forms of practice which re-frame the relation of humans to each other and to the environment. This paper suggests a partnership between such global futuring and marginal architectural practices, seeking to place these into dialog with one another in order to understand how radical, complex possibilities for sustainable futures are already being imagined, and how they might help us think about architectural practice in the present.

As a way of including the field of architecture within radical imaginations of sustainable futures, I am not the first to suggest that architects might learn from reading science fiction (SF) (Butt 2018). However, within SF scholarship, there is a sense that it might be time to retire the denomination "science fiction." As John Reider reminds us, SF is historically entangled with colonialism; it emerges from a set of nineteenth century discourses that, for all their claims to enlightenment rationalism, shelter explicitly racist and exclusionary rhetoric (Rieder 2008). Furthermore, Bodhisattva Chattopadhyay argues that "science fiction" is a recent designation for a type of speculative literature that has existed long before its being named by a western marketplace (Chattopadhyay 2016). Chattopadhyay's ambition to "recenter" SF away from western, Anglo sources is to reclaim speculative literature from its entanglements in colonial, problematically techno-optimistic history, advocating for an attention to futures emerging from the rest of the world (Chattopadhyay 2014).

After Mark Dery coins Afrofuturism in 1994 (Dery, 1994; Eshun 2003), an esthetic expression centering the African diaspora's experience and expectations for the future, more attention has been focused on futurisms from across the globe, now also including South Asian (Banerjee 2020; Khan 2021), Indigenous (Dillon 2012), Gulf (Pinto 2019), Latin American (Brock 2021), or Asian (Chan 2016) futurisms, among many others. Beyond various global cultures, other oft-marginalized voices claim their own spaces in future too, including feminist (Cuboniks), crip (Kafer, 2013), or queer futurisms (Beyond Gender Research Collective 2021). This proliferation of futures does not congeal into a singular mass deserving of a definite article—*the* future. Rather they affirm the possibility of what Chattopadhyay describes as cofutures; cofutures are complex, coeval—they arise separately from different sources, and com-possible—they are possible together, requiring "solidarity" in "recognition of difference" (Chattopadhyay 2020, 2021). While they might bear similarities to SF, these global futures do not rest upon the same epistemological foundations, and as such, open up possibilities foreclosed by a more carefully delineated SF, especially as they develop from marginalized worldviews and suggest more just and sustainable relationships between human communities and with the more-than-human world.

To seek out other futures, we need to seek out other worlds in the present. As anthropologist Arturo Escobar reminds us, we already inhabit a pluriverse (Escobar 2018); "the" world is as much a misnomer as "the" future. Different people have very different histories and everyday lives, and each point in the plurality of the present also harbors and nurtures its ideas about what the future can be:

> To think new thoughts, by implication, requires stepping out of the epistemic space of Western social theory and into the epistemic configurations associated with the multiple relational ontologies

> of worlds in struggle. It is in these spaces that we might find more compelling answers to the strong questions posed by the current conjuncture of modern problems with insufficient modern solutions... can design practice contribute to broadening, and drawing on, the rich spectrum of experiences that should be considered viable alternatives to what exists? (Escobar 2018, pp. 68–69)

While the project of the UN Sustainable Development Goals (SDGs) is the future story occupying the discipline in the present moment, as Escobar reminds us, the idea of development, which we see in such terms as "sustainable development," is often already a colonial project in its imposition of a global order on local practices. Escobar reminds us that we cannot see "sustainablity" as a universal signifier, and that sustainable practices cannot be universalized. However, if we understand the SDGs in their potential to inspire a drastically different relationship with the world and with each other, then, we can see that such hopes are only augmented by their correspondence with other forms of radical futuring.

If the aim of the SDGs is really to "leave no one behind," we cannot only talk about a people's material sufficiency. What good is it to preserve diverse peoples while losing their stories, their worldviews? Following design theorist Tony Fry, Escobar goes so far as to call development practices "defuturing" in that they foreclose possibilities for the future in the erasures of the globalist project in the present (Escobar 2020, pp. 138–148). Within the architectural discipline, we have such defuturing projects as Bjarke Ingels' Masterplanet, like its fore-bearer in Buckminster Fuller's Spaceship Earth (Letkemann, 2021), which proposes a singular worldwide master plan for sustainability. Liam Young rightly calls this project a "continuation of the colonialist project that has already masterplanned the planet in its own image" (Fairs 2021). That is, we have already reached a world determined by hubristic techno-scientific solutions, and the architectural discipline must be careful lest the imagination of architectural sustainability only proceed from more techno-scientific hubris.

Global futurisms help the spatial practitioner to question the epistemic foundations of their own discipline, to unsettle preexisting ideas about the future, and to ask different questions about what architectural sustainability is and what it can be. After all, a discipline exists to discipline its adherents, to reproduce its own ways of knowing (Lykke, 2010, p. 20). For the discipline of architecture to change, the conscientious architect needs to seek out new ways of seeing—the "optical devices" made possible by listening to *other* stories (Haraway 1992). As an extension of the attention and care which architects already direct to "the site," we might also see the stories of and hopes for the future as a part of that "site."

17.2 Vandana Singh's Reunion

Reunion, a novella by New Delhi born Vandana Singh, was first published in 2019 (Singh 2022). This novella supplies radically different perspective of the climate crisis and the role of designer or technologist in the organization of one specific community but also supplies a chance to think past the collapse of the present's socioeconomic framework and find ways of learning from marginalized voices. The sustainable future that Singh imagines is about understanding human relationships with one another, with marginalized people as well as with a marginalized more-than-human world, and about cultivating a relationship to technology that serves community instead of capital.

Singh tells the story through the late-life reminiscences of Mahua, a character descended from Santhal people of eastern India. Through Mahua's eyes and through her actions, we see dramatic social shifts as society moves from what Singh terms the "age of Kuber"[1]—a period "madness of the mid twenty-first century"—through a "Great Turning" in the development of a network of experimental communities to replace the city (Singh 2022, p. 351).

[1] In the Hindu pantheon, Kuber is the divinity associated with wealth and treasure.

Through Manua's eyes, we see our own age as "the old megapolises die through the combined machinations of extreme weather and human greed" (Singh 2022, p. 360). The fears which haunt our own present are a reality for Mahua, as she observes Mumbai becoming an archipelago again:

> the sea has reclaimed the city – fish now swim in what was once Charni Road, and crabs and mussels have taken up residence in the National Stock Exchange. The fisherfolk ply their boats and barges in the watery streets. (Singh 2022, p. 342)

However, the character is instrumental in her future India's adaptation to a changed climate. In her motivation "to move civilization away from self-destruction," Mahua introduces new models for India's climate adaptation, including for architecture and urban infrastructure. In the story, a younger Mahua develops an open data infrastructure for a connected city called "Sensornet." Although recognizable as the ubiquitous sensors and big data we see in other smart city plans, her project arises from very different ambitions; the "embedded intelligence agents" she develops arise out of the "desire to be companionably present with the non-human and the inanimate." In a departure from the smart city as we know it, the open data infrastructure is not about the efficient transfer of goods and capital, but of a sensitive relation to the more-than-human ecologies that also make the city home (Singh 2022, pp. 352, 354, 348).

As the character's life progresses, she grows to question the city entirely. In her pursuit of a balanced ecological community, "...the city isn't the right idea for what we're trying to do." She asks how much of our everyday environments are intended to perpetuate an existing way of living:

> why would we want to live in a city as it is now – when people don't have any time for anything but work? There's constant stress, people don't know each other, don't care either, where democracy is sham?... a megapolis is beyond the scale of human social adaptation. (Singh 2022, p. 354)

In response, Mahua develops an experimental settlement, a "basti," called Ashapur as a model for a type of community living which is more attuned to a sensitive relationship between people and the non-human world. She describes how each of the few houses in the settlement is:

> dome-shaped to reduce the impact of the storms, thick walls of clay, straw and recycled brick.... a marriage of the ancient and the modern. The walkways follow the natural contours of the land. The vegetables cascade off the walls on vines. (Singh 2022, p. 341)

Some of these innovations are familiar to an architectural audience, from traditional clay building techniques to the re-discovery of long-standing Indigenous practices now branded as permaculture; I write this article only a few kilometers from Andels Samfundet, an ecovillage on the outskirts of Hjortshøj, Denmark, which features adobe houses, on-site agriculture, and a well-developed and thriving infrastructure for community (https://www.andelssamfundet.dk/). Like the community in Hjortshøj, Singh devotes considerable attention to the organization of the settlement:

> inner roads for people and bicycles, the outer ones for busses that connected them to the greater city. Here, there was room for groves of jamun and neem trees, for gardens on the building walls and roofs. Each domicile held families related by blood and by choice, up to 50 people under one roof, cooking together in large common kitchens. (Singh 2022, p. 358)

While introducing these architectural practices, Singh pursues a rhetorical twist whereby, rather than seeing these often marginalized practices as of an abandoned past, these be read as *futuring* practices. That is, she is re-framing ancient practices as images of what the future could be.

Singh also introduces the prospect of technological community infrastructures; networked "sun towers" collect and distribute energy to all bastis in the network,[2] and each basti is "embedded with sensors that monitor and report a constant stream of data—temperature, humidity, energy use, carbon storage, chemical contaminants, biodiversity" (Singh 2022, p. 343). Importantly, such data does not feed into the

[2] This technology is better developed in another Ashapur story, Indra's Web (Singh 2018).

algorithmic maw of big tech, but like the character's earlier "Sensornet," is open and shared among other networked bastis in order to monitor and maintain the ecological health of the settlements. Singh also intensifies the different strategies in order for the reader to see how they would develop into global settlements:

> connected by green corridors, each settlement embedded with sensors, farm towers replacing conventional agriculture. Such settlements would spring up in different parts of the country and the world. Former agricultural lands would return to the wilderness, or to subsistence farming, repairing the damage done to the biosphere's life-maintaining systems. (Singh 2022, p. 355)

The Hindi word "basti" means dwelling and refers to any place people might dwell but has gained a pejorative sense in referring to village or slum dwellings.[3] Importantly, however, this pejorative sense is re-appropriated in the new form of settlement. The original occupants of Singh's fictional basti are marginalized people—refugees from Bangladesh fleeing the rising seas. In this sense, the basti is a site of "marginal resistance" (353) in that it learns from tribal societies in India and also grows from broad coalition of people excluded from authority in the present rather than from supposed expert knowledge.

That is to say, Singh's vision of the future is also about agency. In a thinly veiled critique of how our public spaces are increasingly administered by private interest, Singh's novella shows that a sustainable community needs to develop new infrastructures for collaboration, sharing, communication, and to leverage technology for a community's benefit. In the story, the bastis are made possible by new political organizations in concert with the new spatial environments, the scale of each house and village corresponding to scales of political negotiation, while the global scale is managed by the collective digital infrastructure. Notably, in Singh's story, the collective governance is substantially inspired by the Santhal tribal people[4] in governing through consensus, community, and in prioritizing "reverence for the web of life"(Singh 2022, p. 365).

Singh's is one of many imaginations of future community that ask for a different kind of architect and a different kind of architectural practice. She shows one image of what that community might look like, but Singh's story also shows storytelling as a synthetic practice—one that includes architectural speculation alongside political, social, economic, and technical speculations. Taken together, these speculations argue that, if we should drastically alter our relationship to the more-than-human world, we can't live in the same way, nor work from the same assumptions. Singh asks us to reconsider the basis for community, for transportation, and infrastructure for sharing and accountability, even, if necessary, to the extent of questioning the city as the apotheosis of human settlement. Instead, the primary goal of Singh's architecture is being "companionably present" with each other and with the more-than-human world. But Singh does not stop there. Her character Mahua is a model for spatial practitioners too, as much as for what she does as for what she doesn't: Building political alliances with a community means rejecting centers of authority and capital, especially as the representative building of the "Age of Kuber," the Mumbai Stock Exchange, is underwater. Working together with Indigenous knowledge means that the spatial practitioner must understand different ways of valuing the world and of organizing one's place in it.

17.3 Global Futurisms in Architecture

Singh's future, though inspiring, is only one possibility lying latent in the pluriverse of possibilities explored within global futures. Even if we are inspired by a vision of a form of human settlement which takes the well-being of people

[3] As the word comes into English, it only maintains this pejorative sense. My thanks to Bodhisattva Chattopadhyay for a fruitful discussion of this word in Hindi, Bengali, and its derivations from Sanskrit.

[4] The Santhal people are tribal people from eastern India, especially in the Jharkhand, West Bengal, and Odisha states.

and planet as its primary goal, what Escobar reminds us about is that it is not workable to replace one master narrative with another; solutions arrive locally, each within its own context. By advocating for architects to pay greater attention to global futurisms, I am also suggesting that we learn more about the pluriverse of practices already within architecture, many which find resonance with movements in global futurism. The more the discipline empowers such pluralistic voices, the better the discipline evolves new strategies for addressing climate change. While it might be tempting to dismiss speculative storytelling as fantasy, these practices also affirm the many real avenues for futuring within our own discipline.

In one project drawing upon themes in Afrofuturism, Nigerian-born Olalekan Jeyifous produces an imagination of his own neighborhood in Brooklyn, New York as, resisting gentrification, it develops self-sufficient communities around food and water production and exchange (Budds, 2020; Also see: Summers 2021; *Olalekan Jeyifous (@kidcadaver) • Instagram photos and videos*). The project's striking images play upon familiar SF tropes in speculative technology, but what is notable about them is not necessarily their esthetic or technical innovation. Instead, these collages are developed from Jeyifous' own photos of his Brooklyn neighborhood and thus have a unique resonance not as an abstract possibility, but in historical continuity to the present day—we can see how that future might grow out of the material of the present. Though the Black and Latinx communities who have long called Brooklyn home are now facing gentrification, these images reaffirm not only the continued presence of these communities but also show them reclaiming the space in a "post-capitalist" intervention (Budds, 2020); the Brooklyn of the future, they claim, will include a future for people of color, while also hinting at urban agriculture systems with new technologies and systems to capture and store water and to grow and distribute food within the neighborhood. While such a project suggests a radical vision for the sustainable cities and communities championed by the SDGs, the post-capitalist implication also calls into question what economic growth and responsible consumption might look like. The work of the architect in this case, like in Singh's story, is to build and nourish community, rather than to serve capital.

Another vector for pluriversal futures in architecture is the reassertion of Indigenous spatial knowledge in a post-colonial practice. The Gathering Circle by Anishinaabe architect Ryan Gorrie reclaims the shore of Lake Superior as a place of Indigenous gathering, especially poignant in a city and a country which has long suppressed Indigenous knowledge and cultural practices. (*Spirit Garden | Brook McIlroy*) The pavilion takes on a spatial and material order that is in contrast to the colonial powers in Canada, carving out a space unique to the communal and ceremonial practices of the Indigenous people of the region. For these peoples, the circle is a potent form to encourage discussion and building consensus, while the structure itself is an adaptation of traditional wood-bending techniques.

Gorrie also invited Indigenous craftspeople to construct the ceremonial space with rough-hewn wood in a renewal of the Indigenous tectonic culture; material assemblies of the pavilion are marked by strategies of tying and bundling rather than cutting and fitting. In this case, in an echo of Grace Dillon's reminder that Indigenous scientific literacies are shaped by the diverse environments in which they are found (Dillon 2017, p. 472), this project also suggests a tectonic literacy that rejects the alien invasion of Western tectonic and material cultures and which embraces the traditional knowledge of the land while staking a categorical claim to the future of the site. The imagination of sustainability on this site is both a reminder that Indigenous storytelling and futuring traditions proceed from a deep knowledge of the lands in which they are found, the sustainability they represent cannot be universalized but is nested in a deep commitment to care and community with that land.

17.4 Conclusion

Vandana Singh's Ashapur stories join other speculative futuring traditions from across the world to suggest that a pursuit of the SDGs is not a singular pursuit, but one which must be in dialog with local communities, attuned to the unique needs, the unique knowledge, and the hoped-for futures of local communities across the globe. Singh reminds us that architectural sustainability is often not about novelty and technological invention, but rather, that futures might arise equally from traditional forms of knowledge and ways of being. Importantly, she shows that these are incompatible with an infrastructure that is owned and administered by a small elite, or by the imposition of western practices into diverse local contexts. Instead, Singh describes spatial practices that grow from a broad coalition of people excluded from authority in the present rather than from supposed expert knowledge. Her vision of the future is about agency, about spatial practice in new political alignments together with local communities. The sustainable futures imagined from the knowledge of marginalized peoples offer just such a possible partnership for the architect or spatial practitioner, even, as Ryan Gorrie shows, to the level of the material and tectonic cultures appropriate to each place.

If architectural practice is to evolve sustainable possibilities and practices beyond existing disciplinary frameworks, it might be time both to empower the diverse subjects already in practice and to pursue broad coalitions with other ways of knowing about the world. After all, if the goal of sustainable practice is to work toward the prosperity of the more-than-human world as much as supporting all the different lives that humans have, then it is worth probing how much of the knowledge and practices we have are oriented toward preserving a status quo. If we are truly to follow the guiding principle of the SDGs—to "leave no one behind"—then as a discipline, architects must also be willing to embrace other worldviews, to probe their own personal histories rather than the origin myths of modernist practice, and to seek out radical futures to pursue. The storytelling traditions in global futurism are a potent tool-set to do just that.

References

Banerjee S (2020) Indian science fiction: patterns, history and hybridity. University of Wales Press (New dimensions in science fiction), Cardiff

Beyond Gender Research Collective (2021) A future collectively salvaged from the rubble. Architectural Rev

Brock P (2021) Futurism and genre genesis in Brazilian science fiction. Zanzalá: Rev Bras de Estudos sobre Gêneros Cinematográficos e Audiovisuais 8(1):8–18

Budds D (2020) Olalekan jeyifous is imagining an afrofuturist brooklyn, curbed. Retrieved 27 Sept 2022, from https://archive.curbed.com/2020/7/1/21308742/afrofuturism-olalekan-jeyifous-interview

Butt A (2018) "Endless forms, vistas and hues": why architects should read science fiction. Architectural Res Q 22(2):151–160. https://doi.org/10.1017/S1359135518000374

Chan D (2016) Asia-futurism. Artforum. Retrieved 9 Sept 2021 from https://www.artforum.com/print/201606/asia-futurism-60088

Chattopadhyay B (2016) On the mythologerm: kalpavigyan and the question of imperial science. Sci Fiction Stud 43(3):435–458. https://doi.org/10.5621/sciefictstud.43.3.0435

Chattopadhyay B (2020) The pandemic that was always here, and afterward: from futures to CoFutures. Sci Fiction Stud 47(3):338–340. https://doi.org/10.5621/sciefictstud.47.3.0321

Chattopadhyay B (2014) Recentering science fiction and the fantastic: what would a non-anglocentric understanding of science fiction and fantasy look like? In: Grilo A, James T (eds) Speculative fiction 2013: the best online reviews, essays and commentary. Jurassic, London, pp 213–228. http://strangehorizons.com/non-fiction/articles/recentering-science-fiction-and-the-fantastic-what-would-a-non-anglocentric-understanding-of-science-fiction-and-fantasy-look-like/

Chattopadhyay B (2021) Manifestos of futurisms. Found Int Rev Sci Fiction 50(2):8–23

Cuboniks L (n.d.) Xenofeminism: a politics for alienation. Laboria Cuboniks. Retrieved 3 Oct 2021, from https://laboriacuboniks.net/manifesto/xenofeminism-a-politics-for-alienation/

Dery M (1994) Black to the future: interviews with Samuel R. Delany, Greg Tate, and Tricia Rose. In: Dery M (ed) Flame wars: the discourse of cyberculture. Duke University Press Books, Durham, pp 179–222

Dillon GL (2012) Imagining indigenous futurisms. In: Dillon GL (ed) Walking the clouds: an anthology of indigenous science fiction. University of Arizona Press, Tucson (Sun tracks : an American Indian literary series, v. 69), pp 1–12

Dillon GL (2017) Indigenous scientific literacies in Nalo Hopkinson's ceremonial worlds. In: Latham R (ed) Science fiction criticism: an anthology of essential writings. Bloomsbury Academic, London, New York, pp 470–486

Escobar A (2018) Designs for the pluriverse: radical interdependence, autonomy, and the making of worlds. Duke University Press, Durham

Escobar A (2020) Pluriversal politics: the real and the possible. Duke University Press, Durham

Eshun K (2003) Further considerations of afrofuturism. CR New Centennial Rev 3(2):287–302. https://doi.org/10.1353/ncr.2003.0021

Fairs M (2021) Bjarke Ingels' Masterplanet vision is 'a continuation of the colonialist project'. Dezeen. Retrieved 3 Oct 2022, from https://www.dezeen.com/2021/01/06/bjarke-ingels-masterplanet-liam-young/

Garden S and McIlroy B (n.d.). Retrieved 10 Oct 2022, from: https://brookmcilroy.com/projects/spirit-garden/

Haraway D (1992) The promises of monsters: a regenerative politics for inappropriate/d others. In: Grossberg L, Nelson C, Treichler PA (eds) Cultural studies. Routledge, New York, pp 295–336

Jameson F (2007) Archaeologies of the future: The desire called utopia and other science fictions. Verso, London

Kafer A (2013) Feminist, queer, crip. Indiana University Press, Bloomington, Indiana

Khan SA (2021) Star warriors of the modern raj materiality, mythology and technology of Indian science fiction. University of Wales Press, Cardiff

Letkemann JPW (2021) Critical dystopias in the digital project. Folio 36:90–97

Lykke N (2010) Feminist studies: a guide to intersectional theory, methodology and writing. Routledge, New York

Olalekan Jeyifous (@kidcadaver) • Instagram photos and videos (n.d.). Retrieved 10 Oct 2022, from https://www.instagram.com/p/B_zmdXApiyk/

Pinto AT (2019) Alien nations. Mousse (64):166–175

Rieder J (2008) Colonialism and the emergence of science fiction. Wesleyan University Press, Middletown, Conn

Singh V (2018) Indra's Web. In: Easthampton MA (ed) Ambiguity machines: and other stories. Small Beer Press, pp 145–154

Singh V (2022) Reunion. In: Saint TK (ed) New horizons: the Gollancz book of South Asian science fiction. Orion Publishing Group Limited, pp 341–365

Summers BT (2021) Untimely futures. Places J. Retrieved 10 Nov 2021, from https://placesjournal.org/article/black-homelessness-in-oakland/

www.andelssamfundet.dk/english1 Retrieved 10 Oct 2022, from https://www.andelssamfundet.dk/english1

18

Breaking the Rules: Towards an Experimental Design Pedagogy

Camilla Forina, Sofia Leoni, Tommaso Listo, Valeria Federighi, and Edoardo Bruno

Abstract

The *Great Game* is a laboratorial course that problematises the traditional *design studio* approach, seeking for reframed modes of knowledge transmission through interactions among a multiplicity of actors and establishing situatedness as design principle. The didactic experiment of the *Great Game* tries to answer two questions. The first asks whether it is possible to teach an effective way of designing without applying the binary logic of what one (the teacher, a theoretical authority, an ideology) decides wrong (students' knowledge up to that point, a certain way of doing architecture, intentions), and what is decided to be right (new and updated skills, another way of designing, ethically acceptable goals). The second asks whether it is in the form and organisation of the *design studio* that we should intervene (and modify) if we want to produce a change in terms of approaches to design. Logic and structure of the *Great Game* are presented, as well as (graphical) reports on the experiences of two iterations of the course: on these contents is built a correspondence between the pedagogical objectives and the effectiveness of design in intercepting the multiplicity of instances that emerge in any specific place and situation. We conclude by showing how through playful experimentation of architectural education, the *Great Game* tries to incentivise a form of knowledge that derives from nonlinear actions of mutual interlocutions and reciprocal positioning, or rather, the product of multiple instances, detournements, and even errors.

Keywords

Design strategy · Game of strategy · Design practice · Design theory · Design pedagogy

In the world of architectural education, *design studios* are historically the institutional sites for the transmission of knowledge and the training of future architects. Configuring themselves as both a physical place and pedagogical models, they are the places where architectural design is simulated under controlled conditions, recalling what happens in the field of scientific experimentation within the laboratory. Such a simulation implies a series of simplifications assumed to be internal to the didactic object, with the precise intention to mitigate the complexity of the external world to a minimum denominator. The predominant result is hence a detachment of design teaching from the very dynamics of

C. Forina (✉) · S. Leoni · T. Listo · E. Bruno
Politecnico di Torino, Torino, Italy
e-mail: camilla.forina@polito.it

V. Federighi
Politecnico di Torino, Torino, Italy

S. Hilal et al. (eds.), *Design for Partnerships for Change*, Sustainable Development Goals Series,
https://doi.org/10.1007/978-3-031-36993-3_18

professional practice in favour of the transmission of values based on creativity and ideas as absolute truths anchored in certain authorship (Kuhn 2001; Till 2005; Rodriguez et al. 2018).

Nevertheless, acknowledging the complexities that today's challenges dictate (Deamer 2020)—namely a sustainable, accountable, multidisciplinary, participatory, and representative approach to decision-making—we believe that it is (also) in *design studios* that we should look, and it is (also) here that we should try to (re)act to redefine disciplinary positions. Based on these assumptions, in October 2020, the *Great Game* took its first steps (Federighi and Bruno 2022).

Set up as a sort of role-playing game for architects, *Great Game* is a laboratory course held during the third year of a Bachelor programme in Architecture that tries to problematise the more traditional approach of *design studios*. The *Great Game* thus tries to build a simulation that allows you to transmit, develop, and train a set of skills and sensibilities in addition to that of the traditional *design studio*. This construction has been carried out through a different organisation of the students' works review and discussion, according to rules, constraints, and environmental conditions attempting to define a sort of "retrospective thickness", which belongs to the client's mandate and the project site. Dropped into such thickness, students find themselves interacting within a complex and contradictory environment in the role of architects whose clients have a general perspective, rather than a clear idea. In facing the specific request of the mandate, they are thus confronted with different actors and interests to which responding while setting their broader strategy. In its course, the game unfolds in an overlapping of moves, emerging situations, and unexpected irruptions, as opposed to the relative linearity of *design studios*; the result of this setting produced a second type of thickness, which could be called thickness-in-action. The initial intentions of the groups of students are repeatedly deviated as the result of a plurality of actors who have different and changing interests, and not, for example, in the search for any kind of combination between form and function. This also requires that students are aware of their relative position and situatedness (Haraway 1988), and that they organise their actions through a prefiguration of their possible effects.

Wondering how we might effectively contrive a pedagogy of architectural design that ensures responsibility, inclusivity, and participation, the *Great Game* thus seeks to recognise complexity as a category that can be taught to act and navigate with. In this sense, the *Great Game* is an everchanging experiment that continues to investigate new forms of architectural teaching through intellectual and technical tools which hold a large potential in the renewal of the discipline (Figs. 18.1, 18.2, 18.3, 18.4, 18.5, 18.6, 18.7, 18.8, 18.9, 18.10, 18.11, 18.12, 18.13, 18.14, 18.15, and 18.16).

Fig. 18.1 First simplification commonly adopted within design studios consists of the exclusion and reduction of the multiple external realities by adopting the perspective of an absolute subject conveyed through a series of separations in which the didactic object is broken and served into pieces to be manipulated by the will of that same subject: —the project assignment is clear and straightforward;—the project location is problematised in a biunivocal way: there is a problem, summarised by sets of data, that the project must solve; or rather analysis and then synthesis;—teaching staff holds the monopoly of the "crit";—freedom is granted to students to express their idea, that is assumed as an absolute entity itself, apart from its conformity with the critical ideal of the teacher;—the transmission of knowledge is organised in progressive order: first a skill, then the next one which presupposes the first and so on; or rather first knowledge as a finished product, then its application. Diagram by the Authors.

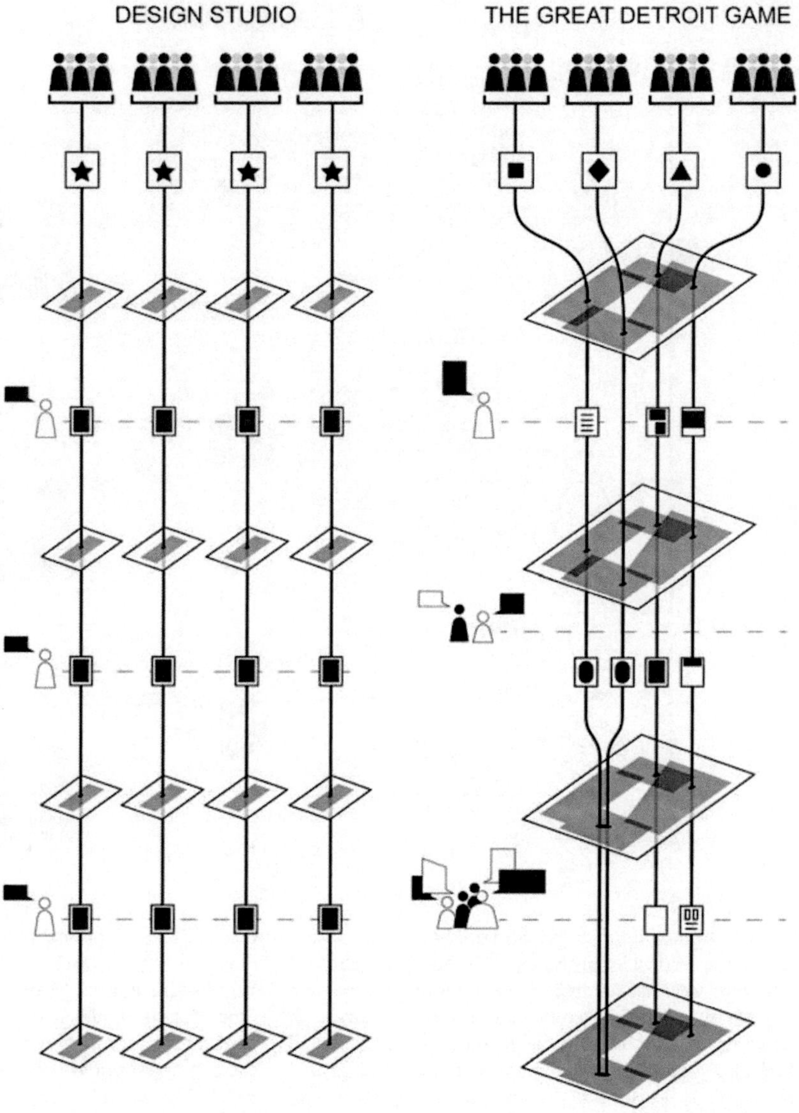

Fig. 18.2 *Great Game* intervenes critically on the conventional workflow of the *design studio* hinged on a dominant authorship model, multiplying both amount/types of interlocutors and documents while trying at the same time to keep the positioning action that constitutes this intervention evident and participatory. The course starts from the assumption that transmission of knowledge does not begin when the critical line is drawn between what one (the teacher, a theoretical authority, an ideology) decides wrong (students' knowledge up to that point, a certain way of doing architecture, intentions) and what is decided to be right (new and updated skills, another way of designing, ethically acceptable goals) but, rather, it consists from beginning to end in the interaction allowing for acts of discernment and reconstitution. *Diagram* by the Authors

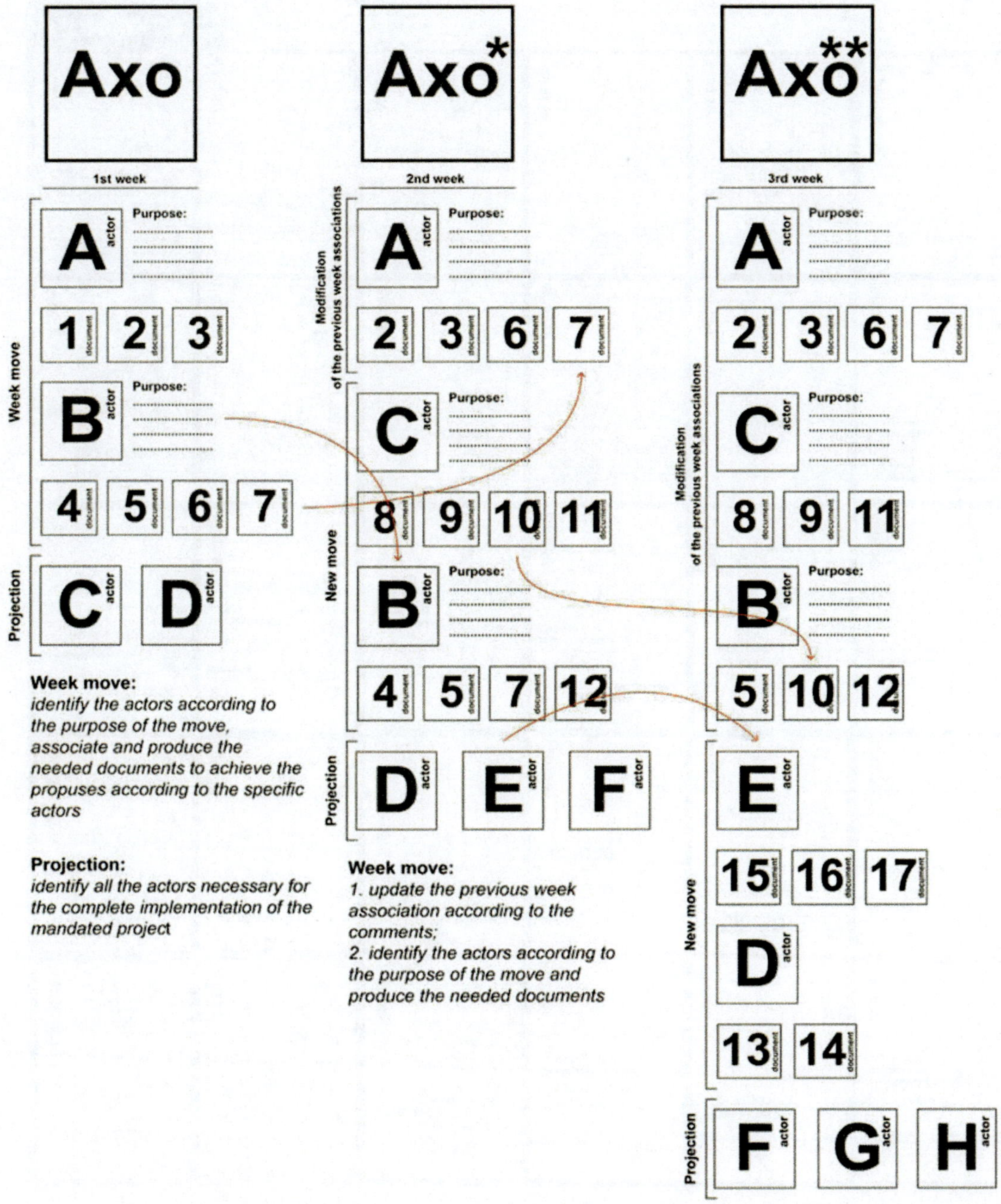

Fig. 18.3 Each week the students foresee the most suitable strategy for the implementation on the project through objectives embedded into a set of interlocutions with possible actors, such confronts are mediated by documents specifically produced for the intent. Students successively modify the associations in the name of the feedbacks received and project their possible future directions. *Diagram* by the Authors

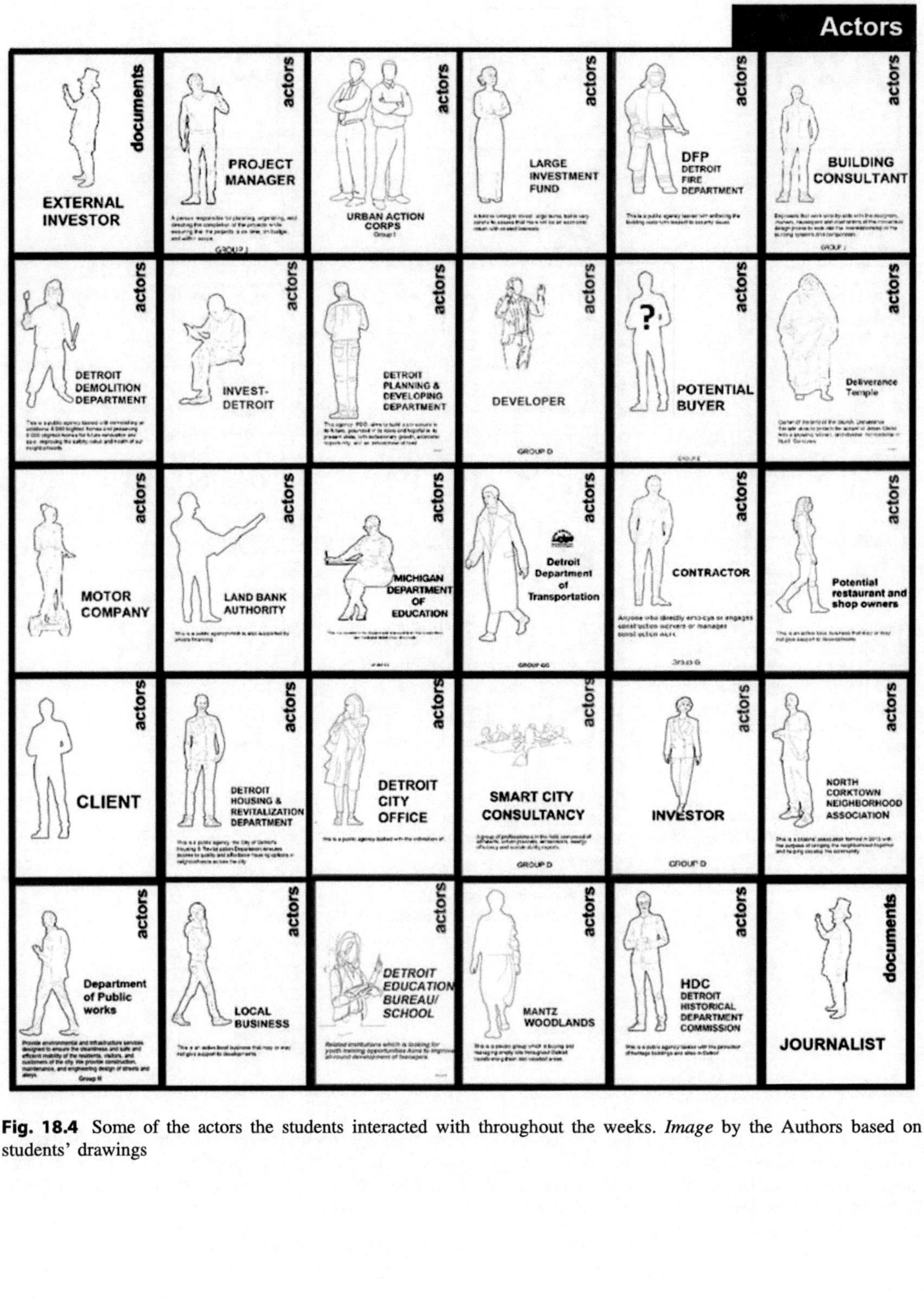

Fig. 18.4 Some of the actors the students interacted with throughout the weeks. *Image* by the Authors based on students' drawings

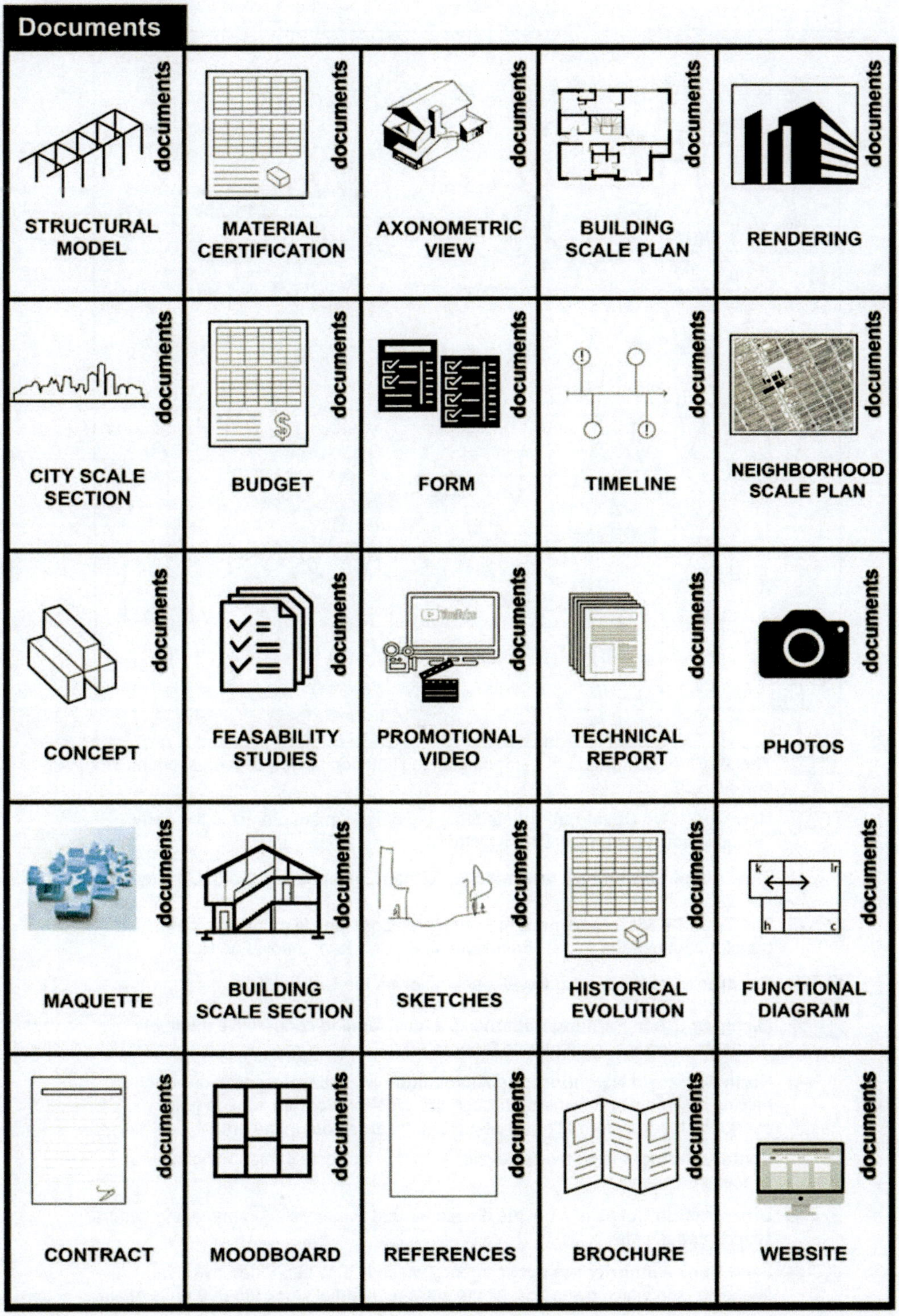

Fig. 18.5 Nature of the documents is up to the groups' own choice: technical analyses, budgets, zoning change forms, technical plans, renderings, city-wide sections, but also websites, promotional videos, and marketing brochures. *Image* by the Authors based on students' drawings

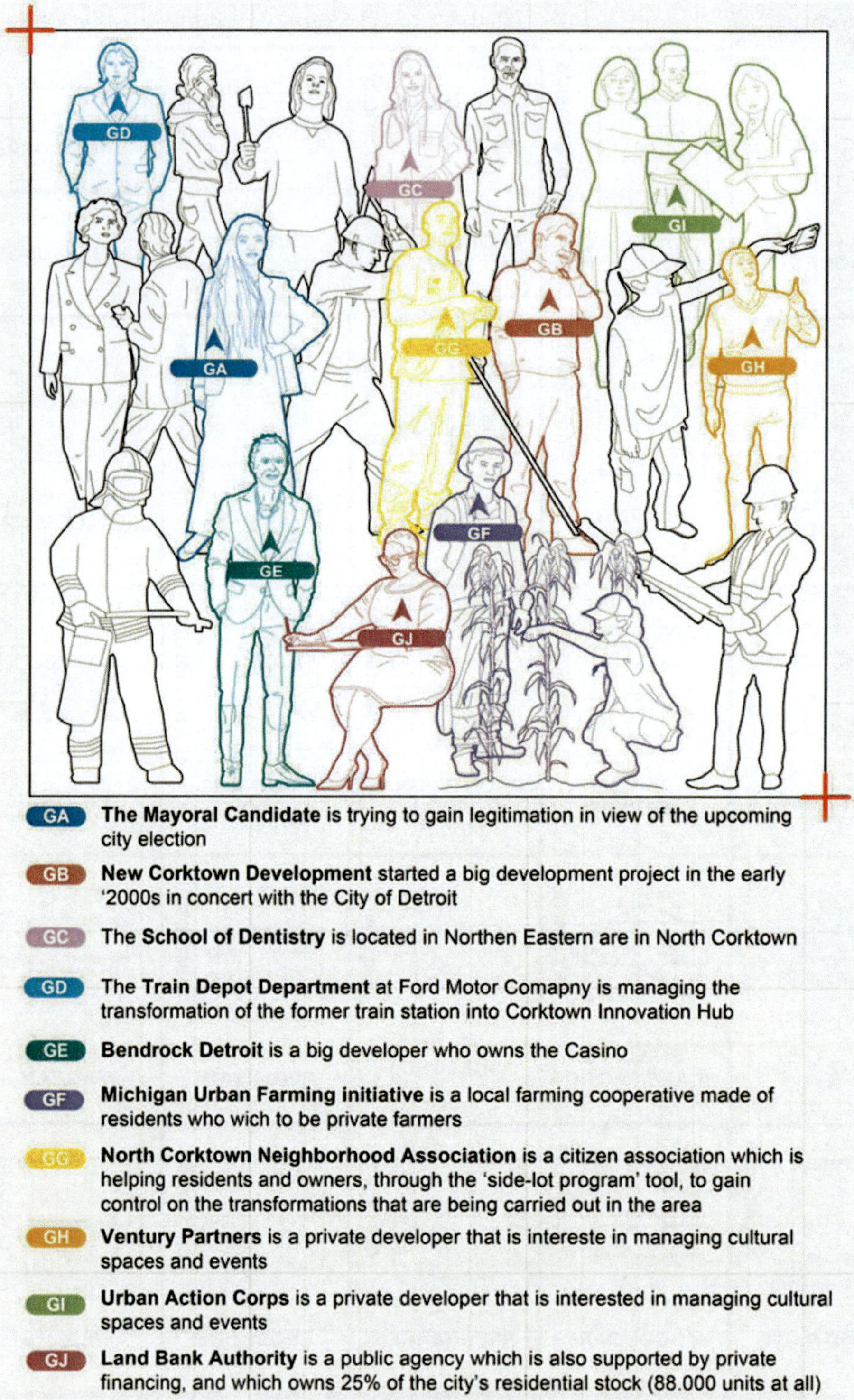

Fig. 18.6 Each group of students has a different client, whose mandates are offered through varied profiles: clients have a story, their intentions are not always clear, they don't know exactly what they want, they can change their mind during the time. *Image* by the Authors

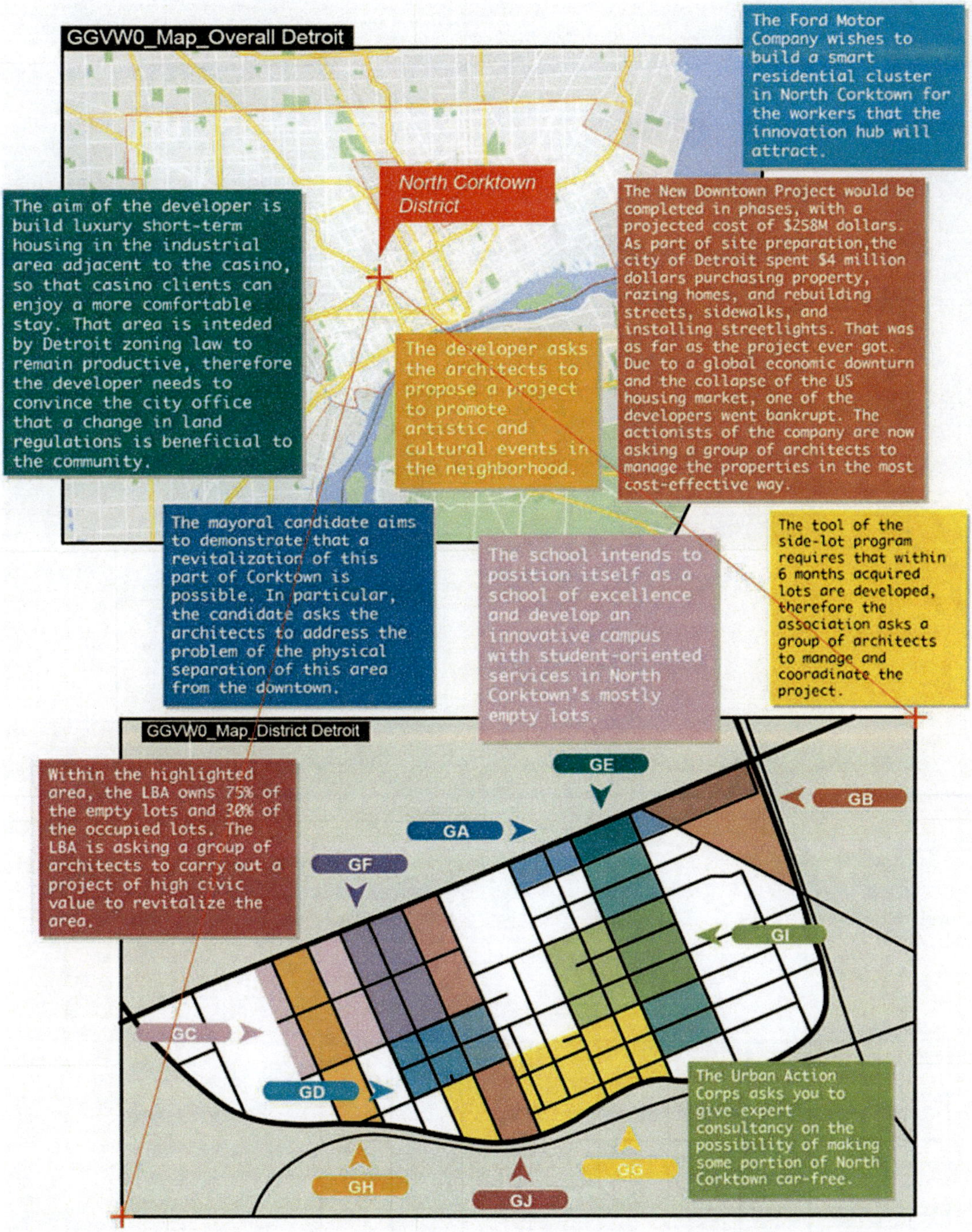

Fig. 18.7 In the first two years of the course the Detroit neighbourhood of North Corktown works as "board of the game"; it is offered to students as animated by opposing interests and controversies, an entanglement in which it is not possible to make easy distinctions between analysis and synthesis, a place of the Latourian matters of concern and not of matters of fact (Latour 2004). *Image* by the Authors based on studio drawings by the students

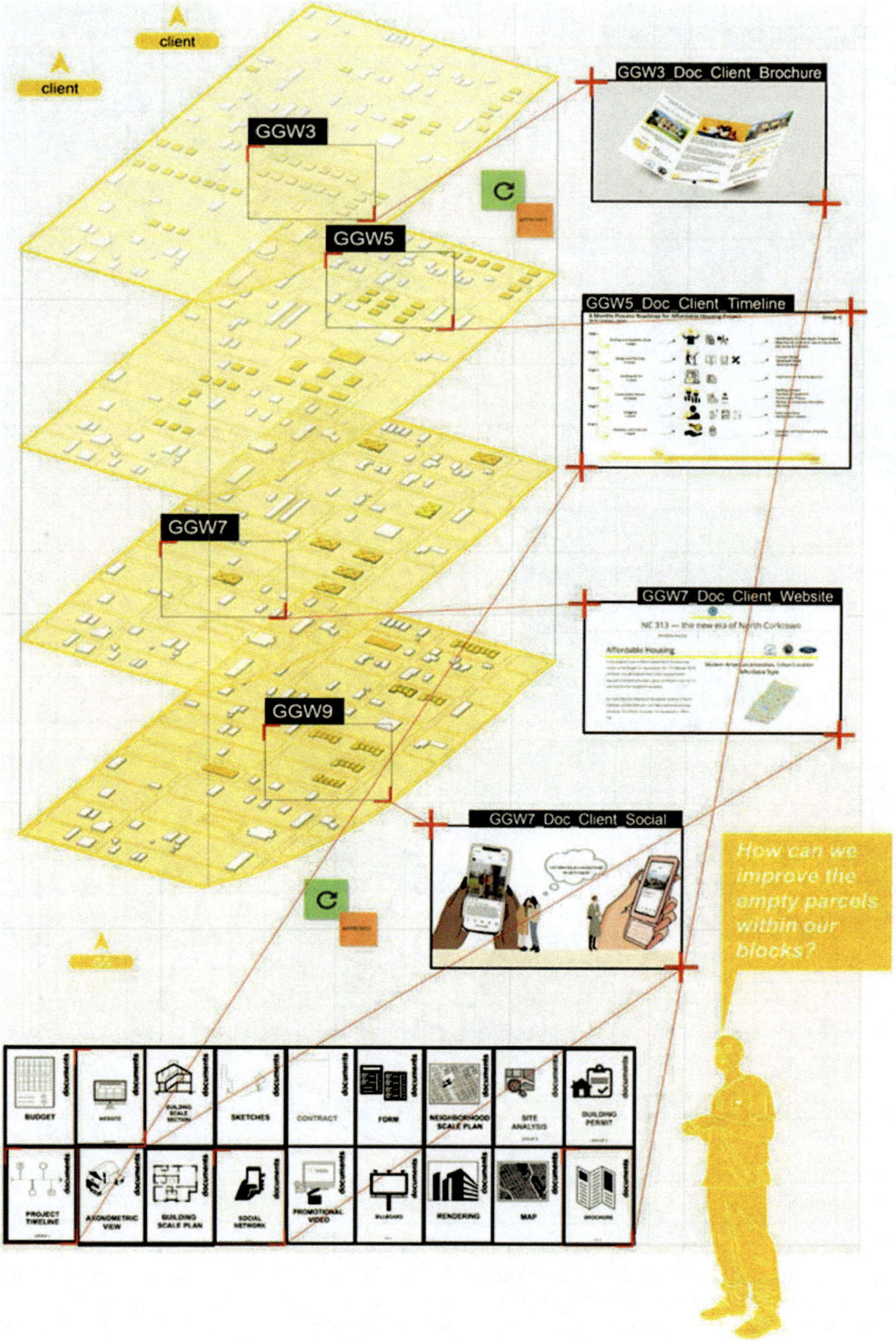

Fig. 18.8 Project consists of specific documents produced to negotiate with a specific interlocutor, in this image a summarisation of the exchanges occurred with the client during the weeks by one group of students. *Image* by the Authors based on students' drawings

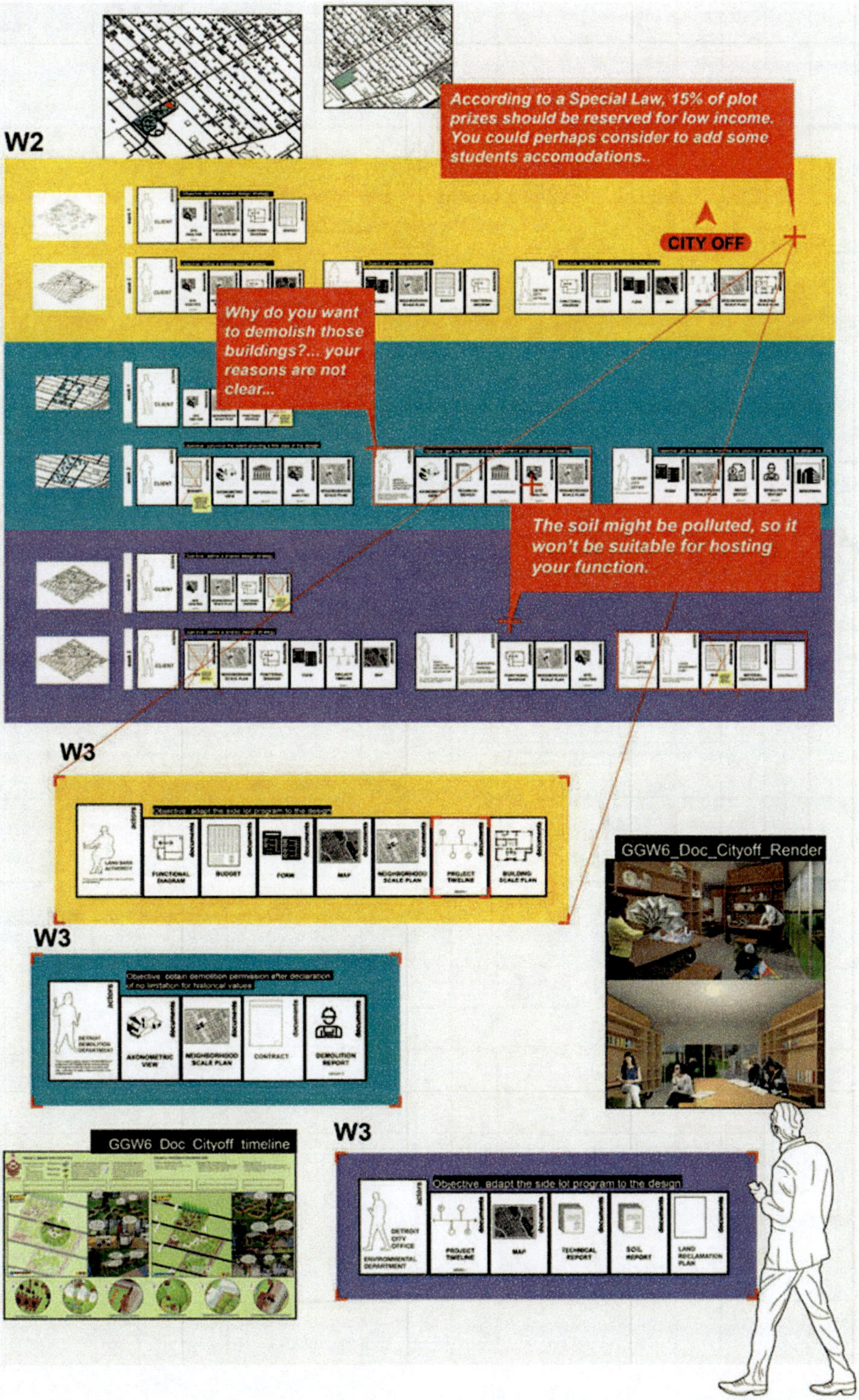

Fig. 18.9 During the course, students interact not only with the client but also with other actors who they wish to associate with to carry out their design strategy. Actors, in turn, support, negotiate or oppose the proposals from their own perspective, whether it is a bureaucratic perspective (as the Detroit City Office represented in this case), or a private one (a citizens' association or an investor). *Image* by the Authors based on students' drawings

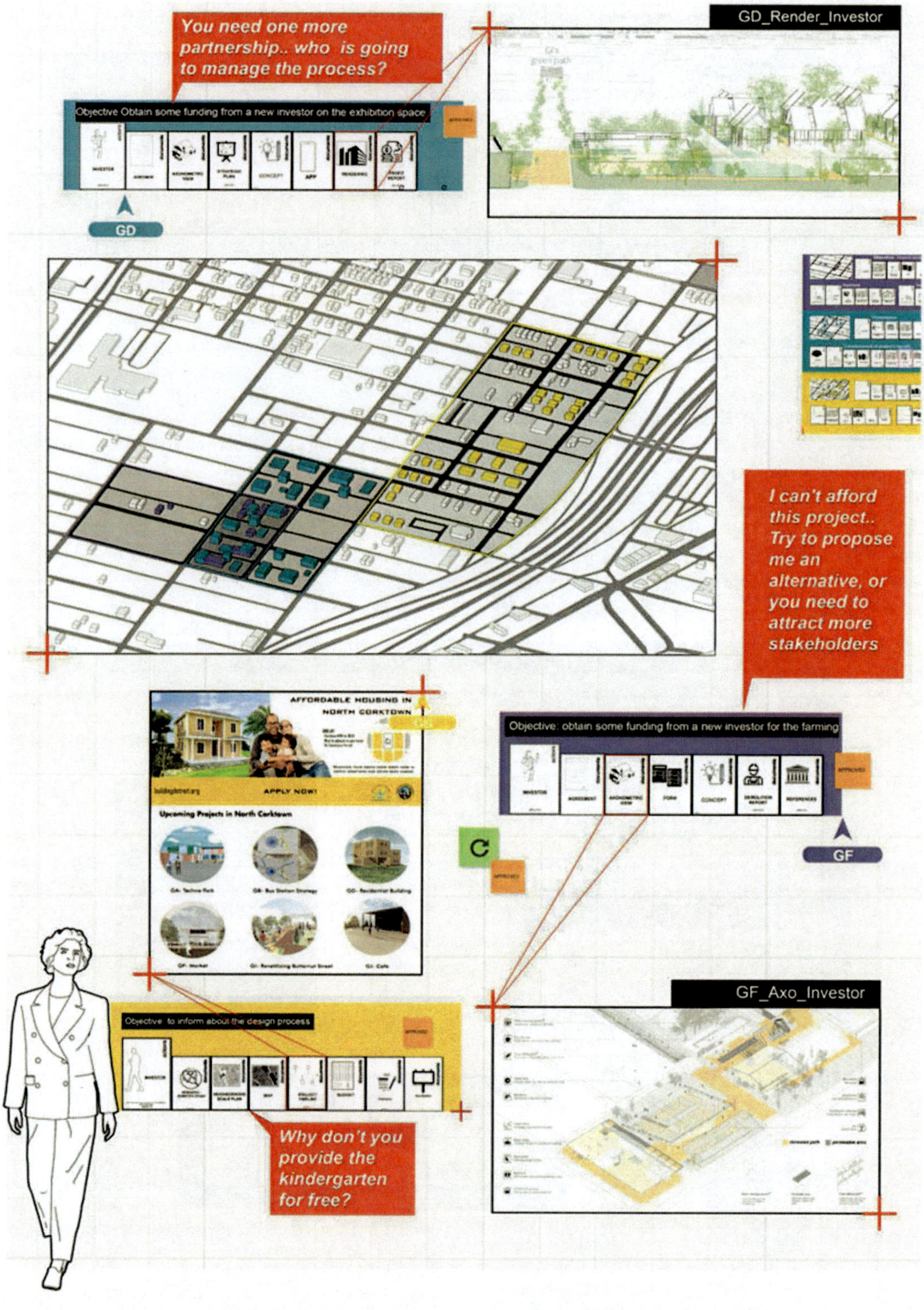

Fig. 18.10 Students received feedback throughout the semester by teachers playing different roles as needed and by a series of guests who interpreted specific actors such as public agencies, citizens associations, developers, and so forth; in this case, the developer. *Image* by the Authors based on students' drawings

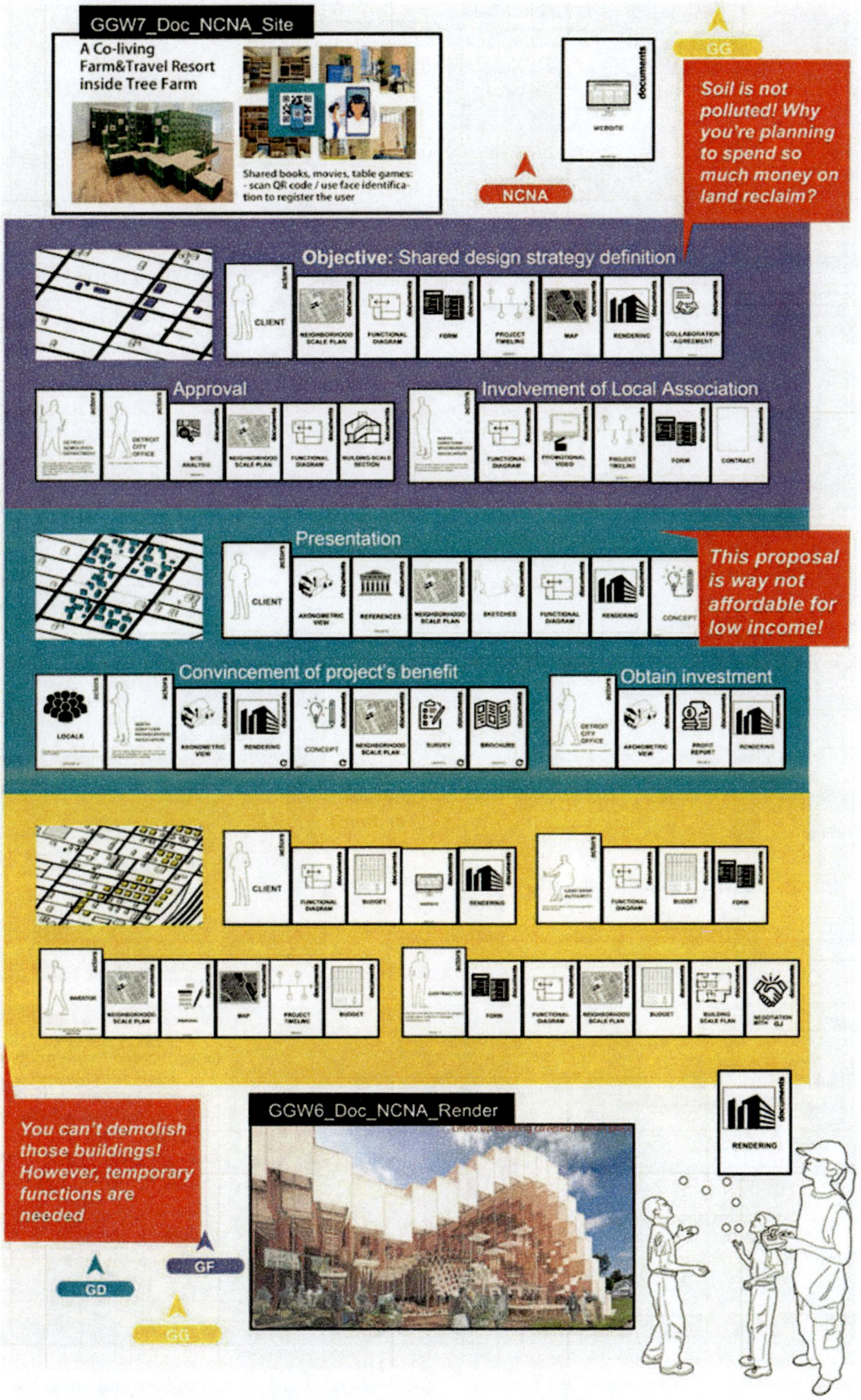

Fig. 18.11 Interlocutors the students interact with can contradict themselves, it is up to the students to navigate them, using the project as a political tool to associate those actors they deem necessary; in this case the Neighbourhood Association. *Image* by the Authors based on students' drawings

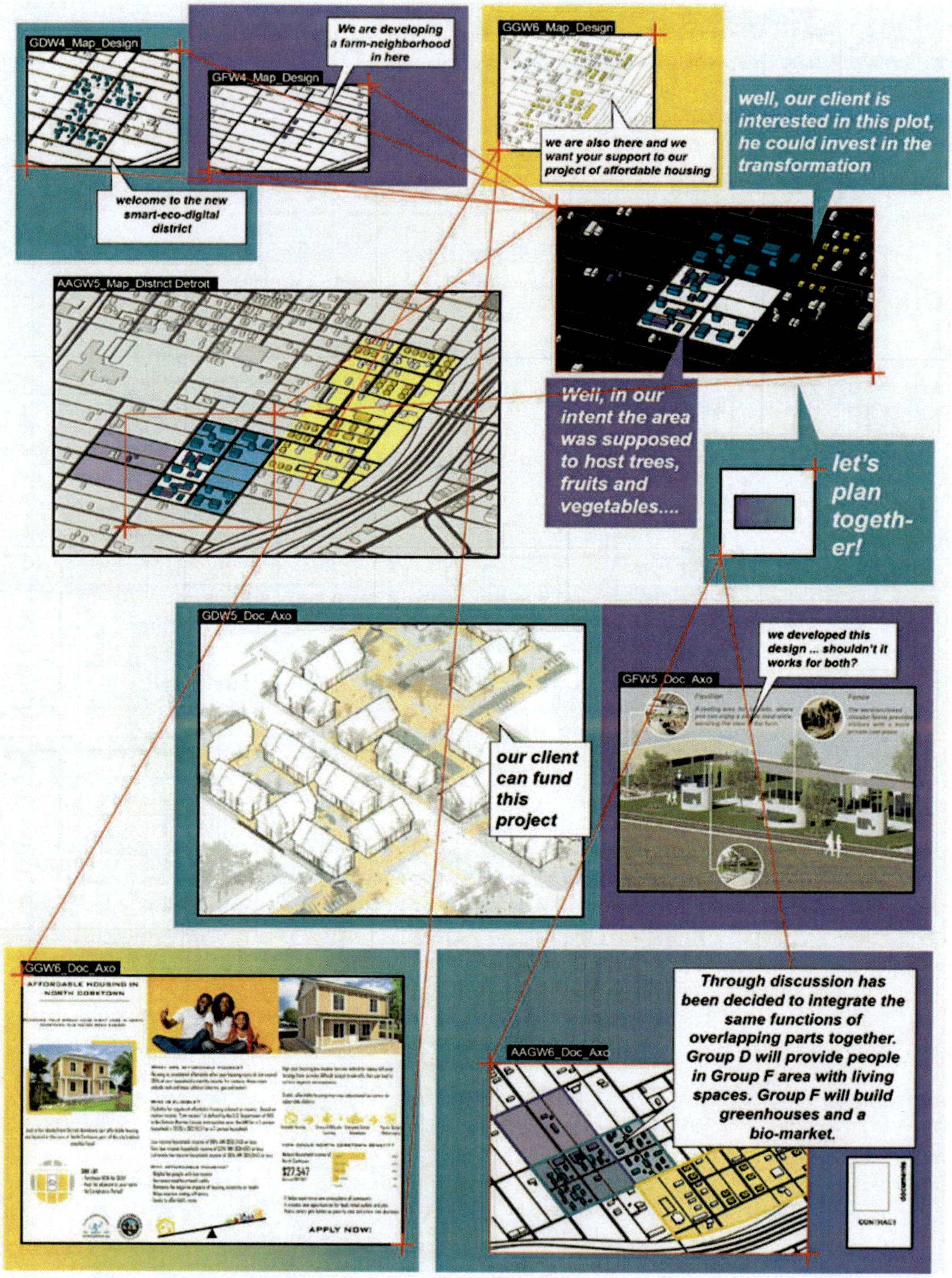

Fig. 18.12 Students do not work in an absolute time and space, but rather at the same time and in the same space where other students work. To foster conditions for interaction, different groups' project areas intentionally overlap, to the effect that the groups step on each other's toes at different times during the course, having to consider each other's perspectives and negotiate a shared strategy on where/when/what/how to carry out their project. In this sense, the groups of students are encouraged to confront each other, rather than act as single units that work independently from the actions of the rest of the class. *Image* by the Authors based on students' drawings

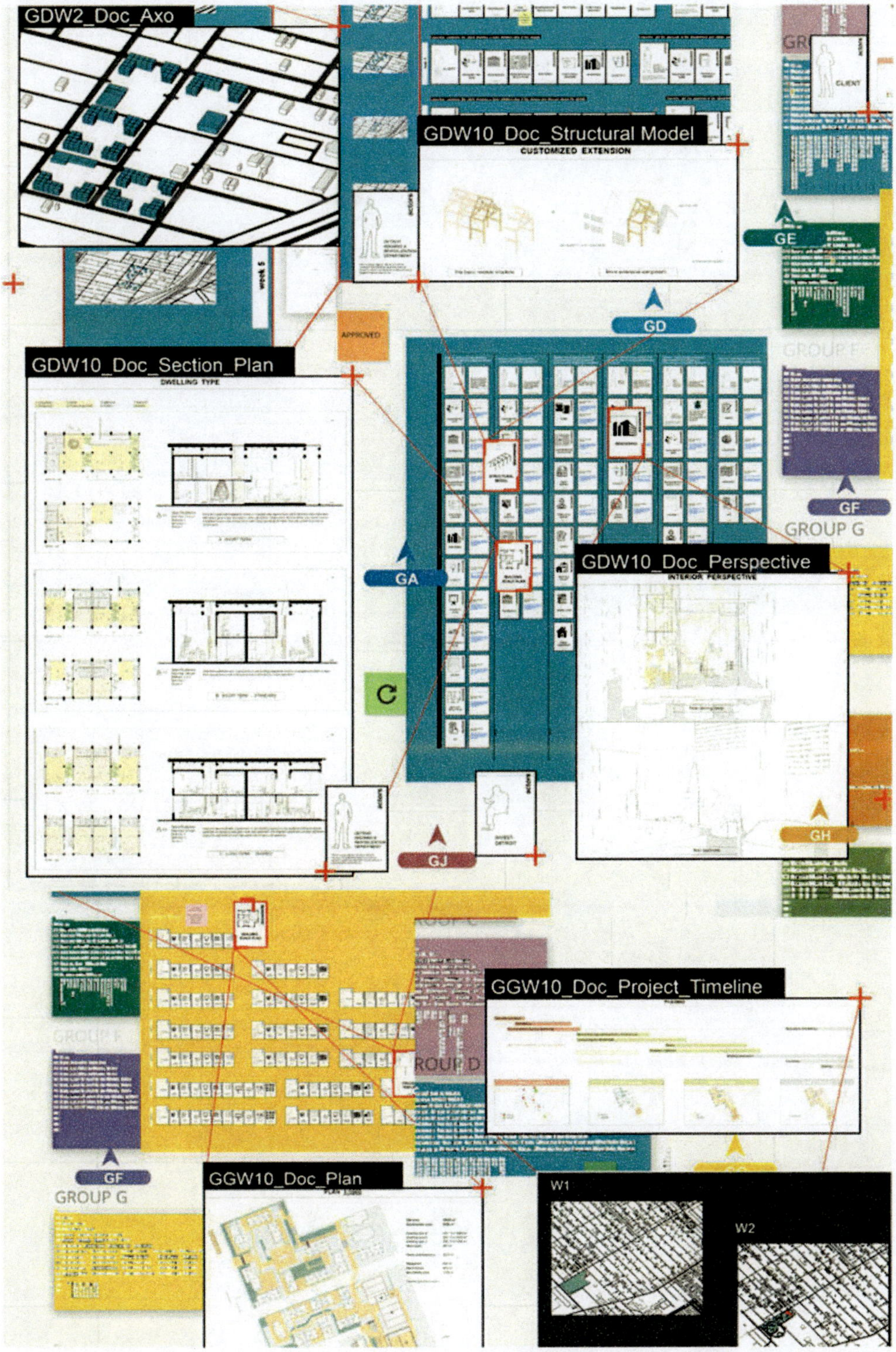

Fig. 18.13 Week after week exchanges and detournements are layered into the game board, producing a thick entanglement of actors, documents, and design proposals. *Diagram* by the Authors

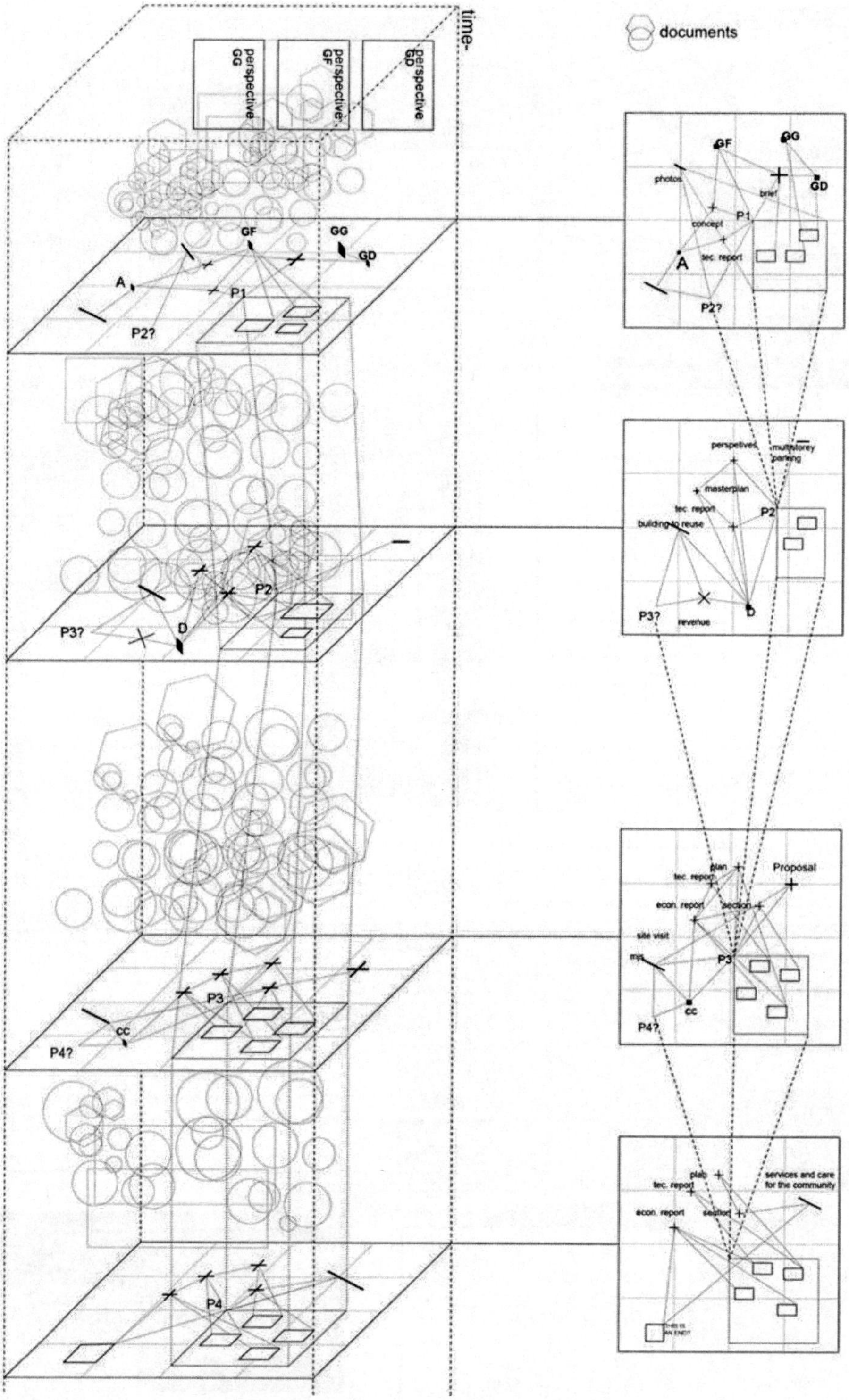

Fig. 18.14 In the conclusion of the course, it is possible to retrospectively enact the succession of each interaction and its spatial repercussions. *Image* by the Authors based on students' drawings

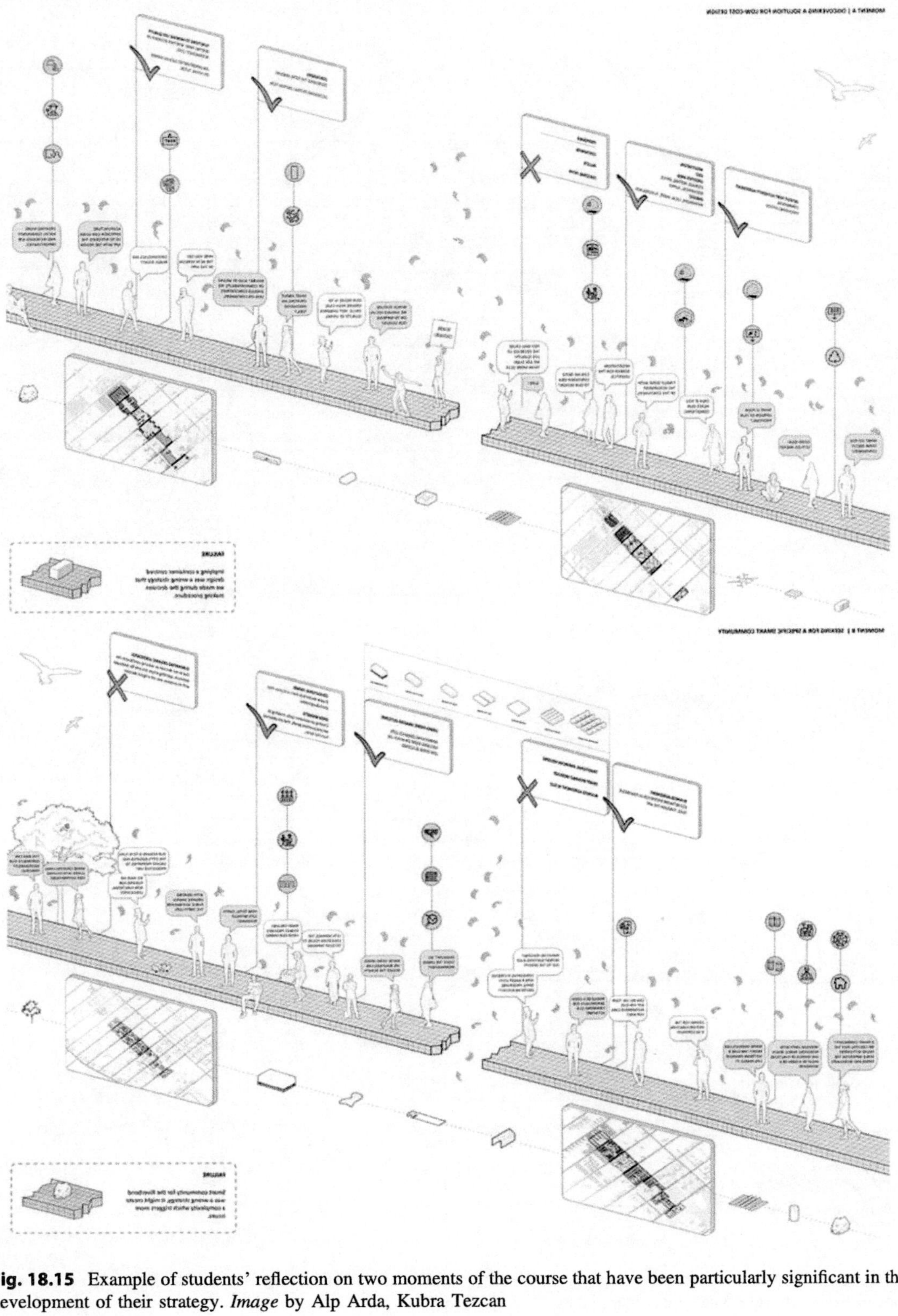

Fig. 18.15 Example of students' reflection on two moments of the course that have been particularly significant in the development of their strategy. *Image* by Alp Arda, Kubra Tezcan

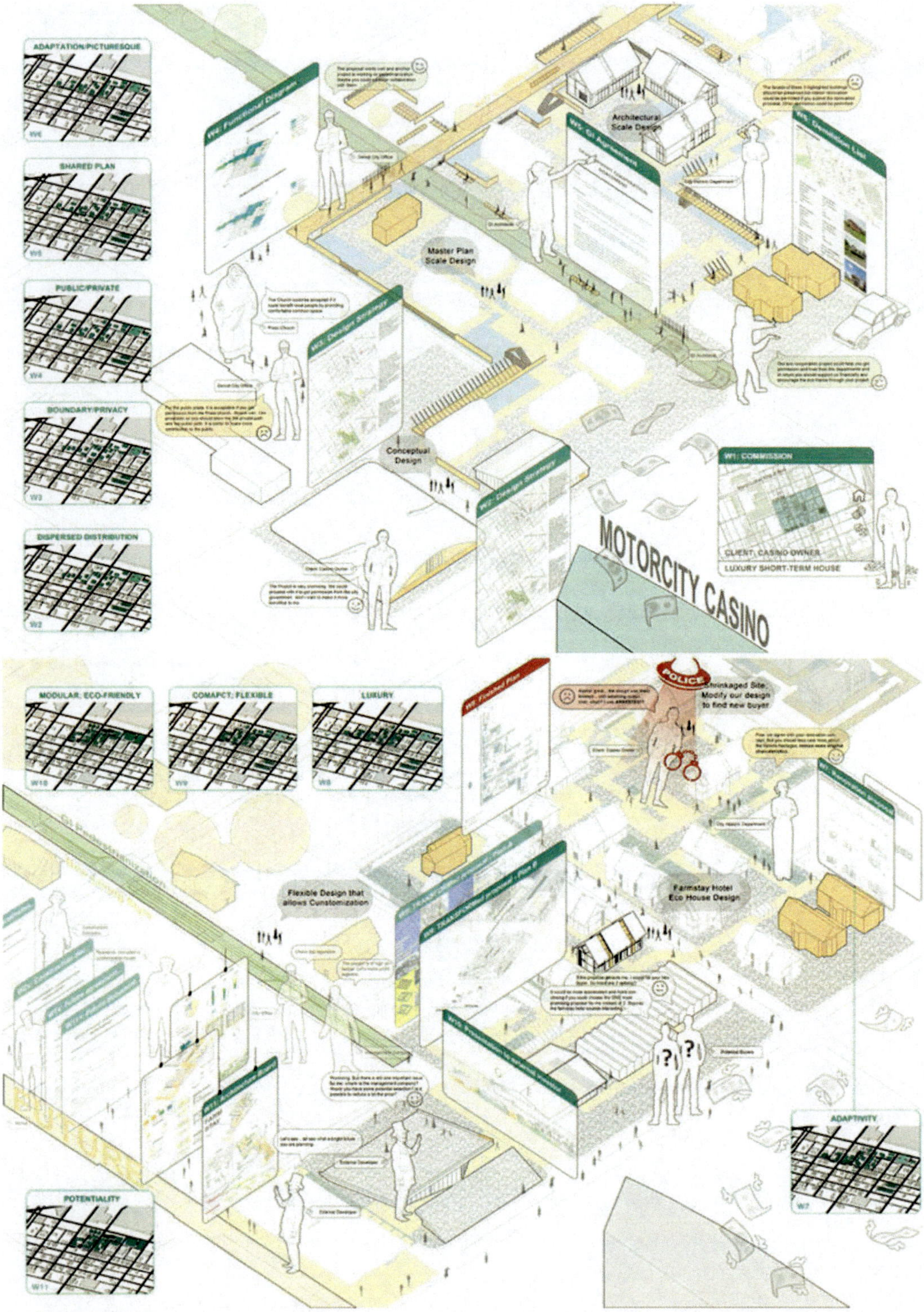

Fig. 18.16 Example of retrospective strategy realised by a group of students. *Image* by Giordana Parisi, Yu Ziyue, Zhong Jihao, Phan Tran Khue Tu, Zhang Xian

Acknowledgements This article traces a pedagogical experiment carried out in the course "Architectural Theory Lab A" 2020–21, 2021–22 at Politecnico di Torino. The Lab was part of a curriculum reform of the Bachelor "Architecture", the result of a three-year effort and debate within the School of Architecture, and took shape through a year-long exchange among faculty and Ph.D. students. Authors wish to specifically thank Professor Giovanni Durbiano for supporting the experiment.Authors want also to thank the students for their effort and for the works they have produced and which are partially shown in the images of this article: Alp Arda, Kubra Tezcan, Mengyao Li, Beicheng, LiuZitong Niu, Zhihao Zhou, Sevjiddulam Erdene-Ochir, Esin Asan, Ege Genc, Irem Yigitoglu, Telatar Beliz, Yagmur Arabaci, Beliz Telatar, Margerita Assi, Yeraz Lora Yenice, Yesim Ergen, Mario Alonso Ortiz, Maria del Pilar Ferreira, Sarah Delphine Georges, Donatella Rossi, Giordana Parisi, Ziyue Yu, Jihao Zhong, Tra Khue Tu Phan, Xian Zhang.

References

Deamer P (2020) Design pedagogy: the new architectural studio and its consequences. Architecture_MPS 18:2–8

Federighi V, Bruno E (2022) The detroit great game. Explorations around architectural design and its agency. AADR -Art Archit Des Res

Haraway D (1988) Situated knowledge: the science question in feminism and the privilege of partial perspectives. Fem Stud 14(3):575–599

Kuhn S (2001) Learning from the architecture studio: implications for project based pedagogy. Int J Eng Educ 17:349–352

Latour B (2004) Why has critique run out of steam? from matters of fact to matters of concern. Crit Inq 30:237–241

Rodriguez C, Hudson R, Niblock C (2018) Collaborative learning in architectural education: Benefits of combining conventional studio, virtual design studio and live projects. Br J Educ Technol 49(3):337–353

Till J (2005) Lost judgement. EAAE Writings Architectural Educ 26:161–181

Speculative Futures: Design for Change

19

Caroline Sohie

Abstract

Communities worldwide are in flux, challenged by uncertainty and volatile pressures to redefine their built, ecological, economic and social existence. It raises fundamental questions about how communities can be enabled to better plan for change and set out a future trajectory that enables them to thrive. The paper proposes a speculative futures model that positions design at the heart of the sustainability debate. It proposes a paradigm shift to meet the complex challenges of the future, underpinned by three sustainable design transitions outlined in the paper, namely systemic sustainable design, regenerative sustainable design and speculative sustainable design. The focus on sustainable transitions enables design to transcend the limitations that come with a problem-solving perspective of sustainability. Within this context, the medium of design, and architecture in particular, is positioned as an active agent of change that can positively influence the dynamic relations that constitute living systems at all scales. It reframes the core focus of sustainable design as enabling continual systemic transformation, where resilience is pursued—not by resisting change—but by designing for change. The author applied the proposed model within the context of an academic research-by-design process. Referring three project case studies, the potential of the approach is examined as a tool to create new entry paths towards critical imaginaries that can engage with the complex challenges of the future while positioning design as a medium to stimulate agency, enabling informed collective decision making.

Keywords

Sustainability transitions · Regenerative design · Speculative design · Systems change · Design agency · Futures

19.1 Introduction

Two irresistible forces—change and complexity—have never been as pertinent as now.

Until recently, the notion of change was an abstract phenomenon, removed from our day-to-day existence. Currently, we keep adjusting our lives, in accordance with fundamental shifts impacting our world. Societies worldwide are in flux, challenged by volatile pressures to redefine their built, ecological, economic and social environment, torn between wide-ranging demands, such as a rapidly urbanising environment, scarcity of resources, climate change and the consequences of rising inequity.

C. Sohie (✉)
Instinct, London, United Kingdom
e-mail: caroline.sohie@instinct.london

S. Hilal et al. (eds.), *Design for Partnerships for Change*, Sustainable Development Goals Series,
https://doi.org/10.1007/978-3-031-36993-3_19

This shifting reality raises fundamental questions about how design, and architecture in particular, can adopt more effective agency in responding to the complex challenges communities face today while unlocking the potential for a sustainable, equitable and inclusive future.

Sustainability, as a science, practice and movement, has expanded over time and made significant inroads in framing a sustainable agenda globally, through the adoption of the UN SDGs (United Nations 2015) by 193 countries and the introduction of the New Urban Agenda (United Nations 2017), introducing the principles of an inclusive roadmap towards a more sustainable future. However, the sustainability field has not led to the fundamental transition of the social-ecological system trajectories towards sustainability, with environmental and social degradation continuing at accelerated pace. The focus on addressing symptoms rather than causes is deemed one of the major factors that continue to stifle the field in achieving not only net-neutral states but the much more comprehensive ambitions of generating thriving and flourishing living systems (Gibbons 2020).

If we are serious about these ambitions and aim to achieve a more regenerative approach to the urban agenda, it is essential to develop design methods that can engage with growing complexity and tools that facilitate critical choices in the present that can influence sustainable future outcomes.

It raises intrinsic questions about how SDGs can be translated on the ground, responding to local and global circumstances, in a world of constant flux. It also highlights the need, at a fundamental level, to exchange important ideas about the future we will be sharing and find collective alignment in order to formulate sustainable responses together.

In view of this context, the paper explores the potential of design as an agent of positive change, reframed by three sustainable design transitions, namely systemic, regenerative and speculative sustainable design.

Section 19.2 describes how the implementation of the concept of sustainability is challenged by grappling with complexity. A systemic design approach, vested in complexity theory, is posited as an approach to engage and meet the layered challenges of the future.

Section 19.3 focuses on the second transition and contextualises how the concept of sustainability has evolved overtime, elevating its paradigms, aims and methodologies to support more holistic regenerative development goals.

Section 19.4 establishes the intrinsic link between speculative design and the sustainability praxis. The third design transition introduces futures thinking as an entry point to reframe the sustainability debate within a context of intergenerational accountability and the need for future-oriented collective narratives.

Section 19.5 introduces a speculative futures research-by-design framework, which assimilates the three transitions. The model is exemplified by three academic architectural design projects, leading to concluding reflections in the last section, on how the model can lead to a paradigm design shift towards sustainability (Fig. 19.1).

19.2 Systemic Sustainable Design

The concept of sustainability has evolved significantly since the publication of the Brundtland Report 'Our Common Future' in 1987 in response to the need for a more comprehensive model akin to the complex nature of the issues the concept intends to address (Adetunji et al 2003). Sustainable development was deemed to reconcile economic development with the protection of social and environmental balance.

It is now acknowledged that the concept of sustainability extends beyond the narrow scope of reductionism and compartmentalised specialisation, implicated by the model. The issues addressed by sustainability are complex with dynamic multidimensional variables and sub-variables. Their complexity cannot be fully comprehended nor engaged with, without a systems approach and strategy (Rose 2001; Dovers and Handmer 1992).

Consequently, a systemic model is essential in achieving sustainability with a more comprehensive focus, through the integration of environmental,

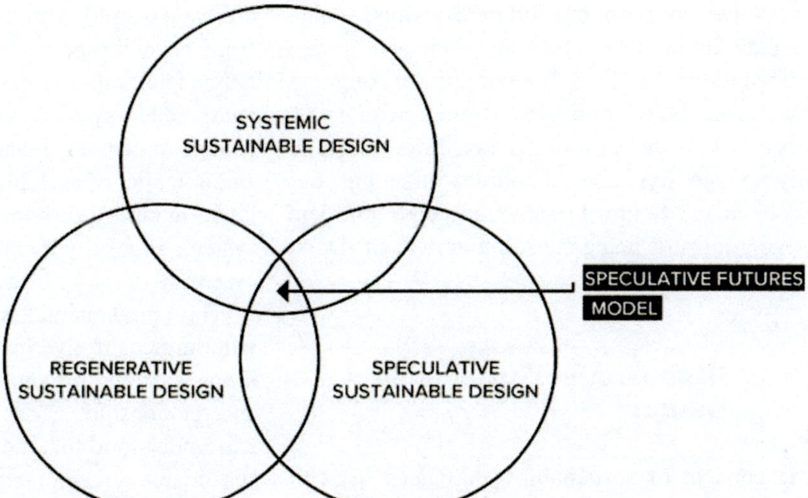

Fig. 19.1 Speculative futures model

social and economic systems to improve the quality of life within earth's carrying, generating and assimilating capacity to ensure intra- and intergenerational equity (Adetunji et al 2003).

This raises the question; if design aims to rise to the challenge of sustainable development, what fundamental shift in approach is needed to engage with those complex demands now and in the future.

Dealing with multifaceted, networked conditions destabilises a traditional design process. Multiple factors make it difficult to read and interpret the situation holistically and challenge the understanding of the role and hierarchy of entangled relationships at play.

The natural human tendency is to counter complexity and reduce uncertainty by jumping to conclusions all too easily, moving away from the openness of design abduction to result-focused reasoning. However, in the complex problem space of sustainability, any attempt to search for 'the' solution would be riddled with assumptions. As such, the design quest is to be repositioned and move away from striving towards an immutable structure, that so-called solution, to keep the design abduction alive in a complex design process (Dorst 2019).

In the light of this, an explorative and reflective design practice is required as the prerequisites for applying a conventional problem-solving approach to designing—first determine the goal, map problem and solution spaces, optimise the search path, create a solution—are not fulfilled in truly complex design situations (Dorst 2019).

We need to move away from design models based on problem solving towards a new paradigm based on complexity theory and systems thinking (Ball 2012; Dorst 2019).

In highly complex systems, fundamental shifts take place due to the emergence of order, rather than goal-directed creation; change is achieved through influencing the system, rather than the implementation of a plan to solve a problem. A new state of relative stability can be achieved by instilling resilience. Viewed through the lens of complexity theory, the way to achieve progress is to create impactful interventions that enable systemic change and move the whole system to a more desired state (Dorst 2019).

To increase resilience, the approach to sustainable design needs to undergo a massive shift by relinquishing the idea of a definable design problem, solution and project and by fully embracing the dynamic nature of a complex systems context. This means that also, the outcome of design needs to be reconceptualised as a highly flexible system with an inherent, transformative purpose that keeps redesigning itself as time goes by and circumstances change.

As we envision our futures; resilience and adaptability are key. Systemic sustainable design interventions have to be primed for ongoing transformation over time in a continuing dialectic process with the various factors, informing and driving the dynamic conditions affecting our world in the future. Design transcends problem solving towards being the solution in itself (Dorst 2019).

19.3 Regenerative Sustainable Design

The concept of sustainability has been used to describe conserving environmental resources for human benefit, popularised by the Bruntlandt Report as meeting current and future needs within environmental limits. This conventional approach to sustainability emphasises that unbound use and depletion of environmental resources is detrimental for continued human existence. The focus is largely anthropocentric reinforcing the question on how to ensure continued economic growth within a context of finite resources (Gibbons 2020).

With the birth of sustainability science, a new type of 'contemporary sustainability' emerged which incorporates considerations of ecosystemic viability, social justice, social-ecological and social-ecological-technical systems, satisfying liveability and normativity (Gibbons 2020). Notwithstanding the more comprehensive approach, the focus is still anthropocentric aiming for human wellbeing now and in the future, within limits, through 'solving' complex 'problems' that are value-laden, contested and locally specific' (Gibbons 2020), introducing goal-oriented concepts such as transition, transformation, leverage points, process and transdisciplinarity. While some scholars motivate that this contemporary strand of sustainability science is gaining maturity, shifting from quantitative to qualitative development processes, others critique it as a paradigm in crisis.

'Regenerative sustainability' is deemed to represent the latest wave of sustainability discourse. It is founded on a holistic worldview and creates a paradigm shift as it moves away entirely from an anthropocentric view of sustainability. Human and other life systems are considered one autopoietic system in which developmental change processes manifest the unique essence and potential of each place or community. Its aim is to manifest thriving and flourishing, living systems in a fully integrated individual-to-global system.

This fundamental shift calls for symbiosis with humans to live in conscious alignment with living systems principles of wholeness, change and relationship, as nature does (Gibbons 2020). It is understood that the move from conventional sustainable development principles to a regenerative focus has the propensity to support a radical transition towards thriving and inclusive communities at all scales.

In a world in which the only constant is change, fostering the capacities of living systems to evolve in a developmental change process is deemed the best strategy to attain thrivability and consequently resilience. In view of this, sustainable development is repositioned as a process for supporting the inherent self-organising, life enhancing, self-inducing capacities of whole living systems (Gibbons 2020).

As living systems change in ways that are not predictable or certain, this sustainable design transition requires us to be comfortable with uncertainty and allow time to let regenerative processes unfold. It focuses on regenerative capacity development rather than achieving fixed goals, catalysing positive transformational change overtime.

19.4 Speculative Sustainable Design

From the onset, the future-oriented principle of intergenerational equity was implicitly encapsulated in the definition of sustainable development (Brundtland 1987) and explicitly stated in the revised definition of the Rio Declaration on Environment and Development 'to equitable meet developmental and environmental needs of present and future generations' (UNCED 1992).

Pearce et al. (1989) proposes three pathways of achieving sustainable development, which includes futurity aside of environment and equity. In this conceptualisation, futurity involves concerns for short and medium as well as longer term that will ultimately impact on the inheritance of future generations and their quality of life. Also, equity places emphasis on a future prospect, as in addition to providing for the needs of the least advantaged within current society, it advocates for intergenerational equity with fair treatment of future generations. In other words, sustainability embodies the promise of societal transition towards a 'more equitable and wealthy world in which the natural environment and our cultural achievements are preserved for the generations to come' (Dyllick and Hockerts 2002, p. 130).

However, when attempting to describe this world, the question remains unanswered in terms of what future desired state we are collectively aiming for. The sustainability agenda lacks a collective vision, or a set of aspirational futures, to work towards (Angheloiu et al 2017).

To accelerate sustainable transition, it is ever more critical to establish a platform that can provoke critical collective reflection, where ideas and ambitions can be shared about the futures that we, as individuals and communities, will be living in together.

Our futures are influenced by the choices we make in the present. Even though we cannot determine which future will eventuate, nevertheless we influence the shape of the forthcoming future by or actions or inactions in the present. Our choices have consequences and so they need to be made wisely (Voros 2001).

Speculative design can be a significant contributor to making these choices responsibly. This strand of design, located at the intersection of critical design research and future studies, is employed to challenge the status quo in order to rethink products, systems and worlds. Speculative design envisages and anticipates the future, while at the same time it helps us to understand and rethink the world of today (Mitrović et al 2021) or as Dunne and Raby (2013) suggest, 'design can allow an individual to open windows on the future in order to better understand the present'.

Future studies have provided a futures taxonomy framework (Hancock and Bezold 1994), as outlined in Fig. 19.2, to speculate about potential alternative futures, distinguishing four classes (Henchey 1978): probable, plausible, possible and preferable futures. Through the lens of 'alternative futures', space is created to discuss and consider possibilities and options, towards imagining and redefining our relation to reality itself. Through radical imagination and by using design as a medium, it propels thinking, raises awareness, questions, provokes action, opens discussions and can offer alternatives that are necessary in today's world (Mitrović et al 2021).

The exploration of future scenarios is used as a tool to help expand the 'possibility space' by encouraging speculation of multiple and wide-ranging alternative futures (Miller 2006). The emerging scenarios can be used as the basis for back casting, a method to construct a plausible causal chain leading from a contemporary context towards a particular future scenario or event.

The unique contribution of future scenarios as a reflective tool lies in the development of a more comprehensive understanding of systemic change. The framework can enable designers and change agents to identify what contemporary strategic agenda can positively impact the sustainability trajectory and influence the current conditions governing tomorrow.

19.5 Design for Change

19.5.1 Introduction

The author developed and led an academic design programme to examine the potential of applying a speculative futures model; a research-by-design methodology which is underpinned by the sustainable design transitions described in the previous sections of the paper.

Set up as an interdisciplinary incubation platform, the programme aimed to provoke informed debate on the broader societal role of architectural practice and gain insight in design's

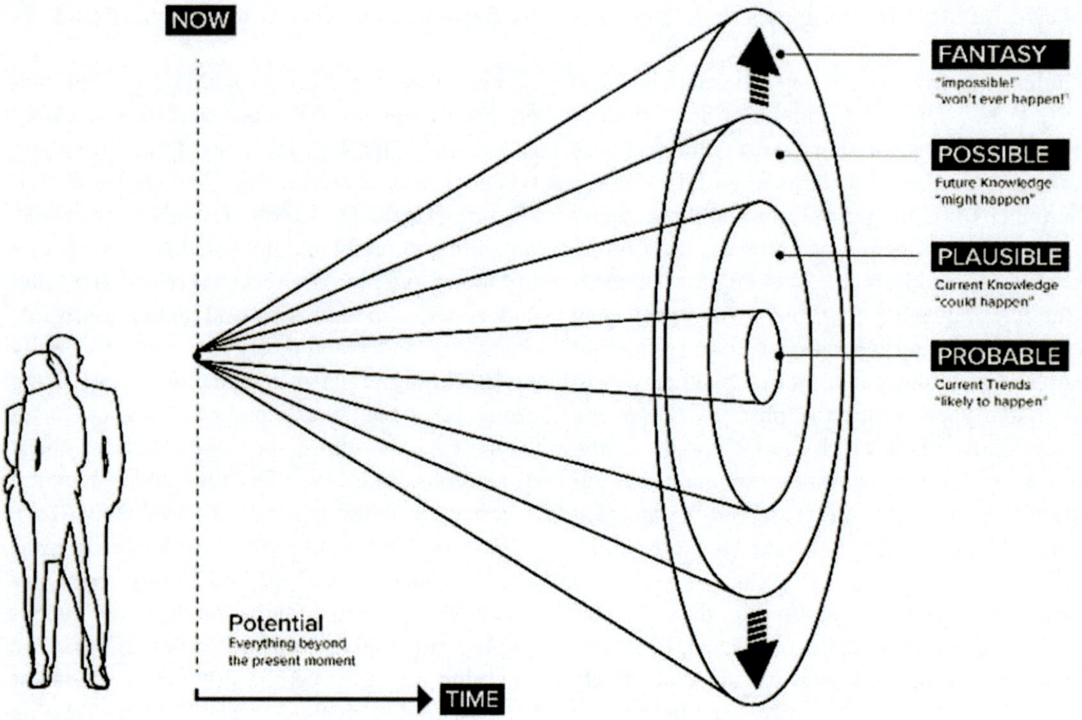

Fig. 19.2 Futures taxonomy. Adapted from Hancock and Bezold (1994)

potential engagement with a changing reality. The focus was to encourage innovative ideas and approaches that challenge the status quo and bring a critical, albeit imaginative perspective towards designing the world of tomorrow.

The study was undertaken as part of an Architecture Master's Design Studio programme at the Faculty of Architecture, KU Leuven University, Belgium. The methodology was first tested during speculative interdisciplinary design workshops held during three consequential years (2016–2018), which consisted of an intensive design week, attended by approximately 100 architecture International Master's students.

The method was further elaborated and refined, leading to an Architecture Master's Thesis Studio Programme (2020–2022).The research-by-design programme was held during one semester and initiated by an intense workshop week introducing the key tools and terms of reference of the speculative design approach, with a total of 11 architecture students completing a final Master thesis.

The studio operated in the space between design, critical urban theory and speculative futures. Conceived as an urban futures think-tank, the studio's framework positioned architecture as a critical medium to articulate the unknown—and in some ways unknowable—futures. The central objective was to identify strategic design agendas and enable alternative—sustainable—development trajectories to leverage empowered responses to twenty-first century global challenges.

19.5.2 Research-By-Design Method

The research-by-design method was formulated around the question on how a design process can inform critical insights and decisions that as a society we need to make now to enable the shift from an extractive to a regenerative future, and secondly, how the medium of architecture can stimulate and contribute to enabling and shaping this changing reality.

Core to the framework was a future-oriented research methodology, used to stimulate strategic thinking and speculative design capabilities. The theoretical lens of 'possible futures' was adopted to enable design and decision-making processes that challenge the status quo and move beyond conventional problem solving; offering innovative imaginative perspectives.

The programme followed a structured approach, enabling the participants to adopt gradually the various research-by-design tools introduced throughout the process, while providing opportunity for micro processes of divergent and convergent design reflection to unfold.

The stages outlined in Fig. 19.3 were used to benchmark layers of knowledge building, as part of a nonlinear reiterative process, leading to insight when design thinking patterns coalesce.

The decision-making processes that underpin the method were not aimed at creating an ideal solution for a specific future context, in accordance with a set of principles. By contrary, the objective was to establish a reflective practice that encourages a profound understanding of the current drivers of change to pre-empt the wide-ranging parameters that might govern our conditions in the future. These were explored by simulating a potential future scenario, that formed part of the 'possible' and 'plausible' spectrum of the futures taxonomy (Hancock and Bezold 1994).

Navigating the complexity of this speculative future by means of design led to critical insights about the potential impact of contemporary decisions on sustainability in the longer term.

In the light of the programme, the students had to select a city as the study area for their research and subsequently, their design intervention.

The research-by-design process, as outlined in Fig. 19.3, consisted of six stages.

19.5.2.1 De-Coding

Inscribed within our cultured spaces are social, technological, economic, ecological and political structures that reflect the tangible and hidden forces that create and sustain them. A strategic scan of prevailing trends and societal undercurrents are undertaken of the study area to identify urgencies and drivers of change that inform the evolution of the built and natural environment.

This includes the search for weak signals, analysis of behaviours and norms within the present that suggest emerging patterns of change impacting at a local and global scale. To aid the

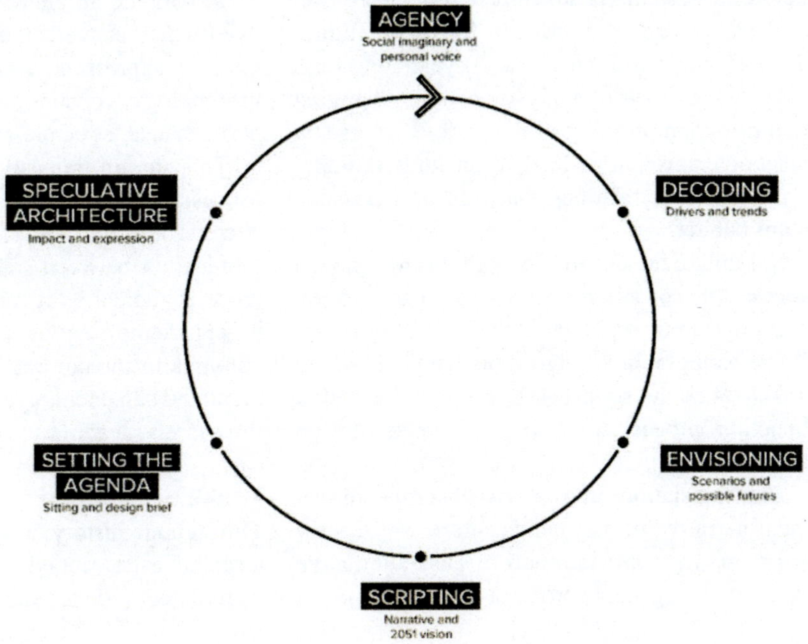

Fig. 19.3 Speculative futures model research-by-design process

process, systems thinking is introduced to dissect complex contexts and de-code active relations and interdependencies.

19.5.2.2 Envisioning

The process shifts to the field of futures research, by adopting the Futures Taxonomy defined by Hancock and Bezold (1994). The programme examines how the speculative lens of 'potential futures' can intensify the design process, by countering bias and conventional thinking patterns, and open up fresh perspectives on a complex problem space.

Using the data attained during the first stage, foresighting (Voros 2001) techniques facilitate the prototyping of a range of possible future scenarios and a number of diverse peak scenarios or sensitivity tested. A reiterative process of analysis, comparison and evaluation enables deepening of knowledge leading to a critical evaluation of the factors and change agents that can have an impact on the human condition. Horizon scanning and back casting techniques assist in unpacking the genesis of patterns of change and their impact on the short- and long-term future of the study area.

19.5.2.3 Scripting

A selected scenario is developed to the next level of depth, creating a critical 'possible future' rationale and narrative for a 30-year long-term horizon. Parametric analysis of the changing conditions and profiling of the various actors and stakeholder groupings lead to an understanding of the dynamic relations that inform a possible future habitat.

Not unlike the development of a screenplay, a process of speculative scripting leads to an emerging narrative. Data visualisation and speculative cartographic design tools are deployed to further refine the script while keeping the design abduction process alive in a complex design process.

The translation into a visual propositional visual narrative of a systemic future world view opens up ways to share and discuss alternative ways of being and provoke reflection on a designer's critical position in relation to this alternative reality.

19.5.2.4 Setting the Agenda

The focus of this stage is to identify a programmatic proposition and urban strategy towards an impactful design agenda, set within a possible future context.

The ambition is to hack the future; through a well-considered architectural programme and strategically chosen site, the design aims to shift the development patterns of the city, leading to a meaningful and transformational outcome. The strategic act of destabilising and reconfiguring the status quo is powerful as it has the propensity to influence the conditions governing tomorrow.

To achieve this, the speculative evolution between the now and the future is interrogated to identify and influence points of inflection, which can fundamentally reframe the future trajectory.

19.5.2.5 Speculative Design

The speculative design process transcends into a spatial proposition situated in the future. Urban design strategies are tested to maximise the project's leverage to create a high impact intervention that can increase community resilience and catalyse positive change.

In parallel, an architectural language is developed which converts the project's ambition into a tangible expression, which includes the study of morphology, poetics of space, programming, performance, structure and materiality.

This step of translation is critical to the entire process as it forces decision-making processes to become specific and create a vehicle for communication between stakeholders. Its mere existence is radical as it raises critical debate about urgent themes, exposes opportunities and shortcomings in the current approach to our environment and can encourage people's imaginations to flow freely, leading to new social imaginaries.

19.5.2.6 Agency

Throughout history, visionaries, thinkers, architects have developed visions of the future and formulated radical ideas, provoking the status quo and levering change in line with alternative value systems they believed in.

It raises critical questions about agency of architecture today and encourages reflection on how design practice can meaningfully engage with a context of uncertainty and give rise to more holistic perspectives on contentious issues of our times, contributing to an alternative social imaginary.

The provocative and potentially disruptive nature of the process opens up the possibility to bring architecture back to the crux of public debate, enabling informed decision-making processes about the world we will be inhabiting in the future.

19.5.3 Case Studies

To ground the potential of the described research-by-design method in enabling sustainable transitions, three examples of the outputs are described in more detail. The architectural reference projects adopted the 6-stage process, leading to highly diverse agendas and design propositions rooted in their respective study area. All generated in their own unique way, catalytic change.

Project 1—**Lima's Block Attack**

Exploring alternative methods of rapid hyper-densification in informal settlement (Farren 2022)

The project responds to the urban crisis affecting the developing world; peak rates of urbanisation combined with reliance on limited financial and infrastructural resources. As one of the megacities in Latin America, the city of Lima, Peru, is used as a case study to investigate alternative approaches to vertical rapid densification to counter urban sprawl.

Lima 2050 evaluates a possible future scenario framed by the core variables of hyper-densification combined with continued degradation of the environmental landscape.

The analysis of intangible and physical attributes of local informal communities led to alternative design parameters which informed an ecosystemic approach to architecture and place-making. This evolved into a new spatial settlement typology, which harnesses social relations and spatial qualities embedded in informal settlements while integrating attributes of low-tech sustainable self-construction.

The project's game changing agenda proposes new pathways to develop dense thriving urban neighbourhoods within transitional environments. It questions new models of governance with a focus on decentralised autonomous communities and creates opportunities for inclusive socio-economic progression through an environment of hybrid economies, integrating informal and formal ecologies of exchange and transaction.

Project 2—**Towards a cross-species symbiosis**

A non-human centred habitat (Koç 2022)

Framed by climate change, the project investigates the alteration of natural ecosystems by man, leading to decreased resilience, resulting in augmented risk of flooding and loss of biodiversity over time. Set in the Belgian city of Liege, the project interrogates the effects of centuries of industrialisation along the city's natural water network, on its physical and socio-economic fabric.

The study explores a possible 2050 future, based on a peak flooding scenario and which expands on the effects of the 2050 European no net land take policy target, reducing land artificialisation and maximising rewilding, characterised by a shift towards non-human centred urban development.

A crisis is repositioned as an opportunity for creating an urbanism of resilience. From the design process, a regenerative model of co-existence emerges which is founded on a non-human centred approach to our habitat, creating an interspecies living system. A sensitive form of regenerative development is proposed which mediates the dynamic relations and demands between habitats, leading towards an architecture of symbiosis.

Project 3—**Tuko Pamoja**

The digital liberalisation of Dar es Salaam empowered by the blockchain technology (Trautmann 2022)

The project interrogates trends of growing informality in the Global South and harnesses existing coping strategies of informal urban dwellers, based on intangible networks and collaborative practices.

In a speculative Dar es Salaam 2051 future, blockchain technology serves as a public ledger; a tool to unlock digital peer-to-peer transactions at grass root level. The scenario reveals tensions and synergies which co-exist in a global city, informed by global trends and local drivers of change. Underpinned by technology, a globalised economic network is marked by potent systems and networks spanning across continents. Locally, the acceleration of urbanisation exhausts resources and outpaces formal planning strategies, leading to a surge in informal living patterns, with often precarious living conditions as a result.

The proposal introduces technology as a disrupting change agent of the status quo, leading to the reframing of Dar es Salaam's sustainable paradigm. It transitions bottom-up socio-economic progression through democratising access to opportunity, by bridging the divide between formal and informal city structures. A counter-narrative is created for a vibrant urban future which is more equitable and inclusive.

19.6 Outcomes and Discussion

The outcomes of the research-by-design method, exemplified by the three case studies, display the following core characteristics;

19.6.1 Systemic Approach to Complexity

Systems thinking was embedded in the entire process—in research, design and project representation. The global and local STEEP analysis of drivers of change and their entangled relations was key to the progression of each project.

The de-coding process at the onset, instilled an acute awareness that the study area, and therefore the context of intervention, was continually in flux, informed by dynamic parameters which the students gradually discovered through their research, and later manipulated by means of their design project.

As a result, the focus moved early on from creating a conventional singular project to a strategic approach towards design to influence key factors, with the objective to disrupt and reconfigure the system to enable positive change. For example, whereas project 2 was primarily driven by factors associated with climate change, project 1 was focused on analysing the complex dynamics informing peak urbanisation in a developing context with limited resource.

19.6.2 Critical Imagination

An inquisitive, reflective approach to design led to unconditioned spaces for design exploration, which were unanticipated at the onset of the creative trajectory. Design abduction was harnessed throughout the entire process; the projects were founded on iterative framing and reframing of the issues until the layering of knowledge and insights led to a natural emergence of a dominant design direction.

Exploring alternative futures through a speculative design lens significantly expanded the 'possibility space'. For example, project 1 investigated a speculative 2050 scenario for the city of Lima defined by peak urbanisation due to urban migration combined with peak environmental degradation, whereas project 3 focused on the intersection of informality and the surge of technological advancements such as blockchain, in a very different context of Dar es Salaam 2051.

Both propositions led to a counter-narrative which reached beyond conventional responses rooted in the status quo. In Lima, the common response of building inadequate standard social housing tower blocks was reimagined as an innovative high-density spatial typology which harnesses the social and spatial qualities embedded in informal settlements. Project 3 created a counter-narrative by bridging the conventional divide between formal and informal city structures, through the introduction of blockchain as a public open-access network, leading to unprecedented opportunities for socio-economic upliftment at grass root level.

19.6.3 Design for Change

The projects succeeded in relinquishing the idea of a definable design problem which needed resolution. This fundamental shift resulted in design outcomes with an inherent resilient systemic quality; the architectural proposition represented an active form with an ability to adapt to new conditions over time.

The designs embodied an ecosystemic strategy, whereby the dynamic exchange between architecture and context could catalyse change far beyond the contours of the immediate intervention. The medium of architecture was used as a strategic act of reconfiguring and reimagining the status quo; with the propensity to influence the conditions governing tomorrow.

For example, the spatial settlement typology proposed by project 1 was envisaged as an intricate ecosystem that could evolve incrementally in line with the needs and demands of its residents. Furthermore, at a macro-level, the proposed typology could lead to a radical reimagining of high-density urban settlements.

The attributes of the outputs indicate the potential of the speculative futures model in enabling design as an active change agent, underpinned by the three sustainability transitions (Table 19.1).

19.7 Concluding Reflections

In this article, I link the design for resilience with a pivoting question embedded in the concept of sustainability; how do we make collective decisions that lead to regenerative futures with the insights of today?

Furthermore, I enquire whether, within the realm of this quest, the role and agency of architecture can be extended through a paradigm design shift to meet the complex challenges of tomorrow.

The paper opens up new pathways in response to this enquiry.

A speculative futures model is introduced that positions the medium of design and architecture in particular, at the heart of the sustainability debate. It proposes to reframe design in its role of challenging the status quo to meet the unprecedented challenges of the future, underpinned by three sustainable design transitions, namely systemic sustainable design, regenerative sustainable design and speculative sustainable design. The focus on sustainable transitions enables design to transcend the limitations that come with an inadequate problem-solving perspective of sustainability. It repositions the core focus of sustainable design as enabling continual systemic transformation, where resilience is pursued—not by resisting change—but by designing for change.

Subsequently, the paper explores the potential of applying the proposed speculative futures model, by translating the approach in a research-by-design method. Referring three project case studies of an Architecture Master's Thesis Studio Programme, the method is examined as a device to unlock new entry paths towards critical imaginaries that can engage with the complex challenges of the future, promote imaginative counter-narratives and catalyse systemic change.

Thirdly, throughout the paper, the importance of design agency is emphasised in the democratisation of sustainability. The holistic and speculative nature of the proposed approach enables discussion and sharing of important ideas about the futures that we will be living in together. In this light, the medium of architecture has an important role to play. By prototyping speculative worlds, and making them concrete and tangible, architecture can provoke new insight, can model a range of possibilities and probe possible futures that might lie ahead of us. Launching these speculative messages might catalyse traction and connect local and global audiences, to have conversations about the critical choices we collectively need to make.

Table 19.1 Speculative futures model indicators distilled from the research-by-design process

Sustainable design transitions	Process	Outcome
Systemic sustainable design	Abductive process	Dynamic sustainable metabolism
	STEEP analysis of drivers of change	Tunning of dynamic informants
	Systems thinking to examine complexity	Agility and embedded resilience
	Defining vulnerabilities and urgencies	Strategic high-impact focus
	Parametric analysis	Inter-scalar integration (local to global scale)
	Mapping relations and interdependencies	Systems design
Regenerative sustainable design	Living systems framework	Holistic world view
	Stimulating a multiplier effect	Regenerative capacity
	Influencing dynamic systemic relations	Evolution
	Multi-functional synergetic approach	Capitalising on symbiotic relations
	Creating capital	Increase in capital (natural, social, human, financial, built/manufactured)
	Change agency	Focus on self-organisation
	Strategic interventions catalysing impact	Adaptation
	Hacking the system	Transformation
	Metabolic foundation	Ecosystemic response
	Engaging with a changing context	Resilience
Speculative sustainable design	Change as positive design agent	Sustainable transition
	Possible futures framework	Action enabled through foresight
	Analysis of patterns and inflection points	Integrated time continuum
	Future scenario planning and sensitivity testing	Accountability and resilience
	Prototyping	Democratising the futures discussion
	Backcasting	Strategic steps and interventions catalysing impact
	Narrative environments and storytelling techniques	A shared futures narrative
	Challenging the status quo	Counter-narrative and new imaginaries

References

Adetunji I, Price A, Fleming P, Kemp P (2003) The application of systems thinking to the concept of sustainability. In: Greenwood DJ (ed) 19th annual ARCOM conference 3–5 Sept

Angheloiu C, Chaudhuri G, Sheldrick L (2017) Future tense: alternative futures as a design method for sustainability transitions. Des J 20(sup1):S3213–S3225, Dig J Moi Med. https://doi.org/10.1080/14606925.2017.1352827

Ball P (2012) Why society is a complex matter: meeting twenty-first century challenges with a new kind of science. Springer Science & Business Media, Heidelberg

Brundtland G (1987) Report of the world commission on environment and development: Our common future. United Nations General Assembly document A/42/427

Dorst K (2019) Design beyond design. In: She Ji: The journal of design, economics, and innovation. Tongiji University Press, Tongji, Dig J Moi Med. https://doi.org/10.1016/j.sheji.2019.05.001

Dover S, Handmer J (1992) Uncertainty, sustainability and change. Glob Environ Chang 2(4):262–276

Dyllick T, Hockerts K (2002) Beyond the business case for corporate sustainability. Bus Strateg Environ 11:130–141

Dunne A, Raby F (2013) Speculative everything, design, fiction, and social dreaming. MIT Press, MA

Farren CP (2022) Lima's block attack. Exploring alternative methods of rapid hyper-densification in informal settlements. Dissertation. KU Leuven, Brussels

Gibbons LV (2020) Regenerative—The new sustainable? Sustainability 12(13):5483, Dig J Moi Med. https://doi.org/10.3390/su12135483.www.mdpi.com/journal/sustainability

Hancock T, Bezold C (1994) Possible futures preferable futures. Healthc Forum J 37(2):23–29

Henchey N (1978) Making sense of futures studies. Alternatives 7:24–29

Koç F (2022) Towards a cross species symbiosis. Dissertation, KU Leuven, Brussels

Miller R (2006) Futures studies, scenarios, and the "possibility-space" approach. Schooling Tomorrow 58

Mitrović V, Auger J, Hanna J, Helgason I (eds) (2021) Beyond speculative design: Past—Present—Future. SpeculativeEdu; Arts Academy, University of Split, Split

Pearce D, Markandya A, Babier E (1989) Blueprint for a green economy. Earthscan, London

Rose JN (2001) A systems views of sustainability. J Committee Monetary Econ Reform 13(2)

Trautmann K (2022) Tuko Pamoja. The digital liberalisation of Dar es Salaam empowered by the blockchain technology. Dissertation, KU Leuven, Brussels

UNCED (1992) Report of the united nations conference on environment and development, (UNCED Report) A/CONF.151/5/Rev.1 13 June 1992

United Nations (2015) Transforming our world: the 2030 agenda for sustainable development the 2030 agenda for sustainable development. Accessed: 25 Sept 2022

United Nations (2017) New urban agenda. https://habitat3.org/wp-content/uploads/NUA-English.pdf. Accessed: 25 Sept 2022

Voros J (2001) A primer on futures studies, foresight and the use of scenarios. In Prospect, the Foresight Bulletin, no. 6 (December 2001). https://thevoroscope.com/publications/foresight-primer/. Accessed: 5 Sept 2022

Check for updates

20 From Past to Future: Learning from Reconstruction of Kasthamandap

Binita Magaiya and Manindra Shrestha

Abstract

Kasthamandap is an iconic monument, from which the city of Kathmandu derives its name and located in Hanumandhoka Durbar Square World Heritage Zone. It collapsed to the ground in 2015 Gorkha Earthquake. After the mega earthquake hits the Kathmandu and neighboring cities, the most affected were the monuments, mostly listed in the World Heritage Monument Zones. Once a cherished public structure, now into ruins, Kasthamandap attracted a dedicated community who advocated and faithfully restored to its monumental glory. The state authorities divided the reconstruction works of national monuments into various funding nations. The most controversial was of Kasthamandap, which then led to large public outcry to protect the traditional building methods, materials and technologies, versus the modern. The indigenous knowledge, the supervision and restrictions, new academic understanding to the practiced oral traditions are explored in the paper. The overall sense of community and belongingness was heightened with massive activism against the prevalent lowest bidding system in heritage conservation, the retention of community pool, the assured integration of intangibles in tangible heritage making in this reconstruction project. This paper intends to bring out the community involvement in various stages of reconstruction of Kasthamandap, which was assured its true restoration value and purpose, also the nearest to the authentic restoration sought in modern times.

Keywords

Heritage · Reconstruction public · Community · Space making

20.1 Introduction

> ... heritage sites are important for community recovery because they serve as tangible anchors that help communities reorient themselves (Daly and Rahmayati, 2012).

Nepal has a long history of earthquakes; the earliest recorded earthquake in the Kathmandu Valley is can be found from 1255 AD. Since six major earthquakes have hit the country with heavy loss of lives and property (Thapa et al. 2017). In the Gorkha Earthquake 2015, almost all of the major monuments within the three cities of Kathmandu Valley suffered damages in varying degrees. However, the self-ordained emergency

B. Magaiya (✉)
Social Science Baha, Research, Kathamandu, Nepal
e-mail: binita.magaiya@gmail.com

M. Shrestha
Kasthamandap Reconstruction Committee, Project, Kathamandu, Nepal

S. Hilal et al. (eds.), *Design for Partnerships for Change*, Sustainable Development Goals Series,
https://doi.org/10.1007/978-3-031-36993-3_20

rescue and salvaging was done with available resources and with spontaneity required by the situation. The community and especially the bystanders were the first responders not to mention the government officials and the security personnel. And eventually, when the issue of rebuilding heritage surfaced, a series of discussions was initiated by the Department of Archeology (hereafter DoA), continued for months. The discussion was aimed to guide the DoA in the forthcoming rebuilding work.

Seven monument zones inside Kathmandu Valley have been enlisted in the World Heritage List under the single ensemble of Kathmandu Valley World Heritage Property. It is in the Kathmandu Durbar Square, within the Kathmandu Metropolitan City (hereafter KMC), lies the Kasthmandap, one of the prominent monuments listed in World Heritage List and after which the city of Kathmandu is named. As stated by Daly and Rahmayati, *It is of the utmost importance to understanding how reconstructing the built environment could have been better synchronized with easing social trauma, and allowing the inhabitants of these regions to have the maximum (albeit under heavily-distressed circumstances) opportunity to re-orientate themselves in a shattered landscape.* (Daly and Rahmayati 2012). The catastrophe which shook the whole nation needed a nationwide recovery project to help people out of the trauma from the disaster and here heritage sites of monumental value addressed in such times of distress.

20.1.1 Introduction to Kasthamandap

Kasthamandap, also locally known as *Marūsata,* has been dated as a 7th cent (circa) structure as per the excavation reports done in 2015–16 (Coningham et al. 2016). Kasthamandap is centrally located in the historic core settlement of Kathmandu. Kasthamandap has its name derived from two words, viz. "*Kastha*" meaning wooden and "*mandapa*" meaning a defined space for ritual or any sacred purpose. There is a hearsay that the structure is built of a single tree named *Kalpabrikshya*. The myth lives on till this date. For centuries, Kasthamandap has remained as a palimpsest in the socio-political and economical history of Nepal (Joshi et al. 2021). The city of Kathmandu has been named after the structure of Kasthamandap. The three-story pavilion collapsed altogether in the 2015 Gorkha Earthquake, killing ten people present in the ground floor. During the reconstruction period 1970s, many erroneous practices were done, which came into light during the recent restoration. These were seen as one of the reasons for its collapse. Since known history, Kasthamandap has been used as a public rest house, which was soon adapted as a temple after the installation of Gorakhnath, presumably a powerful yogi, in 1465 AD. Other idols of four popular Ganesh in the Kathmandu Valley, installed in the inner corners of the quoin wall, were a recent addition in the twentieth century. Among many other adaptive uses, Kasthamandap had performed as royal council hall and coronation pavilion (Slusser and Vajracharya, 1974). Other functions included residence, shops and a daily communal interaction space. The pavilion also acted as a

Fig. 20.1 Kasthamandap after the restoration in 1970's. *Source* National Archives, Department of Archaeology

marketplace as depicted in the old paintings of Rajman Singh Chitrakar (1844 AD) and H.A Oldfield (1880 AD).

20.1.2 The Aftermath of Gorkha Earthquake 2015

As the state party to the Kathmandu Valley World Heritage Site, DoA should have initiated the rebuilding work of Kasthamandap in the aftermath of Gorkha Earthquake 2015. But the National Reconstruction Authority (hereafter NRA) was immediately formed by the government to cope with the crisis situation, and it was authorized to oversee all reconstruction processes of both private buildings and the national level monuments including other heritage sites throughout the country. As far as heritage rehabilitation is concerned, the ad hoc and ill prepared approach in recruiting technical professionals, very few of them trained enough to deal with heritage structures, especially monuments where the "sacred secret" were to be upheld, only led to poor results.

With the upsurge of rebuilding projects, the need to rebuild Kasthamandap was also widely voiced from the general public. It was leveled up to its plinth level; this had led the general public to suspect whether Kasthamandap has no foundation. But the archeological investigation, carried out by the DoA in coordination with the Durham University, revealed that not only the foundation was perfectly laid, but it was also intact and not damaged in the earthquake. As a part of the same study, OSL dating of the central foundation structures revealed them to belong to the seventh century (circa). Similarly C14 dating of one of the wooden capitals atop the main post revealed it to belong to fifth century (circa). These studies only helped to heighten the value of the monument (Coningham et al. 2016).

Traditionally, a systematized community called Guthi was responsible for the upkeep and maintenance of monuments as well as to carry out other socio-cultural affairs. The institution sustained themselves with the income from Guthi lands (land donated for piety by wealthy patrons). But after the land reform act was effected in the early 1970s, most of the *guthi* lands were nationalized. This, in turn, forced the *guthi* institutions to rely on government support for organizing such communal activities. (Pradhananga et al. 2018). Some of them are still active in other monuments and cultural practices. This made it difficult for the community to deal with government authorities and procedures and eventually led them to divide or sell the land parcels and organize a fund in trust for required expenses. This method, however, proved to be unproductive and later the community disintegrated in due time. Hence, the monuments and heritage sites were left without their original custodians, and in recent times, government appointed officials took over the role of custodian. As a result, the traditional knowledge transfer was not possible; therefore, a big gap in the repair and maintenance procedure was eventually observed. The impact of which was felt vividly in 2015 Gorkha Earthquake, when it was seen that the "modern" interventions and negligent maintenance practice were the main cause behind the damages to the monuments and heritage structure. And the argument that the community can act as the most important entity for the sustenance of any heritage structure surfaced. In case of Kasthamandap, eight active guthis were traced but later over the centuries only two had their continued practices.

20.2 Methods

The paper is based on the research process before, during and after the construction period, where the community members played a decisive role in different phases of reconstruction. Taking Kasthamandap as a case study, this paper focuses on overall reconstruction process driven by the community and the challenges faced in heritage reconstruction. The insight is shared through the hands-on experience where the authors themselves had to place themselves in the multiple facets of community activism and participation methods. The paper aims to deliver the new insights and understanding gained by the authors whose role shifted from being activists, researchers, consultants, architects, restorers and finally to a community member. Reviews of different studies carried out during this phase, websites, social media, news items and discussion with local community members, government representatives and other stakeholders have been incorporated in the paper. The authors have closely worked with different community groups, and they were part of the technical team during the reconstruction process.

Kasthamandap is the struggle and success story of the community who adamantly refused to give up against the state that was ill prepared to tackle unforeseen damages to the heritage structure of such prominence. After the 2015 Gorkha Earthquake, the reconstruction of monuments was a huge challenge in itself. After the collapse of Kasthamandap, it took national and international activism to recognize the historical value it possessed. The contestation for its recognition as a national monument itself was a long battle. The start of activism was when the monuments in the Kathmandu Durbar Square were being "distributed" among the "donor countries" for reconstruction. While the national and international donor funded agencies rushed in to get their share of Nepal's heritage and the "contribution" they would be making toward the preservation of Nepal's "heritage," the agencies were themselves not sure of what and how the heritage reconstruction was to be handled. The locals, scholars, researchers, professionals and youths all resisted this act for Kasthamandap put out an iconic message that Nepal's Monument can be rebuilt by the people themselves and with national funds and traditional means. In the later phase, the reconstruction project of Kasthamandap ultimately got support from the local government and earned the reputation of the "National Pride Project," which later was an undertaking from the Kathmandu Metropolitan City (hereafter KMC).

20.2.1 Need for Immediate Reconstruction

Many of the processes through which meaningful discussions about the past, present and future of traumatized people are facilitated not by foreign intervention, but rather by access of the population to the basic culturally appropriate venues and contexts in which different forms of discussion occur. This not only involves community leadership, social networks and hierarchies, but also the material settings within which communities know how to interact (Daly and Rahmayati 2012).

It is very important to start reconstructing the tangible heritages immediately, as it helps people to overcome the sense of loss quickly. The primary stakeholder for the monuments in Nepal is DoA, while the other national and international agencies like the municipalities and UNESCO play an integral role in the decision making process. Before the earthquake, the awareness in the general public about the reconstruction of our temples and other heritages was solely under the guardianship of the DoA. The various municipalities also had their own department dedicated to the preservation of cultural properties. There was not much interest in the community to be actively involved in the construction/reconstruction or preservation of the tangible heritages. The community never felt themselves needed in building or rebuilding heritage structures under the supervision of the experts from DoA. But after the earthquake, the aggressive "build back better" campaign of the NRA was floated which was taken as insensitive approach as far as heritage is concerned. And the general public was very much offended by the insensitive approach taken by DoA/NRA.

20.2.2 Design as Scholarship Where the Process of Design is the Research Method

The initial point was when few locals voiced their knowledge about Kasthamandap. A presentation by Mr. Yagyaman Pati Bajracharya, a renowned Vajrayana teacher and scholar, shared his research claiming his lineage to the builder of Kasthamandap and the symbolism it carried. The research paper titled "Kasthamandap: Tesko Nirmata ra Dhancha Shailee," presented on January 22, 2016; he elaborately described the

Fig. 20.2 Saptabidhanottar Puja ritual being held infront of Kasthamandap. *Source* Shailesh Rajbhandari for CRK

design and theological references to the actual design of the structure. He claimed that the structure was built around 1200 years ago, probably in the seventh to eighth century. He had also highlighted this in his book "Lilavajra: Kasthamandap ka Nirmata" published few years before. The archeological study carried out after the 2015 earthquake by the DoA together with the team from Durham University, dated the foundation of the structure to the seventh century (circa.). He further claimed that the structure itself represents the cosmological embodiment in a shape of mandala and every measurement, elements and entities have their symbolic meanings. Bajracharya further went on to publish his findings in his book titled "Kasthamandap ko Nalibeli," in the year 2020. Many of his claims had been tested and proven correct during the time of reconstruction and had been a guiding factor for some of the crucial measurements. He leads a Vajrayana Scholar group called "Bauddha Darshan Adhyayan Pucha," which has been instrumental in performing intermittent rituals much needed for the structure's sacredness. Many of the design decisions had been made in accordance with ritualistic values. The strict order of numerology and Vajrayana philosophies at times have shaped or restricted the design of Kasthamandap. As shared by Flavier, Jesus and Navarro, the indigenous system is locally bound to a specific area and culture. They use informal knowledge which is orally transmitted and generally not documented. It is dynamic, adaptive and closely related to the survival of many people (Flavier et al. 1995).

In order to transmit the oral knowledge held secret to the Vajracharya priest as part of their ritualistic practices, the group also welcomed the initiations of non Vajracharya students, the technical professionals from the volunteers pool in the campaign, who were active in the Campaign to Rebuild Kasthamandap (hereafter CRK). Interested architects and engineers were trained up to a minimum designated spiritual level, by master Bajracharya, for the better understanding of sacredness of the structure. The reconstruction and reassembling process of Kasthamandap began in collaboration with the technical professionals and spiritual practitioners.

20.2.3 The Design Specifications for Kasthamandap

The specifications appeared in due course during the reconstruction process. There is hearsay about the design of Kasthamandap that the structure is a 36 × 36 × 36 cuboid. A system of hand measurements was prevalent in the Kathmandu valley, and one of the hand measurement units, viz. *sishu hasta* (427 cm)*, equivalent to a length of 24 angula (digit) is taken a length of a hand, i.e., cubit* (Pant and Funo 2007). *This was a key finding from which standardized specification of Kasthamandap was derived. The courses of bricks in the foundation and in the superstructure were all set according to numerical specifications guided by the Vajrayana Principles. The prime number 9 is taken as a factorial unit, and all other values can be ascertained using the various systems of numerology within the Vajrayana principle. A number of posts in each floors, i.e., 100 in the ground floor, 64 in the first floor and 20 in the top floor are all numerologically assigned representing each deity, forming a mandapa in conception. The 100 numbers of floor joists, 10 specifically in the center, running east west were also strictly guided. The height of each floors, the detailing of the niches and the detailing of the pinnacle were hence guided and ritually invoked to bear the sacredness of the mandapa. The general misconception that Kasthamandap was a rest house (sattal) was replaced and was actually proven to be a sacred mandapa in the making. The various layers were enshrined with the holy charya dances after completion of each construction phase. It was seen after the excavations that the original structure of Kasthamandap had been gradually changed over the course of two centuries,* viz. *from seventh to ninth century. Upon closer inspection the defined perimeter of the ninth century foundation lines clearly gave the measure of 36 hasta* (15.3 m). *Kasthamandap is designed to have a cubical volume, hence it has equal length breadth and height. For the height, the measurement is to be taken from the plinth up to the roof right under the pinnacle. Omer's "continuity principle" which "stipulates that through all stages of disaster, management and treatment should aim at preserving and restoring functional, historical and interpersonal continuities, at the individual, family, organizational and community levels."* Omer and Alon (1994) *guides in the practice followed for the reconstruction of Kasthamandap* (Fig. 20.3).

20.3 Results

20.3.1 The Formation of CRK

After much efforts from the voluntary movement, CRK came into formation with a ritual ceremony "*Saptabidhanottar Puja*" organized by the

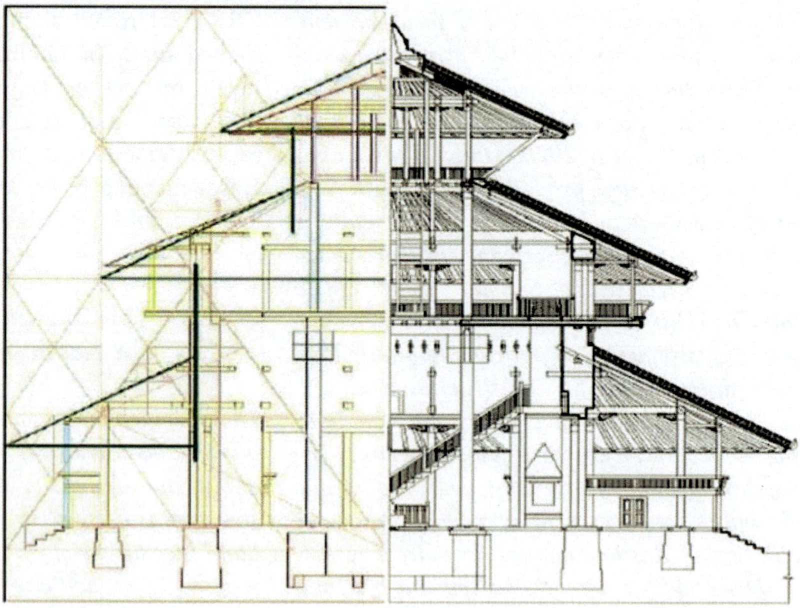

Fig. 20.3 Traditional measurements derived from site measurements and applied in the reconstruction phase. *Source* Authors

locals. Altogether 185 Vajracharya priests came for the praying and purification ceremony in Kasthamandap. This created the momentum much needed to accelerate the creation of office of CRK. The office of CRK signed a four party agreement with the DoA, NRA and the Kathmandu Metropolitan city on May 12, 2017, for taking up the responsibility to rebuild Kasthamandap.

There were a series of presentation by Mr. Dipesh Risal, a heritage enthusiasts on "Why Kasthamandap Matters." There were two presentations, one on top of the debris in Kasthamandap itself on September 25, 2016, and the other in Patan on November 10, 2016. The presentations had fine details of what were the spatial function, tangible and intangible associations it had during nineteenth and twentieth centuries. These were cited in the ancient *thyasafus* (handwritten folios) and *Paubas* as well as in the *Pancarakshya* Manuscript in 1135 CE. The discussion session gave away many interesting facts about Kasthamandap and its functions based on his collection of researches from Mary Slusser, Gautam Vajracharya and H. A Oilfield and many more scholars.

The campaign continued with other activities of lecture series from renowned scholars to ensure community engagement in various levels. The presentations are as follows:

Kasthamandap Punarnirman: Sampadaka Murta ra Amurta Roopharu (Sudarshan Raj Tiwari)— July 25, 2017

Prof. Dr. Sudarshan Raj Tiwari presented his research for Kasthamandap titled "*Kasthamandap Punarnirman: Sampadaka Murta ra Amurta Roopharu*" on July 25, 2017, Kathmandu. It was a focused lecture series with discussions from the audience integrated to see the best possible way out for the reconstruction process. It was concluded that to ensure longevity of the structures in heritage context, the correct methods and materials needed to be used during construction, and monitoring and maintenance of the monument was to be carried out periodically.

Kasthamandapko Avilekhma Kasthamandap (Kashinath Tamot)— August 6, 2017

After the 2015 earthquake in Nepal, only three copper-plate inscriptions (tamrapatra) out of the nine installed were recovered from the ruins. The ones that are recovered are dated Nepal Samvat 444, 499 and 543. Several of these inscriptions have been translated and published by Mr. Kashinath Tamot, the most celebrated historian in Nepal, made public his research in Kasthamandap. In his findings, Kasthamandap was used as a rest house, a public building and a city council hall, over other functions later adapted by the changing needs of the community.

Interaction with Ar. Wolfgang Korn

Ar. Wolfgang Korn, the German architect, who had measured Kasthamandap in 1976 A.D, was invited in the CRK Office on September 22, 2017, for an interaction of his measurement techniques that he followed during his days for the drawing of the structure that was very challenging in itself. The drawings that he had made were submitted to the DoA multiple times. Ar. Wolfgang Korn also handed over his measured drawings on November 7, 2015, asserting its importance to the national identity.

In December 2015, a UNESCO funded archeological team from Durham University excavated on Kasthamandap site, which further helped understand the construction, cause of collapse and further history of Kasthamandap (Coningham et al. 2016). The team of CRK, along with the experts, visited many related sites and tried to gain detailed knowledge of the building and the construction details so that the mechanism proposed comply with traditional system that all had been advocating for. The case visits were conducted mostly in Bhaktapur. Mr. Rabindra Puri, a renowned conservationist in Nepal, and Ar. Rabi Jonchhay took the technical team of CRK around Bhaktapur to help the team gain in-depth knowledge of the traditional construction systems and project management as a whole. Some other projects of the Kathmandu Valley Preservation Trust (KVPT) were also visited, and ongoing construction works in Hanumandhoka Palace complex also provided important ideas for traditional construction and technology.

CRK volunteers team formation

Presented are the key dates and activities that led into formation of the Campagin to rebuild Kasthamandap team later into a non-governmental organization, which challenged the national agencies such as NRA and DoA for their minimalist knowledge of heritage making and the participation from the public mattered the most.

- ***Local Meet-up at Nasal Chowk, Hanumandhoka April 21, 2017***

A local meet up was organized in the Hanumandhoka Palace Museum where individuals from various walks of life, including locals, ward officials, technical experts, intellectuals, social and political leaders—all came together to discuss the rebuilding of Kasthamandap, highlighting the core idea of CRK.

- ***Volunteers' Workshop on April 22, 2017***

An all-day workshop was conducted in Kathmandu, with close to a hundred young enthusiastic volunteers enlisted to brainstorm about various approaches in which the people could participate in the rebuilding of Kasthamandap. The participants were classified into twelve groups depending their area of interest and expertise and the professional areas in which they could contribute. Many participants were students of architecture and engineering who had passion for heritage. Other participants included entrepreneurs, lawyers, social workers, communicators and accountants, all of whom shared the emotional attachment toward Kasthamandap and heritage in general.

- ***Oath Taking Ceremony on April 25, 2017 (12_{th} Baisakh 2074 B.S)***

On April 25, 2017, the memorial day of 2015 Gorkha Earthquake, there was a public event organized by the team of CRK in front of the Kumari Temple in Kathmandu Durbar Square, where all the present dignitaries local community members, professional and social organization representatives and individuals from all walks of society took an oath to rebuild Kasthamandap keeping intact the original intangible and tangible values and integrity of the monument.

- ***Workshop with the Guthis Associated with Kasthamandap October 26, 2017***

Eight different community guthis in direct association with Kasthamandap were identified. Three major guthis of Kasthamandap were solely intended for the diligent upkeep of the structure as well as the cultural traditions. Ta: Chatan guthi and Sa: guthi were the only two guthis that had continued the cultural practices relating to Kasthamandap. The other guthis involved in Kasthamandap were Manandhar Guthi, Gorakhnath Community, Kapali Guthi, Mahabali Guthi, Maru Kishan Guthi and Pachali Bhairav Guthi. A discussion session was held with the guthis where the CRK team tried to get an in-depth knowledge of the spatial functions in the earlier days before the collapse of Kasthamandap. It was an interesting experience to share, as the participating communal guthis had their specific rituals and timings that were sometimes dependent on each other. At other times, there were correlations between the guthis, which were sorted out with dialogue and coordination with each other.

- ***Workshop with the Technical Experts Team October 27, 2017***

The workshop was held with the technical experts including scholars, professors, engineers and practitioners who had been in continuous dialogue with the team of CRK during the development of the proposed drawings. The details of the proposed drawings were minutely discussed. The experts presented their views of the reconstruction practices and challenges in the design and the implementation phases, their approach methods and challenges ahead.

– ***Public Interaction and feedback collection October 28, 2017***

The three-day workshop concluded with the public display of the proposed drawings and the results of the findings from the workshops. This public event was periphery of Kasthamandap. The interaction between the team of CRK and the enthusiastic audiences from the community and the public was met with overwhelming response, with many suggestions put forth. It was a whole day event, and everyone from the public could view the proposed drawings or could interact with the volunteers from CRK for the work process. The suggestions were recorded and used in the revision of the drawings.

Other supporting events such as oral story collections, archival researches, interaction with artists and painters, prominent linguists, community groups, technical schools and colleges were conducted in all three cities of the Kathmandu Valley. Numerous consultations were made, and finally, the produced outcome was the drawing proposal with the community's ownership. The efforts from the CRK team paid off and the continued support from the community poured in.

20.3.2 The Local Election and the Rebuilding Process

The local elections in April 2017 put forward a very challenging dynamic for the reconstruction of Kasthamandap. There was a four party MoU signed by CRK, KMC, DoA and the NRA that gave CRK an authority to rebuild Kasthamandap in a unique Public–Private Partnership model. But after the KMC's elected officials took charges, they saw this as an opportunity to reclaim the jurisdiction over. The argument was that the CRK office didn't have enough funds to handle such a project of "national pride" and were concerned that the donation drive won't be sufficient enough to raise such a huge amount of funds. The MoU was scrapped and later the KMC was given responsibility to rebuild Kasthamandap. There was also the issue of who had jurisdiction over what areas of the Durbar square, all the confusion regarding who had control over which area of the durbar square. The turmoil was evident. The ward offices, DoA, NRA, KMC all were at odds with each other. Once again, the discussions of a tender system was on the table, but this time, the KMC were convinced of a larger Public–Private Partnership and support for a sixty plus member committee "Kasthamandap Reconstruction

Fig. 20.4 Presentation to the community. *Source* Sandesh Munikar for CRK

Committee" (hereafter KRC) was formed under the steering committee of the Mayor. As argued by Daly and Rahmayati, [...] Vividly demonstrate how monuments, buildings, city walls and so forth have become sites of intense political and symbolic contestation through the convergence of developmental and governance frameworks that oscillate between past and future, local and global. (Daly and Rahmayati 2012). *This was clearly the case in the reconstruction process.*

20.3.3 KRC Technical Team

The community support went to KRC as soon as it was formed; there were many subcommittees formed to look into various aspects of reconstruction and the community engagement process. The technical committee was formed with renowned experts in architecture and heritage, structural engineering, archeology and other technical disciplines. An in-house technical team, in consultation with the technical committee, supervised the construction of Kasthamandap. With the funds from KMC, KRC was guided by the "Adhar Patra (Principle Guidelines)" that has been unanimously adopted by all the committee members. The whole reconstruction of Kasthamadap was executed by this committee within 3.5 years with the final expenditure amounting to approx. 120 million. As anticipated by the community earlier, the cost of reconstruction came down significantly assuring both quality and continuity of the traditional practice. Such model of community participation should be an example for other heritage structures too where complexity of ownership matters.

20.4 Discussion

20.4.1 A Final Note: Returning Back to Community?

Now, the post reconstruction, the handing over process back to the community is in a standstill because of a court case, where few individuals had filed a complaint regarding the reinstatement of new Gorakhnath statue. The old statue, which was broken in pieces during the earthquake, was not reinstated because of religious reasons and a new replica was made and installed reserving the old one for the museum. The arguments caused stifle, and hence, the site has not been opened for public till the time this paper is written. This has caused the issue of ownership. After a long community support and partnership, how can a organization single handedly decide, when to return to the public? What are the measures and management plans that need to be implemented to give it back to the community. Does the conservation action plan actually work while dealing with communities in a living heritage site? These questions are to be resolved as Kasthamandap awaits to be open to all. The structure once a public monument without any inherent religious value and purpose suddenly gaining a religious significance with the installation of a deity figure and later reclaiming its originality as a public space is to be heard, as the reconstruction process itself showed the public sentiment, regardless of its religious status. Had it been a religious structure, the reconstruction and the public support would have taken a different discourse. Hence, the value creation is the most important debate that could spark the idea of identity and space making entangled with heritage and culture.

Kasthamandap unfolded itself with new meaning and created dialogue that was heard and diligently responded by everyone involved. The underpinning guide and ever present support to its meaningful reconstruction was always found within the folds of community. The memory of which disappeared was so strong that craving for the associations to its earlier physical form was always felt which gradually subsided as Kasthamandap slowly took shape. Now that Kasthamandap stands where it has ever stood before, a new debate can be initiated—do we always need shape, size and color and even the sense of touch to relate to the manifestation of reality? After all, everything is delusional in spirit world. Community remained as ever present custodian to the monument and its livingness.

Acknowledgements Authors acknowledge the contribution of all the volunteers, community members, experts and people involved during the period of both CRK and KRC. We also acknowledge *guthi* members, officials from NRA, DoA, KMC and other institutions for their support and coordination during reconstruction.

Declaration This research did not receive any specific grant from funding agencies in the public, commercial or not-for-profit sectors.

References

(n.d.). Retrieved from https://kathmandupost.com/miscellaneous/2017/04/26/locals-seize-kasthamandap-rebuild-initiative

(n.d.). Retrieved from https://rebuildkasthamandap.com/

(n.d.). Retrieved from https://kathmandupost.com/miscellaneous/2017/07/01/the-tussle-for-kasthamandap

(n.d.). Retrieved from https://caravanmagazine.in/reportage/storeyed-past

(n.d.). Retrieved from https://old.risingnepaldaily.com/nation/sc-order-to-be-followed-to-place-old-idols-at-kasthamandap

(n.d.). Retrieved from https://english.onlinekhabar.com/kasthamandap-statue-debate.html

(n.d.). Retrieved from https://thehimalayantimes.com/nepal/locals-protest-plan-to-instal-new-idol-in-kasthamandap

(n.d.). Retrieved from https://www.bbc.com/news/world-asia-56873438

(n.d.). Retrieved from https://english.onlinekhabar.com/the-tale-of-two-dharaharas-destroyed-by-two-earthquakes-at-a-100-year-interval.html

Coningham R, Acharya K, Davis C, Kunwar R, Simpson I, Schmidt A et al. (2016) Preliminary results of post-disaster archaeological investigations at the Kasthamandap and within Hanuman Dhoka, Kathmandu Valley UNESCO World Heritage Property (Nepal). Ancient Nepal 28–51

Daly P, Rahmayati Y (2012) Cultural heritage and community recoveery in post-tsunami Aceh. In: Daly P, Feener RM, Reid A (eds) From the ground up: perspectives on post-tsunami and post-conflict Aceh. ISEAS Press, Singapore, pp 57–78

Flavier JM, Jesus AD, Navarro CS (1995) The cultural dimension of development: indigenous knowledge systems. In Warren DM, Slikkerveer LJ, Brokensha D (eds) The regional program foor the promotion of indegenous knowledge in Asia. Intermediate Technology Publications Ltd (ITP), London, pp 479–487

Gallacher P (2005) Everyday Spaces—the potential of neighbourhood space. Thomas Telford, London

Joshi R, Tamrakar A, Magaiya B (2021) Community-based participatory approach in cultural heritage reconstruction: a case study of Kasthamandap. Prog Disaster Sci

Keates J(1991) Italian journeys. Picador

Locke JK (1997) Where have all the guthis gone? Face Face Mag Dev. Kathmandu

Omer H, Alon N (1994) The continuity principle: a unified approach to disaster and Trauma. Am J Community Psychol 22:273–287

Pant M, Funo S (2007) Stupa and Swastika: Historical Urban planning principles in Nepal's Kathmandu Valley. NUS Press

Pradhananga N, Shrestha KK, Dee J (2018) Retrieved 10 Jan 2022, from guthiaustralia.org: https://guthiaustralia.org/newa-culture/newa-social-system-guthi/

Slusser M, Vajracharya G (1974) Two Medieval Nepalese buildings: an architectural and cultural study. Artibus Asiae 169–218

Thapa DR, Wang GX, Fan F (2017) Deterministic seismic hazard assessment for Nepal. In: Sixteenth world conference on earthquake engineering. Santiago, Chile, pp 1–8

21 The House of People

Irina Teodora Comanita

Abstract

The Palace of the Parliament located in Bucharest, Romania, commonly known as 'The House of People' is a building with an area of 333.000 square meters, being the world's second-largest building by surface area, recorded in the Guinness Book of World Records in the section Administrative Buildings (Urbact in A controversy in the heart of Bucharest: The Parliament Palace. Retrieved October 12, 2022, from https://www.blog.urbact.eu/2011/01/the-romanian-parliament-palace-a-controversy-in-the-heart-of-bucharest/). This facility is the legacy of the communist era, a symbol of the Dictator Nicolae Ceausescu's power, being designed to host the most important institutions of Romania as well as Ceausescu's home. A place of contrast some might say, as it was rated as "the most beautiful" but also "the ugliest" building of the city in a survey conducted by the National School for Political Studies. At the same time, the building represented a mutilation of the Uranus neighborhood, as in order to build it, Ceausescu demolished an area of seven square kilometers and displaced 40 000 people, their homes, and their stories. (Pelehatai in Uranus atunci, Uranus acum: Razbunarea fantomelor orasului. Scena 9, 2019. Retrieved October 12, 2022, from https://www.scena9.ro/article/uranus-acum-expozitie-mnac-casa-poporului). This dialogue is an invitation to explore the following question: if a building could speak, what would you ask? I imagined having a dialogue with the House of People to 'understand' its life and purpose as a building before giving up on it or transforming it. In this dialogue, I envision the role of the architect (I) as being a therapist for buildings, not only listening but helping the building regain its (self) esteem and further impact the life around it (the neighborhood, the city, the country). How can we understand before we demolish? How can we keep a sense of place if we have not come to peace with our past yet?

Keywords

Architecture · Community · Neighborhood · History · Communism · Identity

21.1 The Talk

"How many did you kill?" I asked, while looking up at its tall, imposing body that would often make me feel so small like a grain of rice. This time I was there to listen and help and that gave me power.

I. T. Comanita (✉)
Independent Researcher, Copenhagen, Denmark
e-mail: irinateodorac@gmail.com

S. Hilal et al. (eds.), *Partnerships for Change*, Sustainable Development Goals Series,
https://doi.org/10.1007/978-3-031-36993-3_21

Fig. 21.1 Collage of the Uranus neighborhood (gray) and the current existing buildings (red). Collage created by Radu Manelici for OAR Bucharest

"I think almost 9000… there were supposed to be 7000 but in that chaos 2000 more fell. And then there were those 20 churches, maybe that's why I am cursed. I stand on holy ground but I am anything but holy."

"And the people?", I asked impatiently.

"Well… first you kill their homes and then their lives are starting to fall apart as well. 40,000 people have perished in that madness. Old people, young people, a piece of their life has disappeared like dust in the wind" (Fig. 21.1).

"Do you remember how it happened?"

"What do I know? Amnesia - seems to be the disease of the century. I don't remember much, but word goes that it was brutal. It was a massacre. There might even be photos of the buildings while being pulled out from the ground, while black smoke and tears covered the entire place rapidly. And then, it all came to dust. The air was unbreathable for a while, I guess they would have probably disappeared deprived of oxygen anyways."

You could watch straight into its soul through its dark glass eyes. There was not much happiness in there. Tears were gushing down its pale yellow cheeks. Its voice was cracking while describing its story, its beginnings. Its face looked sad and I didn't have any good words to say to make it better.

"Have you met any of them ever since?"

"Sadly, yes. Some of them are still alive today and they even come and take pictures of me. Some come with their grandchildren and they start telling their story. It breaks my heart. They know every street that used to be here, every house, every family… I remember one time this woman came with her granddaughter and she was pointing at the place that she used to call home. The little girl was fascinated about this new discovery, that her city used to be different before, and soon they both started talking about how life looked like back then.

"We inherited this house from your great grandfather when he came back from war. I was a student in the 80's and I was right in the middle of my winter exams when, one day, we received a letter in the mailbox. We have been informed that we needed to evacuate the building within three days. My father managed to extend the deadline with two more weeks and then, in one single day, we were gone. I thought my life was never going to be happy again. I was very attached to our home, to our garden, the smell of freshly baked bread from the bakery around the corner, the streets where I used to play with my little brothers, the bench in front of our door where we used to gather in summertime with all the neighbors living on our street and dance and laugh and talk until late night. We used to be like a big family, one big community. I vividly remember this episode when they came to tear down this gorgeous house, built in a neo-Romanian style. I panicked, it felt so tragic and in that moment I didn't know what else to do, so I took out my tape recorder, turned the volume high and I played a piano song, all while tears were falling down my cheeks as I was watching that house falling apart" (Fig. 21.2).

"Did anyone try to stop any of it?"

Fig. 21.2 Collage superpositioning two different realities: the 'lost' reality and today's reality. Designed by Radu Manelici in collaboration with Andrei Bîrsan and Ştefan Tuchilă

"There were many of them questioning if this was the only way... but no one really dared saying out loud how they felt, you could have been killed for only saying something bad about the regime and the decisions they took. There was also a lot of uncertainty, people didn't really know if the demolishment rumors were true or not until they saw the bulldozers coming. "Is it going to be us today?" Many people refused to leave their homes and stayed there until the very last day. They would see their roof being taken down, and then the walls and then their electricity and water being cut. They lived without a roof over their head for a while. Other people have invested everything in their homes. Their home was their life and so, when the home was taken away from them, they also decided to take their own lives. I have their blood on my hands...

We were all actors, I guess, playing our roles in a movie that no one knew much about. I am not trying to get away with it... but I also didn't have much of a choice. I didn't choose this life; I didn't choose the place I was 'born' nor my creator. If you asked me, I wished I was one of those imposing opera houses, made of curves and glass and steel, looking down at the blue sea watching my inner dazzling light reflecting in it. And there would be elegant people coming in the evening to admire me and listen to concerts and music that fills their mind and hearts with joy, while I stand there, living in the moment. But that is a dream. I am what you see, what they see. I am both a victim of faith and a perpetrator. And both were a given."

"You will not heal until you learn how to forgive yourself. You are being very harsh and still clinging to the past. I am not here to judge you, but rather to listen and understand. Maybe it's not anymore about who you were, but rather who you want to be. How do you feel now, after so many years? I asked.

I must say, at that point, those words came to me like a bullet. I was thinking of how the life of a building and my life could resemble. We were both thrown into the world and now trying to figure not only who we were but who we want to be. I can be a therapist for it, I have the skills to listen, design and transform it. I could have even been its creator in a past life. But then who can transform me? And will that mean that the person who impacts me, will also have an impact on this building? What a circle, we are all entangled and twisted.

"How do I feel? Why would anyone care? They point fingers at me all the time. "The murderer!", "The ugliest in the world!" ... As if pretty can erase what I've done.

"See, I get from you an immense sense of shame, that's been lingering around for a while. Shame is an emotion that we learn, that sometimes has been induced to us and it makes us paralyzed. There is no growth coming out of shame. I see something inside you that wants to change, that wants to challenge this life that was given to you. I see a common fault, a complex mix of factors that led up to this situation. And you happened to be the result of it. But accepting your faith as doomed to fail is not an option."

"You have a point about my shame but I am tired. I believe I will end up smashed from head to toe like my predecessors. I am just waiting for them to come after me and give me the same treatment."

"Don't get too defensive. I am sure you have paid the price of your existence already. I was just wondering if you could leave a lesson behind, you know, a good learning drawn from this tragedy. And even if they knock you down, there's still life after death for a building, so you should consider this option. The fact that we're talking now makes me think you are."

"I am. I want to be different, I want to bring joy! I need you to help me, to guide me somehow. I feel completely lost. I am sad, cold, and alone. There was a time when I couldn't understand who I was and I just didn't seem to fit in. I was too much, completely out of scale. But then I shined for a while, I was the pride of the country and everyone in the world knew who I was. The tallest, the greatest...the symbol of socialism, and that made me confident. But then so were they... the ones who fell, I mean. 9000 wonderful stories, 9000 beautiful bodies turned into skeletons. Some people still remember them and how they looked, but the young ones don't even know the history. The only thing they see now is me, shining like a piece of jewelry in the middle of a crime scene. They think I became a symbol of democracy now but I don't feel this way.

I know who I've been but who am I now? They still call me 'the house of people'... yet no people find a home in me. I am empty on the inside and my body is aching... My back skin is wrinkled by the sunsets I covered. I don't know how long I can last like this. I need someone to cure me and give me a purpose to live. Otherwise, they might as well smash my body to the ground and turn it into 7000 houses for 40,000 people.

"That's why I am here now, to tell your story at last. The future can't start until we learn from the past."

References

Pelehatai I (2019) Uranus atunci, Uranus acum: Razbunarea fantomelor orasului. Scena 9. Retrieved 12 Oct 2022, from https://www.scena9.ro/article/uranus-acum-expozitie-mnac-casa-poporului

Urbact (2011) A controversy in the heart of Bucharest: the parliament palace. Retrieved 12 Oct 2022, from https://www.blog.urbact.eu/2011/01/the-romanian-parliament-palace-a-controversy-in-the-heart-of-bucharest/

22 Individual Actions in Collective Processes and Forms: Collaboration for Change Across Education, Practice and Public Debate?

Marie Frier Hvejsel, Wiebke Ahues, and Karin Hartmann

Abstract

The diversity of UN's 17 Sustainable Development Goals (SDG's) unfolds a series of intertwined environmental, social, and economic imbalances calling upon us to succeed in activating our individual competences and perspectives in partnerships for change. This argumentative essay serves as a mutual self-reflection towards this end, aimed at supporting the identification of resourceful individual action in partnerships across each of our respective fields of action through architecture; in education, practice, and public debate. Motivated by a re-reading of Maki's investigations in collective form (Nurturing dreams—collected essays on architecture and the city, MIT Press, Cambridge, 2008 [1964]), the paper identifies a correspondence between the multidimensional challenges reflected in the SDG's and Maki's call for collective forms of architecture. Consequently, the paper scrutinizes collective processes for collective forms by proposing a translation from 'form' to 'process' in Fumihiko Maki's original definition of collective form. Doing so, we can observe that *investigation into the collective* <u>*process*</u> *is important because it forces us to reexamine the entire theory and vocabulary of architecture, the one of single* <u>*authorship*</u> {(Maki 2008 [1964], p. 57) translated by the authors}. Hence, through translation we uncover a critical linkage between the current quest for accommodation of plural competences within the architectural process and the changes that architecture must necessarily undergo as a physical form to support viable change of the built environment. In conclusion, the paper discusses the implications hereof for our actions in each of our individual contexts.

Keywords

Collective form · Collective process · Self-reflective action

M. F. Hvejsel (✉)
Aarhus School of Architecture, Aarhus, Denmark
e-mail: mfh@aarch.dk

W. Ahues
Studio Director Berlin, Henning Larsen Architects, Architects Chamber Berlin, Board, Berlin, Germany
e-mail: wiebke.ahues@henninglarsen.com

K. Hartmann
Association of Female Architects, Bonn, Germany
e-mail: karin.hartmann@khap.de

22.1 A Shared Motivation

As visualized in the diversity of UN's 17 SDG's, a series of intertwined imbalances call upon us to succeed in establishing partnerships for change in order to contribute to a viable development of the built environment (UN 2022 [2015]). Entailing

S. Hilal et al. (eds.), *Design for Partnerships for Change*, Sustainable Development Goals Series,
https://doi.org/10.1007/978-3-031-36993-3_22

changes across the environmental, social, and economic dimensions of our behaviour, such partnerships must necessarily be built upon care for individuals and resources locally and globally (Meadows et al. 1972; Rosling et al. 2019; Krasny 2019). Daily, we ask ourselves as individuals acting in architectural education, practice, and public debate, how to utilize each of our individual resources and privileges in the establishment of such partnerships for the collective good? Scrutinizing collective processes for collective forms, this argumentative essay serves as a mutual self-reflection in this matter, aimed at supporting the identification of individual actions to be pursued in each of our contexts.

For architectural form as such, viable changes of the built environment can be observed to entail a simultaneous minimizing the environmental footprint and cost of our constructions while maximizing their social capacity and value as lived space (Hvejsel 2022). The complexity of this challenge requires us to rethink the still prevailing description of the architectural work as an artistic entity in itself, due to its tendency to leave the urban context and lived reality of many out of reference. Simultaneously, it reflects as a shortcoming in our understanding of the architectural process. A process, in which the association of the solitary work translates into an idolization of solitary authorship that tends to compromise the move that we need to make towards a collective conception of architecture. This association compromises the charting of the combination of diverse individual views and competencies that the intertwined imbalances stated in the 17 SDG's entail (Hartmann 2022, p. 47–49). Hence, this assumed unity of work and author does not match the pressing need for an inclusive, pluralistic vision of society as described in SDG 11, nor a sufficient coverage of the impact of the forms we construct as entailed in SDG 13 (UN 2022 [2015]). Clearly, the forms we construct must be developed and valued upon their collective potentials in response to the before mentioned imbalances, calling for a conceptual change in the processes from which these constructions result, a change from an individual to a collective conception of form and process.

In the early 1960s, Fumihiko Maki collaborated with Masato Ohtaka and Jerry Goldberg on a series of 'Investigations in Collective Form' (Maki 2008 [1964]). They did so in response to the urban challenges that became visible in the implementation of universal modernism. Whereas Jacobs (1961) fought these challenges head-on in the streets of her neighbourhood, and Gehl (1971) described them from the point of view of urban space and—design in 'Life Between Buildings', Maki's investigations read as an attempt at clarifying the physical role of architectural form to this end. These investigations aimed at describing an architectural form capable of supporting a collective experience and development of *'the city as a pattern of events'* (Maki 2008 [1964], p. 62). *'Investigation of the collective form is important because it forces us to reexamine the entire theory and vocabulary of architecture, the one of single buildings',* Maki claimed (Maki 2008 [1964], p. 57). Given the before mentioned intertwined challenges that govern a viable development of the built environment, these investigations appear increasingly relevant today. It is our idea that Maki's investigations offer a direction for investigations into the nature of the processes and ways of working that the necessary expansion of the collective capacity of architectural form entails. Consequently, this essay is structured around an attempt at translation from collective architectural 'form' to 'process' suggested in a rewriting of Maki's original motivation. Hence, to us, *investigation into the collective process is important because it forces us to reexamine the entire theory and vocabulary of architecture, the one of single authorship.* ((Maki 2008 [1964], p. 57) translated by the authors by exchanging aspects of 'form' with aspects of 'process' in the quote).

22.2 The Shortcomings of Individual Authorship

Contemporary to Maki's initial investigations in collective form, Kay Fisker wrote about 'Idolization and Anonymity' stating that' *those architects that succeed in ordering our urban spaces and our landscape and that succeed in creating a human environment ... are more valuable to society than those that create individual and sensational artwork"* (Fisker 1964, p. 526). In combination, the title of Fisker's essay and to call to value architecture upon its contribution our collective life form suggested a necessary change in the self-image of the architect corresponding with Maki's description of the collective form. In today's discourse, Fisker's call resonates in public debate, where it is amplified in a critique of single authorship for its association with marketing logics rather than design reality (Hartmann 2022). Furthermore, it is being identified as a key hindrance in arriving at a sustainable development of the built environment, because it implies compromising inclusion and equality in the architectural process as specifically addressed in SDG 5 goal (UN 2022 [2015]). As stated by Elke Krasny in her Essay 'Architecture and Care', the historical development of the architect's profile deeply ingrains the gendered binarity opposition (Krasny 2019, p. 39) reflecting in a compromised ability of the resulting architectural form to unfold the necessary spatial care for individuals and resources locally and globally. Hence, the still prevailing tendency to maintain the image of a solitary author, concerned with productive work while leaving topics related to inclusive, intergenerational collaboration in reproductive work out of the professional focus is deeply problematic (Krasny 2019). In our view, we are still short of principles to value the collective form and to credit the architects that succeed in contributing collectively to the social capacity of architecture as called for by Fisker. Seemingly, the answer is neither idolization nor anonymity, but a question of valuing collaborations that succeed in enabling a mutual recognition of the individual views and competencies in the eventual spatial capacities of the forms resulting from our design processes.

As argued above, viable development of the built environment calls for major expansion of the collective spatial capacity of the forms we construct while minimizing their environmental footprint. Likewise, it appears that the production of such collective capacities in built form necessitates a simultaneous redefinition of the underlying processes from which they result. Almost 60 years after publication of his first studies in collective form, Maki picked up this thread in a panel discussion at the *1st International Conference on Structures and Architecture*, ICSA2010, by adding collaborative processes and communication as measures of collective forms. Hereby Maki questioned the single authorship within a sustainability framework and raised a call for the development of collective processes, stating that *'the question is not about sustainability of each individual building, but also how we can make a more sustainable community and environment in larger scale... the collective form can have an important place as an object where planners, architects, engineers, and others might be able to work together in a more creative way addressing this challenge. It is less a question of independent effort, somehow the situation suggests something collaborative that we still have not explored fully'* (Maki 2010, 00:16:00). If following Maki's line of thought, there is an opportunity to link the social demand for diversity, inclusivity, and equality to innovations in the environmental and economic dimensions of sustainable construction of architectural form. Seen in this light, the introduction of instruments against structural discrimination translates into necessary working tools to be implemented in the architectural process. Hence, upon Maki's invitation to collaborate across generations and disciplines, which we see mirrored in the targets of the SDG's, we find a shared motivation and direction for pursuing principles for a collective conception of architecture as process.

22.3 Collective Processes for Collective Forms? An Attempt at Translation

In 'Investigations in collective form', Maki stresses the importance of 'linkages' as the foundation for collective forms and hereby for establishing a collective development and experience of the city: *'Linkage is simply the glue of the city. It is the act by which we unite all the layers of activity and resulting physical form in the city. Insofar as linkage is successful, the city is a recognizable and humanly understandable entity. We are at home with it. We depend on understanding how two events within a city are combined to make a living sequence'* (Maki 2008 [1964], p. 62). In this matter, Maki lists a series of physical operations characterizing the collective form; to mediate, define, repeat, make a sequential path, and to select, but observes that the act of linking is in principle *'a singular one: that of making a comprehensible and humanly evocative environment... once a link is established for any reason, it takes on a complicated secondary system of meanings and uses'* (Maki 2008 [1964], p. 62). In this physical description of how architectural form itself can help to support our ability to coexist, communicate, and eventually collaborate, we find a foreground for Krasny's work on care in and of architecture (Krasny 2019). Following this line of thought, a potential to translate from Maki's description of linkage as the physical glue of the city to an understanding of linkage as the collaborative glue of the design process opens up. A potential allows us to read linkage as *'the glue of collective process'* and as *'the act by which we unite all the resources of the individual parties and resulting physical form in the city. Insofar as linkage is successful, the collective process is a recognizable and humanly understandable entity. We are at home with it. We depend on understanding how two views within a collective process are combined to make a living sequence'* (Maki 2008 [1964], p. 62, translated by the authors of this essay by exchanging aspects of 'form' with aspects of 'process' in the quote).

'Modern Individualism is an imaginary structure—this is why it fails' Aldo van Eyck argued in 1962, stating the insufficiency of the individual, and continued by stating the collective's ditto: *'Collectivism is the final barrier man has thrown up against himself as a substitute'* (Eyck 2008 [1962], p. 54). By exchanging form with process in Maki's description of linkage above, a potential to identify a principle for the collective process appears, shedding new historical light on this mutual insufficiency of individual and collective that reflect in both form and process. Instead, a unified conception of individual and collective is implied in which form and process are valued as a joint expression of the multiple individual views and resources acting in its realization. Hence, as insinuated by Eyck, a strong and caring collective must necessarily be integral with a diverse, inclusive, and plural capacity on behalf of the individuals constituting its parts. Likewise, this diversity must consequently reflect in the physical capacity of the built form resulting from it. *'...individualism and collectivism cannot be reconciled as abstractions or absolutes since only what is real can shake hands and acquire ambivalent meaning—it needs real hands to really shake hands. The real third is a real dialogue, a real embrace, a real dual between real people'* (Eyck 2008 [1962], p. 54). In this way, we are not searching for a collective process as such, but a collective process that cannot be reduced to the sum of the competencies and views acting in it. Using the expression of Eyck, it must form an 'in-between realm' in which form and process are mutually dependent reflections of the realities of each of the individual views and parts it consists of. A realm requiring of us to question, negotiate, balance, and amplify our individual contributions together.

22.4 Statements for Collective Authorship of Collective Forms

As stated in the introduction, the scope of this essay is to engage in a mutual self-reflection aimed at supporting the identification of actions to be pursued in each of our everyday contexts across education, practice, and public debate. Hence, this section of the essay is an attempt at applying the above translation in the formulation of three statements for individual action (together).

22.4.1 Education

'Builders of the cities we admire—cities that we sense are good environments—have generally been generations of men and women working over decades, even centuries. We perceive what they have done in our limited span of study' (Maki 2008 [1964], p. 59). As an educator and researcher, focused at developing tectonic methods for improving the resourcefulness of architectural form through a conception of construction technologies as potential spatial gestures (Hvejsel 2022), Maki's studies provide a central perspective for the improvement of my practice. By pointing simultaneously to the confined frame of action that condition the work of the individual architect and the magnitude of complexity that the work entail, the current foundation of architectural education is put into question. It requires us to strengthen each student in identifying and developing their individual artistic sensibility while simultaneously aiding them in contributing to a collective technological uncovering of the scientific consequences and value potentials of architectural form.

If applied within the current context of architectural education, Maki's linkage of form and process implies profound changes in architectural curricula. From it, we can understand that the positive impact potential of the structural changes rightfully called for by students demanding a safe, plural, and equal learning environment (Hartmann 2022, pp. 58–60) and can be multiplied if successfully linked to the changes we need to accommodate in architectural form itself. In my view, the workshop pictured in Fig. 22.1 exemplifies a direction to follow in this matter as it reflects the full challenge of piecing together the trash we are leaving behind in order to form a new viable collective reality in both form and process.

In my work, I consider that of didactically reflecting the processual realities of architectural practice in a way that enables students to critically develop their discipline towards resourceful practical action as my main responsibility. If incorporating the linkage of process and form implied by Maki, our teaching must continuously explore collective processes enabling us to create and value architecture upon its spatial capacity to engage positive environmental, social, and economic change. Not only does this call for my action in nurturing a shift from a hierarchical student-teacher relation to a shared care for mutual learning processes: Simultaneously, it points to a need to act in a re-conception of architectural research as the backbone of lifelong learning processes, rather than as an isolated academic discipline.

22.4.2 Practice

'It is worth noting that group form generally evolves from society rather than from powerful leadership' (Maki 2008 [1964], p. 53) As argued by Maki, forms for our challenging situation should be created in collective processes, but from my experience as studio director and leading architect in several acknowledged architecture studios, the planning industry is still very homogeneous. How can collectiveness develop in a homogeneous environment, I ask myself? We see that diversity in our planning industry is becoming an existential necessity to come to collective forms that reflect our society.

While we complain a shortage of applicants it seems difficult to hold certain people in the profession. For example, in terms of gender distribution in Germany, we have 59% female graduates while only 35% will eventually be chamber-registered architects and only 2% are working in leadership positions (Bundesarchitektenkammer 2020). Keeping diverse profiles in our profession and support them to go for leadership positions can be a lever to promote Maki's approach of collective processes, and hereby to arrive at more collective forms. In this way, leadership for collective processes means to ensure that each individual team member can fully use and develop their potential and thereby contribute to the success of the project. Leaders must recognize the potentials, promote them, help to compensate for weaknesses, and mediate within the team. In addition, an important task is to absorb the pressure from outside and not to

Fig. 22.1 How to expand our collective range of action in form and process? The photos are from a new workshop launched at Aarhus School of Architecture developed by Associate Professors Jonathan Foote and Anders Gammelgaard Nielsen. The students are deconstructing an old chair from bulky waste in order to expand its social capacity by offering a seat, not only for one but for two people. Simultaneously, the students are taking the leap from single authorship to the formation of a collective process. Photo credits: Jonathan Foote

pass it on unreflectively. Structures and a culture of communication must be established within the team to enable the goal of personal responsibility and initiative. Leading persons act and communicate in an exemplary manner for the way the members deal with each other. Feeling responsible not only for your own communication but also for communication problems within the team and trying to intervene in a clarifying way is an important attitude. Criticism should generally be regarded in the team as something that promotes process and is detached from personal value.

Identifying care as a crucial aspect of leadership seems essential for people to grow and feel belonging. The evolution of leadership from team leading to team caring in response to the SDG's, inherently entails not leaving any marginalized perspectives and competences behind in our design process if we are to succeed in creating collective forms in practice.

22.4.3 Public Debate

'*Can we, then, create meaningful group forms in our society? The answer is not a simple one. It requires new concepts and attitudes of design. It also requires the participation of cities and their social institutions*' (Maki 2008 [1964], p. 53). In public debate, the invisible and often not precisely articulated canon of universal modernism forms the key criteria for valuing architecture influencing its processes profoundly. Therefore, Maki's call for a new attitude including more voices as the participation of institutions is central to me as an architect and author.

In co-curating the Netherlands' pavilion at the 2021 Venice Architecture Biennale, architect Afaina de Jong responded to the claim proposed by the General Commissioner, 'How will we live together?' asking; 'Who is We?'. Her project The Multiplicity of Other, explored the many perspectives inherent in a 'we' through a variety of interdisciplinary approaches. In public debate and its media, architecture is contrarily mostly regarded and rewarded for itself: '*[This] is likewise seen in representations of architecture with as few traces of user paraphernalia as possible and flares up in discussions when award-winning architecture, once in operation, is 'not understood' by users.*' (Hartmann 2022, p. 73) How could the public way of debating and valuing architecture be expanded, I ask myself? How could the mainstream public debate address the narrative of single authorship as a crucial challenge in fulfilling the SDG's within in the field of architecture?

The often unconscious, yet inherent, impact of modernism tends to be amplified in the emphasis of single authorship that dominates at the expense of the possible collective potential of design in form and process. To open up the framework and enable collective processes, the canon of universal modernism could be re-framed as *one* possible foundation, opening up to 'other methods' taking into account the lived experience of many. Referring to Maki, the implementation of a new approach in public debate based on questions may form a central 'linkage' in this matter. Asking questions referring to the concerns of care, user-friendliness, re-framing design as a viable instrument to amplify people-centred approaches. The attitude of questioning the status quo could be less about adding new perspectives to a 'state of the art' but to aim to make architecture whole by giving it the role of serving the needs of many valuing it upon its ability to engage. Understanding the needs of many necessitates open caring debates across all parties involved in the production of architecture, including cities, and their social institutions.

22.5 Self-Reflection for Resourceful Action

In summary, this mutual self-reflection for individual action has uncovered a need to diversify perspectives to contribute to viable change in each of our respective contexts in response to the SDG's. Mediated by a re-reading of Maki, Ohtaka, and Goldberg's investigations in collective form, we have discovered a correspondence between the SDG's and their call for collective forms that must necessarily reflect in our actions within the architectural process, entailing of each of us to move;

- from teaching and research to facilitation of mutual learning processes
- from team leading to team caring
- from a canon-based debate to caring perspectives for the voices of many.

In this matter, we have found that the first 'linkage' to be established, using the terminology of Maki, is the link between collective form and process: The physical and conceptual recognition that transition towards a viable development of the built environment cannot be covered by an account for the environmental footprint of the forms we construct alone, but must be integral to the development of the collective processes through which these forms are necessarily produced and eventually valued. We are determined to continue our learning path towards putting our individual resources and privileges to their maximum use in making this linkage, and we wish for this paper to read as a gesture to anyone wanting to join us in the process.

Acknowledgements Finally, we want to express our sincere gratitude to the students at Aarhus School of Architecture and to Associate Professors Jonathan Foote and Anders Gammelgaard Nielsen for their generosity in letting us present their work as an example for inspiration in the essay.

References

Bundesarchitektenkammer (2020) www.bak.de. Accessed 14 Oct 2022
Fisker K (1964) Persondyrkelse og anonymitet. Arkitekten 26:522–526
Gehl J (2001 [1971]) Life between buildings: using public space. The Danish Architectural Press, Cph
Gehl J (2010) Cities for people. Island Press, Washington, DC

Hartmann K (2022) Black turtleneck, round glasses: expanding planning culture perspectives. Jovis Verlag, Berlin

Hvejsel M (2022) What gestures (can we) afford? On the resourcefulness of tectonics in architecture and engineering. In: Djebbara Z (ed) Affordances in everyday life. Springer, Cham. https://doi.org/10.1007/978-3-031-08629-8_8

Jacobs J (1961) Death and life of great American cities. Random House, New York

Krasny E (2019) Architecture and care. In: Fitz A, Krasny E (eds) Critical care: architecture and urbanism for a broken planet. Architekturzentrum Wien/Cambridge, MA, Wien and MIT Press, London

Maki F (2008 [1964]) Investigations in collective form. In: Mulligan M (ed) Nurturing dreams—collected essays on architecture and the city. MIT Press, Cambridge

Maki F (2010) Video of panel discussion on the occasion of the special seminar on 'mega structures' at the first International conference on structures & architecture, Guimarães, Portugal

Meadows DH et al (1972) The limits to growth: a report for the Club of Rome's project on the predicament of mankind. Earth Island

Rosling H, Rosling O, Rönnlund AR (2019) Factfulness: ten reasons we're wrong about the world—and why things are better than you think. Sceptre, London

UN (2022 [2015]) The United Nations' 17 sustainable development goals. https://sdgs.un.org/goals. Accessed 14 Oct 2022

van Eyck A (2008 [1962]) The child, the city and the artist—an essay on architecture, the in-between realm. In: Ligtelijn V, Strauven F (eds) Aldo van Eyck writings. SUN Publishers, Amsterdam

Part IV

Re-framing Agency

23 Re-framing Agencies: 'Production' of Non-human Subjects in Contested Territories: Counteracting Ecological Deprivation

Gisle Løkken and Magdalena Haggärde

Abstract

The divide between the human 'subject' and the non-human 'object' since enlightenment is an important cause for today's deprivation of natural landscapes, ecologies, and the escalating loss of biodiversity. Consequently, there is an increasing recognition of the nature crises and the climate change as interrelated and needed to be solved reciprocally, as parts of a complex common ecology. However, at the same time in the political planning system, there is a predominant highly instrumental and reductionist practice considering natural landscapes, not least along the shoreline, as commodities open for industrialisation, urbanisation, and privatisation. In the case from Lofoten in 2017, a master studio at Bergen School of Architecture, a different approach was experimented by the production of 'Deleuzeguattarian' maps of new figurations of *vulnerability* and *imbrication* of layers of social activity, geology and biology, and in particular in the article; the sea birds' *lines of flight*. The article discusses the material and mental context of the studio and the possibility of using artistic means and methods in the 're-framing' of (non-human) agencies—to make planning more resilient and ecological consistent.

Keywords

Rhizome · Actor network · Artistic methods · Non-human subject · Agency

23.1 Introduction

Loss of biodiversity and the threat of mass extinction of non-human species, animals and plants, are increasingly considered an integral part of the climate change and consequently a crucial element in the global environmental crisis.[1] The threat to vulnerable animals is

G. Løkken · M. Haggärde (✉)
70°N arkitektur, Tromsø, Norway
e-mail: magdalena.haggarde@70n.no

G. Løkken
e-mail: gisle.lokken@70n.no

G. Løkken
Norwegian University of Science and Technology, Architecture and Planning, Trondheim, Norway

M. Haggärde
The Arctic University of Norway, Academy of Fine Arts, Landscape Architecture Program, Tromsø, Norway

[1] The IPCC latest report; *Climate Change 2022. Impacts, Adaptation and Vulnerability. Summary for Policymakers*, states in SPM.D.4 (2022: 35): "Safeguarding biodiversity and ecosystems is fundamental to climate resilient development, in light of the threats climate change poses to them and their roles in adaptation and mitigation (*very high confidence*). Recent analyses, drawing on a range of lines of evidence, suggest that maintaining the resilience

S. Hilal et al. (eds.), *Design for Partnerships for Change*, Sustainable Development Goals Series,
https://doi.org/10.1007/978-3-031-36993-3_23

compounded, but is largely due to overexploitation of natural systems and reduced access to food resources. But as pressing is the rapid destruction of habitats where natural landscapes and seascapes more often are taken over by industrialisation, urbanisation, and leisure use, where today "one-third of all land is degraded" (Guterres 2022).[2] The severity of biodiversity loss and environmental deprivation, with subsequent devastating consequences for human existence, has been highlighted for decades—from Alfred North Whitehead (1925/2022: para. 155) claiming that; "[a]ny physical object which by its influence deteriorates its environment, commits suicide" to Gregory Bateson (1972: 483); "We are learning by bitter experience that the organism which destroys its environment, destroys itself"—to a numerous researchers, governmental and non-governmental organisations today adducing evidence for the same[3]—where the latest report from the most prominent global authority, IPCC, clearly states the entanglement of environmental issues: "Safeguarding biodiversity and ecosystems is fundamental to climate-resilient development, in the light of the threats climate change poses to them and their roles in adaption and migration" (IPCC 2022: 35).

However, since René Descartes in the enlightenment reasoned in his text *Animals Are Machines,* that the body should be "regarded as a machine [...] made by the hands of God" (Descartes 1989: 14), and marked the distinction between humans as "both mind and body" (Powell 1971: 209), and animals being bodies but no minds, the view on animals and nature in western societies has been highly instrumental. This seals the perception of the non-human not having subjective 'rights' on their own; further laid out by Allan White (1989: 120), that "only a person can logically have a right because only a person can be subject of such predications". The inherited notion from Descartes, coincides with the Christian orthodoxy of a clear distinction between 'man'[4] and nature, reinforced by the supreme right of 'men' to "treat them [the animals] in certain ways" (White 1989: 120).

Today the conception that nature exists in its own right is by all means more recognised, by ecologists and environmentalists,[5] and not least through the more recent development of circular and eco-economy (Brown 2001; Ingebrigtsen and Jacobsen 2007; Raworth 2017; Dasgupta 2021) based on the premise that nature cannot be a 'free asset'. However, it is still not fully reflected in realpolitik, where, e.g. architectural production and planning frequently serve as tools for an often instrumentally aimed political system. However, as architects and planners, we are reminded by Bateson (1972: 512) that; "We are not outside the ecology for which we plan—we are always and inevitably a part of it". And as any issue regarding ecology and survival of endangered species (and consequently humanity) is time critical, the need for 're-framing of agency' in planning and decision-making is imminent and demands radical reconsiderations and actions with other means than those coming from the systemic thinking that caused the problems. In the article, it is argued that this

of biodiversity and ecosystem services at a global scale depends on effective and equitable conservation of approximately 30–50% of Earth's land, freshwater, and ocean areas, including currently near-natural ecosystems (*high confidence*)."

[2] Secretary-General António Guterres' opening remarks at the UN Biodiversity Conference—COP15 (2022, December 6) states clearly the devastating prospects of biodiversity loss; "A million species teeter on the brink", and the "loss of nature and biodiversity comes with a steep human cost."

[3] See e.g. Dasgupta (2021): *The Economics of Biodiversity: The Dasgupta Review. Abridged Version*, stating clearly the failure in the capitalist economic system by considering nature as a 'free asset' in the economic models.

[4] The concept of 'man' has until far into the twentieth century been considered equivalent with 'the white man of a certain prosperity and position, leaving out children, women, poor people, and any person outside the western realm.

[5] The IPCC 2020 report is the sixth assessment report of the United Nations Intergovernmental panel on Climate Change (IPCC), engaging a large amount of scientists worldwide. The first part (working group one) published in 2021, as an example, was made by 234 scientists from 66 countries and built on more than 14,000 scientific papers and approved by 195 governments (https://en.wikipedia.org/wiki/IPCC_Sixth_Assessment_Report).

implies more open and adaptive approaches with explorative mapping and artistic means.

23.2 Approach and Theory

The article takes these realities as a point of departure and discusses the possibility of a more open, non-biased, and less mechanical, instrumental and reductionist approach towards the complexity of the field we are planning in. The article explores means, methods, and work from a master studio conducted by the authors of the article, at Bergen School of Architecture during the spring semester of 2017 (BAS 2017), the way it is presented and discussed in the subsequent book made on the basis of the studio; *Layered Landscapes Lofoten—understanding of complexity, otherness, and change* (Haggärde and Løkken 2018: 17). The aim of the book is described to critically "challenge internalised concepts about planning and investigation, which today often are too predefined, limiting and simplifying". The open approach in the master studio is *rhizomatic*[6] and connects findings and experiences of various characteristics and of different substances. Rhizomatic thinking does not differ between what is assumed objective, or what will normally be considered subjective and consequently not regarded valid in planning. Neither does it distinguish between the agencies of the human and the non-human, as it is necessary to be informed by the depth of complexity the landscape contains—in time and space.

The open and non-biased conception of the Actor Network Theory (ANT), the way it is elaborated by Bruno Latour (1996), Anne Marie Mol (2010) and others, is introduced as a theoretical framework for a critical discussion of hierarchisation of the 'human' and 'non-human' 'rights' in planning. The ANT is not considered a complete or closed theory on how actors 'interact', but follows a rhizomatic logic that is "acentered, non-hierarchical, nonsignifying" (Deleuze and Guattari 2004: 23)—always open for the unexpected and undefined. Likewise, will the concept of 'making of subjects' through subjection in power-processes, from Michel Foucault (1980, 1982) and Judith Butler (1997), be discussed as applicable for re-framing of agencies in planning. The introduction of 'new subjects' in the form of non-human entities, forms parts in a Foucauldian 'productive discourse' (Butler 1997: 2) with a critical approach to prevalent structures and institutions.

23.3 Materials and Methods

The case from the master studio is centred around Lofoten in Northern Norway, an archipelago of mountainous islands pointing south-westwards from the Norwegian mainland out in the North Atlantic. The studio searched for 'conceptual ecologies' (Cole and Rafe 2017: 12) by inquiring information from the landscape that is not immediately accessible. The aim of the work was to construct new concepts "ecologically, ethically and aesthetically", hence, according to Hans-Georg Gadamer "there is no essential difference between understanding and interpretation" (Bernstein 1983: 138) and Gilles Deleuze (2007: 325) states that "[i]nformation is a set of imperatives, slogans, directions—order-words" we can use for new action and "appropriation" (Bernstein 1983: 145). This means that in a landscape like Lofoten, with a history of 'living by nature' through centuries but today facing increased pressure of mechanisation, industrialisation, and commodification, it is an imperative to challenge any conception of human activity in the littoral zones of the entire archipelago.

Existence in these coastal landscapes has always been conditioned by an ever-shifting nature, and even if Lofoten is located literally *in* the centre of one of the world's richest cod fisheries, survival has been predicated on knowledge about the changes and the ability to deal with unpredictability. The history of the

[6] The concept of the *rhizome* is coined by Deleuze and Guattari (2004, p. 21) from the organic root system of rhizome, as a non-linear, "acentered" and sturdy structure radically differing from a hierarchical and vulnerable tree structure; "the rhizome is made only of lines: lines of flight or deterritorialization [... and] operates by variation, expansion, conquest, capture, offshoots."

seasonal cod fisheries in Lofoten, 'the Lofotfiske' is a story about exploitation of natural resources in a profitable business since medieval times,[7] which has had a significant impact on national economy and the foundation of trading cities like Bergen, exporting dried fish to an international market. As such have the forces of globalisation impacted Lofoten for centuries.

In a local context, the fisheries have, through history, altered the demography and patterns of settlement and shaped the physical features of the islands along with modernisation, mechanisation, and industrialisation. From medieval times, where small fishing boats were hauled and launched at the very shore in the littoral zone, a development towards larger and motorised vessels in the twentieth century resulted in new needs for harbours and quays. In the continuation, today the fish is to a large extent caught by huge pelagic trawlers flagged to other places. This means that the fisheries in reality have become independent of 'onshore facilities' and the remaining processing facilities in Lofoten, are highly mechanised and located in larger factories. This development has caused a demand for ever deeper and larger harbours—and equally larger industrial areas on shore. In the Lofoten topography, the land between the ocean and the mountains is of limited extent, and the expanding industry and urbanisation has been solved by land-reclamation in the littoral zone and in shallow waters.[8] In the article, *The Two Landscapes of Northern Norway* the Northern Norwegian philosopher Meløe (1988: 387) elaborates the concepts, conditions, and connections of the mountain based landscapes for reindeer herding and the costal based fishing and argues that our understanding of concepts comes "from our common activities in the world" and consequently; "Without coastal fishing, or seafaring, in boats too large for their crews to draw them ashore, there is no place for the concept of a harbour". Through increased mechanisation and less dependence of knowledge about natural conditions, flow and ebb, weather conditions, other species or seasonal changes, amplified by the Cartesian notion of nature being a machine, there are thus fewer 'common activities' on which concepts about birds and wildlife; their conditions for survival or importance for our existence, can be created.

23.3.1 Human and Non-Human—Power Dynamics and Hierarchies

By accepting the premises of today's ecological crisis, we acknowledge the need for new methods of exploration with awareness for the complexity of connections and dependencies, preconditioned in an ecological paradigm. In history, the strict division between the subject being human, and the object being non-human (Wilding 2010: 21), has sealed a hierarchy where the non-human has been open for exploitation and extinction. However, in an 'Actor-Network Theory' (ANT), according to Mol (2010: 255), things get more 'blurry', as what counts as an 'actor' is not restrained to be 'the conscious human', but can be "different things". As such, the ANT is a conception of the world that counteracts strong and oppressive power relations and what Foucault (2000: xiil) describes as "subdivision and pyramidal hierarchization", consistent enough to stimulate "action, thought, and desire by proliferation, juxtaposition, and disjunction". Mol claims that the ANT is not only stating the obvious that "an actor acts" (2010: 255), but also that an ""“actor” is made to be", which marks a "shift from the active to the passive". An essential

[7] The 'Lofotfiske' is the annual seasonal fishery from January to April on the migrating pelagic codfish 'skrei' sprawling in the waters of Lofoten. The amount of fish caught annually is fluctuating and in 2020 the total amount of cod fish from the 'Lofotfiske' was approximately 110,000 tons with a total value of NOK 2.5 billion. Approximately 40% of the fish was delivered locally in Lofoten (NRK 2020).

[8] The issue is exigent in any costal settlement, and Fig. 23.1 shows an example from the fishing village of Ramberg where a large part of the tidal zone was filled out for industrial use in the 1990s (B), and there have been several plans for extension of the harbor in the adjacent areas (A and C)—which is considered very rich in wildlife with 43 registered species of aquatic birds, more than any other tidal zone in Lofoten, and with 9 red-listed species (D) (Wilson and Eggen 2015, p. 4).

Fig. 23.1 Aerial photo from Ramberg, Lofoten, showing already existing landfill (**B**) for industrial use, and new planned ones (**A**, **C**), threatening the tidal zones crucial for the birds' survival (**D**). Illustration made by the authors with information from Wilson and Eggen (2015). See also note 8

aspect of the ANT, and what she finds "one of the pleasures of engaging", is that it offers the "possibility of seeing, hearing, sensing, and then analysing the social life of things—and thus of caring about, rather than neglecting them" (Mol 2010: 255). The act of 'caring' makes a difference and establishes an ethical stand for an active engagement and resistance against what is systemic and hegemonic. Hence, Deleuze (2007: 327) claims that "[c]ounter information is only effective when it becomes an act of resistance", which is what a critical-creative approach in its nature will aim for.

In the ANT, it is possible to map what at the same time is both material, like between things and physical appearances, and semiotic in the sense of abstractions and concepts—and all relations and phenomenon transcending and transgressing the borders between fixed categories. Analysing in the spirit of ANT is open explorative and rhizomatic the way Deleuze and Guattari (2004: 21) describe the rhizome to connect "any point to any other point, and its traits are not necessarily linked to traits of the same nature". As such, this open approach creates an awareness for the multiple and for the possibility that there are layers of information that are hidden and easily can be overlooked and neglected. The important reminder is that the "multiple must be made, not by always adding a higher dimension, but rather in the simplest of ways, by dint of sobriety, with the number of dimensions one already has available—always $n - 1$" (Deleuze and Guattari 2004: 21).[9]

[9] The '$n - 1$' principle indicates a sturdy and resistant system, which will not collapse even if a component or unit falls out.

23.3.2 Artistic Approach, Mapping, and the Making of Concepts

The master studio in Lofoten aimed to create an open testing ground for the students to experiment with progressive mapping. As mapping can be a strong political tool either restricting or liberating, it is accordingly crucial to develop a clear consciousness about the intentions of mapping, and also by which methods we are working. According to Deleuze and Guattari (2004), we should therefor make maps that are experimental to be able to find the hidden and unexpected. This means that the mapping should be complex and artistic—to liberate new knowledge and creativity, and also because there is "a fundamental affinity between a work of art and an act of resistance" (Deleuze 2007, p. 327).

By an 'actor-network' approach and equally by an "acentered, non-hierarchical, nonsignifying" (Deleuze and Guattari 2004: 21) rhizomatic approach, progressive mapping will be the opposite of a mechanical 'tracing' of an apparent reality or an "object of reproduction". It will rather be something that always "must be produced, constructed, a map that is always detachable, connectable, reversible, modifiable" (Deleuze and Guattari 2004: 21). These preconditions were fundamental for the studio assignments (BAS 2017; Haggärde and Løkken 2018) and the composition of different issues to be investigated in the landscape—revolving around the concepts of *complexity, imbrication, vulnerability, fieldwork, flexibility,* and *reorientation*. The concepts formed clear theoretical frameworks and working tasks, but were at the same time open for exploration, interference, and alteration—over again from ever new perspectives. Hence, the making of concepts is a creative act as they, according to Deleuze (2007: 318) "do not exist ready-made" just to take over, but "have to be produced".

Further, Deleuze and Guattari (1994: 15) state that; "There are no simple concepts [...]. It is a multiplicity", and "every concept is at least double or triple" because no concept can possess "every component, since this would be chaos". Consequently, this means that the concepts we make have to be worked on and interpreted into new overlapping ideas[10] and concepts containing fragments, new aspects and becomings, which involve "relationship with concepts situated on the same plan [...which] link up with each other, support one another, coordinate their contours, articulate their respective problems, and belong to the same philosophy, even if they have different history" (Deleuze and Guattari 1994: 18).

The concepts and terms were introduced in an interdisciplinary mode, which means that the curriculum and the lectures were influenced by other disciplines and subjects like anthropology, sociology, philosophy, engineering, visual arts as well as architecture and landscape architecture and consequently were the assignments formulated to stimulate heterogeneous and conceptual thinking, and creation of new maps in the Deleuzeguattarian (2004: 24) meaning of liberating new knowledge by 'making a map and not a tracing'.

23.3.3 Short Introduction to the Initial Concepts (BAS 2017)

Complexity; opens for infinite connections of what is past and what is to become—regardless of limitations in time and space—revealing endless possibilities and new beginnings "yet to be made, juxtapositions yet to flower into interaction" (Massey 2005: 11).

Imbrication;[11] juxtaposes a multiplicity of layers and concepts that are constituting the landscape and the society—of material nature and coming out of "a broad social science [and] philosophical approach" (Kofman and Lebas 2008: 23).

---

[10] According to Deleuze (2007: 323), "An idea is very simple. It is not a concept; it is not philosophy. Even if one may be able to draw a concept from every idea."

[11] *Imbrication* is originally a concept coined by Henri Lefebvre in the planning competition *New Belgrade* in 1968, which he defines as "the combinatory mode, the situational placement of different elements called to constitute, over a period of time, the city" (2009: 25).

Vulnerability; is about understanding the conditions for life and survival, and the consequences of human activity—with the potential of making changes for vulnerable species and environments.

Fieldwork; is essential for any architectural process "understood as a form of reality check, a means of adjusting one's thinking and practice to the exterior world" (Gabrielsson 2011: 35). It opens for experimental investigation on landscapes, practices, and events as an "embodied engagement and enquiry and as analytical trope" (Daniels et al. 2010: 1)—as a *critical-creative* confrontation of conceptions we have formed.

Flexibility; constitutes the need for resilience and adaptation to change—where in history the most specialized species and societies often are most vulnerable for external influence and environmental changes.

Reorientation; summarises the learnings from the semester and marks a new start and an opportunity to make things differently, based on new or deeper knowledge about the reality of the landscape in time and space.

23.4 Results

"In this studio *Layered Landscapes Lofoten*, we have studied the landscape from many angles and through a wide spectre of concepts. We have walked the landscape and encountered different practices unfolding [...]. We have dug into the historical shifts and natural conditions for the habitation of Lofoten, and we have confronted new challenges and forces that are at stake, both from within and from outside." (Introduction to the concept of *Reorientation*, Haggärde and Løkken 2018: 288).

23.4.1 Imbricating Activities and Vulnerability

The short excerpt from the *Layered Landscapes Lofoten* studio (BAS 2017) and the related book (Haggärde and Løkken 2018) used in this article unfolds under the initial concept of *Imbrication*. But in accordance with the Deleuzeguattarian (1994: 15) 'conceptual multiplicity', it brings in and overlaps essential elements from other concepts and ideas about *Vulnerability* and *Complexity*—and not least the existential dimensions of *Reorientation* as a new *becoming*. In the introduction text to *Imbrication*, John Pløger (2018: 93) explains that the "word imbrication makes one think of layers, connectivities, relations and the like, but the point might be that the world also forces one to take the contextuality of body-mind, material-immateriality, and doings-undoings seriously". The example is directly related to the initial aspect concerning incommensurable interests and activities in the littoral zone in Lofoten, and the exploration of layers of non-human practices which constitutes new maps conditioned on seabirds' behaviour and needs for existence. It is an example of how we —when investigating by other means—always will reveal new and important knowledge. The vulnerability of the birds is obvious—but still their territorial needs are commonly neglected under the pressure of urbanisation and industrialisation along a narrow shoreline (see Fig. 23.1).[12] An isolated tidal zone or a small littoral area is easily sacrificed for 'sake of modernity' and progress. But the birds' map of the territory is complex and consists of nesting and breeding areas, different feeding areas depending on season and tidal conditions, areas for protection from bad weather and heavy winds from varying directions—and from people and predators. The obliteration of even small parts of these patterns cuts off the migration routes and patterns the birds are operating along—and in total it puts at risk entire species. The threat to the birds is an example of what Latour (2016) points at as the major problem in planning: the lack of ecological holistic thinking, where "the problem is precisely to try not connect the dots, that is to

[12] Latest numbers given for construction along the shoreline, from Statistics Norway (SSB 2022, July 5), show a decrease in 'not-built' shoreline from 70% in 2005 to 68.4% in 2022, and as an example, the Lofoten municipality, Vågan, in 2021, approved 34 (out of 36) applications for dispensation—as one of the most extensive on national basis.

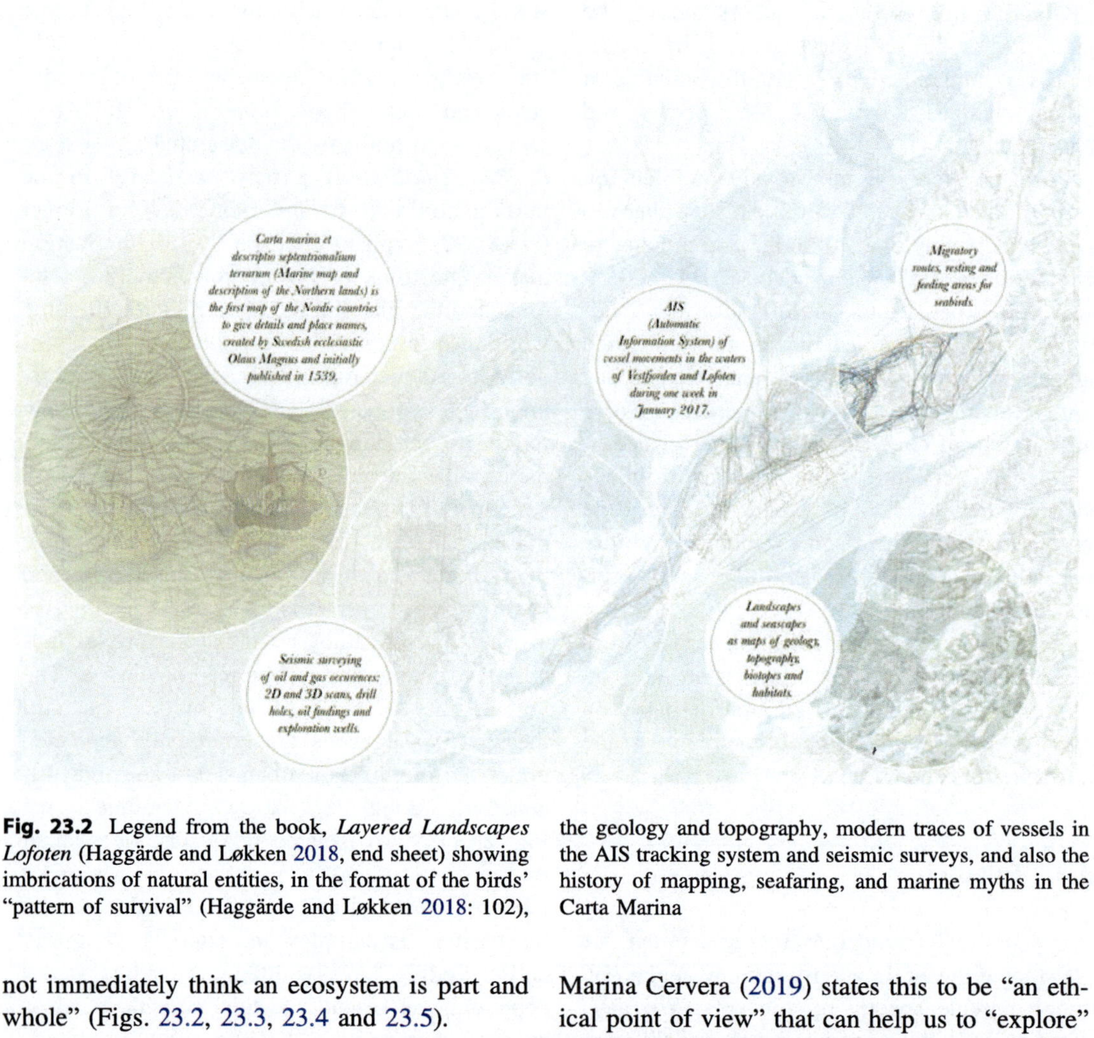

Fig. 23.2 Legend from the book, *Layered Landscapes Lofoten* (Haggärde and Løkken 2018, end sheet) showing imbrications of natural entities, in the format of the birds' "pattern of survival" (Haggärde and Løkken 2018: 102), the geology and topography, modern traces of vessels in the AIS tracking system and seismic surveys, and also the history of mapping, seafaring, and marine myths in the Carta Marina

not immediately think an ecosystem is part and whole" (Figs. 23.2, 23.3, 23.4 and 23.5).

23.5 Discussion

In the introduction to the concept of *Reorientation*, in the book *Layered Landscapes Lofoten* (Haggärde and Løkken 2018: 287–288), Herman Melville's character Bartleby is paraphrased on his resistance to accept any work he does not find meaningful. By claiming; "I would prefer not to", he establishes a "critical consciousness about why" we do things, for "whom we work", for the premises we work on, and "certainly for the impact of our work". In her opening remark at the book launch for *Layered Landscapes Lofoten* at El Col·legi d'Arquitectes de Catalunya, Marina Cervera (2019) states this to be "an ethical point of view" that can help us to "explore" and find the limits for justification of our interventions, and how the concept of *Imbrication* can be a means to "understand the complexity" and the "materiality of the place".

As such, *Imbrication* is not accumulation of data according to a linear and mechanistic logic of "interference of all the lines of imitation" (Latour and Lépinay 2009: 37), but a creative process and folding[13] of information, knowledge and ideas which accept chance and the unexpected.

[13] 'Folding' is a concept (from Deleuze: *le pli*) developed by Jean Hillier (2007, p. 232); "Spatial planning practice can be regarded as a performance of folding. Plans can be folded, metaphorically and literally, in many ways. In planning as becoming, there would be no predetermined style of folding, unfolding or refolding."

Fig. 23.3 Photos of traditional (partly abandoned) and modern industrial harbours in Lofoten and Vesterålen. By Anna Liisa Saavaste, Lassi Tuulonen and Magdalena Haggärde (in Haggärde and Løkken 2018)

Imbrication is spatial and follows a continuous process according to Henri Lefebvre's theory of a "socially produced spatio-temporal configuration" (Schmid 2022: 14). Translated to the Lofoten context, in the logic of the ANT, the birds' 'spatio-temporal configuration' of a new map of Lofoten constitutes a social 'interrelational space' existential to their survival.

23.5.1 Place, Space, and New Subjects

"The fields we are concerned with are, first, the *physical*—nature, the Cosmos; secondly, the *mental*, including logical and formal abstractions; and, thirdly, the *social*. In other words, we are concerned with logico-epistemological space, the space of social practice, the space occupied by sensory phenomena, including products of the imagination such as projects and projections, symbols and utopias." (Henri Lefebvre, *The Production of Space* (1991: 12).

When Pløger (2018: 95) states that; "Space is lived [...]. Place is taking place, it depends on the social forces (en)acting it", it is an analysis of how the landscape is constituted through presence and practices. And when Foucault claims that; "Space is fundamental in any exercise of power" (Pløger 2018: 97), it calls for our analytical approach to be sensitive to internalised hierarchies and power structures. According to Foucault (1982: 792), power is not a "fact" or an "institutional right" nor a "structure", but "it is elaborated, transformed, organized", and "adjusted to the situation". By accepting Foucault's analysis of power—it is possible to transform and "bring into play" the internalised power relations as "action in a field of possibilities", which calls for another approach to power than as an unambiguously subjugating force. Rather Foucault (1980: 97) prescribes to "grasp subjection in its material instance as a constitution of subjects"—by Butler (1997: 2) explained to be the paradoxical

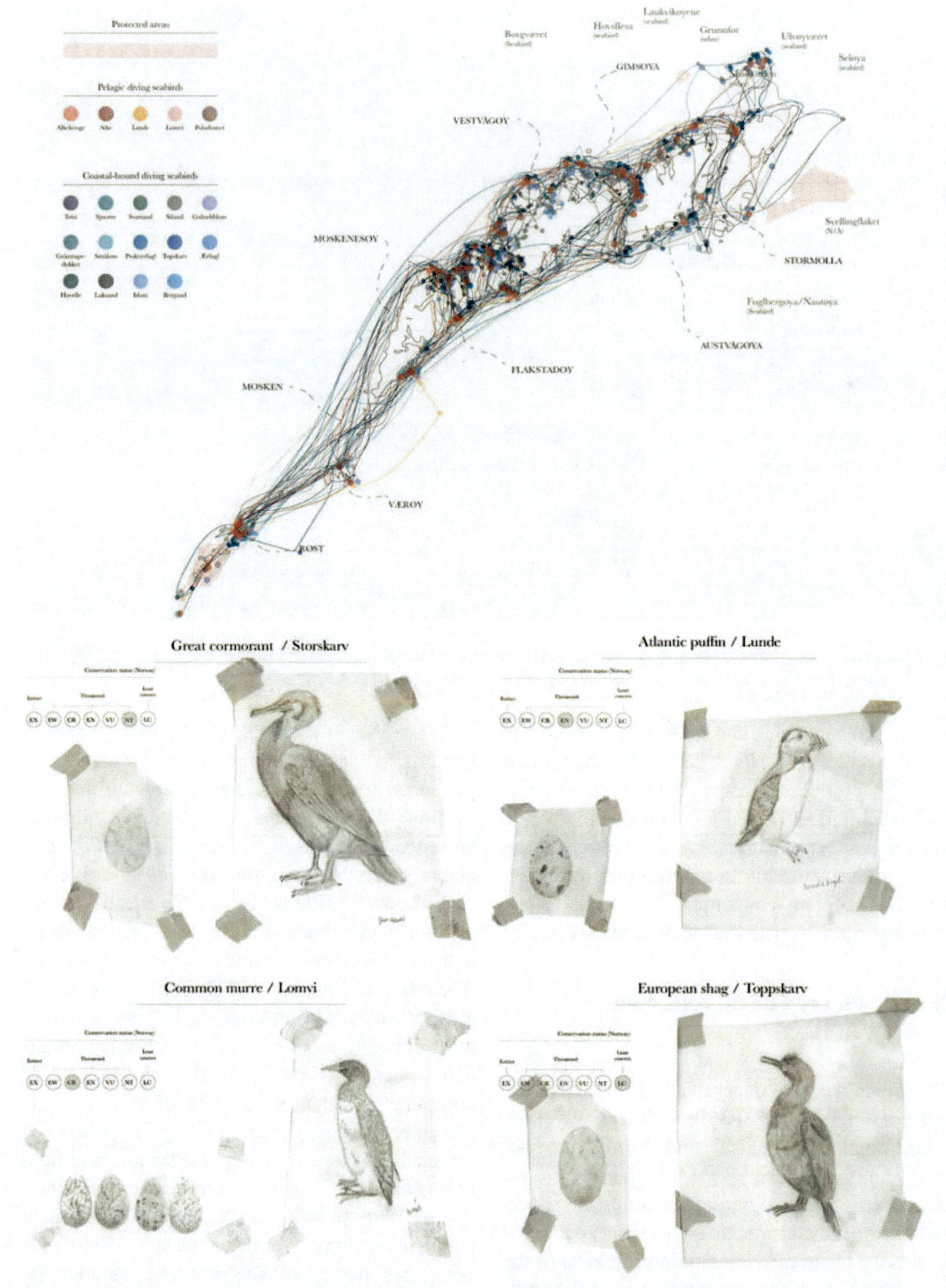

Fig. 23.4 Investigations of the 'conditions' for seabirds; *murres, puffins,* and *cormorants*—creating 'spatial–temporal' (Schmid 2022: 14) maps based on migratory routes, resting and feeding areas, where each spot and connection might be crucial for the species' viability. These beautiful 'patterns of survival' (Haggärde and Løkken 2018: 102) are invisible in most maps, as they are non-existent as physical imprint, but give important meaning and awareness to the understanding of the landscape. "Ornithological Map" by Victoria Helene Haukøya Storemyr (2018: 102–105)

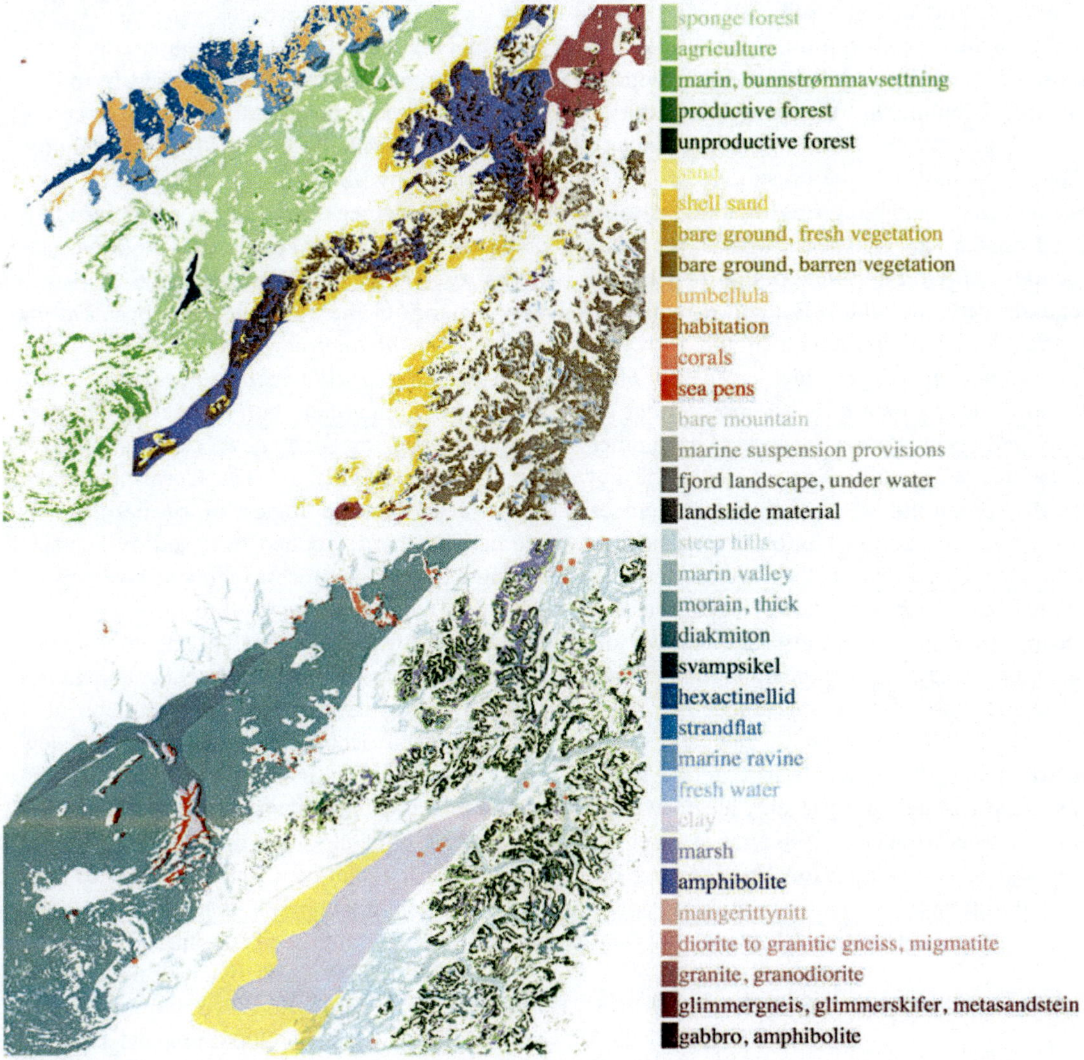

Fig. 23.5 Mapping of landscapes and seascapes with the same intensity; geology, biotopes, flora, habitats, etc., creates a continuous map giving new meanings, showing interconnections transcending the shoreline. Where most maps stop at the coastline, neglecting the importance of the sub-sea landscape, this map shows present, continuous, limitless connections. "Diversity of Nature" by Ingeleiv Andrea Utne Midtun (2018: 106–109)

nature of subjugation being both a signification of the oppressive nature of power and the process of "becoming a subject" by the Foucaultian "discursive productivity".

If we acknowledge the fact that the 'Anthropocene' pressure on the coastal and littoral ecology of Lofoten is severe and escalating, we can use the concept of the sea birds and the notion of a Foucaultian 'productive' discourse to constitute new subjects on an equal level. The discourse in this sense is not an exchange of words but implies a critical analysis of the "instances of control" and the "discursive regularities" forming them (Foucault 1981: 72). It should therefore, according to Foucault, be analysed in its "conditions, its actions and its effect" (Foucault 1981: 66) and "call into question our will to truth, restore to discourse its character as an event, and [...] throw off the sovereignty of the signifier".

Foucault put up some principles for a more open analytical approach of "possibilities" as opposed to "significations" which break with the 'principles' dominating "the traditional history of ideas" (Foucault 1981: 68). For the landscapes of Lofoten, it emphasises the imperative of finding, analysing, and applying information with an ethical standard as 'resistance' against prevailing structures and with a critical notion of what has been considered as 'truth'. Further, it indicates a materiality in the discourse—in the format of "discontinuous practices, which cross each other, are sometimes juxtaposed with another, but can just as well exclude or be unaware of each other" (Foucault 1981: p. 67). Consequently, the new subjects 'produced' in the discursive process possess, so far, unknown information and connections which alter internalised world views and systems, and constitutes new 'regularities', as Foucault says that "we must not resolve discourse into a play of pre-existing significations". And finally, based on the appearance and 'regularity' of the "discourse itself", Foucault claims that we should, "go towards its external conditions of possibility" and use the "aleatory series of these events" (Foucault 1981: p. 67) to set new expectations, in this case; for the becoming of a new and different plan.

23.5.2 Mapping and Artistic Methods

Hence, mapping by artistic means and methods is not strictly framed in any 'regularity' of the past or the present. It works without a predefined expectation to what the mapping should contain or provide, constitutes a possible 'speech act' and creates its own discourse of becoming. According to Deleuze (2007: 319) it is not "much opposition between the sciences and the arts" when it comes to being creative and inventive. Accordingly, the legitimacy of the artistic approach in the format of an open and un-biased mapping comes out of an acknowledgement that the 'artistic method' is not anti-scientific or 'relativistic' in the meaning of dissociation from fact or empiricism, but rather a grasp of the possibility there is to gain knowledge by means that are more flexible and explorative than a positivist approach can be. Hence, Gadamer (2004: 37) claims that "conceptual knowledge" can as well be possessed from "the work of art". As such the artistic approach comes closer to the *abductive* logics of modern ethnography which by Michael Agar (2006: para. 60) is explained to deprecate traditional *deductive* logic, as trying to find "new conclusions from old premises" or *inductive* logics trying to see "how well new material fit the available concepts". On the contrary, an *abductive* approach, which means 'lead away' (Agar 2006: para. 59), is not bound to predefined regularities, but is "taking surprises seriously and creating new explanations for them" (Agar 2006: para. 82). Correspondingly, Agar prescribes the logics to be *iterative*, as "something that is applied over and over again" during the working process, and also *recursive* as a logic that is "calling on itself to solve a problem that comes up even as it is solving a problem" (Agar 2006: para 82). Altogether this constitutes an open and dynamic process with the ability of adjusting itself according to 'learnings' or new knowledge evolving in the process, and it indicates a process contradictory to a positivist notion in planning with the hope of conducting 'exact science' or 'rationality' (Toulmin 2003: 67) but with an imminent 'risk' that "rigour might degenerate into rigidity" (Toulmin 2003: 41).

The validity of the artistic mapping presupposes a considerate process "concerned with knowledge and with truth" (Gadamer 2004: xx), and a call for a more 'flexible' and 'creative' knowledge production with a more 'nuanced' and "non-scripted process wherein the researcher responds to the particularities of what he or she is examining" (Cerwonca and Malkki 2007: 22). In other words, is the mapping of the sea birds a response to the 'situated particularities' which are invisible in the existing planning database but becomes a part of "a pattern of survival" (Haggärde and Løkken 2018: 102) for not only the birds as an entity, but for the whole ecological complexity.

The artistic approach of mapping the subconscious, the subjugated, the subjective, and the practices of entities not being predominant in the discourse, presupposes an approach which is open and not confined within a predefined framing of means or methods. Rather, it constitutes a 'point of departure' for a 'hybridisation' of the planning process taking into account variables, new events and concepts; defining new 'regularities' for the plan which are not easily numerically quantifiable. Latour and Lépinary (2009: 36) state that; "If accumulation is not the relevant point of entry to understand the dynamics of the economy, one must look elsewhere". As for the planning process, it will be more fruitful to call for an imbricated "interference and intersection of the paths of desire", which can be considered a strategy for resilience providing better information of the "probability" of things taking a different direction (Latour and Lépinary 2009: 37) and consequently lead to more appropriate actions and decisions. The mapping of vulnerable ecosystems with means and features that today are alien to an instrumentally aimed, and rationally claimed, planning system, evokes consequently the notion of a new "Ecocene" as a "curative catalyst for cultural change necessary to survive the 'Anthropocene'" (Boehnert 2018: 11)—which has to be created and "designed" in the planning process.

As such, the mapping of the seabirds' conditional existential behaviour constitutes a layer, or a rhizomatic 'plateau'—as a "rhizome is made of plateaus" (Deleuze and Guattari 2004: 24), in an imbricated reality of the coastal landscapes of Lofoten. The 'seabird-map' is an elicitation of 'something' that is materially present but at the same time 'invisible' for the institutionalised planning the way it is conducted in the prevalent modernist tradition –within the Anthropocene paradigm. The birds' needs and their existence 'draw a valid map' first when they are represented and advocated for, and imprinted in an open network structure as legitimate subjective actors, with "respect of specificities" (Lefebvre 2009: 27). The case stands as an example on how architects and planners need to re-orientate the methods and the means in planning to gain knowledge that is essential and existential for the 'unified' human / non-human ecology. And, equally how we need to approach planning by the creation of new concepts and ideas, to adequately map the landscape without bias or constraints—for 'connecting the dots' of a complex ecology, to avoid destructive simplification and deterioration of our common environment.

23.6 Conclusion

Despite well-documented connections between climate change and loss of biodiversity, the destruction of ecosystems and natural landscapes continues with an, in general, too instrumental and reductionist economic policy—with little ability for holistic thinking. Consequently, in the article, it is argued for a rethinking and a re-framing of the internalised historical division between the non-human as object and the human subject—a dialectic which has been problematised since Antiquity and ever stronger during the latest century. The current state comes with a call from Latour (2004: 57) for a "reunification of things and people, objects and subjects", as a matter of survival also for humanity, and as a possibility for construction of a new strong 'partnership for change' by more open explorative and non-hierarchical approaches, through the spirit of ANT and Deleuzeguattarian progressive mapping.

References

Agar M (2006) An ethnography by any other name. Forum Qualitative Sozialforschung/Forum: Qual Social Res 7(4):Art. 36

BAS (2017) Layered landscapes Lofoten. Master studio at Bergen School of Architecture, spring 2017, conducted by M. Haggärde & G. Løkken

Bateson G (1972) Steps to an ecology of mind. The University of Chicago Press

Bernstein RJ (1983) Beyond objectivism and relativism: science, hermeneutics, and praxis. University of Pennsylvania Press

Brown LR (2001) Eco-Economy: building an economy for the Earth. W. W. Norton & Company

Boehnert J (2018) Design, ecology, politics: towards the ecocene. Bloomsbury

Butler J (1997) The psychic life of power, theories in subjection. Stanford University Press

Cervera M (2019) Presentació del libre «Layered Landscapes Lofoten». At El Col·legi d'Arquitectes de Catalunya. Transcription from recorded lecture retrieved from https://arquitectes.cat/content/presentació-del-llibre-layered-landscapes-lofoten

Cerwonka A, Malkki H (2007) Improvising theory. Process and temporality in ethnographic fieldwork. The University of Chicago Press

Cole DR, Rafe MM (2017) Conceptual ecologies *for* educational research through Deleuze, Guattari and Whitehead. Int J Qual Stud Educ 30(9):849–862. https://doi.org/10.1080/09518398.2017.1336805

Daniels S, Pearson M, Roms H (2010) Editorial. Perform Res 15(4):1–4. https://doi.org/10.1080/13528165.2010.539871

Dasgupta P (2021) The economics of biodiversity: the Dasgupta review. Abridged version.HM Treasury, London. www.gov.uk/official-documents

Deleuze G (2007) Two regimes of madness. Texts and interviews 1975–1995 (trans: Hodges A, Taormina M). In: Lapoujade D (ed). Semiotext(e) (original work published 2001)

Deleuze G, Guattari F (1994) What is philosophy? (trans: Tomlinson H, Burchell G). Columbia University Press (original work published in 1991)

Deleuze G, Guattari F (2004) A thousand plateaus. Capitalism and schizophrenia (trans: Massumi B). Continuum (original work published 1987)

Descartes R (1989) Animals are machines. In: Regan T, Singer P (eds) Animal rights and human obligations, 2nd edn. Prentice Hall, pp 13–19 (original work published 1637/1646/1649)

Foucault M (1980) Power/knowledge: selected interviews and other writings 1972–1977 (trans: Gordon C, Marshall L, Mepham J, Soper K). In: Gordon C (ed) Pantheon Books

Foucault M (1981) The order of discourse (trans: McLeod I). In: Young R (ed) Untying the text: a post-structuralist reader. Routledge & Kegan Paul, pp 51–78

Foucault M (1982) The subject and power. Crit Inq 8 (4):777–795

Foucault M (2000) Preface to Deleuze G, Guattari F, Anti oedipus. Capitalism and schizophrenia (trans: Hurley R, Seem M, Lane HL). University of Minnesota Press

Gabrielsson C (2011) Inside the cave, outside the discipline. In: Ewing S, McGowan J, Speed C, Bernie V (eds) Architecture and field/work. Routledge, pp 35-42

Gadamer HG (2004) Truth and method (trans: Weinsheimer J, Marshal DG), 2nd edn. Continuum (original work published 1960)

Guterres A (2022) Secretary-general's opening remarks at the UN Biodiversity Conference—COP15

Haggärde M, Løkken G (2018) Layered landscapes Lofoten: understanding of complexity, otherness and change. Actar

Haukøya Storemyr VH (2018) Ornithological map. In: Haggärde M, Løkken G (eds) Layered landscapes Lofoten: understanding of complexity, otherness and change. Actar, pp 102–105

Hillier J (2007) Stretching beyond the horizon, a multiplanar theory of spatial planning and governance. Ashgate

Ingebrigtsen S, Jacobsen O (2007) Circulation economics. Theory and Practice. Perer Lang AG

IPCC (2022) Climate change 2022. Impacts, adaptation and vulnerability. Summary for policymakers. https://report.ipcc.ch/ar6wg2/pdf/IPCC_AR6_WGII_SummaryForPolicymakers.pdf

Kaufman E, Lebas W (2008) Introduction to Lefebvre H, writings on cities. In: Kofman E, Lebas W (eds and trans). Blackwell Publishing (original work published 1996)

Latour B (1996) On actor-network theory. A few clarifications plus more than a few complications. Soziale Welt 47:369–381

Latour B (2004) Politics of nature, how to bring the science into democracy (trans: Porter C). Harvard University Press

Latour B, Lépinay VA (2009) The science of passionate interests: an introduction to Gabriel Tarde's economic anthropology. Prickly Paradigm Press

Latour B (2016) On not joining the dots. Recorded lecture at Radcliffe Institute. https://www.youtube.com/watch?v=wTvbK10ABPI. Accessed 22 Nov 2016

Lefebvre H (1991) The production of space (trans: Nicholson-Smith D). Blackwell (original work published 1974)

Lefebvre H (2009) Introduction. In: Bitter S, Weber H (eds) Autogestion, or Henri Lefebvre in new Belgrade. Sternberg Press, pp 3–32 (original work published 1968)

Massey D (2005) For space. SAGE

Meløe J (1988) The two landscapes of northern Norway. Inquiry Interdisc J Philos 31(3):387–440

Mol A (2010) Actor-network theory: sensitive terms and enduring tensions. Kölner Zeitschrift für Soziologie und Sozialpsychologie. Sonderheft 50(1):253–269

NRK (2020) Årets skreifiske kan håve inn totalt én milliard kroner. https://www.nrk.no/nordland/tror-lofotfisket-2020-kan-ende-pa-en-milliard-kroner-_-og-eksport-av-klippfisk-oker-1.14970896

Pløger J (2018) Imbrication: the connections of city and landscape. In: Haggärde M, Løkken G (eds) Layered landscapes Lofoten: understanding of complexity, otherness and change. Actar, pp 93–99

Powell B (1971) Descartes' machines. Proc Aristot Soc 71:209–222

Raworth K (2017) Doughnut economics: seven ways to think like a 21st century economist. White River Junction, Vermont, Chelsea Green Publishing

Schmid C (2022) Henri Lefebvre and the theory of the production of space (trans: Murphy King Z). Verso

SSB (2022) Byggeaktivitet I strandsonen. https://www.ssb.no/natur-og-miljo/areal/statistikk/byggeaktivitet-i-strandsonen

Toulmin S (2003) Return to reason. Harvard University Press

Utne Midtun IA (2018) Diversity of nature. In: Haggärde M, Løkken G (eds) Layered landscapes Lofoten: understanding of complexity, otherness and change. Actar, pp 106–109

Wilson J, Eggen, M (2015) Ramberg havneutbygging og virkning på Sandnesbukta. Foreslått utbygging av Ramberg havn, et verneverdig område i Flakstad kommune. Report

White A (1989) Why animals cannot have rights. In: Regan T, Singer P (eds) Animal rights and human obligations, 2nd edn. Prentice Hall, pp 119-121

Whitehead AN (2022) Science and the modern world. Lowell Lectures 1925. The Project Gutenberg eBook https://www.gutenberg.org/files/68611/68611-h/68611-h.htm (original work published 1925)

Wilding A (2010) Naturphilosophie redivivus: on bruno latour's 'political ecology'. Cosmos History J Nat Soc Philos 6(1):18-32

Decolonizing Narratives Through Ecological, Regenerative and Biophilic Design

24

Nadine Samaha and Sarah Naarden

Abstract

Australia's First Peoples are recognized as the oldest living continuous culture in the world. Embedded within their traditional ecological knowledge are bioregional foundations that underpin: biophilic design, regenerative design and development, circular economies, biodiversity restoration and climate change strategies for land and built-form management. Through generations of custodianship, these knowledge systems have been held in Aboriginal Law. With the looming ecological crisis that the world is facing, Indigenous Elders are generously transferring their knowledge to the wider community. Yet, they do so with great caution after 230 years of disrespectful relationships brought about by colonial settlers' sense of entitlement. This presents a challenge for Australian architects, who desire to meaningfully participate in caring for Country whilst respectfully acknowledging the injustices of colonization and the ongoing marginalization of First Peoples. This paper traces the pathways of two socio-ecological-focused architects who come together to co-create methodologies for intercultural collaboration with First Peoples. Presented in this paper is the case study: 'Ghost in the Nets' which was recognized with a 'World Architecture Community Award for Unbuilt Work' in 2021. Initially proposed for the 2020 National Gallery of Victoria competition, it is being conceptually adapted for a new home at the First Peoples Pathway at Initiatives of Change Australia, a Centre of Excellence in Reconciliation, Education and Healing (Fig. 24.1).

Keywords

Ecology · Regenerative design · Biophilic design · Decolonization

24.1 Introduction

'Wominjeka' means 'welcome, come *with* purpose' in the languages of the Wurundjeri, Boon Wurrung and Bunurong People of the Eastern Kulin Nation-The First Peoples in Melbourne. The purpose of this paper is to inspire the architectural profession to develop cultural intelligence with their First Peoples. A modus operandi for any design brief and not just the sites which have an Aboriginal heritage overlay with a public interface. This process requires a pause in the everyday 'practice as usual' to create space to unlearn colonial history and consider

N. Samaha (✉) · S. Naarden
School of Design, Rmit University, Melbourne, VIC, Australia
e-mail: nadine@level-ak.com

S. Hilal et al. (eds.), *Design for Partnerships for Change*, Sustainable Development Goals Series,
https://doi.org/10.1007/978-3-031-36993-3_24

actively participating in social and environmental justice with First Peoples. Rather than a tokenistic gesture, the potential of a unique Australian architecture not only responds to bioregional vernacular but empowers First Peoples with social procurement pathways for Indigenous learning and healing restoring biodiversity with custodian responsibilities. The challenge is to make the hidden narratives of place visible translating the act of decolonization into built form (Fig. 24.1).

This paper reframes the idea of inclusion and marginalization in the built environment. Rather than viewing First Peoples' problems through a deficit lens, it repositions their perspectives as a key resource. First Peoples have responded to climate change with place-based innovation for over 65,000 years. Fragments of this knowledge can be identified through cultural heritage markers, artefacts and ceremonial practice. Evidence can be traced through oral storytelling, artwork and the journals of colonial settlers as

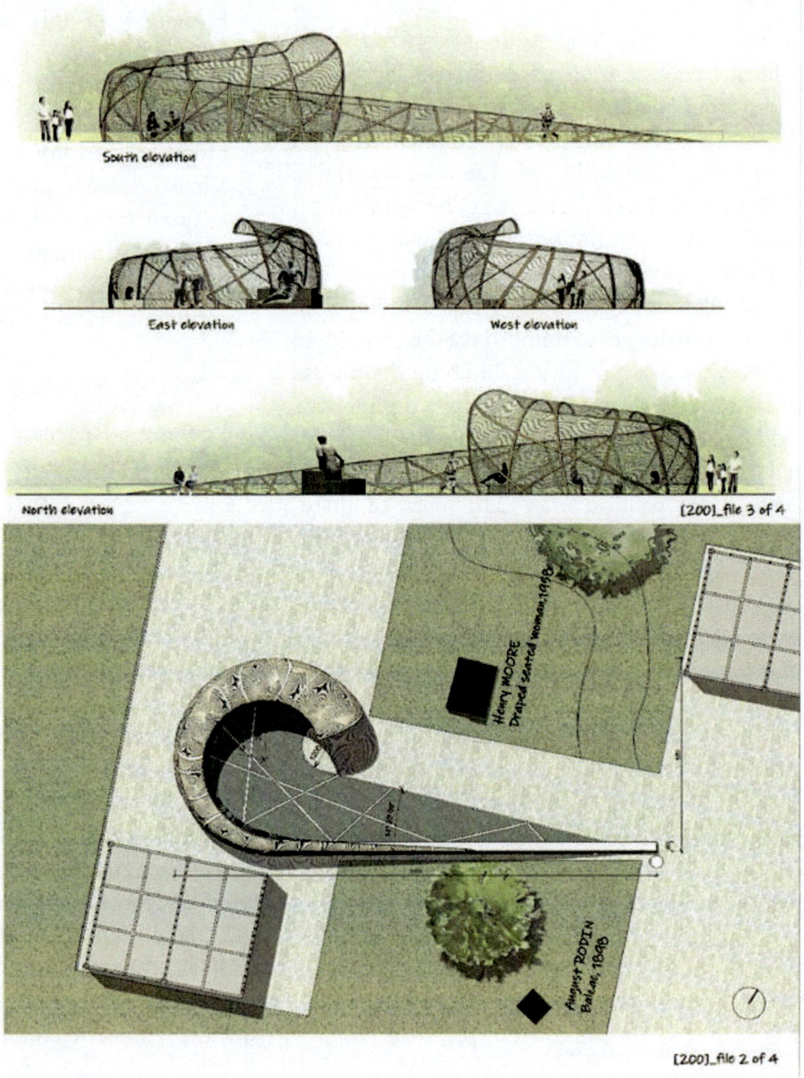

Fig. 24.1 Ghosts in the net-NGV competition entry 2020

referred to by Bruce Pascoe in his Dark Emu book (Pascoe 2018).

24.2 Materials, Methods and Methodologies

This research paper follows a qualitative approach that relies on key methodologies that can assist in engaging, disseminating and acting on produced knowledge. The aim is to establish trust-building with First Nations peoples through truth-telling, deep listening and deep understanding of caring for Country. These practice foundations underpin:

- The proposed '*Ghost in the Nets'* case study as a placemaking for reconciliation and connections.
- Connections between the various multi-disciplinary stakeholders.
- Collection of feedback through surveys to measure dissemination, action and reciprocity of knowledge.
- Creation of a formal code of conduct and reporting system.

This research follows the successful 'Regenerative Futures' pilot event/workshop that was hosted at the Royal Melbourne Institute of Technology (RMIT University) in 2022 (Fig. 24.6). The event was supported by Initiatives of Change Australia (IofCA), the Sustainable Architecture Forum at the Australian Institute of Architects (SAF@AIA), the National Sustainable Living Festival (SLF) and RMIT University. It hosted a Welcome to Country and didgeridoo performance by Uncle Shane Charles and was complimented by an Indigenous mindful movement practice called Wayapa Wuurk (https://wayapa.com/) led by Torres Strait Island woman Alana Marsh. This practice reconnected and grounded participants to place, spirit, true self and community. We then made our way up to the sustainably built Garden Building at RMIT where the emerging Indigenous leader and award-winning film producer Jai Allan Wright, descendant of the Mununjali and Wanggerriburra people of Yugambeh Nation, introduced the event. He was joined on the panel by regenerative design architect Dr. Dominique Hes and regenerative landscape architect and writer Dr. Maya Ward. After three inspiring keynote presentations, we served afternoon tea with scones comprising traditional Indigenous wattle seed and lemon myrtle. After this, each speaker led a workshop group speculating a Regenerative Future using different ideation tools. We finalized the event by sharing the key points from each group's findings. This created a platform to connect First Peoples' knowledge and regenerative design architecture with public participation. Connections were made through networking and continue to this day. Several participants joined the Turruk Yarning circles to deepen their relationships with the First Nations Peoples. Other participants opted in for more Regenerative Futures workshops. As feedback was positive, we decided to follow up with this research paper and hone better our skills to make better use of the results in our future events and projects.

Instead of viewing First Peoples' problems through a deficit lens, we are repositioning their perspectives as a key resource. Our hope is to expose ecological and cultural gaps so that caring for Country is continuous and not limited by project boundaries, budgets and timelines. Open dialogues were created for built environment opportunities with Australian Local Government strategies, biodiversity masterplans, Indigenous engagement and community partnerships.

Otto Scharmer states in his text: *Leading from the Emerging Future,* as designing with ecosystems rather than for an 'ego system' (Scharmer and Kaufer 2013). Shifting our anthropocentric views and destruction of biodiversity to a more eco-centric focus is transcended as Biophilia. *'Biophilia Hypothesis'* was popularized by Wilson in 1984 where he describes that the inner need of human beings is to connect with all living species (Wilson 1984). Many authors, architects and urban designers afterwards have explored how building occupants can connect to other living species and to the natural environment. This became known as *biophilic design.* Kellert defined two basic principles of biophilic design, which are (1) *organic or naturalistic dimension* and (2) *place-based or vernacular* (Kellert et al. 2008). These basic dimensions are related to six biophilic design elements that are very relevant to our events, masterclasses and case studies:

1. Environmental features
2. Natural shapes and forms
3. Natural patterns and processes
4. Light and space
5. Place-based relationships
6. Evolved human–nature relationships

However, First Nations Peoples have always known these principles and have always been intimately connected to their surrounding environment and other living species. Knowing Country and how to care for it is at the heart of their practice.

In line with the above, our following methodologies are demonstrated in partnerships for change practices through:

- Engagement
- Education
- Action

Engagement as the agency of the knowledge holder architect is transferred to First Nations Peoples through a codesign ethos: designing places 'with' community rather than designing 'for' community. In this instance, decolonization is practiced through *Turruk* an intercultural intelligence workshop series and community of practice led by First Nations Peoples (Fig. 24.2. *Turruk* is the Wurundjeri word for river reeds found in the local Yarra River and was used in ceremonial dances of Wurundjeri women in pre-colonial times. *Turruk* is a key program as well informing the 'Ghost in the Nets' case study pavilion (Fig. 24.1).

One of the central aims of Turruk is breaking the *'Great Australian Silence'* (Clarke 2018) about hidden colonial injustices, which is usually beyond the scope of an architectural brief. It requires architects to make space for courageous truth-telling attuning an empathetic ear to the stories of First Nation leaders and grassroots communities. In this process, truth hearing can be transformed into deep listening for a purposeful call to action. Senior Aboriginal elder and writer Miriam-Rose Ungunmerr-Baumann describes this as still inner awareness; '*Dadirri: recognising the deep spring that is inside us. We call on it and it calls to us*' (Ungunmerr 1988).

Daidirri was a key motivation in co-creating Turruk at IofCA in Melbourne. Here the cultural Intelligence program was co-designed with Uncle Shane Charles, a proud Yorta Yorta, Wurundjeri and Boon Wurrung man, and co-chair of Reconciliation Victoria, and First Nation Leaders and Allies. Since 2020, Turruk has run twice a year with over 700 participants across Australia and the globe. In 2021, it won a 'United Nations Intercultural Innovation Award' as part of the IofC Trust-building program which was piloted in 11 different countries over 3 years. Essentially, Turruk provides a culturally safe place to personalize stories beyond media stereotypes and break the stigma around intergenerational trauma. Delicate issues are navigated with the support of trauma-aware counsellors. Turruk typically comprises 6 weekly immersive yarning circles, ceremonies and masterclasses exploring themes that IofCA calls Trust-building. These include:

- Belonging
- Authentic leadership
- Inclusive dialogues
- Healing historical wounds
- Diverse collaboration
- Trust-building futures

The Turruk program responds to the Australian Government's recent commitment to the National Treaty through a process known as the *Uluru Statement of the Heart* (2022). It enables a practical understanding of a 'once in a generation change' and how this might look and feel in future. Unlike Canada and New Zealand, Australia's First Peoples were not invited to negotiate a Treaty at the time of colonial settlement. Still to this day, sovereignty has not been ceded. Progress has been made to create constitutional

Fig. 24.2 Turruk sessions 2022

change for an Indigenous Voice in Parliament through a referendum in 2023. However, it will take some time before this results in a Treaty with practical everyday implications for architectural business. In the meantime, many larger architectural practices work with the framework established through 'Reconciliation Action Plans' (RAP). These plans aim to increase Indigenous cultural awareness, employment and social procurement. They also hope to reduce significant gaps in Indigenous health, education, incarceration and child protection. RAPs are considered a costly and timely investment that only larger architectural practices can afford. Smaller architecture practices without a RAP, can still commit to unlearning colonial history and taking meaningful action. Until recently, architects had very little tertiary education about First Peoples' architecture and the impacts of colonization.

Many Australians were taught in primary and high school: *'Captain Cook discovered Australia in 1788'*. The principle of *Terra Nullius*: Latin for *'land belonging to no one'* shamefully declared *'Aboriginal People as nomadic and in absence of 'civilized' people capable of land ownership'* (McBride and Smith 2021). This was the legal basis for British settlement implemented in Australian law in 1835. It was only overturned in 1992 with the *Mabo Case,* a High Court case instating Indigenous rights to the 'Native Title' claim. After this, the language of *'Traditional Owners'* has taken precedence over *'Traditional Custodians'*. Many First People believe the term 'custodian' better reflects the responsibilities of looking after the land, the community, and all living sentiments. Indigenous academic and artist, proud Wiradjuri man Uncle Glenn Loughrey explains that we don't own the land, *'the land owns us'* (Loughrey 2020). Another Indigenous academic from the Apalech Clan in far north Queensland, Tyson Yunkaporta reinforces this sentiment in his text; *Sand Talk* (Yunkaporta 2019). He makes the provocation that if *'Indigenous thinking can save the world'*, then the dominant narratives around the land *'belonging to us'* need significant disruption. The challenge is to reframe our thinking and ask how do we participate in a regenerative urban context and *belong to the land.*

Education is used as a dissemination tool, through the creation of events, forums and masterclasses to guide through First Nations, biophilic, ecological and regenerative design knowledge. *Indigenous Ecosystem Corridors + Nodes (IEC + N)* was established through a joint venture between the *International Union of Architects (UIA)* and the *International Federation of Landscape Architects (IFLA)* (Rodger and Williams 2020) to restore the ecological damage that was inflicted by human settlement and population growth. Masterclasses started in 2020 and were prepared and presented by a multidisciplinary group online by an AIA working group (Fig. 24.3). They will be expanding in 2023 in collaboration with the Australian Institute of Landscape Architects (AILA) to include workshops and further case studies.

The purpose of these masterclasses is to educate architects, landscape architects and designers on the importance of incorporating biophilic and ecological interventions in the built and non-built environment to boost biodiversity, reduce the heat island effect and mitigate climate change. The subjects include an understanding of the Indigenous Connection to Country, the built environment's impact on biodiversity, the principles of biophilic ecological and regenerative design, and the relationship to SDGs measures.

Garrard et al. in their proposed project 'Beyond green façade: integrating ecology and architecture' (Garrard et al. 2023), seek an ecologically driven architecture approach that will have quantitative ecology and decision analysis integrated into architecture design. We expect our educational events to reinforce this approach.

Furthermore, through these masterclasses and other educational events, we aim to create awareness and establish connections between stakeholders around the current ecosystem corridors being proposed in Melbourne which are concentrated around waterways and transport lines, such as the Greenline. Many projects are shrouded in confidentiality with little public communication about how each project could be interconnected to improve biodiversity targets in future.

Action on the ground entails local groups actively working in the field, understanding the urgency of climate change and how best to mitigate it.

Fig. 24.3 IEC + N

Fig. 24.4 Rewilding Stonington working bees. 2020, 2021 and 2022

Rewilding Stonnington, a local group that started in 2020 by rewilding one nature strip as a pilot project, went on rewilding a corridor along Tooronga railway station through a self-funded process. This gained a quick interest from the Local Government and the community. Local Government is now supporting RS to create an urban forest and an Indigenous heritage walk in consultation with the local First Nations Peoples in 2023 (Fig. 24.4). The group continues to reintroduce Indigenous and endemic plants on nature strips and parks to the Stonnington area, connecting with the pre-colonial heritage and creating knowledge communities. In line with SDG 11 'Sustainable cities and communities', the group aims are to:

- improve our biophilic connection to nature,
- boost biodiversity and attract pollinators,

Fig. 24.5 Inspiration images and context plans

- reduce the urban heat island effect and stormwater runoff,
- respect and advocate for Aboriginal ways of caring for Country,
- consult with local traditional owners,
- create interest and diversity in the landscape,
- improve health and wellbeing,
- reduce noise pollution from lawnmowers,
- reduce the need for artificial fertilizers and weed killers.

These aims correlate with Mills et al. (2020) research study that was conducted at the University of Adelaide in 2020. They found that revegetating urban green spaces with endemic planting could rewild the soil microbiotas. This diversity in the soil is more likely found in woodlands than in grass nature strips and parks. They found that microbial communities and their ecosystems function better when plant species are more diverse. They believe that this soil microbiome that boosts biodiversity seems to improve the health, wellbeing and immune system of humans and all living species.

Ghost in the nets Case Study

As part of these proposed ecosystem corridors, 'Ghost in the Nets' could be considered a 'node' for Indigenous social and ecological knowledge exchange between the public and the private realms of land ownership. The proposed site is at the First Peoples Pathway Centre at IofCA. The site offers a unique spirit of place with 65 years of building reciprocal relationships with First Peoples, through education, reconciliation and healing. Currently, the centre hosts Uncle Shane Charles as a cultural resident, and Dylan Charles as an Indigenous artist in residence. The First Peoples Pathway is in the process of being formed with Indigenous academics, educators, ecologists, social enterprises, philanthropic organizations and employment advisors.

The intention of the 'Ghost in the Nets' project is to co-create community partnerships with opportunities for the western understanding of biophilia with local Indigenous ways of knowing. Some Western ways of knowing separate disciplines, using the mind through quantitative and empirical research and results. It tends to be more anthropocentric in its approach (Levac et al. 2018). In comparison, Gregory Cajete's in his book '*Look to the Mountain: An ecology of Indigenous Education* describes Indigenous perspectives as an invitation to a holistic understanding of all aspects of your being including mind, body, emotion and spirit (Cajete 1994).

The pavilion forms part of an intertwined Indigenous healing garden around a yarning circle with the aim to rediscover the sense of wonder/inner reflection and being connected to all living species and the natural environment. It will host cultural ceremonies, education forums and programs like Turruk around a cultural fire. It will become a place of intercultural exchange for bush medicine and bush tucker and biodiversity knowledge. The invitation will be extended to all community stakeholders including architects, artists, government bodies, NGOs, students and sustainability groups with opportunities for pro bono partnerships. It will be as well the host for global Indigenous dialogues, sustainable community gatherings, intercultural musical performances, weaving/carving workshops and bush tucker.

The 'Ghost in the Nets' pavilion design was inspired by a coiled cochlear ear drum inviting deep listening to ancient wisdom and an embracing arm for intercultural connection. Through its pedagogy of assemblage, the structure and the layering of skins, a placed-based vernacular will emerge. Foraged bio-based materials gathered from the local river and sea could tell a local story about lost pre-colonial landscapes and cultural practices. Through participatory weaving, tree branch frames, local bark and Turruk river reeds could be intertwined through the nets (Fig. 24.5). The structure could hold *'Ghost Nets'* (https://www.sea.museum/2019/01/30/ghost-nets) discarded fishing nets salvaged from beaches preventing sea life strangulation. First Nations weavers could invite Second Peoples; descendants of settlers and Third Peoples; more recent migrants. We expect a different tone of truth-telling might emerge. One that not only reconciles our differences but reconciles our relationship to the environment in the act of making together.

Our hope is that this purpose-built Yarning circle provides a process pilot for others across Australia; a process that privileges the sharing of oral wisdom that connect to time and place through their outdoor settings. By using the stars as maps to tell information stories about the seasons. Observing the patterns in the sky, clouds, sounds and scents of Country with bio-regional insights encodes place knowledge in our memory. Learning about Restoring the land, Bush Tucker and Medicine in situ reinforces oral wisdom. These types of nuanced immersive experiences have less impact if taught indoors via screen technologies. It also encourages curiosity in asking open questions where conversation in today's is becoming more fragmented through social media landscapes. The neuroscience of place-based learning suggests memories are encoded for improved social coherence and sense of inclusion by learning together in nature, which First Nations Peoples have done for millennia as reiterated in Lynn Kelly and Margo Neale Songlines' book (Neale and Lynne 2020). It also enables social ecological cultural barriers to be broken down, through biophilic principles whereby our vital signs synergize with our circadian rhythms, sounds and patterns in nature. Biologically we are attuned to each other for deeper listening which is a key ingredient of establishing cultures of care.

24.3 Results

Positive feedback came from the Regenerative Future workshop as it responded to the need for architects and all other stakeholders to develop cultural intelligence with their First Nations Peoples. This encouraged us to pursue a holistic program that can connect all stakeholders in future projects and forge partnerships between Indigenous and interdisciplinary game changers through engagement, education and actionable projects.

The feedback from Turruk workshops identified that participants felt confident to embark on a reconciliation journey and begin respectful relationships with Indigenous Elders to understand issues more holistically with multi-sector communities including health and education sectors.

The IEC + N educational masterclasses feedback identified architects' eagerness to connect with First Nations peoples wisdom, the principles of biophilic, ecological and regenerative design,

Fig. 24.6 Regenerative Futures at RMIT University February 2022

and SDGs so that they can apply it in their built and non-built environment design. It taught them the importance designing for biodiversity.

The feedback from Rewilding Stonnington, the local action group has proven our bottom-up approach of immediate action gets the attention of Local Government and builds momentum with increased community participation. We noticed during our working bees that rewilding urban areas created communities of care addressing issues of loneliness.

Our methodologies of engaging, educating and acting through First Peoples voices, multidisciplinary and multicultural community of practice is assisting us in articulating a more impactful agency of architecture that can respond better to the 17SDGs. In this instance, we are responding to the following United Nations Sustainable Development Goals (SDGs) (United Nations Sustainable Development Goals Website 2022):

Figure 24.2: Good Health and Wellbeing: Biophilic connections to nature and culture.

Figure 24.2: Quality education: Masterclasses, forums and workshops are available for everyone.

Figure 24.2: Industry, innovation and infrastructure: Partnering with the community to create vernacular structures.

Figure 24.3: To reduce inequalities: Having Indigenous voices to be heard.

Figure 24.4: Sustainable cities and communities: Creating sustainable and beautiful corridors to establish well-connected communities. Improve the built environment.

Figure 24.4: Climate action: Reducing heat island effect and air pollution in urban to mitigate climate change and improve air quality.

Figure 24.4: Life on land: Creating Indigenous ecosystem corridors and nodes to boost biodiversity, improve habitat and create better connections between fauna and flora.

Figure 24.4: Partnerships for the Goals: Working in partnership for a sustainable built and non-built environment.

24.4 Discussion and Conclusions

In summary, this paper restores these Indigenous pedagogies highlighted by biophilic immersive experiences. It reimagines design by transferring and engaging the agency of the architect to the community of practice with reciprocal knowledge and cultivation of culture of care through story sharing. After several years of COVID lockdowns, sharing stories about overcoming adversity, and being active in climate change solutions with First Peoples enables new stories of place. In effect, the proposal addresses two key challenges the world faces today: the loneliness epidemic and the ecological crisis.

The themes in this paper will be explored in *Regenerative Futures*: a series of public think tanks at Sustainable Living and Design Festivals. We ourselves are seeking to become regenerative practitioners, trying to develop our capability in ourselves using our potential and actioning them in projects and communities. We are inspired by the Regenesis principles (Mang et al. 2016), and the Inner Development Goals (IDGs) (2022) framework that states to achieve regenerative development we need to transform our skills and qualities through five dimensions: being, thinking, relating, collaborating and acting.

Since 2020, the discussions started enabling pathways for biophilic-informed cities to be inclusive of First Peoples' voices about the future of our cities. With the looming ecological crisis that the world is facing, elders are more keen to transfer their knowledge to the wider community through an understanding of caring for Country.

The 'First Nations Wisdom for Regenerative Futures' yarning event will be held on the 19th of February 2023. It will host a special First Nations delegation from Canada led by Lewis Cardinal Co-Chair of the Aboriginal Commission for Human Rights and Justice in dialogue with Tyson Yakanporta, Uncle Shane Charles and Uncle Glen Loughrey. Through the theme, Knowing Country, as part of the Sustainable Living Festival in 2023, this event is a collaboration between IofCA and SAF@AIA. Architects, landscape architects, ecologists, project managers, academics and engineers leaders are invited to engage in the dialogue. The objective of this collaboration is to decolonize the narratives through truth-telling, deep listening and reconciliation. Through weaving First Nations' wisdom, biophilic, ecological and regenerative design, we hope the storytelling and the dialogues will consolidate deeper relationships and facilitate connections between all stakeholders.

This yarning event together with ongoing IEC + N masterclasses hopes to cross-pollinate First Peoples knowledge solutions with leaders advancing UN Sustainable Design Goals. This pollination will feedback into other conversations and interventions around zero carbon, regenerative ecological design, circular economy, biophilic design, biodiversity, community partnership, sustainable development goals and affordable housing. The intention of this initial workshop is to bring a bigger picture into focus, with the aim to expose gaps and opportunities so that restorative landscape projects are continuous and not limited by project boundaries, budgets and timelines. This opens the dialogue for Local Government strategies, biodiversity masterplans with Indigenous engagement and community partnership opportunities. This will be the first in a series to create a platform for future place-making projects such as our 'Ghost in the nets' case study and Indigenous ecosystem corridors and nodes projects (Fig. 24.6).

Acknowledgements We acknowledge the Indigenous wisdom shared by Uncle Shane Charles, Uncle Glen Loughrey, Alana Marsh, Jai Allan Wright and Dr. Tyson Yunkaporta. We are grateful for the insight of Professor Sarah Bekessy, Dr. Dominique Hes, Dr. Maya Wardand Jane Toner in regard to their ecological, biomimicry, biophilic and regenerative knowledge. We are grateful as well for the support of IofCA and AIA in all our event endeavours. We have been fortunate to have had the late Professor Allan Rodger as mentor and as one of the founders of IEC+N whose knowledge and experience inspired us in our regenerative design journey.

References

Cajete G (1994) Look to the mountain: an ecology of indigenous education. Kivaki press, Asheville

Clarke A (2018) The great Australian silence, 50 years on. The conversation. https://theconversation.com/friday-essay-the-great-australian-silence-50-years-on-100737 . Accessed 29 Sept 2022

Garrard G, Baracco M, Bekessy S, Weisser W (2023) Beyond green façade: integrating ecology and architecture. https://dataportal.arc.gov.au/NCGP/Web/Grant/Grant/DP210103787. Accessed 13 Jan 2023

Inner Development Goals (2022) https://www.innerdevelopmentgoals.org/. Accessed 4 Oct 2022

Kellert RS, Judith HH, Mador M (2008) Biophilic design: biophilic design: the theory, science, and practice of bringing buildings to life. John Wiley & Sons Inc., New Jersey

Levac L, McMurtry L, Stienstra D, Baikie G, Hanson C, Mucina D (2018) Learning across Indigenous and Western knowledge systems and intersectionality: reconciling social science research approaches. University of Guelph.

Loughrey G (2020) On Being Blackfellas Young Fella, pg 68 Coventry Press, Australia.

Mang P, Haggard B, Regenesis (2016) Regenerative development and design : a framework for evolving sustainability. Wiley, New Jersey

McBride L, Smith M (2021) Terra Nullius. The Australian Museum. https://australian.museum/learn/first-nations/unsettled/recognising-invasions/terra-nullius/. Accessed 30 Sept 2022

Mills JG, Bissett A, Gellie C, Lowe J, Selway A, Thomas T, Weinstein P, Weyrich S, Breed F (2020) Revegetation of urban green space rewilds soil microbiotas with implications for human health and urban design. Restor Ecol 28(S4):S322–S334 (Web)

Neale M, Lynne K (2020) Songlines: the power and promise. Thames & Hudson, La Vergne

Pascoe B (2018) Dark emu: aboriginal Australia and the birth of agriculture, new ed. Magabala Books, Western Australia

Rodger A, Williams T (2020) Available via UIA website. https://www.uia-architectes.org/en/resource/indigenous-ecosystem-corridors-and-nodes. Accessed 7 Oct 2022

Scharmer O, Kaufer K (2013) Leading from the emerging future from ego-system to eco-system economies, 1st edn. Berrett-Koehler Publishers, Inc

Sea Museum Website. https://www.sea.museum/2019/01/30/ghost-nets. Accessed 10 Sept 2019

The Heart Website (2022) https://fromtheheart.com.au/uluru-statement/what-does-it-mean/. Accessed 1 Oct 2022

Ungunmerr MR (1988) Daidirri, Miriam Rose foundation. https://www.miriamrosefoundation.org.au/dadirri/. Accessed 29 Sept 2022

United Nations Sustainable Development Goals Website (2022) https://sdgs.un.org/goals. Accessed 28 Sept 2022

Wayapa® Wuurrk. https://wayapa.com/. Accessed 10 Sept 2021

Wilson EO (1984) Biophilia. Harvard University Press, England

Yunkaporta T (2019) Sand talk: how indigenous thinking can save the world. Text Publishing, Victoria

Check for updates

25 Strengthening the Roles and Responsibilities of Architects as Civil Society Organization (CSO) Partner to Local Government Units (LGUs) for Community and Urban Development

Mary Kristine Argao Segovia,
Romeo Vinco Baesa Jr., and Joan DeLeon-Tabinas

Abstract

As professionals, Filipino architects are not unfamiliar with the idea or initiative of carrying out corporate social responsibility (CSR) initiatives for diverse communities. Despite the professional upheaval, the disorganized workloads, and the project deadlines, some were already involved in civic work. It demonstrates how fervently we still believe in the significance of our collapsing profession to our nation as the current state in the Philippines concerning the practice of the architecture profession. Despite the initiatives, efforts, and partnerships made through these various CSR projects through our professional organization, the United Architects of the Philippines (UAP) as the integrated and accredited professional organization for architects in the Philippines exert efforts to maintain the profession's integrity and importance. Furthermore, the non-recognition of the architecture profession by local government units (LGU)/ authorities, as well as the frequent disregard for its relevance in nation-building and development, has been a growing challenge. Many people seemed to have forgotten the value of our technical knowledge and expertise in community development. The architecture profession in the Philippines is currently undervalued and underutilized in terms of its critical role in community development, which has severe effects. Furthermore, architects' technical skills frequently overlap with that of other professions such as interior design and engineering. However, gaining approval from a particular LGU to become a partner through Civil Society Organization (CSO) accreditation is both a responsibility and an opportunity for Filipino architects to defend our profession. The UAP as an organization has taken on the challenge of conducting corporate social responsibilities directly to each community, serving as an avenue for Filipino architects seeking recognition in practicing the profession. The Philippine Architecture Act (RA9266) is becoming less widely implemented at the local and national government levels. To address this issue, architects may endeavor to strengthen the act's implementation by participating directly as CSO partners and by playing a

M. K. A. Segovia (✉) · R. V. Baesa Jr.
Commission on Governmental and External Affairs (CGEA), United Architects of the Philippines, Scout Rallos Street, 1103 Quezon City, Metro Manila, Philippines
e-mail: mksegovia@eccdcouncil.gov.ph

J. DeLeon-Tabinas
Professional Development Commission, United Architects of the Philippines, Scout Rallos Street, 1103 Quezon City, Metro Manila, Philippines

S. Hilal et al. (eds.), *Design for Partnerships for Change*, Sustainable Development Goals Series,
https://doi.org/10.1007/978-3-031-36993-3_25

role in the government structure, forming alliances with various government agencies through professional and community engagements. This approach not only supports in the execution of projects that benefit the community, but it also acknowledges architects as true builders and rebuilders of hope for a more sustainable future. Furthermore, involving architects as CSO partner could be an effective way to dispel misconceptions about the profession being expensive and fancy. This will encourage other countries in comparable circumstances to follow suit, fostering a deeper understanding of the vital role of architects in community and urban development.

Keywords

Partnership for change · UIA2023 congress · Science track · CSO partnership · UAP · Changing agency mindset · CSR activities · UN SDG 17 goals

The roles and duties of architects are to build, enhance, and maintain the built environment. Whenever architects design a structure, it also comes with a great responsibility to the community. The social responsibility of architects is relatively on the conviction that design can create better communities, influence society, and even play a part in bringing about civilization via improving livability. This mindset is considered the norms, having architects' involvement and recognizing its important in community development is essential to the grand design and universal structure. Unfortunately, this is not the case in the Philippines. Despite the fact that Filipino architects are being recognized for creativity and ingenuity around the globe, Filipino architects are not being fairly recognized or acknowledged in terms of exceptional skill and expertise in urban and/or local community development.

In March 2004, the Republic Act No. 9266 or the Philippine Architecture Act was passed regulating the practice of the architecture profession in the country. This was created to prevent disputes and control the overlap practice of the civil engineering and architecture professions in the Philippines. However, the nationwide implementation of this law is not even meeting half of its planned execution after almost two decades from the time it was enacted. Since civil engineers make up the majority of the building officials in-charged with ensuring that numerous national laws, standards, and codes for urban planning, building design, and construction are followed.

The Philippines although considered a developing country in South East Asia still belongs to the list of the so-called *third world countries by the World Bank*. These countries are infamous for having their political and socioeconomic position more in a bad array of factors that turns disparaging in nature instead of exhibiting their strength and advancement. Swarming infrastructure projects are, as we all know, one of the primary indicators of progress. However, there is a false indication that having an expanding number of buildings and other service facilities, including those that are uncontrolled, is acceptable. Each local government unit having a qualified professional as an employee could help reduce the tumult in the bureaucratic and chaotic infrastructure project implementation and have regulated land use and urban planning and development as well as public building construction. Urban development has the potential to positively transform a community into a more sustainable environment responsive to the challenges and opportunities of changing times. But with unplanned and uncontrolled expansion, development due to poor governance could bring unbalance distribution of resources and undelivered basic services to the community.

Moreover, our country suffers from numerous inconsistent rules in this field. The dismissive attitude displayed by certain local leaders toward addressing this issue is deeply concerning. Political interference, a lack of administrative commitment, unequal financial allocation, and personal agendas further exacerbate the problem.

Implementing this vision faces challenges when legislation supporting the hiring of architects for government roles and the practice of architecture in the nation is weak or hindered by building officials, despite being mandated. How does advocating for architects' significance in urban and planning development connect to the concept of being a CSO partner in local government units? Why did the UAP, as the accredited professional organization for architects, seize the opportunity to bolster the enforcement of the architecture law through engagement with various LGUs?

25.1 Analysis and Conceptualization

In 2020, 58.93 million or 54.0 percent of the total 109.03 million population of the Philippines lived in urban barangays (Urban Population of the Philippines 2020 Census). Based on the World Population Review (2022), the total population of the Philippines is now 115,559,009 with a 0.699 Human Development Index with a 1.93% increase from 2021 data. With this number, we have a total of 81 provinces with 146 highly urbanized, independent, and component cities; 1488 municipalities; 42,046. The concept of decentralization and empowering local government units to manage administratively on their own and with the national agency concerned (Department of Interior and Local Government [DILG]) devolving its mandate in regulating the administrative control over the local chief executives within their jurisdiction had an unsatisfactory impact on urban planning and development.

Hence, the imperative for Civil Society Organizations (CSOs) to actively engage as accredited partners with local government units (LGUs) is being encouraged. CSO's participation is crucial in augmenting and providing diverse forms of support, along with sharing technical expertise. This collaboration aims to formulate effective mechanisms and attain targeted goals for community development.

What does Civil Society Organization (CSO) engagement entail, and how does this partnership relate to our profession? Why do we recognize its potential as a powerful avenue for the comprehensive implementation of the Philippine Architecture Act of 2004?

According to World Bank, a civil society is often called or known as the "third sector", "social sector", or "volunteerland" (after government and commerce). CSOs have the power to influence the actions of elected policymakers and businesses through proper forum and legislative discussions. But the CSO's nature is, "*what it is and what it does—is evolving, in response to both technological developments and more nuanced changes within societies*".

Several regulations and protocols have been enforced requiring government agencies to collaborate with Civil Society Organizations (CSOs) in the planning, budgeting, and execution of initiatives initiated and sponsored by government units at various levels. Power dynamics and government control are being reorganized as a result of more substantial civil society engagement, ensuring the effective, efficient, and timely utilization of government resources for community development. A study by Ragrario (2013) demonstrated that LGU-CSO partnerships have been instrumental in driving effective local government reforms and addressing local issues and concerns. Additionally, the proposed House Bill 230, authored by Congressman Alfred Vargas, titled "*An Act Institutionalizing the Participation of Civil Society Organizations in the Preparation and Authorization Process of the Annual National Budget, Providing Mechanisms Therefore, and for Other Purposes*". Further, the institutionalization of CSO participation in shaping and authorizing the annual national budget emphasizes its importance in aligning budget appropriations with national objectives and ensuring the well-being of the people. Responding to the government's call, numerous CSOs have engaged in partnerships duration for three–five years, leveraging their capacities from a wide range of fields. In collaboration with the government, industry partners, and business leaders, CSOs work to identify and advocate solutions for local, national, and global challenges, ensuring their impactful involvement throughout the partnership.

The UAP, as a professional organization for architects, saw these aspects as a leverage in collaborating with LGUs. Using technical skills and expertise, architects are dedicated and committed in looking for possible solutions in helping our community progress and develop in becoming sustainable as possible and to adapt to a new reality of rapidly changing world despite the interconnecting problems involving our professional practice. Innovation, creativity, and transformation are crucial in addressing significant challenges and serving as advocates for urban development. We are aware that it is not an immediate mandate to recognize the architecture profession or enforce government positions in planning and design units at the local level. However, we aim to gradually establish a pathway for such recognition and opportunities.

25.2 Issues and Concerns

Even from a professional standpoint, there were some regional problems and challenges that needed to be resolved. The Philippine Professional Regulation Commission reports that there are 31,091 active professional architects in the country in 2022 (UAP-PRBOA Competitive Roadmap 2022). In contrast to the total number of registered civil engineers in the country, architects make up only 23%. Consequently, it is infeasible for Filipino architects to be present and provide our expertise in every area. As a result, civil engineers are increasingly taking on architectural conceptual design and planning responsibilities. Additionally, establishing architectural planning and design posts and units within LGUs, despite the enactment of RA9266, involves lengthy procedures and significant long-term financial commitments. The cost of engaging an architect is a substantial barrier to the execution of the aforementioned law. The lack of architects in rural areas, as well as the notion of high costs associated with hiring and commissioning one, contributes to this challenge.

25.3 Resolving the Bottleneck

The issuance of DILG Department Memorandum Circular (MC) No. 2022-083 provided comprehensive guidelines on the application and accreditation of CSOs and the selection of representatives to the local special bodies. This MC represents the latest updated issuances concerning these endeavors, outlining roles, responsibilities, specific areas, and the duration of CSO engagements. Section 67 of Republic Act No. 11518 (General Appropriations Act of 2021) explicitly permits government agencies to collaborate with CSOs in implementing designated programs and projects. The transfer of government funds to a CSO is permissible under the FY2021 GAA, subject to the CSO's certification by the national government agency and compliance with relevant laws and regulations. Furthermore, item 1.4 of the aforementioned MC recognizes the valuable role of CSOs in providing services, particularly in the most impoverished and disadvantaged communities.

CSOs play a vital role in fostering citizen engagement, promoting transparency, and mobilizing support for government programs and projects. As an accredited CSO partner, the UAP recognizes the significant impact of this collaboration. Its objective is to reshape local leaders' perception by emphasizing the feasibility of involving architects in technical tasks for planning and designing urban development projects. This transformation leads to improved efficiency and positive societal change.

The directives and engagement include the following areas:

1. **Feedbacking on programs and projects**

CSOs may join in any of the feedback mechanisms that shall be established by the department, e.g., Community-Based Monitoring Program (CBMP), Third Party Evaluation of the Local Government Support Fund-Assistance to Municipalities (LGSFAM), and Local Development Council (LDC) sub-project monitoring.

As an accredited CSO partner, the UAP offers prompt feedback to project implementers on both national and local projects. UAP representatives actively contribute to the Local Development Council, providing insights during consultative meetings on urban planning and development projects. Moreover, UAP members conduct facility audits, research, and technical studies to support relevant programs. This proactive engagement enhances the UAP's role in project monitoring and aids in facilitating effective decision-making.

2. **Pushing for various advocacies**

CSOs may contribute to the implementation of commitments to the Open Government Partnership and Participatory Governance cluster, advocate for the passage of a local ordinance on freedom of information (FOI), participate in the consultation pertaining to Constitutional Reform, participate in initiatives of the local Anti-Drug Abuse Council (ADAC), support the Philippine Development Plan (PDP) and the attainment of the Sustainable Development Goals (SDG), and get involved in the local disaster preparedness program, among others.

UAP Chapters are conducting CSR activities addressing SDGs for recipient communities in collaboration with LGUs, industry partners, and other stakeholders. Some of the SDG addressed are ***SDG 1 (No to Poverty).*** *Philanthropic activities involve donating funds, resources, and expertise to support causes like poverty alleviation;* ***SDG 2 (Zero Hunger)*** *collab with government agencies on projects like "Agri-tektura" supporting farming and fishery, rooftop gardens, vertical farming promoting food self-sufficiency;* ***SDG 3 (Good Health and Well Being)*** *provision of pro-bono designs for healthcare facilities during the pandemic such as temporary isolation facility, barangay swab testing units, design parameters for mixed and adaptive reuse of buildings for COVID by the Department of Health;* ***SDG 4 (Quality Education)*** *provision of back-to-school supplies supports the implementation of the "Brigada Eskwela"—repair, rehabilitation, and improvement of school facilities for the resumption of classes, construction of temporary schools/tents for areas affected by disasters,* etc.*;* ***SDG 6 (Clean Water and Sanitation)*** *provision of design and establishment of handwashing and toilet facility, rainwater collection in rural areas and far-flanged schools;* ***SDG 9 (Industry Innovation and Infrastructure)*** *promoting sustainable building design, adaptive reuse,* etc.*;* ***SDG 11 (Sustainable Cities and Communities)*** *formulating local ordinances for heritage structures, Heritage Mapping with the National Commission for Culture and the Arts, urban planning and design for rural towns, construction of local services facility for rural areas such as fishermen docking point, coastal path walks, and waiting sheds;* ***SDG 13 (Climate Action)*** *tree and mangrove planting, coastal cleanup, advocacy campaign seminars,* etc.*, with the Department of Environment and Natural Resources;* ***SDG 15 (Life on Land)*** *forest restoration initiatives,* etc.*; and* ***SDG 17 (Partnership for the Goals)*** *collaboration with the community with industry partners, professionals, and more for other program and projects.*

3. **Formulating plans, policies, and issuances**

CSOs may participate in consultations or fora for the formulation of plans, policies, guidelines, and issuances. The CSOs can also be involved in policy dialogues in local governance, law and order, and public safety.

UAP, as CSO, also engaged in various programs and projects spearheaded by the LGUs on peace and order like Anti-drug campaign; provision of support for out of school youth, PWD facility audit, etc.

4. **Implementing capacity development programs and other activities**

CSOs may conduct capacity-building programs that assist local governments improve their program implementation and service delivery. Furthermore, CSO participation involves research, surveys, studies, assessments, and monitoring and evaluation activities. This multifaceted engagement increases CSOs' and local governments' capacity to effectively address local issues and achieve positive change.

25.4 Types of Engagement (as Stated on DILG MC No. 2022–083)

1. **Voluntary work**

The department can engage the CSO in a volunteer capacity without any cost to the government to provide feedback on programs and projects; support in pushing for various advocacies; assist in formulating plans, policies, and issuances; and implement programs, projects, or activities.

UAP's CSR projects are voluntary and encompass a range of initiatives, including pro-bono services and active participation in LGU programs. Architects are viewed by the LGUs and the community as dependable technical support, offering diverse forms of assistance.

2. **Voluntary work with cost**

The CSO can also be engaged to assist or participate in a volunteer capacity but shall be reimbursed for applicable cost to provide feedback on programs and projects, support in pushing for various advocacies, assistance in formulating plans, policies, and issuances, and implementation of capacity development programs.

The applicable cost/s shall be subject to existing budgeting, accounting, and existing COA auditing rules and regulations and given for any of the following expenses: (a) actual transportation expenses; (b) per diems; (c) honoraria; and (d) other incidental expenses incurred in the performance of duties.

The UAP engages in collaborative partnerships for special infrastructure projects, providing pro-bono designs for specific buildings or structures. While the reproduction of plans and drawings is covered by the LGU, the cost of establishment or construction falls under their funding responsibility. However, the UAP may also offer free supervision and monitoring services to ensure compliance with building specifications.

3. **Provider of goods and services**

The CSO can serve as a provider of goods and/or services with remuneration or payment for the services rendered. Their services can be procured either to provide feedback on programs and projects, push for various advocacies, formulate plans, policies and issuances, or implement capacity development programs and/or other activities. In this regard, applicable rules and regulations pursuant to Appendix 14 of the 2016 Revised Implementing Rules and Regulations of Republic Act No. 9184 and similar general provisions shall be observed.

25.5 The UAP as Civil Society Organization (CSO)

The United Architects of the Philippines (UAP), as the integrated and accredited professional organization for architects, with its latest analysis with regard to the visibility of the architects addressing the needs or requirements for professional services and engagement in the community or the main market for architectural services in the domestic private sector is not sufficient. Currently, we have 31,091 active architects as of 22 September and as far as an engaging partnership with the community is a concern, we have 1:8 (Sac, PRC PRBoA 2022), and there is one local UAP Chapter that could have government engagement or partnership with eight cities/municipalities within their jurisdiction.

Although Filipino architects through UAP Chapters are conducting various CSR projects in the community without even being accredited as CSO partners by the LGUs within their respective localities, the significance of the profession in the community is still low. With the current flow of events in the practice of the profession, there are still some conflicts, malpractices, and overlapping practice with other profession and this is despite the existence of the Philippine Architecture Act; the presence of architects in respective LGU offices is found to be effective to strongly regulate this law and support its full implementation.

As architects, we play an important part in urban planning and development as we consider long-term variables that affect community growth alongside urban planners. Our technical competence extends beyond the interests of building owners and land developers, and we engage in a significant part in designing and implementing public policies. We work collaboratively with various stakeholders, such as government officials and policy influencers, to establish responsive and sustainable communities through building and reconstruction programs.

These factors should consider variables on demographics, economic trends, and infrastructure elements that are needed to support growth. However, there are some local government units with low initiatives in these areas. Most are not also concerned about proper planning, maximizing the utilization of spaces, and designing infrastructures that could be or great help in providing basic services to the community. The significant roles played by architects in cultivating and achieving this aspect are not being appreciated or acknowledged fully by some local leaders.

In response to the government's call, the UAP's CGEA-Committee on National and Civic Affairs conducted an orientation briefing on July 9, 2022. The briefing aimed to guide newly elected UAP Chapter Presidents and Secretaries in the synchronized application for accreditation of UAP Chapters as LGU Civil Society Organizations in their respective localities. The objective was to familiarize UAP Chapters with the accreditation guidelines and procedures. This LGU-CSO accreditation initiative aims to strengthen partnerships and linkages between UAP, LGUs, and other stakeholders, enabling the implementation of diverse CSR projects within their respective localities.

The UAP through its local chapters advocates to inform and educate the communities to recognize the crucial role played by architects in catalyzing transformative social and economic change and developing the ideas and strategies that are needed to address the challenges surmounting the organization and the profession, motivating civic engagement at the local level through strong engagement with LGUs, NGAs, and NGOs.

Accredited CSO UAP Chapters are actively collaborating with LGUs and industry partners to implement a range of projects. Approximately, 90% of UAP Chapters are engaged in CSR initiatives within their local communities. Notably, several chapters showcased their pandemic-related efforts in 2021, submitting them to the UN SDG during World Architecture Day celebrations. Furthermore, there are annual implementations of sustainable programs and projects in the community that complement the UN 17 SDG.

This unified response of architects at the local level gathered awareness which resulted in the extended partnership through a Memorandum of Agreements with the following governmental groups' leaders (*Union of Local Authorities of the Philippines, the Philippine Councilors League, Vice Mayors League of the Philippines, Provincial Board Members League of the Philippines and League of Vice Governors of the Philippines*) and National Government Agencies (*Department of Interior and Local Government, Bureau of Fire Protection, Department of Education and the National Commission for Culture and the Arts*).

The engagement entails providing prototype designs for vital service facilities and structures, serving as a reference guide for planning and establishing public markets, materials and recovery facilities, public transport terminals, evacuation centers, public libraries, outdoor play facilities, child development centers, and more. The reciprocal objective of this partnership is to strengthen the significance of the architectural profession within the community. This is accomplished through the enactment of local ordinances that mandate the rigorous enforcement and implementation of RA 9266 and its IRR, as well as the Revised IRR of the National Building Code. These ordinances explicitly require that all architectural plans, designs, and documents be prepared, signed, and sealed by duly licensed architects from their respective LGUs or by any duly accredited/registered architects in the Philippines.

The agenda was to eradicate the overlapping professional services and practices of architects with civil engineers in the country. To recall, during the rehabilitation period after World War II, architects were significantly outnumbered, leading to civil engineers taking on a prominent role in supporting the repair and reconstruction of public structures. The unintended long-term consequence of these directives is that civil engineers are now designing, signing, and sealing architectural plans with the assistance of draftsmen. Despite the disarray, most public building officials (primarily civil engineers) are aware of this professional overlap and unlawful conduct. The full implementation of RA9266 to regulate the practice of the profession remained stagnated triggering civil cases prompted by civil engineers for a temporary restraining order relying not on the law itself, but on some of the conditions not even stipulated by the legislation.

As Filipino architects, we aspire to be recognized as designers, project and construction managers, and innovators collaborating with other professions and fields in engineering and architecture. We aim to create innovative structures that address poverty, and environmental concerns, and promote socioeconomic balance through ingenious and artistic approaches. Despite apprehensions about professional engagement and misconceptions about our capacity and integrity, we are still dedicated to supporting this advocacy, actively participate in civil society projects, selflessly share technical expertise and leadership skills, and are committed to providing essential services to the community.

Through our strong partnership and active involvement as CSO partners, we aim to transform the mindset of local chief executives and leaders. This collaboration will enable us to work together toward building a better Philippines, developing socially responsive urban areas that embrace modern technologies, prioritize sustainability, promote environmental friendliness, ensure safety and accessibility for all, and, most importantly, are designed by architects (Figs. 25.1 and 25.2).

Fig. 25.1 Local UAP Chapters participating in various consultative meetings and conducting various CSR projects as Civil Society Organization (CSO) partners of LGUs in their respective localities

Fig. 25.2 Forging of commitment between the UAP and Department of Interior and Local Government Leagues of Local Officials for the partnership and advocacy campaign for the implementation of the Architecture Act in the Philippines

References

DILG (2022) Memorandum circular no. 2022-83. Guidelines on the Accreditation of CSO and Selection of Representatives to LSB. https://dilg.gov.ph/issuances/mc/Guidelines-on-the-Accreditation-of-Civil-Society-Organizations-and-Selection-of-Representatives-to-the-Local-Special-Bodies/3585. Accessed 15 June 2022

UAP-PRBOA Competitive Roadmap (2022) United architects of the Philippines-Commission on Government and External Affairs. Quezon City

UNOPS Cities Alliance (2008) New urban planning strategies of the Philippines. https://www.citiesalliance.org/resources/publications/project-case-studies/new-urban-planning-strategies-philippines. Accessed 9 Oct 2022

Urban Population of the Philippines 2020 Census (2022) Philippine Statistic Authority Website. https://psa.gov.ph/population-and-housing/node/167692. Accessed 5 July 2022

26 Assembly Architecture

Ariel Hernán Jacubovich

Abstract

Assembly architecture describes a new way of producing architecture that transforms inhabitable environments through networked assemblages. The figure of the "assembly" is a means to open up architecture's design-production laboratory to the scenarios that sustain and drive those designs. Groups of architects around the world and in Latin America in particular have been deploying complex strategies to implement forms of collaboration with a multiplicity of actors in the territory. Design is understood as a process that accompanies the transformation of specific places over the course of a more or less extended period of time. What this process yields is not solely a design proposal in the sense of the construction or visualization of a possible transformation, but also a prefiguration of alternative forms of life and spaces. Complex urban ecologies, socio-technical clusters, project assemblies, objects of consensus, vectors of re-urbanization, and association-based urbanism are some of the terms that arise in the dialogue between the project practices that we have been implementing and the attendant research and conceptual speculations. This paper sets out to explain and flesh out those terms as the concept tools at stake in the design experiences of the assembly architecture we have been developing.

Keywords

Assembly architecture · Project assemblies · Complex urban ecologies · Socio-technical networks · Self-management · Participation · Right to the city · Horizontality · Objects of consensus · Architectural devices · Vectors of re-urbanization · Association-based urbanism

26.1 Objects of Consensus and Association-Based Urbanism in Complex Urban Ecologies

For over ten years, **Ariel Jacubovich Oficina de Arquitectura**, from the platform **CAPA** Colectivo de Arquitectura Pública Asamblearia, has been engaged in assembly architecture, a research practice to carry out architecture projects in conjunction with organizations.

Assembly architecture binds in multiple ways the design-production system and unconventional approaches to the disputes at play in the inhabitable environment. More specifically, it

A. H. Jacubovich (✉)
Oficina de Arquitectura, Buenos Aires, Argentina
e-mail: ajacubovich@yahoo.es

S. Hilal et al. (eds.), *Design for Partnerships for Change*, Sustainable Development Goals Series,
https://doi.org/10.1007/978-3-031-36993-3_26

acts in territories where the tension around flows and places cluster in "nodes" in which discords over urban territories and socio-environmental conflicts are manifested. A crucial actor in these scenarios are movements of resistance largely driven by associations and organizations with a collective aspiration to generate new modes of inhabitation.

A series of incipient experiences that, though distributed around the globe (Raumlabour 2017) have been clustered in Latin America in recent years, evidences a change in architectonic culture. (Laddaga 2006).

To describe and participate with the particular agency of architecture in what is happening, we must focus on associations and ties in never stable, but constantly shifting networks.

To insert ourselves in these processes we adopt the assembly—the tool for decision making used by many grassroots organizations —as an instrument with which to make design decisions. While initially a means of generating trust between the architects and the community or of tuning into the community's methods, the assembly ended up transforming how we produce design. That is, it ceased to be a way to respond to existing issues and came to be a means of cooperative coproduction between different groups working together. (Stulwark 2019) The issues or disputes do not predate the design, but rather emerge with it. In other words, the conditions that enable an architectural project to take shape are generated alongside that project.

26.2 From Laboratory to Expanded Network

After the economic and social crisis that gripped Argentina in 2001, the social organizations that had emerged through protest against the economic, social, and institutional decline brought on by the neoliberal policies implemented in the nineteen-nineties turned into social movements committed to the transformation of living conditions in slums (Lewcoyich 2006). Some of those movements were at the center of territorial disputes that, it would turn out, prefigured the construction of city.[1]

In early 2009, several years into the unauthorized occupation of an abandoned industrial complex called Roca Negra—a complex of nearly three-hectare lot of land located in the southern portion of Lanús county on the outskirts of Buenos Aires—we began an architectonic experiment in conjunction with a collective organized under the umbrella of the Movimiento de Trabajadores Desocupados (MTD, for the acronym in Spanish) (Fig. 26.1).

The Ciudad Roca Negra project, which developed over the course of more than four years, came to encompass a series of experiences in architectural design assemblies—a practice our firm continues into the present along with research on emerging forms of producing architecture collectively capable of contributing to an urban or territorial transformation.[2]

Thanks to assemblies, architecture can operate as a series of "objects of consensus" that

[1] Diego Stulwark uses the figure of the *bricoleur* developed by Levi-Stauss in *The Savage Mind* to capture this relationship between prefiguration and alternative forms in contexts of crisis:"Subjectivities of crisis are like the *bricoleur* who does not distinguish between piece and tool or between formal criterion and concrete movement. The *escraches* [public shaming practices] by H.I.J.O.S. [acronym for Hijos e Hijas por la Identidad y la Justicia contra el Olvido y el Silencio, an organization of the children of those disappeared in the most recent Argentine dictatorship (1976–1983)], barter clubs, neighborhood assemblies, worker-run factories, and *piquetes* [protests of laid-off workers blocking highways] were prefigurations of *poststateness*." From *LA OFENSIVA SENSIBLE Neoliberalismo, populismo y el reverso de lo político*, Caja Negra Press, Buenos Aires, 2019 (Stulwark 2019).

[2] Marina Sitrin describes the relationship between the horizontal and the territory in her observation of assembly processes during the crisis."In Argentina, the use of space and the concept of territory were also central to neighborhood assemblies, the movements of laid-off workers, and worker-run businesses and factories. People spoke of new places to meet, places outside the forms of institutional powers. One participant in the assemblies described it as follows:"I understand horizontality as a metaphor for territory, as a way to practice politics through the construction of a territory. That is the foundation of politics; it has to do with direct democracy. It is as if we needed to occupy a space." Horizontalidad y territorio. De Atenas a Buenos Aires, un fantasma recorre el mundo: el fantasma de la horizontalidad. See www.horizontal.mx.

Fig. 26.1 Above, protest in the center of Buenos Aires. Below, project assembly in Ciudad Roca Negra, 2009

articulates interests and divergencies in a complex context. At stake are architectures that act as vectors of urbanization insofar as they are capable, when inserted in those contexts, of channeling disputes toward a process of transformation: a networked urbanization or an "urbanism of associations."

26.3 Project Assemblies

On the basis of Ciudad Roca Negra we undertook a series of projects of varying size and type, all of which begin with a foundational assembly instance and revolve around conflicts and disputes, as well as the shared desire to generate other forms of inhabitation and urban coexistence. Some of our assembly based design projects include

Ciudad Roca Negra: An experiment in the production of urban conditions through self-managed designs with functions reached by consensus.

Estación Kosteki y Santillán: A community space housed in the train station where two activists were killed during a police crackdown; the space is dedicated to their memory.

Plaza del Pan: A sports area adjacent to a local bread factory that provides bread to the community.

Centro comunitario La fe: A community center in a slum in the outskirts of Buenos Aires.

Biblioteca Caminante: A mobile infrastructure to generate spontaneous spaces of cultural interaction.

EMCUR: A memorial space for victims of the dictatorship in Argentina (1976–1983) in the university complex in Rosario.

Biarritz: A thirty-five-unit housing complex for a teachers' union cooperative.

Espacio de Paz Valle del Pino: A community space in Venezuela developed by and for local residents.

The following questions arose over the course of these projects: How can transformation be effected in complex ecologies,[3] that is, in places beyond the reach of public policies and processes of re-urbanization? How can an architecture be at once public and self-managed? How can participation be generated in even the most architecture-specific instances of a project? How can the state or governmental programs be brought into a network of associations as just another player, with no more weight than any of the others? (Figs. 26.2–26.9).

Assembly architecture is a way of continuing with these questions. But design assemblies are not homogenous; they do not adhere to a pre-tested method. There are different types of design assemblies, and their dynamics and objectives change over time, as do the actors involved. Design assemblies take the objects produced to another field, sometimes to a more playful place or a place of material experimentation.[4] (Simone 2004).

The proposals we have formulated include not only deliberative or discursive instances but also workshops in drawing and model-making to produce architectonic documents, building workshops, surveying workshops, carpentry workshops, as well as screenings of shadow puppetry and meetings of writing groups (Escobar 2017). Each of those proposals both produces something and plays a part in determining the architectural design.

26.4 Hybrid Forums with Non-human Specialists

This conception of design requires new protocols of collective production. At stake are multiple players' spheres of expertise open to hybrid forums with decision-making systems not controlled solely by the architects. Have the skills of architecture shifted away from the design process, understood as the production of a pre-visualization, and toward the configuration of parliaments and construction of collective procedures?

The participants in these parliaments can be human persons or objects. (Latour, Nunca Fuimos Modernos 2007) Meaningful "things" that, thanks to their power to mobilize, can act as

[3] I understand the precarious areas where the projects take place as "complex urban ecologies." Ecologies insofar as they function like systems where different parties interact and affect one another; complex because precarity, as absence, is not what determines the state of affairs—at stake is not, in other words, the absence of a principal actor who could unfetter an entrenched situation. The situation is determined, rather, by the absence of agreements between the parties that could enable a change in the state of affairs. Those parties or actors are usually articulated; they are organized like hybrid networks in formation that bind sociotechnical entities formed by groups and associations, material infrastructures, and forms of organization. Those networks of which these projects form part are not what architecture would call the context; they are, rather, the communities that construct scenarios of transformation. In other words, assembly architecture is integrated into the networks as another actor, as—in the best case—a hub. At stake in its participation is both sustaining the connections that make up the network *and* bringing to the table its own interest as discipline.

[4] Interesting regarding the relationship between the absence of material infrastructures and the emergence of popular organizations is AbdouMaliq Simone's vision of the actions that implicate a variety of groups. Together, those groups end up acting as infrastructure insofar as that concept is expanded to include the activity of persons in cities. See *People as Infrastructure: Intersecting Fragments in Johannesburg*. (Simone 2004) Simone, AbdouMaliq. 2004. Public Culture, Goldsmiths Research Online. Available at http://research.gold.ac.uk/1946/.

Fig. 26.2 How to effect transformation in places not reached by public policies and re-urbanization processes?

Fig. 26.3 How to engage in public architecture that is also community run?

Fig. 26.4 How to incorporate a participatory process in the design phase of projects?

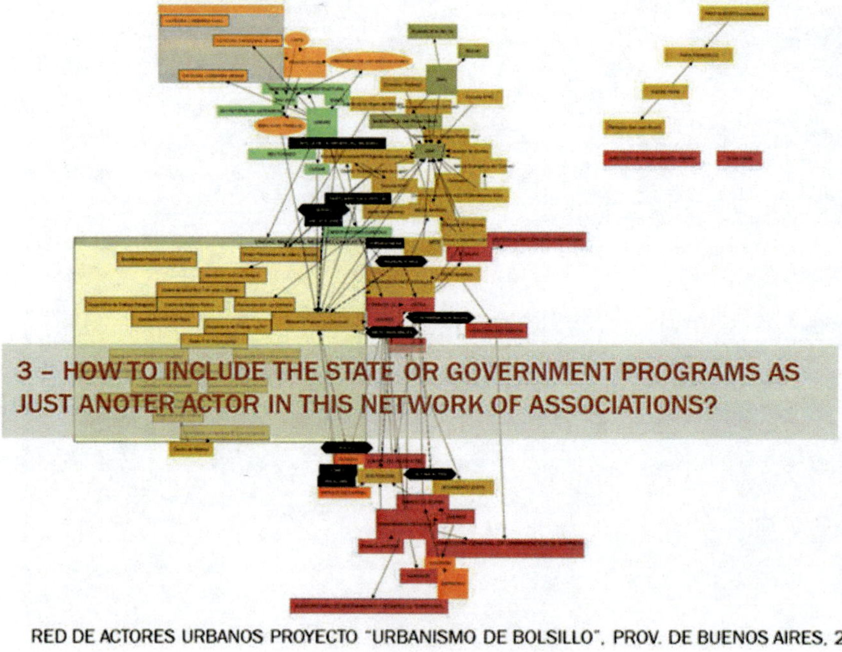

Fig. 26.5 How to include the state or government programs as just another actor in this network of associations?

Fig. 26.6 Right to the city as the right to transform the cities (Harvey 2013)

Fig. 26.7 Public-community-run architecture as vector in re-urbanization process

Fig. 26.8 Assembly based architecture: participation in projects starting at the beginning of the design process

Fig. 26.9 Architecture as objet of consensus: serve to hold together the network in which the architecture is immersed as well as to activate participation in that network. The design not only represents a transformation but is also the representative of a situated socio-technical network

Fig. 26.10 "Poder popular" at the Bloquera Darío Santillán

agents in a possible scenario of transformation. (GPA, Gris Público Americano 2010) These non-human entities play the role of the specialist since their material constitution and their behavior contain a specific knowledge that is essential to the situation.

The concrete blocks produced by activists as well as the commemorative images provided by artists and activists are examples of the "things" that have participated in our projects. Giving voice to each of these non-human actors requires recognizing how other actors act on them. (Ingold 2010) Facilitating their participation means producing a non-discursive consensus through the collective production of graphic documents that will make up the body of the project.

Insofar as a hybrid forum, the "Project Assembly" is a device that relates these heterogenous actors in the system of material transformation that is the project.

26.5 Toward an Assembly Architecture in Ciudad Roca Negra

Our first experience with this assemble practice was in Ciudad Roca Negra. A series of "design assemblies" was held over the course of almost a year. Participation in them was mediated by instruments and procedures we designed beforehand.

Certain "thing-actors" also began to have voice, chief among them the concrete blocks produced by the Bloquera Darío Santillán. Its considerable presence in the assembly was symmetrical to its material productivity, on the one hand, and to its symbolic weight as an element that transforms life on the ground in the slums and surroundings, on the other (Fig. 26.10).

These non-human actors play the role of the specialist since they encapsulate a specificity in which the entire group of participants trusts.[5]

26.6 Assembling the Assembly

To complete the network capable of sustaining the project, Fig. 26.11 new actors were brought in: a construction cooperative, a vocational high

[5] On actor-things, see the chapter *Third Source of Uncertainty: Objects too Have Agency*, where Bruno Latour writes:*...then any thing that does modify a state of affairs by making a difference is an actor—or, if it has no figuration yet, an actant. Thus, the questions to ask about any agent are simply the following: Does it make a difference in the course of some other agent's action or not? Is there some trial that allows someone to detect this difference?...things might authorize, allow, afford, encourage, permit, suggest, influence, block, render possible, forbid, and so on (Latour* 2005).

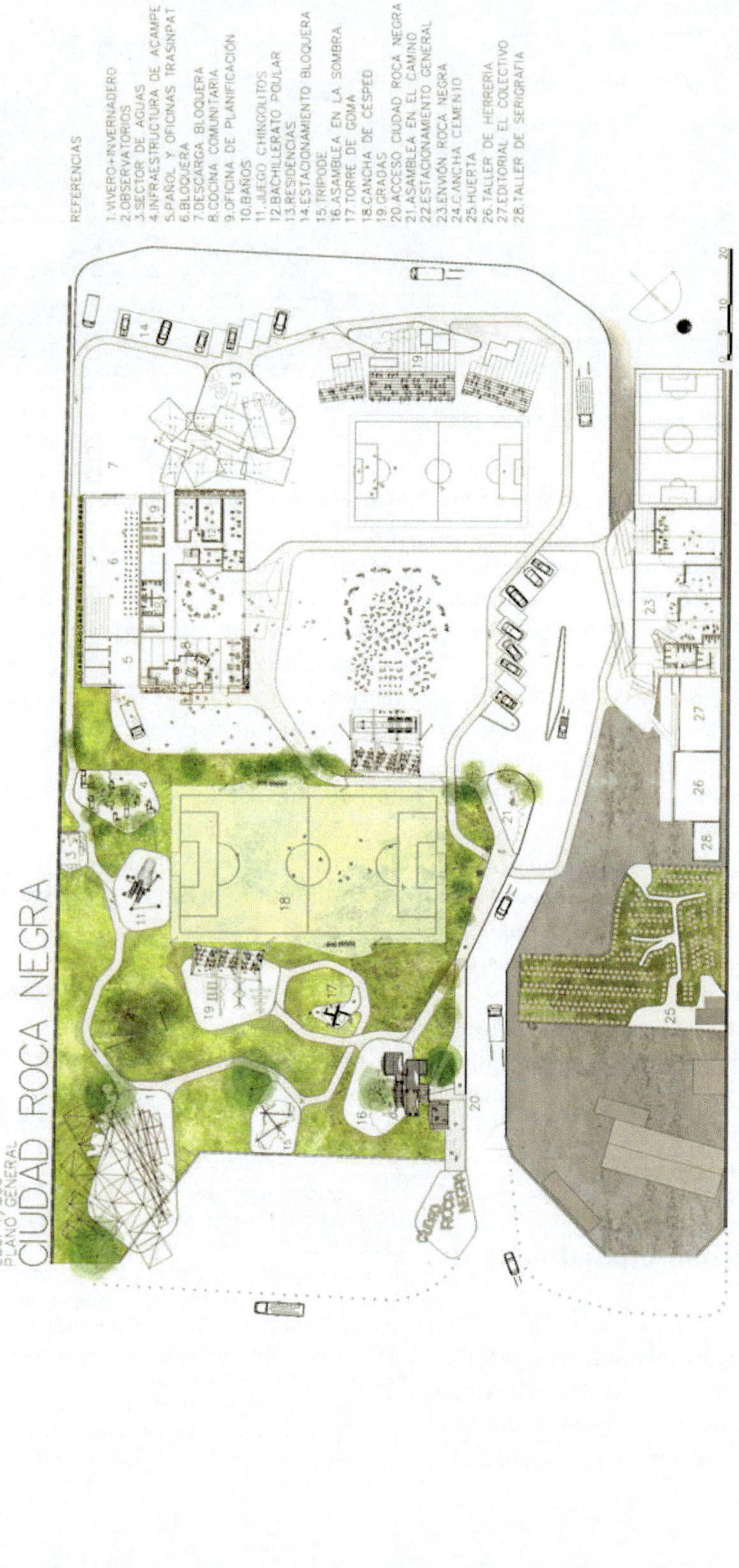

Fig. 26.11 Project plan, achieved by consensus in an assembly process

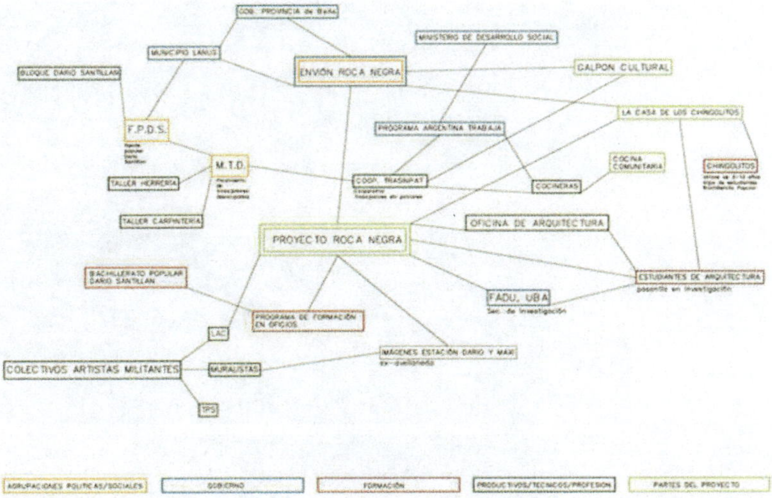

Fig. 26.12 Ciudad Roca Negra actors map

school equivalency program, as well as governmental programs. Civic organizations (soccer clubs, groups of farmworkers, groups of visual artists, and others) also joined the assembly. Since we were researchers affiliated with the Architecture School at the Universidad de Buenos Aires, students there were able to do internships in the territory (Fig. 26.12).

To develop the parts of the project (Fig. 26.13), we organized specific assemblies with the group of cooks at the community kitchen in Roca Negra and with children (Fig. 26.14). The objects produced by those groups served as material for other workshops[6] and then as the basis for a playground that was constructed jointly by teenagers in the community and architecture students[7] (Fig. 26.15).

To assemble assembly means to relate actors who are already involved in the transformation, as well as to bring in actors who complement the network and to give voice to actors who, for a variety of reasons, do not participate in the decision making. (Simone and Pieterse 2017) At stake is a task of re-assemblage that has been repeated in a number of projects and designs we developed after this experience.

26.7 Translating the Assembly

The material produced in the assemblies often has to be translated to be used as material for the design process. In Venezuela in 2015, we were invited to participate in the second edition of Espacios de Paz. We developed a project for Valle del Pino, a neighborhood in the coastal mountains that was lacking in infrastructure and fraught with social problems (Fig. 26.16).

We had only five weeks to design and construct the project in conjunction with the local

---

[6] Lombardi (2014) "El arquitecto como representante," Modos 4 magazine.

"…at stake is not quickly modeling predictable options, but rather re-constructing representation procedures usually part and parcel of the discipline of architecture so that they operate in other conditions; they cease to be the figurations of an author presented to a client and become instead structures that underpin an open decision-making system.

"… If there is one thing that Ciudad Roca Negra has been able to effect with undeniable intensity, it is the reorientation of architecture from the production of predictable—and unresearched—infrastructure models in order to identify and solve collective problems to making the unpredictable surface consistent and collectively."

[7] See the LAC Colectivo Audiovisual Comunitario's video produced for the Festival de Arquitectura Ciudad Roca Negra: https://www.youtube.com/watch?v=ohZRn6_9kQI.

Fig. 26.13 Parts of the project built by the construction cooperative in Ciudad Roca Negra

Fig. 26.14 Specific assemblies with children for a playground design

Fig. 26.15 Playground "La casa de los Chingolitos"

community. In the week we dedicated to the design—the week between the day we arrived and the beginning of construction—we held a specific project assembly with a different group every day.

We organized assemblies with locals to decide on the uses of the project, with transport workers to lay out the pedestrian crossings, with graffiti artists to decide which symbols should appear on the murals. In addition to the assemblies, we organized workshops in drawings and shadow theater (Fig. 26.17). The construction itself could be envisioned as a series of vocational training "assemblies" for the transmission of knowledge from master builders to apprentices (Fig. 26.18).

In a drawing workshop, children were asked to draw their "future home" (Fig. 26.19). The information in the drawings would prove key to the design. The house in many of their drawings was surrounded by large trees. That was how we learned that the 1999 landslide, a catastrophic collapse of the mountain into the sea that literally covered the town, "had taken away the shade." Trees could be seen on the mountains in the distance, but there were none in the neighborhood. Shade was, in fact, scarce.

The trees the children had drawn would become the central figure that organized the fundamental design decisions—among them the decision that the roof that provides shade to the elevated plaza be held up by columns that branched out from four pillars (Fig. 26.20). Our task consisted of translating these assemblies into the design of a common project (Fig. 26.21).

26.8 Construction of Devices

We were invited to design a mobile library for a school near Villa 31, one of the oldest and largest informal settlements (so-called shantytowns) in Buenos Aires. The library was to be built by young people from the community as part of a vocational training program (Fig. 26.22). Each of the participants in the workshop designed a possible library (Fig. 26.23) that was then constructed by all the participants in the training

Fig. 26.16 Before and after in Espacios de Paz Valle del Pino, Venezuela, 2015

using the remains of school desks and other elements in disuse found on the school's premises (Fig. 26.24). That was how a "device"—an object capable of sustaining and articulating practices in a network (Agamben 2014)—was produced[8] (Fig. 26.25).

Once built, the Walking Library acted as the support for the activities (Simone 2004) of not only the different groups that participated in its

[8] Agamben cites Foucault looking for a definition of the devise: "What I'm trying to single out with this term is, first and foremost, a thoroughly heterogeneous set consisting of discourses, institutions, architectural forms, regulatory decisions, laws, administrative measures, scientific statements, philosophical, moral, and philanthropic propositions—in short, the said as much as the unsaid.Such are the elements of the apparatus. The apparatus itself is the network that can be established between these elements…" Giorgio Agamben, What is an apparatus? and other essays. Stanford University Press Stanford. California, 2009.

Fig. 26.17 Workshops in drawings and shadow theater

construction but of others as well.[9] It was the site of storytelling events, shows of fanzines, projections of audiovisual pieces produced by young people, kamishibai performances, hip hop shows, exhibitions (Fig. 26.26), and siestas and reading during school recess.[10]

26.9 Social Urbanism or Association-Based Urbanism

The aim of assembly architecture is always active participation in the configuration of scenarios of transformation. Association-based urbanism, unlike "social urbanism," (Latour 2008) focuses on strengthening the ties between things that seem disassociated but that, in fact, work in concert all the time (Bender 2010). Neither top down nor bottom up.[11] Despite its small scale,

[9] Harvey (2012)."The right to the city is, therefore, for more than the right a right of individual or group access to the resources that the city embodies: it is a right to change and reinvent the city more after our hearts' desire. It is, moreover, a collective rather than an individual right, since reinventing the city inevitably depends upon the exercise of a collective power over the processes of urbanization."

[10] http://arieljacubovich.com.ar/biblioteca-caminante/.

[11] In *Reassembling the Social,* Bruno Latour asks "What is the social made up of?" and answers, "Nothing…" "there is no social dimension of any sort, no 'social context', no distinct domain of reality to which the label 'social' or 'society' could be attributed." We can only define the social once we have described the "associations"—the ties and relations—between the entities that participate in it. Latour calls this the "sociology of associations," as opposed to the traditional "sociology of the social.""Social urbanism" focuses its work on a very specific environment—social emergencies—and makes use of very specific tools—ones that enable "inclusion." That means that its practices are deployed in territories determined by categories projected onto them (poverty, precarity, exclusion, marginality, informality). At stake in social urbanism are top-down processes in which the state is the predominant actor. A double standard is in effect: urban socialism is for the poor and urbanism for everyone else.An "urbanism of associations" would undertake to construct hegemony by strengthening social ties; that strengthening is possible by relating those ties. It proposes describing the heterogenous relations that ensue in a city by tracing associations and relations that already exist or could exist. At play are very specific connections that lead us to describe the urban as shaped by a system of networks that articulate all sort of agents (technological, legal, organizational, political, scientific, material, etc.) as they unfold.

Fig. 26.18 Transmission of knowledge from master builders to apprentices

this conception of an architecture developed in conjunction with organizations can act as a vector of re-urbanization in complex urban ecologies: By intensifying and completing the extensive network required for assembly architecture, a machinery of continuous transformation is, potentially, unleased in a process that ensues on a neighborhood or municipal scale.

One project in progress that illustrates how assembly architecture operates on multiple scale, spanning from the scale of the building to the scale of the city is Biarritz, Fig. 26.27 a collective housing complex for a teachers' union cooperative in Buenos Aires province. Figure 26.28 What began as a commission from that cooperative to design thirty-five housing units currently

Fig. 26.19 Children participating in the decision making at EDPVP

Fig. 26.20 Trees structure

Fig. 26.21 The elevated plaza

Fig. 26.22 Builders an lector

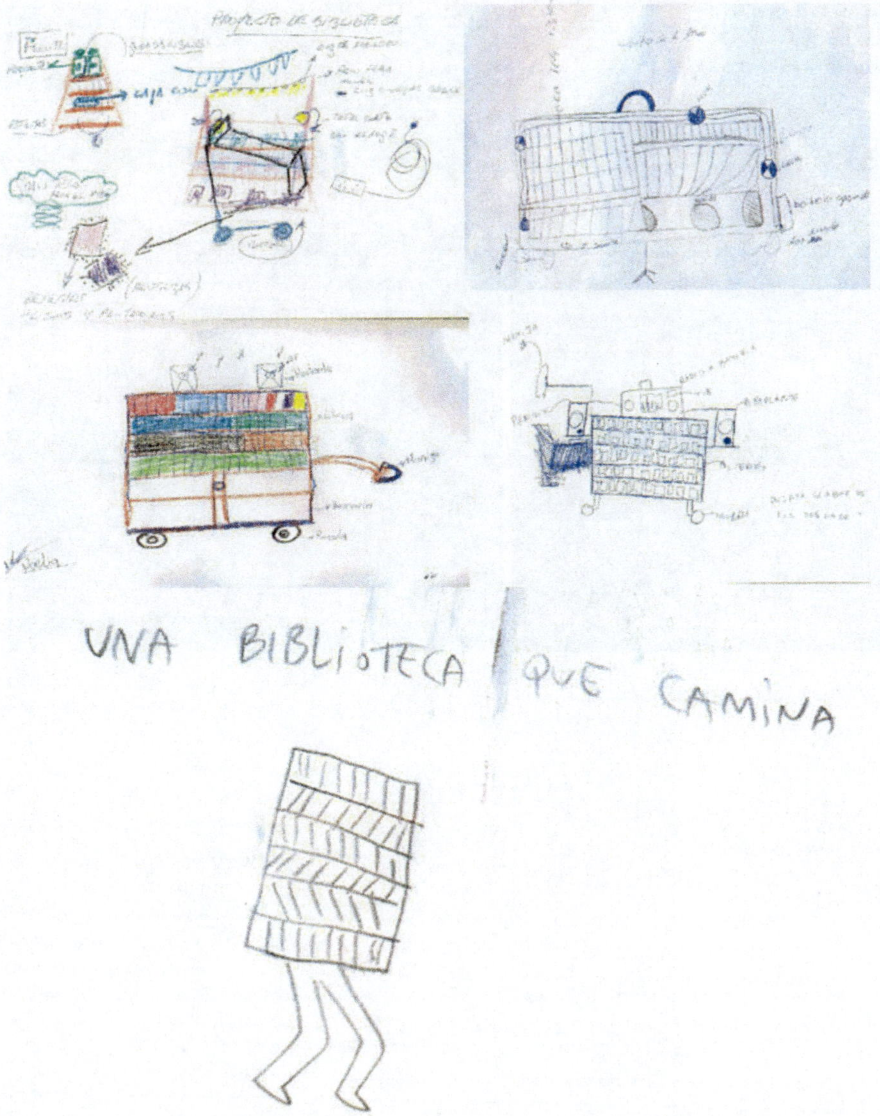

Fig. 26.23 Sketches for the Walking Library made by young people in design workshop and the Walking Library logo

Fig. 26.24 Walking Library map at school

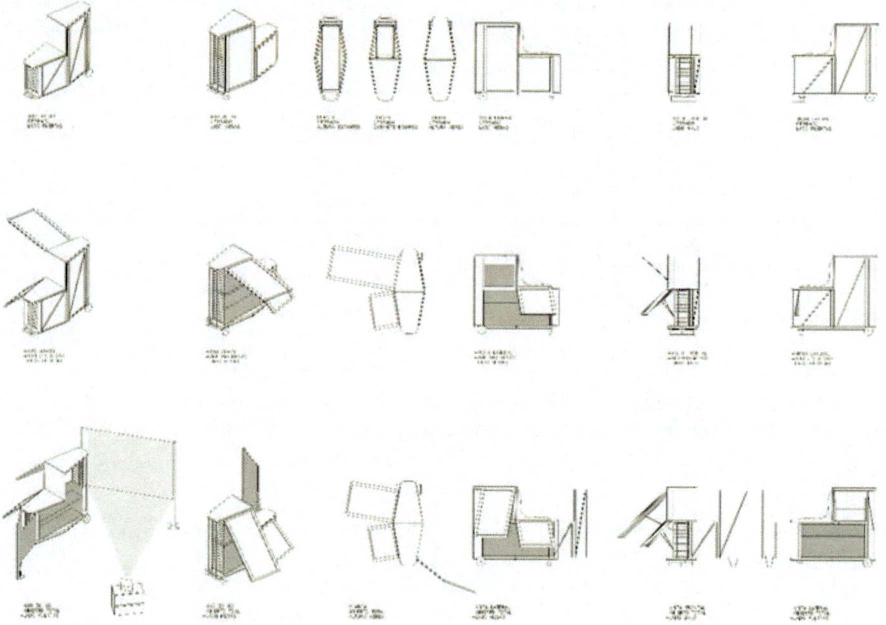

Fig. 26.25 Device: an object capable of sustaining and articulating practices in a network

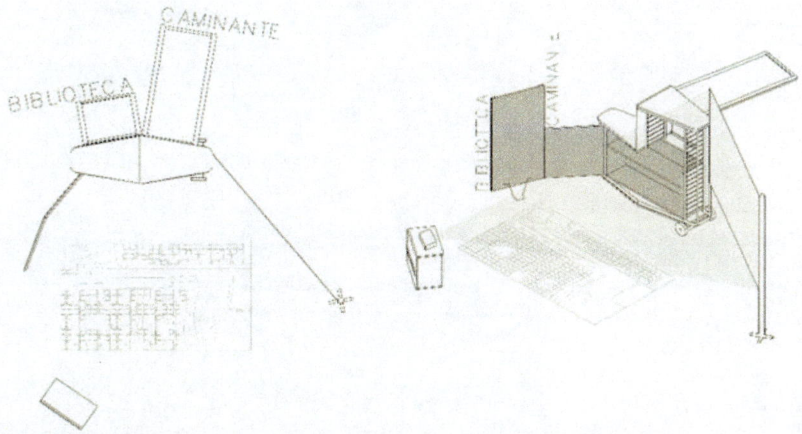

Fig. 26.25 (continued)

Fig. 26.26 At the Centro Cultural Haroldo Conti, an exhibition center in another section of the city

under construction (Fig. 26.29) led to the conformation of a committee of grassroots organizations, technical teams, and the city government. The committee set out to write and present a law that protects access to and democratic production of habitat on the basis of decisions made in conjunction with the municipal authorities, as opposed to the wholesale application of urban policy devised by the city government alone.

That law would permit multiple self-managed experiences in the territory to break out of their isolation. Their transformative potential played a part in shaping a regulation to promote radicalized democratic practices.

26.10 Objects of Consensus

Assembly Architecture, then, opens up an expanded field for an activist practice. It is not limited to the technical certainties of the discipline of architecture, but rather empowered in the potentials of complex networks. At the same time, it modifies existing conditions for the reproduction of the urban.

Fig. 26.27 Under construction: visit of university students to the Biarritz building

Fig. 26.28 Project assembly and pre-adjudication event

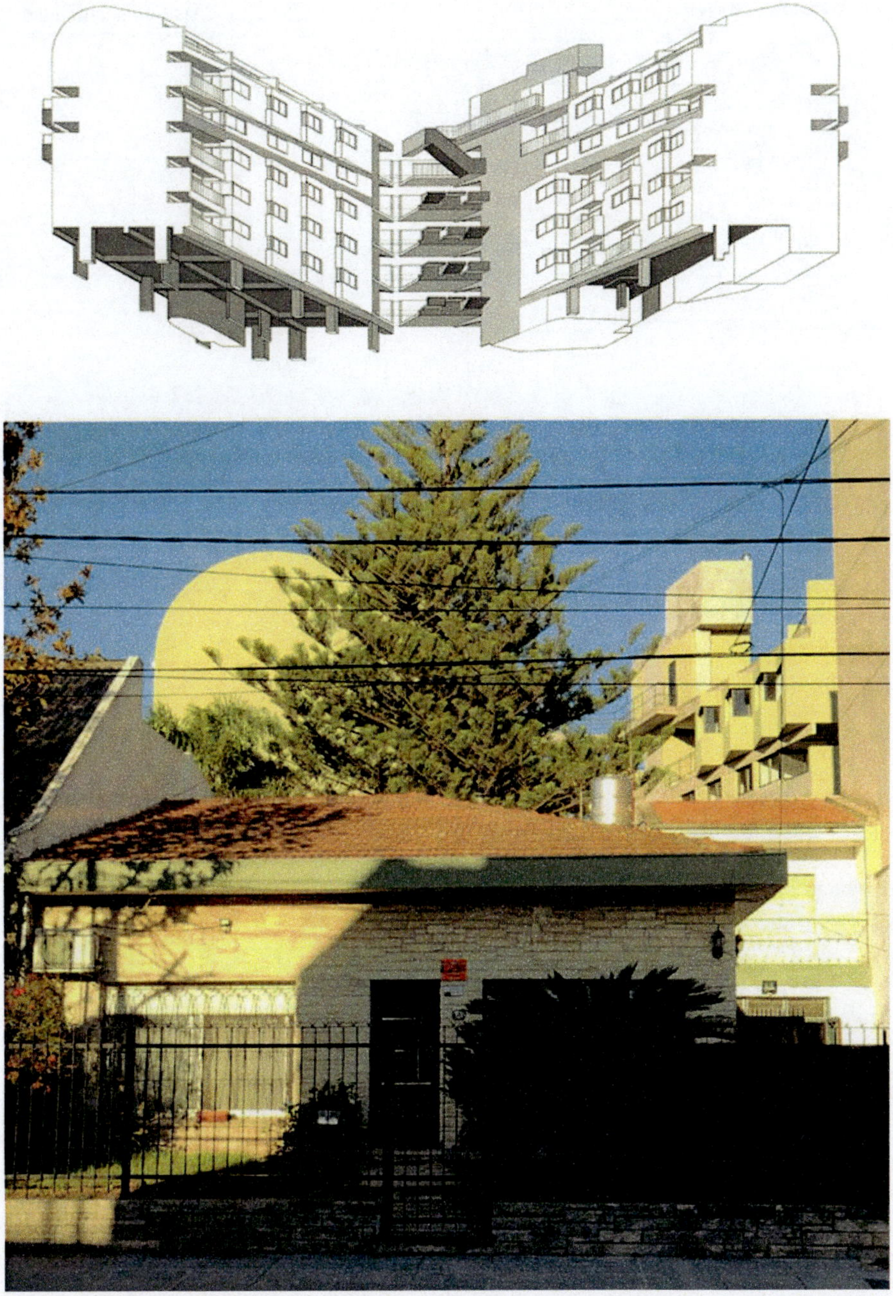

Fig. 26.29 Central patio Axo and Biarritz building seen across the neighborhood

Through this practice, we are able to interrogate the relationships between two things: On the one hand, what is produced in the making and doing of architecture (representations, images, technical documents, and constructions) (Allen 2009); on the other, the forms of organization

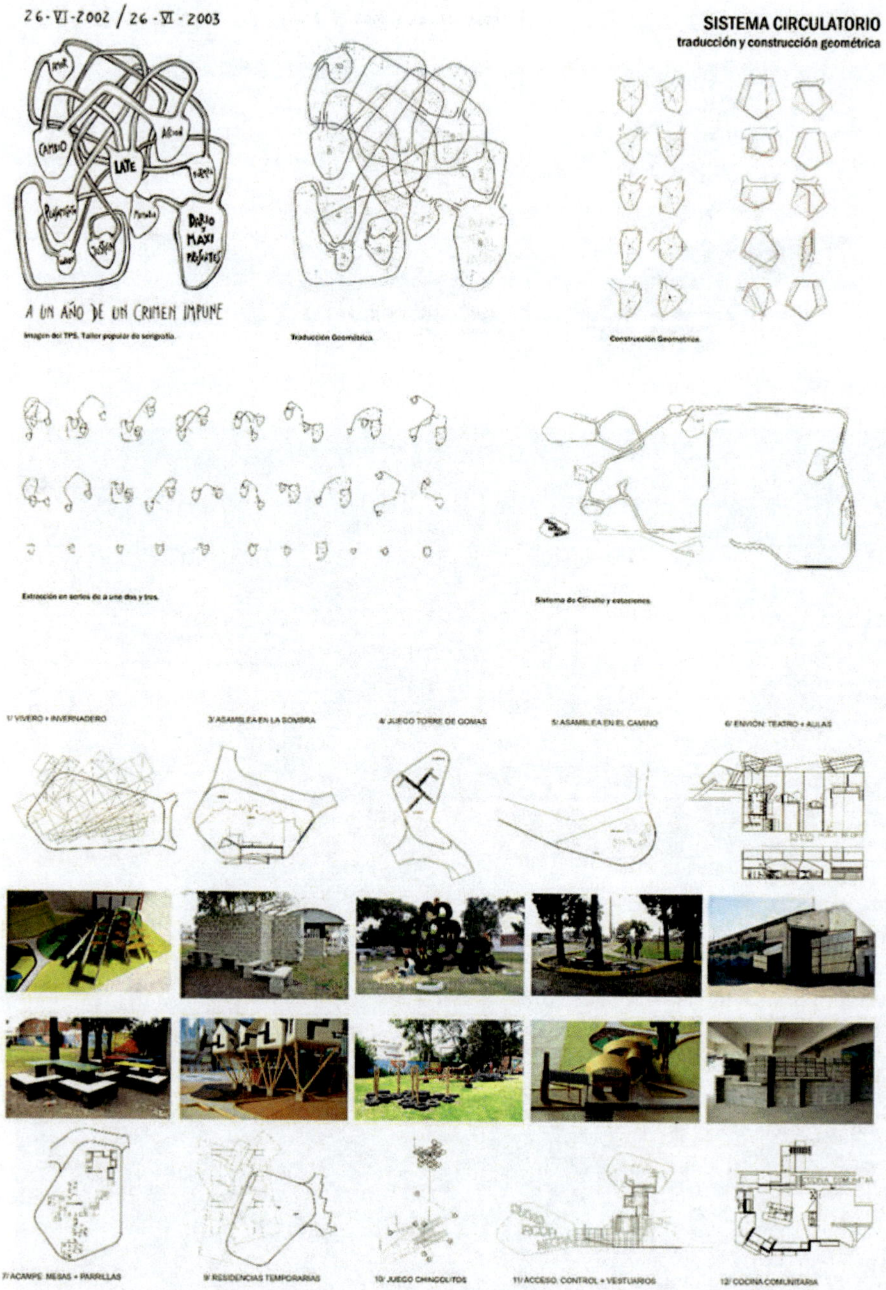

Fig. 26.30 Image by the Taller Popular de Serigrafía. The translation process, and parts of Ciudad Roca Negra project

and representation adopted by the communities where these self-managed architectures arise.[12] But these processes of articulation rely on more than the device of the assembly.

Once a degree of consensus has been reached, other tasks part and parcel of design are performed in the conventional settings of architecture. Work on technical questions and questions of architectonic representation must not lose sight of the sociopolitical representation of the community—that is, the entire set of actors—that participates. What is produced must serve to hold together the network in which the architecture is immersed as well as to activate participation in that network on the part of its human and non-human members (Fig. 26.30).

That is what we call "architecture as object of consensus": The materials produced, whether constructions or representations, maintain a symmetrical double link between the divergent interests of the different actors involved (Latour 1998). Agreements are not produced in discourse, not in speech, but through the material or technical qualities of what is produced.[13]

That final point is key to understanding how assembly architecture works. Insofar as conventional architecture's representation is replaced by the notion of representative (Lombardi 2014), the design not only represents a transformation but is also the representative of a situated socio-technical network—and therein lies assembly architecture's transformative power (Fig. 26.16). At stake is an architecture that participates on the basis of its ability to forge ties and activate constituent powers rather than to respond to constituted powers. A new role for the architect is thus prefigured.

[12] Stan Allen sees this duality as a problem of mapping contemporary urban realities and of the representation of design. He asks the following question in his article *Notation and Diagrams: Mapping the Intangible*:A "CRISIS" OF REPRESENTATION?"To appeal to notational systems in architecture and urbanism is not to suggest a return to perfect transparency of meaning and the smooth implementation of functionality [...] Rather they propose a series of open-ended strategies to work within the indeterminate field of the contemporary city. They propose new scenarios, provoke unanticipated combinations, and allow incremental adjustments over time. They leave space for tactical improvisation of the user in the field. Whatever coherence is attained is always a provisional stabilization of the mobile forces of the city, not set in advance, but developed in practice."In this context it is crucial to remember that the problem of representation in architecture is always double. It is necessary to distinguish carefully between *techniques* of representation: mapping, projection, or notation and the idea that architecture itself functions as a representational system." (Allen 2009)

[13] Simondon addresses this dual capacity of images in relation to objects: On the one hand, images act as memory that brings to the present past experiences, and on the other they are a path toward the object. (Simondon 2013) (English title: Imagination and Invention).

References

Agamben G (2014) *Qué es un dispositivo*. Adriana Hidalgo editora, Buenos Aires

Allen S (2009) Practice: architecture, technique + representation. Routledge, Avingdon

Bender F (2010) Urban assemblages, how actor-network theory changes urban studies. Routledge, UK

Escobar A (2017) Autonomía y Diseño: la realización de lo comunal. Tinta Limón, Buenos Aires

GPA, Gris Público Americano (2010) In: Centro cultural de España en Buenos Aires (ed) Parafomal. Ecologías Urbanas. Buenos Aires

Harvey D (2012) Rebel cities: from the right to the city to the urban revolution. Ed Verso, London-New York

Harvey D (2013) Ciudades rebeldes. Del derecho de la ciudad a la revolución urbana. Ediciones Akal, Madrid

Ingold, T (2010) Bringing things to life: creative entanglements in a world of materials. Realities. Working Paper 15, University of Manchester

Laddaga R (2006) Estética de la Emergencia. Adriana Hidalgo Editora, Buenos Aires

Latour B (2005) Reassembling the social, an introduction to actor-network-theory. Oxford University Press Inc, New York

Latour B (2007) Nunca Fuimos Modernos. Siglo XXI Editores, Buenos Aires

Latour B (2008) Reensamblar lo social, una introducción a la teoría del Actor-red. Manantial, Buenos Aires

Latour B (1998) La Tecnología es la Sociedad hecha para que dure. En M. T. Domènech, Sociología simétrica. Gedisa, Barcelona

Lewcovich I (2006) Pensar sin estado. La subjetividad en la era de la fluidez. Paidós, Buenos Aires

Lombardi R (2014) El arquitecto como representante. In: Fernandez R (ed) Modos 4

Raumlabour (2017) In: Abmann K, Bader M, Shipwriaght F, Talevi R (eds) Urban school Ruhr series. Erschienen im Verlag, Barcelona

Simondon G (2013) Imaginación e Invención. Cactus, Buenos Aires

Simone A, Pieterse AS (2017). New urban worlds: inhabiting dissonant times. Polity Press, Cambridge

Simone A (2004) People as infrastructure: intersecting fragments in Johannesburg. Obtenido de Public Culture, Goldsmiths Research Online. http://research.gold.ac.uk/1946/

Stulwark D (2019) LA OFENSIVA SENSIBLE Neoliberalismo, populismo y el reverso de lo político. Caja Negra, Buenos Aires

27 Architectural Competitions in a Maturing Milieu: Mapping the Agency of Actors

Nezih Burak Bican and Gizem Guneri

Abstract

The last three decades, overwhelmed with capitalist competitive urbanisms, have witnessed a renewal of interest in the relationship between urban form and "the right to just cities," mainly upon discussions over diversity and participation. The same period also witnessed the triumph of architectural competitions both as effectual tools of agonism and as potential means of cultivating participatory cultures in the production of urban space. Recent discussions, hereof, suggest that competitions' agonistic vocations counteracted the pluralist. To solidify this critical rhetoric, withal, there is insufficient research on power dynamics, structures, and implications of competitions. Herein, this study, in its broadest sense, investigates the potency of architectural competitions in allocating spatial, social, political, and economic resources and capacities. The work critically and creatively maps the dynamics and structures of competitions and further testifies (potential) agencies of multifarious actors among these, specifically delving into the emergent local governance structures of two cities from a developing country: Istanbul and Ankara. With divergent spatial development histories and trajectories, the two become complementary cases, novel local governance models of which fundamentally operationalize architectural competitions as means of commoning and register a radical paradigm shift. From a critical and comparative framework, the research reviews literature and relevant data and relies on semi-structured interviews with local government officials, city councilors, competition organizers, jurors, and critics. The findings highlight repressive processual components and potential means and methods of cultivating public agency and a solid and longstanding tradition of public participation in urban space production via architectural competitions.

Keywords

Architectural competition · Participation · Stakeholder mapping · Agency mapping · Conflict · Dialogue · Operation

27.1 Prologue

The thickening of social, political, economic, and spatial polarities and inequities, as both residues and triggers of capitalist competitive urbanisms, especially over the last three decades, specifically rendered questions of diversity and participation

N. B. Bican · G. Guneri (✉)
Atilim University, Architecture, Ankara, Turkey
e-mail: denizguneri@gmail.com;
gizem.guneri@atilim.edu.tr

N. B. Bican
e-mail: burak.bican@atilim.edu.tr

S. Hilal et al. (eds.), *Design for Partnerships for Change*, Sustainable Development Goals Series,
https://doi.org/10.1007/978-3-031-36993-3_27

momentous for practices of urbanization. Even though compartmentalizations among design disciplines had long dominated the fields of spatial production and arrested their capacities to respond, the turn of the century marked a significant shift. The exponential deliberations of relationalities and reciprocities among an extensive array of domains radicalized how design disciplines define their agencies. The understanding of the "urban" expanded far beyond the confines of the "artifact." This, on one side, inaugurated interrogations into architecture's exclusive/introverted modalities of operation and validation. On the other, it reiterated the discipline's capacity as a questioning, interpretive and reflexive apparatus.

For their potentialities to challenge, within a confined scale and frame, procedural, democratic, and participatory methods/aspects and as powerful agents of decision-making and validation, architectural competitions therein came to the fore as trans-disciplinary test bases. Literature, however, indicates that their critical pluralist vocations are still significantly rare in the face of the agonistic.

This work in progress departs from questions that intend to identify potential spots of acupuncture to challenge established balances among operations and operators and aims to unravel the potency of architectural competitions in allocating spatial, social, political, and economic resources and capacities. In line with the 2030 Agenda and the Sustainable Development Goals of the UN (particularly SDG 17—Partnerships for the goals) call for new collaborative ways of working (Stibbe and Prescott 2020), it eventually intends to carve out ground for possible high-impact multi-stakeholder context-responsive (non-generic) partnerships. Specifically dwelling on the maturing urban condition in two Turkish cities: Istanbul and Ankara, the inquiry expands over a review of the literature and legal regulations and in-depth interviews with ten key stakeholders in competition processes spanning the last four years (Table 27.1).

27.1.1 Architecture, Competitions, and Participation

The fact that architecture, by essence, eventually and inevitably targets *the artifact* positions it in a dichotomy where closure into static form with an author and a clientele contradicts the fluidity and plurality of social, cultural, political, and economic constituents of urban space (Owen and Dovey 2008). For that reason, not only the discourse on participation in architecture, except from early and pioneering texts such as de Carlo's (1970, 1972) and Albrecht's (1988) but also exemplary practices remained significantly limited until the last two decades, despite the climax of participatory matters in planning as early as the 1960s (Gusevich 1991; Lane 2005). Within recent discourse, however, architecture's responsive, and responsible efficacy in the complex and politicized urban scene escalated relevance (Jenkins and Forsyth 2009; Jones et al. 2013; Hofmann 2014).

Competitions as integral parts of the discipline and as *fields of research, experiment, and negotiation* (Rönn et al. 2013) constitute a niche yet influential vein within this literature (among others). Throughout this critical vein, competitions are scrutinized outside the boundaries of their inner workings and within a broader socio-political context. (Bern 2018; Bern and Røe 2022; Danielsen 2010; Zettersten 2010). In that, the questions of consensus politics and procedural justice, or in other words, the question of establishing participatory opportunities for those considered exclusive of the classical decision-making circles of architectural competitions, constitute the core (Bern and Røe 2022).

There are two fundamental rifts within this emergent literature. First and foremost is the lack of studies that expressly chart agents and processes operative in participatory examples to potentially open up room for comparative readings and discussions to follow. The second is the fact that the literature is heavily west-centric. Intending to span these rifts, this work critically

Table 27.1 List of interviewees

Interviewee	Background	Association	Discussion context
I1	Architect, former online architecture platform owner	External Consultancy	İstanbul
I2	City planner, academic	City Council (Deputy Chair), Academic Board	Ankara
I3	Architect, academic	City Council (Vice Chair), Academic Board	Ankara
I4	Architect, academic	Competitor, Critic, Jury Member	Ankara-İstanbul
I5	Architect, adjunct instructor	Competitor, Jury Member	Ankara-İstanbul
I6	Architect, academic, former international award institution secretary general	Critic, Jury Member	Ankara-İstanbul
I7	Architect, adjunct instructor	City Council (Chair)	İstanbul
I8	Public administration specialist, former county mayor	Metropolitan Municipality (Department Head)	Ankara
I9	Architect, academic	Metropolitan Municipality (Department Head)	İstanbul
I10	Architect, adjunct instructor; architect; architect	Professional Association	Ankara-İstanbul

maps the dynamics of architectural competitions and further testifies (potential) agencies of multifarious actors among these, specifically dwelling on two cities from Turkiye—a developing milieu that portrays varieties of democracy that are not limited to the western—which changed hands in favor of the opposition in the latest municipal elections.

27.1.2 Architecture, Competitions, and Participation in a *Maturing Milieu*

Architectural competitions have been embraced as public procurement tools, particularly for large-scale investments, since the early republican period of Modern Turkiye. Major urban projects and public buildings were realized through competitions, including the plan of Ankara—Hermann Jansen, 1928. Particularly in the 1970s and 80s, local and central administrations recruited architects through competitions. However, the 1990s, associated with political unbalance and economic instability, and the post-millennial period, with the divergence from inclusive politics, became periods of gradual decline in competitions, eventually leading almost to evanescence until the last few years. Even though the previous two decades under the reign of the Justice and Development Party had been a relatively progress-based period consolidating its base on a nationwide construction-led economy, the majority of the public procurements had been realized through bidding or direct supply. During that time, bids of local municipalities were no exception. In Ankara, for instance, the most recent competition before the latest election dated back to 2003.

The year 2019 represents a periodic shift in the practice of the two metropolitan municipalities under examination. Following the election of the year through which the offices of both were replaced with social-democratic counterparts, both municipalities commenced holding competitions at varying scales of architecture and urbanism [15 competitions in Istanbul (*konkur. istanbul*); 5 competitions in Ankara (*yarismayla. ankara.bel.tr*)] alongside others for artifacts, signage, and cemeteries.

The competitions have been appreciated by several stakeholders for their contribution to urban quality and aesthetics. However, they have also received harsh criticism regarding an array of issues, including the processes embraced for public participation, site selections, jury assessments, low winner prizes, conflicts of authorship, and delayed implementations. Among these, particularly, the public participation mechanisms applied received the harshest opposition from professionals. (i.e., Acar 2023; Cengizkan 2023; Gülsün 2022; TMMOB Mimarlar Odası Ankara Şubesi 2022; TMMOB Mimarlar Odası İstanbul Büyükkent Şubesi 2020) These involved accusations of populism and provoked tough questions regarding the validity of the procedural designs of these competitions. Nevertheless, the actors are still struggling to rally and learn from these contested processes and from each other in a milieu where urban, spatial, professional, and administrative contexts remain immature.

27.1.3 Actors and Agencies in a Nutshell: Mapping the Current Scene

The current political milieu of polarities in Turkiye puts opposition party mayors, who are also attributed to be potential presidential candidates and their administrations, under close monitoring of the central government. The new left-wing administrations which have taken the posts in 2019, therein, have been keen on leaving no space for (publicly) invalidated implementations. They argued that they would maximize the public benefit and value acquired by municipal investments by devoting specialized knowledge to spatial implementations and joining forces with the public.

Competitions have been embraced by the administrations in line with that approach. External consultants (I1) to Istanbul Metropolitan Municipality (IMM) and "the academic board" (I2, I3) advising the Ankara Metropolitan Municipality (AMM) became the primary facilitating actors catalyzing the recent competition initiatives. Furthermore, both municipalities have received the support of the city councils, which have become active public bodies following the election, and the boards of which consist of high-profile professionals and academics (I2, I3, I7). In Ankara, a group of city council members offers further voluntary support as part of the academic board. The consultants and the academic board guide the municipal bureaucrats and the mayors in selecting critical sites, preparing the outlines of building/urban programs, and deciding the jury compositions for the competitions.

It may, therefore, be suggested that outside the rather conventional body of competition stakeholders who had been operational on the Turkish scene—who are primarily actors that are involved in the constitution of the legal framework—the lawmakers, in demand—the administrative body, in the selection—the jury, in the fair sustenance of the processes—the professional body, as participants—the latest competitions also involved actors operational on the facilitation of competition processes—external consultants in Istanbul and the academic board in Ankara and the voice of the public (Fig. 27.1).

So far, the competitions have initiated a keen interest and appreciation among spatial design professionals, as they encouraged the involvement of a plural voice.

27.1.4 Reflections on the Map: On Ruptures of Agency

Although several competitions were completed despite the unprecedented conditions of COVID-19, none have been implemented yet. Parallel to interest and appreciation, this also brought about public criticism that majorly highlighted conflicts on the scene. The initial outcomes of the current research reveal that, as constituents of the conflictive scene, there are rifts of dialogue among/within actors/stakeholders on several occasions; obstacles of economics, politics, and

Fig. 27.1 Role chart. Actors and stages of competitions

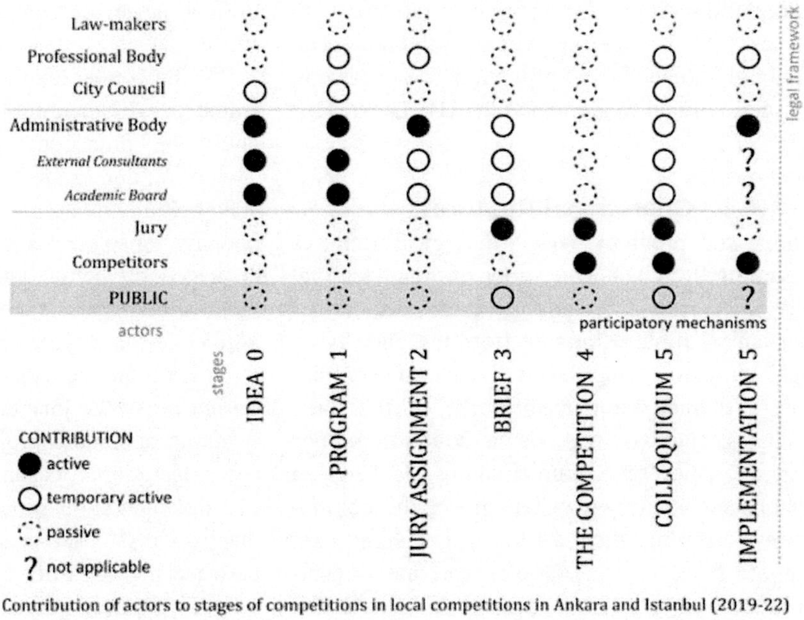

administrative procedures; and other flaws/malfunctions that arise from time pressures and related undertakings to minimize losses.

The work exhibits three significant and intermingled domains of rupture—as also potential nodes of acupuncture—within the operative framework of participation in architectural competitions in the specified milieu. These are namely: ***dialogic***, ***operational***, and ***temporal*** ruptures. The first refers to discrepancies in dialogue among and within actor bodies, the second to inoperative fragments of the competition processes, and the third to complications generated by time-dependent dynamics.

27.1.4.1 Dialogic Ruptures

Although the history of competitions in Turkiye goes back to the early republican period, local and national authorities did not have sufficient opportunities to develop an extended knowledge base and build foundational institutions to organize fluent competition processes. (I6) This significantly owed to the discontinuity of the previous attempts. The administrative bodies, lacking that sound experience, encounter difficulties selecting sites, developing briefs, determining jury members, deciding the modes of participation, and coordinating the implementation phase (I2, I3, I4).

To illustrate, acute ruptures regarding site selection took place in some critical contexts (i.e., Taksim Square in Istanbul), leading to considerable implementation delays. In some cases, bypassed or shortcutted grounds of discussion among the municipal authority, jury members, professional chambers, and the preservation councils seem to embed crookedness to the processes right in the beginning (i.e., Çaldağ Hill in Ankara) (I2, I10). Restricted dialogues between the administration and jury members throughout the processes inevitably result in disappointment of multiple parties, as timing or modes of several participatory methods impair the validity of jury verdict (i.e., *100. Yıl Çarşısı* in Ankara). In other cases, skipping potential dialogues prevent better formulations. For instance, as revealed by interviews, although the professional body has a detailed list of jury members and a database to avoid repetitive appointments at close intervals, they did not have the opportunity to discuss such potentials with the authorities (I10). In some cases (F.I. Büyükada Fayton Square) within which such dialogues were not bypassed, however, the competition processes came to a closure with success thanks to the involvement of experienced staff and bureaucrats with previous experience in competition and implementation. (I1, I7, I9, I10) (Fig. 27.2).

27.1.4.2 Operational Ruptures

Unlike the previous set of dialogical ruptures stemming from lack/restrictions of organizational and interpersonal relations, there are rather mechanical ruptures arising from the underlying legal structures—legislations—lack of formulations to define flexible frameworks, or routinized practices, such as methods of colloquium. For instance, the two municipalities lacked an established institution to administer the competitions, and thus they explicitly got significant support from external bodies (private consultancy or an academic board). This led the administrative bodies or their sub-divisions to be bounded to the frameworks defined by other actors (I1, I3, I4, I7). Furthermore, current legislation based on the public procurement law not only assigns the nonmaterial rights of the author to the administrative body but also does not fully define the paths to set agreement between the parties. This malfunction, for instance, stays as a critical barrier in front of implementations (110, I9, I7, I8).

On the other hand, the legislation does not frame how/if/when public participatory processes could be practiced, modified, or scheduled. Therefore, due to the lack/limits of experience of stakeholders, recent processes tended to delay or become obsolete. Similarly, colloquiums are criticized for being limited organizations where the awarded projects are introduced after the final verdict of the jury; verbal discussions take place—which is appreciated by the interviewees—but, do not affect the jury decisions or lead to modifications of winner projects (I7). There are also other operational reasons that impair competition or implementation processes, such as a lack of mechanisms to set healthy negotiation grounds between the authority and the winning competitors (I5); unresolved authority conflicts before/after competitions (I7, I10); or lack of the legislation in defining rules for brief-development (I7) (Fig. 27.3).

27.1.4.3 Temporal Ruptures

The electoral motives of the Turkish political ecology of stringent polarities generate significant time-dependent/borne pressures, as well,

Fig. 27.2 Dialogical connections and ruptures in dialogues

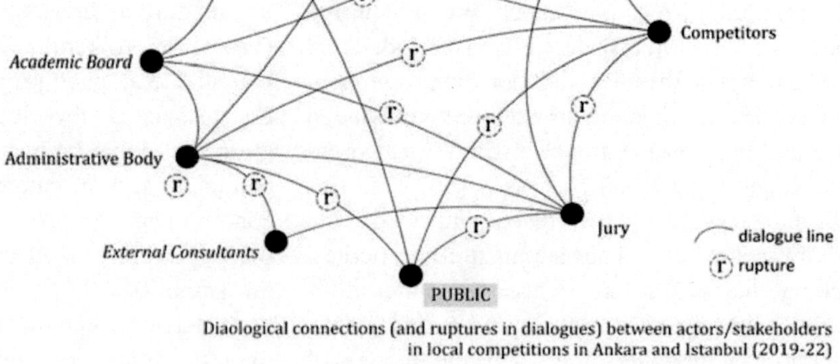

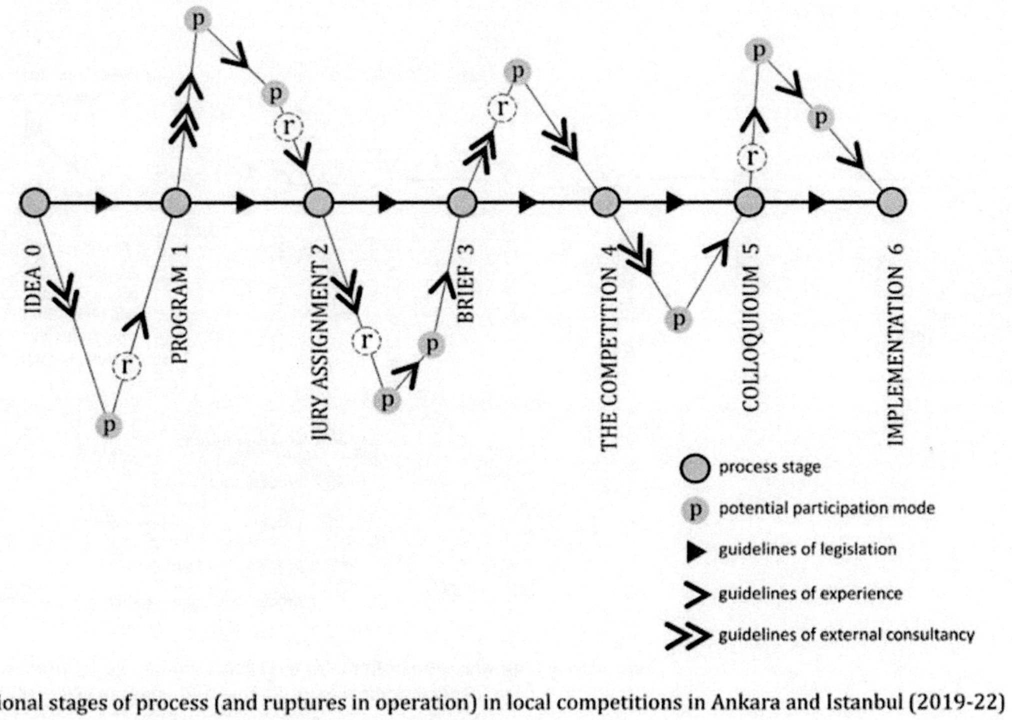

Fig. 27.3 Operational stages of process and ruptures in operation

over spatial implementations that pave the way for ***temporal ruptures*** in the disposal of participatory mechanisms within the continuum of competition processes (I2, I5, I7, I9). Local governments experience the condition in its extreme as their interactions with citizens are comparably more direct than those of the central. (I1, I3, I7, I9) Urbanites expecting rapid results in their living environments tend to call municipalities to account for any retardations. (I1, I7, I9) However, the processes that adopt participatory motives involve a wide array of agents and processual fragments. Yet, the initial findings of the study suggest that the time pressure is majorly breasted by the municipality, which is, to a great degree, compelled, in the end, to some sort of a final result rather than a complete execution of all the program components. The process, vulnerable in the face of political manipulations due to polarities between local and central authorities, seems to slow down when both parties need to collaborate. This detention entails the elimination of certain processual fragments as well as actors to approximate the initial intentions, as discussed with interviewees majorly over the case of the Istanbul Taksim Square Competition (I1, I7, I9) (Fig. 27.4).

Both metropolitan municipalities are known to have experienced interruptions in the processes. This, indeed, complicates transitioning from heavily exploited (especially before 2009) means of closed circle express bids in obtaining spatial design projects to more transparent and comprehensive modes of bidding founded on public negotiation—like architectural competitions engaging participatory processes (I3, I4, I9). Constituting and implementing a fruitful participatory competition process requires time. This timeframe, however, is not only susceptible to the influence of political dynamics but also generates and deepens economic ruptures (rapid rises in prices due to instabilities in the economy) that further obstruct the implementation of decisions due to unignorable budget restrictions (I8, I9).

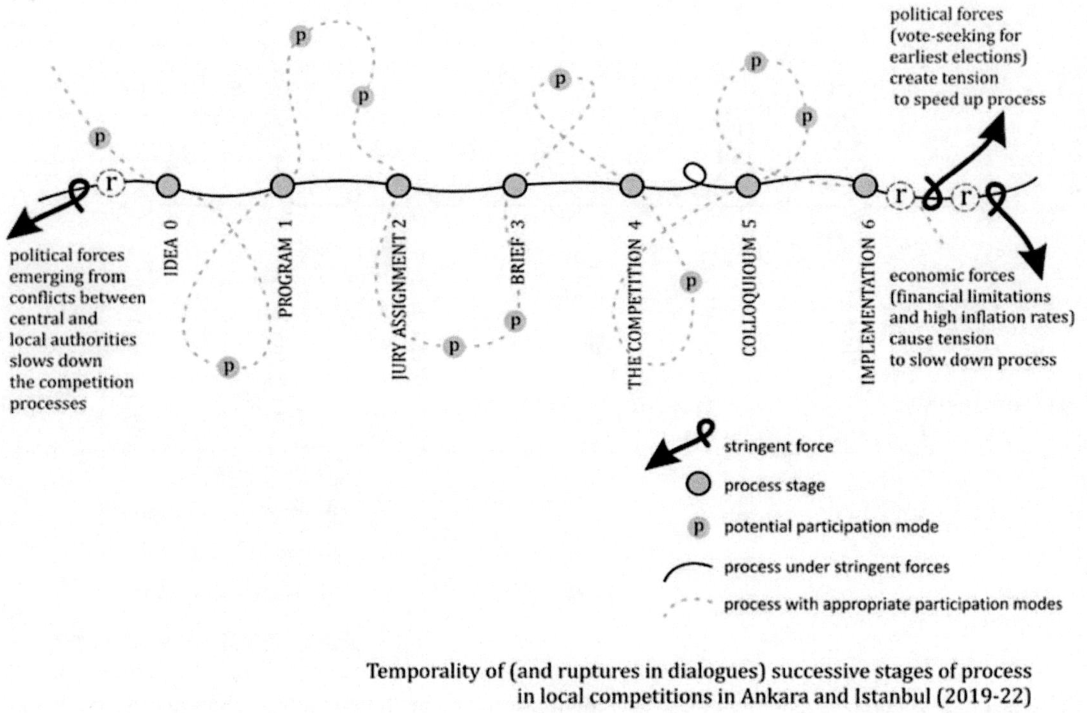

Fig. 27.4 Temporality of/and ruptures of successive stages of process

27.2 Epilogue

- What kind of *spaces of agreement* might be generated through architectural competitions?
- Who might/should be the shareholders in that scene?
- Where would be the best locales of operation?
- Which ruptures block the closure of such processes?
- When would these experiments turn into a culture of space production?
- How would that change the overall urban condition?

The initial findings of this study, unfolding on the bases of these questions in relevance to the undulating and yet resilient democratic ecology of Turkiye, indicate that the contested conditions of experienced participatory processes in the nutshell of competitions (as test grounds) provide a sound discussion domain through which dialogic, operational, and temporal ruptures in a broader picture may be appraised. These findings majorly highlight that (1) the system is significantly complex and top-down and short-term approaches are insufficient in delivering lasting impacts, (2) there is an urgent demand for facilitating mechanisms (i.e., frameworks, agents), and (3) the institutional memories and repertoire of operative tools are insufficient to steer inclusive processes of urban production.

The ruptures identified in the study illustrate domains of considerable blockages that slow down—if not totally prevent—partnership development. These domains are accentuated to inform organizational agendas to be revised to optimize and sustain partnering support institutionally. They are also to inform all parties about their/others' roles and capacities in maximizing value creation to attain SDGs via combining their complementary resources and competencies for the benefit of all. This informative attempt tends not only to shed light on the much-debated arena of architectural competitions, and consequently maximize the quality of interventions in the built environments, but also to present a wholistic framework of/awaken an awareness of partnering systems to the private and institutional bodies within such a maturing milieu.

The formative impacts of competitions on establishing a democratic culture of implementing public projects, therefore, deserve normative research backing, especially for geographies within which the political and economic climates only allow for minor test grounds. This initial study which intended to identify potential spots of acupuncture to unsettle established mechanisms and power relations in the production of urban space will further be complemented via focus group studies involving staged workshops with the participation of stakeholders from various domains to obtain reflexive input on the initial findings and expand the initial diagrammatic mapping to challenge how to engage and bring together agents, institutions and resources in novel and potentially transformational ways.

References

Acar Y (2023) Popülizm ve Katılımcılık: Türkiye'de Son Dönem Deneyimleri. Mimarlık 427:12–15

Albrecht J (1988) Towards a theory of participation in architecture—an examination of humanistic planning theories. J Archit Educ 42(1):24–31

Bern A (2018) Architecture competitions in an urban planning context. J Urban Des 23(2):239–256

Bern A, Røe PG (2022) Architectural competitions and public participation. Cities 127:1–12

Cengizkan A (2023) Ankara'nın Hafızası, Ulus 100. Yıl Çarşısı Yıkılacak mı? Yapı [online]. Available at: https://yapidergisi.com/ankaranin-hafizasi-ulus-100-yil-carsisi-yikilacak-mi/. Accessed: 9 Oct 2022

Danielsen T (2010) New architectural competitions: communication and dialogue. In: Rönn M, Kazemian R, Andersson JE (eds) The architectural competition: research inquiries and experiences. Axl Books, Stockholm, pp 19–36

De Carlo G (1970) Architecture's public. Parametro (5)

De Carlo G (1972) An architecture of participation. Royal Australian Institute of Architects, Melbourne

Gülsün B (2022) Kentsel Tasarım Yarışmalarında "Katılımcılık". Yapı [online]. https://yapidergisi.com/kentsel-tasarim-yarismalarinda-katilimcilik/. Accessed: 8 Oct 2022

Gusevich M (1991) The architecture of criticism. In: Kahn A (ed) Drawing, building, text. Princeton Architectural Press, New York, pp 8–24

Hofmann S (2014) Architecture is participation. In: Architecture is participation. JOVIS Verlag GmbH.

Jenkins P, Forsyth L (2009) Architecture, participation, and society. Routledge, New York

Jones PB, Petrescu D, Till J (2013) Architecture and participation. Routledge, New York

Lane MB (2005) Public participation in planning: an intellectual history. Aust Geogr 36(3):283–299

Owen C, Dovey K (2008) Fields of sustainable architecture. J Archit 13(1):9–21

Rönn M, Andersson JE, Bloxham Zettersten G (2013) Architectural competitions—histories and practice. KTH Royal Institute of Technology, Zurich

Stibbe D, Prescott D (2020) The SDG partnership guidebook: a practical guide to building high impact multi-stakeholder partnerships for the sustainable development goals. The Partnering Initiative and UN DESA

TMMOB Mimarlar Odası Ankara Şubesi (2022) 100. Yıl Çarşısı Ve Ulus'un Geleceği Anketle Değil, Bilimsel Düşünceyle Belirlenir [online]. Available at: http://www.mimarlarodasiankara.org/index.php?Did=12424 . Accessed: 10 Oct 2022

TMMOB Mimarlar Odası İstanbul Büyükkent Şubesi (2020) Taksim Meydanı Yarışma Sürecine ve Sonrasına İlişkin Zorunlu Açıklama [online]. Available at: http://www.mimarist.org/taksim-meydani-yarisma-surecine-ve-sonrasina-iliskin-zorunlu-aciklama/. Accessed: 10 Oct 2022

Zettersten GB (2010) The building of visions and the municipal client's role? In: Rönn M, Kazemian R, Andersson JE (eds) The architectural competition—research inquiries and experiences. Axl Books, Stockholm, pp 373–393

How Alternative Governance Models Can Help the Design Community Combat Forced Labor

28

Leslie Louie, Brandon Cuffy, Ryan Welch, and Billie Faircloth

Abstract

Forced labor is a widespread problem embedded in the built environment's materials supply chains. The design community's awareness of this problem has led to industry-wide calls to action centered on selecting building materials to promote supply chain equity (SCE). While these efforts represent a significant step forward, the industry must confront the power dynamics driving modern slavery. By examining current global trading dynamics and the drawbacks of corporate-led supply chain governance approaches, we argue that any attempt to address SCE will only be successful with workers' and local communities' guidance. Lacking this understanding, architects risk repeating the failures experienced by other industries that have attempted but were unable to eradicate forced labor in their supply chains. The design community can learn from bottom-up, worker-driven programs such as the Fair Food Program, created by the Coalition of Immokalee Workers (CIW). These exemplars of worker-led initiatives can inspire new and future design, computation, and material delivery strategies.

L. Louie · B. Cuffy · R. Welch (✉) · B. Faircloth
KieranTimberlake, Philadelphia, USA
e-mail: rwelch@kierantimberlake.com

Keywords

Modern slavery · Supply chain equity · Corporate-led · Worker-led · Building materials supply chain

28.1 Introduction

The architecture, engineering, and construction (AEC) industry's awareness of forced labor in the extraction, processing, and manufacturing of building materials is growing, and there is an urgent call to address this issue (CIOB 2016; RIBA 2020; Jolliffe 2017; Prince et al. 2020; ArchitectureAU 2021; AIA 2022). *Modern slavery* is defined by the International Labor Organization (ILO) as "situations of exploitation that a person cannot refuse or leave because of threats, violence, coercion, deception, and/or abuse of power" and includes child labor, forced labor, debt bondage, human trafficking, and forced marriage (ILO and Walk Free 2017, p. 9). Modern slavery is endemic to the global and local flow of raw and processed materials and is inherently organized through building materials supply chains. The United Nations *Sustainable Development Goals* (SDGs) demand eradicating modern slavery to achieve Goal 8, *Decent Work and Economic Growth;* Goal 10, *Reduced Inequalities*; and Goal 13, *Climate Action* (UN 2022). As architects organize to increase their labor and supply chain literacy and

S. Hilal et al. (eds.), *Design for Partnerships for Change*, Sustainable Development Goals Series,
https://doi.org/10.1007/978-3-031-36993-3_28

demonstrate design, computation, and project delivery strategies to eradicate modern slavery and drive *supply chain equity* (SCE), we will continue struggling to navigate a path forward. This is partly due to a limited understanding of the problem of labor exploitation, root causes, and overly deterministic views of the solution space—namely an over-reliance on top-down, corporate-led strategies.

In this paper, we examine the nascent practice of SCE, the global trading and labor relationships that make it challenging for architects to combat forced labor, the limitations of specific approaches, and what we can learn from a successful program (Fig. 28.1). Our transdisciplinary team represents the architecture and research practice of KieranTimberlake, where we routinely engage processes for evaluating, selecting, and specifying building materials. Our argument is informed by our team's knowledge of architectural design and detailing, environmental- and social-life cycle assessment (E-LCA and S-LCA), Fair Trade practices, computation, and machine learning. It is also informed by our experience developing and maintaining Tally®, the first building information modeling-life cycle assessment tool (BIM-LCA), which helps designers evaluate the environmental impact of building material selections (Bates et al. 2013). Finally, our recent participation in the *Buy Clean and Buy Fair Washington Pilot Study* afforded us limited insight into the current state of AEC-side barriers to SCE and the potential role of policy (CLF et al. 2022).

The numerous efforts underway to eradicate forced labor across industries tend to be guided by industry leaders, high-level supply chain managers, and governments (USGBC n.d.; Prince et al. 2020; AIA 2022). They represent a top-down approach to address modern slavery where a corporation's or industry's senior leaders determine goals, projects, and tasks. We argue that a bottom-up model may be an unfamiliar but more effective path forward in the AEC industry. For instance, Fair Trade is defined as "a trading partnership, based on dialogue, transparency, and respect, that seeks greater equity in international trade," a mission supported by *10 Principles of Fair Trade* and a quality assurance system *World Fair Trade Guarantee System* (WFTO, 2019; USGBC n.d.; WFTO 2022). In supply chain management, *worker-driven supply chain governance* is where laborers, their leaders, or worker-led organizations guide goals, projects, and tasks (Reinecke and Donaghey 2021). As architects wrestle with how to practice SCE through the trade and procurement of building materials, our industry's attempts will be severely hampered or even flawed if we discount our ability to conceive of and participate in industry-wide solutions that prioritize worker participation and direct connections to issues on the ground.

Fig. 28.1 Argumentation logic for a worker-led approach to supply chain governance is situated in a deeper understanding of global market dynamics that drive exploitation, workers' expertise, and experience. Diagram ©KT

28.2 Market Dynamics that Make Labor Exploitation Inevitable

Forced labor is an endemic problem in many industries despite the plethora of sustainability standards and commitments from multinational corporations to eradicate it. This is because decision-makers in conventional, corporate-led approaches prioritize financial business objectives even if they conflict with the well-being of laborers. Companies are controlled by a board of directors, executive management, and shareholders (MSI Integrity 2020). They are the decision-makers in corporate-led approaches, and they are incentivized to maximize shareholder profit (Simpson et al. 2021). Legally, boards are prohibited from placing societal interests above the financial interests of shareholders (MSI Integrity 2020). Fundamentally, in a corporate-led model, workers' interests cannot align with management's interests (Reinecke and Donaghey 2021). Increased market consolidation over the past few decades has further exacerbated this problem. Three-quarters of industries have seen an increase in market concentration between the late 1990s and early 2010s (SPERI et al. 2021). There are now fewer, larger companies who have the power to effectively oppose strong regulations, limiting the influence of other actors to advocate for public well-being. With the way global markets are structured, there is a compelling incentive to prioritize the needs of business and to devalue the cost of labor to procure a high quantity of cheap materials while maintaining high profits (The Changing Markets Foundation 2018).

The design community must confront the sobering reality of industrialized building culture—high demand for plentiful, cheap materials drives poor working conditions. It becomes clear that the issue of modern slavery is not caused by individual bad actors or shadowy criminals. Rather, it is a systemic problem of contemporary business practices and global trade (SPERI et al. 2021). Criminal or bad actor activity thrives in this environment as current market dynamics place the needs of international trade and consumerism above human rights and dignity. Demand from multinational corporations guides and dominates market relationships. Other entities, such as state actors and workers, have significantly less influence. Thus, while SCE strategies introduced by corporate-led supply chain governance mechanisms may contribute to marginal improvement, labor exploitation remains a main feature of our supply chains.

28.3 Labor and Supply Chain Literacy in Architecture

An architect's approach to tackling modern slavery must address the AEC industry's tendency to use corporate-led methods when tackling intractable problems.

These frameworks could reinforce our tendency to be:

- Physically and psychologically distant from the issue of modern slavery (Simpson et al. 2021)
- Focused on symptoms instead of root causes that drive the prevalence of forced labor
- Ignore market dynamics, specifically the downward pressure for low-cost materials that artificially devalues labor.

These conditions inhibit the effectiveness of regulations or technological solutions intended to reduce the incidence of labor exploitation.

Whereas a bottom-up, worker-led approach:

- Leverages the knowledge, leadership, and drive of workers who deal with these conditions daily
- Provides dynamic solutions that respect the complex conditions that give rise to the prevalence of modern slavery
- Directly addresses the downward pressure for low-cost materials and recommends solutions to subvert that systemic dynamic.

Before we propose a way forward, we must generally acknowledge our industry's limitations: architectural professionals lack full comprehension of the dynamics that give rise to the "culture of building", including the legal framework governing the building materials supply chain (Davis 2006, p. 5; King and deBaca 2020). In our firms experience, our ethical frameworks, contractual conventions, and project delivery methods hinder modern slavery eradication efforts from the outset. In the USA, our professional covenants do not explicitly extend to individuals laboring to mine, process, transport raw materials, or manufacture building products (AIA 2020; Slade 2020). As many studies demonstrate, what constitutes "construction labor" is not confined to a building's site and is instead spatially and temporally diverse, occurring in many jurisdictions, communities, and sites globally, far removed from a building site (Glocal Research and India Committee of the Netherlands 2015; Anti-Slavery International 2017; Bales and Sovacool 2021; Rozani 2022). Our profession may feel disadvantaged in coordinating a response to modern slavery because we have not learned how labor is organized formally and informally, nor do we know how to account for labor in its totality. These circumstances present serious challenges to any attempt to evaluate and mitigate risks that forced laborers are used in the service of architectural projects.

Acknowledging these limitations requires us to admit that the industry is mainly inept at tackling the issue of modern slavery. The AEC community needs to explore beyond conventional networks in our space and seek out organizations that have sought labor justice and advocated for workers' voices. These organizations are more intimately aware of the dynamics that drive labor exploitation. They confront exploitation daily and historically have sacrificed and risked their lives to fight for their rights (IndustriALL 2021, 2022). Many successful attempts to drive ethical sourcing in global supply chains owe their achievements to workers who were deeply involved in the development and implementation of SCE programs (Pike 2020; Fair Food Program 2021; Reinecke and Donaghey 2021; WFTO 2022). Thus, as the design community becomes acquainted with the problem of forced labor, we must steer away from corporate-led approaches and learn from successful, worker-driven programs in other industries. Learning more from workers' experiences will enable architects to close the *psychological distance* between themselves and this pressing issue, putting us in a better position to address it.

28.4 The Limitations of a Corporate-Led Approach to Address Modern Slavery

There is growing evidence that corporate-led approaches are ineffective in addressing modern slavery in supply chains (MSI Integrity 2020; Reinecke and Donaghey 2021). This is mainly because they do not acknowledge a significant root cause of labor exploitation—the power imbalance between corporations and workers (MSI Integrity 2020; Fair Food Program 2021). Instead, corporate-led approaches focus on the needs of the industry (Reinecke and Donaghey 2021). As a result, when they attempt to tackle modern slavery, they seek solutions prioritizing the assurance of ethical sourcing to consumers to remain competitive in their markets and maximize profitability. Often, they create lenient standards and certifications that can validate a high volume of material to quickly assure purchasers of ethical sourcing (The Changing Markets Foundation 2018). Corporate-led approaches also tend to focus on symptoms and are reactive to the presence of forced labor. They neglect to explore the context in which labor exploitation and abuse occur. Without understanding these conditions, they often implement misguided solutions that harm the very people they intend to protect (Rozani 2022). This lack of understanding also leaves technologies created to track forced labor or estimate modern slavery risk severely inhibited (Fig. 28.2).

28.4.1 Problems with Certifications

Often, the language employed in certifications allows for corporations to continue business as

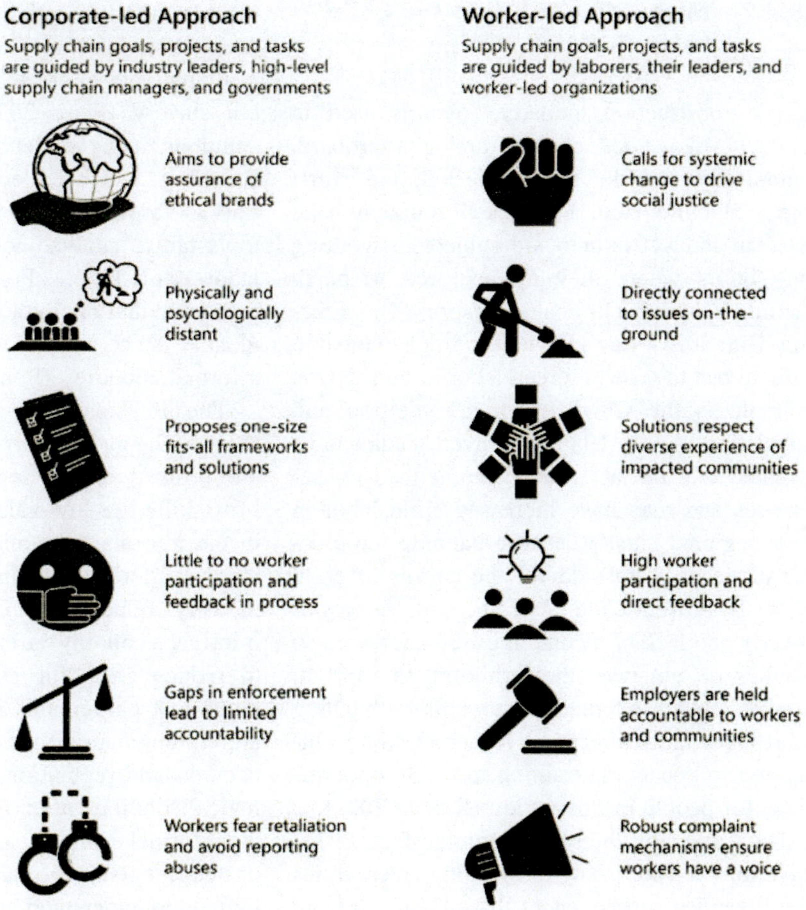

Fig. 28.2 Attributes of a corporate-led and worker-led approach to supply chain equity (SCE). A bottom-up, worker-led approach will help architects close the psychological distance between themselves, people, and communities impacted by modern slavery. Diagram ©KT

usual (The Changing Markets Foundation 2018). For example, the Concrete Sustainability Council's (CSC) technical manual states that a plant must fulfill certain prerequisites to obtain certification. Under the criterion of Human Rights, the manual indicates that the "organization must declare that all efforts have been made that may reasonably be expected of the organization in order to ensure that all of its operations comply with the Universal Declaration of Human Rights (UDHR)."

The required evidence for this criterion is one of the following options:

- A "written declaration by senior management satisfying the requirements"
- "An SA8000 certificate covers the scope of this certification, not older than three years"
- "Publicly available company commitment to follow the OECD guidelines for multinational enterprises" (CSC 2021).

This wording and required evidence give significant latitude to corporations seeking certification. A written declaration provides zero transparency into the inner operations of an organization. They also explicitly state that the company must declare efforts made that can "reasonably be expected of the organization." This vague language could permit leniency for companies that don't have stringent rules to protect workers, easily allowing exploitation to continue. Thus, in the worst-case scenario, corporate-led mechanisms may facilitate a false sense of assurance of ethical sourcing to architects.

28.4.2 Misguided Solution for Child Labor in Mica Mining

In the construction industry, mica is used in products like fiber cement and plasterboard. Among the world's top producers, the worst forms of child labor have been found to take place in India. Children are subject to working long hours where they are exposed to harsh chemicals and back-breaking labor (The Freedom Hub 2021). Laws have been implemented in India to ban the use of forced labor in mines. One example is the Child Labor (Prohibition and Regulation) Act in 1986. However, studies have revealed that the act caused unintended consequences and may have increased child labor in these regions. This is because banning the practice alone does not address the causes of child labor. To prevent child labor, the main factors of poverty and lack of income must be addressed so families do not need their children to work to survive. Better alternatives to simply banning child labor are reallocating resources to schooling and to vocational training to create opportunities for people in this region (Rozani 2022).

Understanding the socioeconomic factors that drive forced labor is key to avoiding externalities or unintended consequences that may result from corporate-led management approaches. The physical and psychological distance can cause those who seek to mitigate modern slavery to have a vague understanding of the problem. Thus, they are prone to adopt efforts that may exacerbate the issue (Simpson et al. 2021).

28.4.3 Limitations with Data-Driven Solutions

To address modern slavery, the design community is seeking data-driven solutions (Bernstein et al. 2020; Flare 2022). Some proposed solutions are modeled against other supply chain governance issues tackled in different industries. For instance, Flare, a machine learning tool, employs an algorithm to detect leading indicators that food or medicine has been purposefully adulterated and applies a similar approach to identify the potential of forced labor being embedded in materials. They pull publicly available data to create their indicators for modern slavery (Flare 2022). Other technological solutions focus on enhancing traceability, like Evrythng, which collects data that traces materials to farms (Evrything 2021). Many of these tools fail to analyze conditions that give rise to labor exploitation. For instance, they do not recognize that exploitation can be challenging to identify since most forced labor exists in the informal economy (Simpson et al. 2021).

The informal economy represents a large share of the global workforce but remains outside the protection and regulation of state actors. Low-skilled or low-valued workers are involved in the informal economy and can participate in many different sectors, including the AEC industry (Chen 2012). Some argue that the informal economy is exploitative and functions to reduce costs for large corporations; others argue that participants in the informal economy are entrepreneurial and benefit from the lack of taxation and regulation (WIEGO n.d.). Technological solutions often ignore the involvement of the informal economy and thus can be dismissive of nuances related to labor abuse. It is incredibly difficult to understand how the informal economy influences supply chain dynamics without actively seeking feedback from worker communities.

Data-driven solutions have employed open-sourced datasets to estimate the risk of forced labor (Flare 2022). These comprehensive datasets are often available only at a country or regional level. When examining data at a coarse level, it is difficult to determine findings confidently due to an inability to verify data with local sources. This ambiguity can obscure our understanding of exploited populations as factors like demographics, population density, and resource allocation are likely not homogenous over large geographic regions. Ultimately, this level of data can only describe part of the labor exploitation narrative.

There is a high barrier to collecting data regarding labor exploitation in contexts where forced labor is likely to exist. This type of missing data can be classified as "Not Missing at Random," suggesting a causality between external circumstances to the dataset and the desired target variable. One of the major barriers to collecting reliable data is the fear of retaliation associated with speaking out against labor abuse (Harvey 2014). Consequences of speaking up can range from task reassignment and firings to death threats and bodily harm. Therefore, workers are reticent to speak out if it risks their economic security or severe reprisal (Pike 2020). Further, this type of missing data does not have corollary features that can assist in the estimation of the missing information with high probability. Even when using other datasets as a proxy to infer missing information, conclusions may be biased in unknown ways (Little et al. 2007).

Consequently, data incompleteness, skewness, and overconfidence are significant barriers to performing accurate and precise analyses of conditions related to labor exploitation. Many of these problems with labor exploitation data can be overcome through various imputation methods. However, the most effective means of improving data collection is creating safe and discrete methods for reporting among worker populations.

28.4.4 Case Study: Worker-Driven Social Responsibility and the Fair Food Program

The Coalition of Immokalee Workers (CIW), the human rights organization that developed the US-based Fair Food Program, asserts that the root cause of labor abuse is the vast imbalance of power between farmworkers and their employers. They contend that high demand for low-priced produce creates a downward cost pressure, which is passed down to workers as low or non-existent wages. These economic conditions lead to frequent labor abuse and trafficking as a means of maintaining artificially low labor costs (Fair Food Program 2021; WSR 2019).

Launched in 2001, the CIW's Campaign for Fair Food aimed to address this power imbalance by asking corporations to use their market power as a force for good. They sought to achieve this in two ways:

- A price premium, specifically a "penny per lb." added to farm workers' regular paycheck.
- A commitment from companies to buy from those who have adopted a human rights-based Code of Conduct (Fair Food Program 2021).

The FFP model, which embraces a worker-driven social responsibility (WSR) approach, has seen reasonable success. In addition to several multinational corporations agreeing to participate, 90% of tomato production in Florida, major tomato operations in five other states, and pepper operations in Florida have agreed to adopt the Fair Food Code of Conduct. The program engages and empowers farmworkers who participate actively in committees and surveys. Because of their feedback, farmworkers have access to necessities that were often denied before the implementation of FFP. They can work in an environment without tolerance for sexual harassment, discrimination, or verbal abuse. Furthermore, they can make complaints without fear of retaliation. As a result, they've recovered almost $500,000 of wages and increased wages for many workers by 10% (Fair Food Program 2021).

A 10-year study of the 40 best known social responsibility programs revealed that the Fair Food Program provides unprecedented transparency into the food supply chain, mainly due to its worker-centered mechanisms (MSI Integrity 2020). The experience of FFP is unfortunately the exception and not the rule. However, it is a compelling example of how worker leadership and participation can help transform a supply chain.

28.4.5 Near Term Next Steps

The design community must take incremental steps to help build new frameworks and approaches to counteract the systemic dynamics that have exploited laborers for generations. This is undoubtedly a daunting task, but on a positive note, in recent years, the design community has adopted new practices to address environmental sustainability. In light of this, E-LCA may provide a helpful roadmap to also address social sustainability. In scarcely a decade, E-LCA has grown from a niche practice into the mainstream thanks to incentive programs, government mandates, and a rich ecosystem of tools and data tailored to the design community's needs. With little knowledge of the extraction, processing, and manufacturing processes that go into building products, architects can now evaluate a material's environmental impact and compare that against a functionally equivalent alternative. As manufacturers have recognized the market value of transparency through the development of product-specific environmental product declarations, architects and builders can continue finding ways to reduce environmental impact in the procurement process (USGBC n.d.; CLF n.d.).

There is an opportunity to develop analogous methods for assessing the social dimension of our product and material selections through all phases of design. E-LCA has given us a head start, providing a simple means of setting up functionally equivalent comparisons and helping us quantify materials that go into our buildings. What remains is associating materials and products with representative data on labor exploitation and providing guidance on how to interpret results to guide design decisions. This aspect will evolve dramatically in the coming years as data and methods mature. Ideally, architects should seek data from organizations that engage and leverage the expertise and knowledge of workers impacted by labor abuse. In the meantime, it would be advisable for the design community to evaluate available social certifications that pertain to building materials critically. A deeper analysis of why approaches like the Fair Food Program are more effective than corporate-led approaches should be pursued. This exercise can help architects decide what to indicate on specifications.

Ultimately, to be in a better position to address modern slavery in supply chains, the design community should:

- Partner with organizations that work closely with worker populations to inform technological solutions and LCA practice
- Assess applicable social certifications against worker-driven models like the FFP to encourage the adoption of more robust standards
- Explore applications of Fair Trade practices and *worker-driven supply chain governance* to the extraction, processing, and manufacturing of building materials (WFTO, n.d.; Reinecke and Donaghey 2021)
- Pursue further education regarding conditions that drive labor exploitation.

28.5 Conclusion

It is understandable that as the design community's awareness of modern slavery increases, there is a clarion call to eradicate this despicable practice. However, modern slavery is a deeply embedded problem with no easy, effective solution. This issue must be handled with thoughtfulness and delicacy, as acting too quickly without the right information may allow forced labor to continue with impunity or possibly exacerbate the problem. Acknowledging this reality may inspire hopelessness and frustration, but there is still a path forward. Architects should advocate for a worker-led approach to supply chain governance. The design community should place less consequence on corporate-led approaches and should seek expertise from worker populations on how to tackle forced labor in their communities. Ultimately, respecting a worker-led approach to supply chain governance leverages the knowledge and experience of workers who are intimately familiar with the problem of labor exploitation. This may allow designers to develop creative and dynamic methods to eventually eradicate modern slavery from building materials supply chains.

References

AIA (2020) 2020 Code of ethics and professional conduct. Retrieved 13 Jan 2023, from https://content.aia.org/sites/default/files/2020-08/2020_Code_of_Ethics.pdf

AIA (2022) AIA and mindful MATERIALS partner to advance holistic product specification. Retrieved 13 Jan 2023, from https://www.aia.org/press-releases/6521249-aia-and-mindful-materials-partner-to-advan

Anti-Slavery International (2017) Slavery in India's brick kilns & the payment system. Retrieved 13 Jan 2023, from https://www.antislavery.org/wp-content/uploads/2017/09/Slavery-In-Indias-Brick-Kilns-The-Payment-System.pdf

ArchitectureAU (2021) Products and materials library to help architects specify safely, sustainably. Retrieved 13 Jan 2023, from https://architectureau.com/articles/institute-product-and-materials-library/

Bales K, Sovacool B (2021) From forests to factories: how modern slavery deepens the crisis of climate change. Energy Res Soc Sci 77. https://doi.org/10.1016/j.erss.2021.102096

Bates R, Carlisle S, Faircloth B, Welch R (2013) Quantifying the embodied environmental impact of building materials during design: a building information modeling based methodology. Retrieved 13 Jan 2023, from https://api.semanticscholar.org/CorpusID:110258910

Bernstein P, Sharples C, Ulicny B (2020) Big data and other tools needed to move from disaggregation to aggregation and a slave-free supply chain from the design for freedom report. Grace Farms Foundation, New Canaan, CT

Chen MA (2012) The informal economy: definitions, theories and policies. Retrieved 12 Oct 2022, from https://www.wiego.org/sites/default/files/publications/files/Chen_WIEGO_WP1.pdf

Concrete Sustainability Standard (CSC) (2021) Concrete sustainability council technical manual version 2.1. Retrieved 12 Oct 2022, from https://csc.eco/wp-content/uploads/2022/04/Technical-Manual-CSC-Technical-Manual-Version-2.1_2021-11-22_compressed.pdf

Davis H (2006) The culture of building. Oxford University Press, New York City

Evrything (2021) Evrything product cloud for apparel. Retrieved 12 Oct 2022, from https://cdn2.hubspot.net/hubfs/2577446/ACTIVE_WHITEPAPER_PDF/EVRYTHNG-Product-Cloud-for-Apparel_whitepaper.pdf

Fair Food Program (FFP) (2021) Fair food program 2021. Retrieved 12 Oct 2022, from https://fairfoodprogram.org/results/

Flare (2022) Flare. Retrieved 12 Oct 2022, from https://www.gfems.org/flare/

Glocal Research and India Committee of the Netherlands (2015) Rock bottom modern slavery and child labour in south Indian granite quarries. Retrieved 13 Jan 2023, from http://www.indianet.nl/pdf/RockBottom.pdf

Grace Farms (2022) Design for freedom toolkit. Retrieved 12 Oct 2022, https://www.designforfreedom.org/home/design-for-freedom-toolkit/

Harvey B (2014) Supply chain safety emerging initiatives in the aftermath of Rana Plaza in Bangladesh. Retrieved 12 Oct 2022, from http://mhssn.igc.org/Prof%20Safety%20-%20May%202014%20-%20Bangladesh.pdf

IndustriALL (2021) Trade unionist killed in the Philippines. Retrieved 12 Oct 2022, from https://www.industriall-union.org/trade-unionist-killed-in-the-philippines

IndustriALL (2022) Violent attack against Myanmar unionists. Retrieved 12 Oct 2022, from https://www.industriall-union.org/trade-unionist-killed-in-the-philippines

International Labour Organization (ILO) and Walk Free Foundation (2017) Global estimates of modern slavery. Retrieved 12 Oct 2022, from https://www.ilo.org/wcmsp5/groups/public/@dgreports/@dcomm/documents/publication/wcms_575479.pdf

Jolliffe E (2017) Architects must take a stand on modern slavery. Retrieved 13 Jan 2023, from https://www.bdonline.co.uk/opinion/architects-must-take-a-stand-on-modern-slavery/5090665.article

King L, deBaca L (2020) An ancient problem, a new awareness: slavery in construction materials supply chains. Retrieved 13 Jan 2023, from https://www.americanbar.org/groups/construction_industry/publications/under_construction/2020/spring2020/slavery-construction-materials-supply/

Little T, Bovaird J, Card N (2007) Modeling contextual effects in longitudinal studies. Lawrence Erlbaum Associates Inc., Mahwah, NJ, pp 15–29

MSI Integrity (2020) Not fit-for-purpose. Retrieved 12 Oct 2022, from https://www.msi-integrity.org/wp-content/uploads/2020/07/MSI_Not_Fit_For_Purpose_FORWEBSITE.FINAL_.pdf

Pike K (2020) Voice in supply chains: does the better work program lead to improvements in labor standards compliance. ILRReview 73(4):913–938. https://doi.org/10.1177/0019793920911905

Prince S, deBaca LC, Thatcher C (eds) (2020) Design for freedom. Grace Farms Foundation, New Canaan, CT

Reinecke J, Donaghey J (2021) Towards worker-driven supply chain governance: developing decent work through democratic worker-participation. J Supply Chain Manag 57(2):14–25. https://doi.org/10.1111/jscm.12250

Royal Institute of British Architects (RIBA) (2020) RIBA modern slavery statement. Retrieved 13 Jan 2023, from https://www.architecture.com/about/modern-slavery-statement-2019

Rozani AS (2022) Child labour in mining in India and the DRC: two case studies. Retrieved 12 Oct 2022, from file:///C:/Users/705/Downloads/rozani_child_2022%20(3).pdf

Simpson D, Segrave M, Quarshie A, Kach A, Handfield R, Panas G, Moore H (2021) The role of psychological distance in organizational responses to modern slavery risk in supply chain. J Oper Manag 67(8). https://doi.org/10.1002/joom.1157

Slade H (2020) Now is the time to include slave-free criteria in our industry code of ethics. Retrieved 15 Jan 2023, from https://www.designforfreedom.org/take-action/now-is-the-time-to-include-slave-free-criteria-in-our-industry-code-of-ethics/

SPERI, Stanford and Yale (2021) Labour share and value distribution. Retrieved 12 Oct 2022, from https://static1.squarespace.com/static/6055c0601c885456ba8c962a/t/61b3a4a8ab5752443ff63e70/1639163049070/ReStructureLab_LabourShareandValueDistribution_December2021_AW.pdf

The Carbon Leadership Forum (CLF) (n.d.) Embodied carbon in construction calculator (EC3). Retrieved 12 Oct 2022, from https://carbonleadershipforum.org/ec3-tool/

The Carbon Leadership Forum (CLF), Department of Commerce, & Meserow Design (2022) Buy clean buy fair Washington pilot study. Retrieved 13 Jan 2023, from https://carbonleadershipforum.org/wp-content/uploads/2022/11/CLF-Commerce-Buy-Clean-Buy-Fair-Final-Report-Nov2022.pdf

The Changing Markets Foundation (2018) The false promise of certification. Retrieved 12 Oct 2022, from https://changingmarkets.org/wp-content/uploads/2018/05/False-promise_full-report-ENG.pdf

The Chartered Institute of Building (CIOB) (2016) Building a fairer system: tackling modern slavery in construction. Retrieved 12 Jan 2023, from file:///C:/Users/705/Downloads/CIOB_Modern_Day_Slavery_WEB.pdf

The Freedom Hub (2021) Child mica mining. Retrieved 12 Oct 2022, from https://thefreedomhub.org/wp-content/uploads/2021/07/TFH-Child-Mica-Mining-Report.pdf

The World Fair Trade Organization (WFTO) (n.d.) Definition of Fair Trade. Retrieved 15 Jan 2023, from https://wfto.com/who-we-are#our-grassroots-history

The World Fair Trade Organization (WFTO) (n.d.) Our impact. Retrieved 12 Oct 2022, from https://wfto.com/

The World Fair Trade Organization (WFTO) (2019) Guarantee system handbook. Retrieved n.d. Mar 2022, from https://wfto.com/sites/default/files/WFTO%20Guarantee%20System%20Handbook%20-%20updated%20August%202019.pdf

United Nations (UN) (2022) Transforming our world: the 2030 agenda for sustainable development. Retrieved 13 Jan 2023, from https://sdgs.un.org/2030agenda

USGBC (n.d.) LEED rating system. Retrieved 12 Oct 2022a, from https://www.usgbc.org/leed

USGBC (n.d.) Social equity within the supply chain. Retrieved 12 Oct 2022b, from https://www.usgbc.org/credits/new-construction-core-and-shell-schools-new-construction-retail-new-construction-data-22?return=/pilotcredits/New-Construction/all

Women in Informal Employment Globalizing and Organization (WIEGO) (n.d.) Informal economy debates: dominant schools of thought. Retrieved 12 Oct 2022, from https://www.wiego.org/sites/default/files/resources/files/WIEGO_IE_Dominant_schools_of_thought.pdf

Worker-Driven Social Responsibility Network (WSR) (2019) Building dignity and respect. Retrieved 12 Oct 2022, from https://wsr-network.org/wp-content/uploads/2019/12/Building-Dignity-and-Respect-english.pdf

Part V

Re-thinking Land

29 Transportation Infrastructure in Yunnan: Repercussions on Cultural Diversity

Vanessa Ma

Abstract

China's transportation network has burgeoned over the twentieth century, motivated by an ambition to boost its economy with the formation of an extensive high-speed transportation network across the country. As a whole, this initiative has been claimed to be a means of prioritising development in the hinterland and improving the livelihood of communities in the rural areas. In reality, however, the building of such a network bypasses these communities. In their analysis of networked infrastructure, Graham and Marvin (Splintering urbanism: networked infrastructures, technological mobilities and the urban condition. Routledge, London, 2001) suggest that in the course of unbundling infrastructure to suit specific modernisation goals, socio-spatial relationships will inevitably be reshaped, with impacts that vary at different levels across different economic systems and social parameters. Often, a flattened cultural landscape is observed subsequent to the development. This research uses the roads connecting Yunnan to Tibet, a historically controversial link, as a case study based on the discussion points posited in Graham and Marvin (ibid) (Splintering urbanism: networked infrastructures, technological mobilities and the urban condition. Routledge, London, 2001) thesis *Splintering Urbanism.* The content of this paper comprises a contextual examination of the dilemma aforementioned under a comparative framework which draws upon existing theories and contemporary scholarly discussions on the regional transformations in the past decades. This is aimed at evaluating the discourse on the formation of 'urban enclaves' in rural China which has been pointed to be the culprit of cultural segregation and dilution over the course of modernisation. In doing so, insights into inclusive development may also be gathered.

Keywords

Splintering urbanism · Roadology · China · High-speed transportation

29.1 Introduction

As one of the world's largest countries by landmass, China has a long-standing history of being culturally diverse. Yet, the country's history of modern economic development is comparatively short. The World Health Organisation (WHO) and the United States (US) had recognised it as a developing country until 2020 when it was crossed off the list for its spontaneous

V. Ma (✉)
Project: As If, Hong Kong, China
e-mail: ma.changling@gmail.com

S. Hilal et al. (eds.), *Design for Partnerships for Change*, Sustainable Development Goals Series,
https://doi.org/10.1007/978-3-031-36993-3_29

upward gross domestic product (GDP) growth. Since 2010, China's GDP has come close to that of the USA.

The scale and extent of infrastructure network expansion is commonly used as an index of development. 'To build wealth, first build roads' [要致富，先修路]. Aligned with this Chinese proverb, the expansion of transportation networks has been a major development policy of the Government of China. This is seen to be particularly true in the lesser developed provinces in the outskirts. Admittedly, the fact that a settlement is accessible via a transportation network alone cannot be a comprehensive indicator of development without considering the availability of other basic infrastructures. Carelessly, however, the conception that building new roads equals a developed city echoes how the Chinese government communicates development.

There exists a disparity between the level of development of the transportation network and of other basic infrastructures. Confronting pressures of globalisation and digitalisation, and the increasing pace of life, time is of the essence. Sacrifices in development are inevitable, most evidently perhaps is the flattening of the cultural landscape observable across the country.

The generic development model in the West centuries ago evinced a migration from a 'modern infrastructural ideal' to a 'liberated infrastructural reform', as proposed in *Splintering Urbanism*. With this shift, an unbundling process takes place. This occurrence will then lead to a socio-spatial splintering of modern cities, giving rise to access inequality. The thesis also explains the imbalanced cultural development.

However, the presumed existence of a 'modern infrastructural ideal' in developing cities is not indisputable for the case of Mumbai, let alone the subsequential introduction of privatised, specialised service providers. In contrast, the generic development model adopted in Chinese cities is controlled by the 'capitalism' of a communist state where no infrastructural service providers are permitted to join the market as an independent enterprise.

It is therefore our intention to examine and discuss the degree to which *Splintering Urbanism*, in its reasoning of the splintering process of urbanism in a few cities of study, may be used to describe China's imbalanced cultural development. This has a significance in illuminating the way cultural diversity can be preserved in face of rural modernisation.

29.2 Methodology

Following the publication of Graham and Marvin's thesis, considerable discussions on its universal applicability have ensued. In November 2008, a series of articles were published in GeoForum under the section *Placing Splintering Urbanism* (Coutard 2008). It consisted of a collection of case studies on various cities around the world in which responses to the thesis were offered.

This paper situates *Splintering Urbanism* in rural China at the emergence of roadology study [路學] which sprouted in 2010. This is done by juxtaposing Graham and Marvin's (2001) concept of 'urban enclaves' alongside several theories and narratives.

29.2.1 Roadology

Roadology is an interdisciplinary study initially developed by Zhou (2010) who wanted to establish a framework for the analysis of the production, usage, construction, and consumption of roads. Similar to the intention behind the synthesis of Graham and Marvin's (2001) 'critical networked urbanism', the emergence of roadology can be viewed as a local attempt to formulate a dialogue on the transportation network in China and beyond. It encompasses research methods across multiple disciplines: road ecology, economics, politics, architecture, and anthropology.

This study is aimed at filling up the gaps in existing dialogues regarding the significance of road construction, and other related aspects undermined by the discourses in more developed countries which are presumably equipped with a wider set of infrastructures that includes road, power, etc. (Fig. 29.1).

Fig. 29.1 Shaded area on the map above shows Yunnan Province. It is located at the south-western border of China, bounded by the Tibet Autonomous Region, Sichuan Province, Guizhou Province, Guangxi Province, Vietnam, Laos, and Myanmar

29.2.2 Yunnan

The cultural make-up that exists in Yunnan is evidently more diverse and multifarious as compared to other provinces in China which typically have a large Han-majority and a small composition of other ethnic minorities. Here, although the Han community still forms the majority, there is a greater proportion of ethnic minorities. This composition is noteworthy in informing our discussion on transportation networks as it impacts urbanism and spatial planning, etc. This is one of the reasons why the high-speed transportation network connecting Yunnan and Tibet has been chosen to be the subject of this case study.

Rather than a single city or suburb, the case chosen is a segment of the infrastructure network linking dozens of townships, big and small, across thousands of kilometres. Therefore, granting a potent illustration of the effect of time–space compression (Harvey 1990) which has an evident cultural impact. Through this, we are able to demonstrate the transformation of the multifarious cultural layers shaped by its complex ethnic composition (i.e. greater proportion of ethnic minorities as compared to other provinces in China) and the varied geographical proximity of the townships to the border.

Furthermore, in comparison with the water and power networks discussed in some of the case studies in *Placing Splintering Urbanism* (Coutard 2008), the construction of a transportation network is spatial. It also covers vast distances across lands and is a direct result of urban politics which significantly impacts the internal and external flow of residents and travellers that in turn drive the import and export of culture.

Evolving from an 'open spatial' entity which can be entered and exited at any point (Fig. 29.2a, b) to a 'closed premium networked space' one with discreetly controlled access (Fig. 29.3a, b), it is argued that the construction of modern transportation networks and hubs in modern China has led to the segregation and dilution of local cultures resulting from the introduction of a bypassing and closed system.

This paper begins with an analysis of the importance of the transportation network in the region between Yunnan and Tibet based on its geographical and historical context. It then discusses the factors affecting its unprecedented and exponential expansion in recent decades. This discussion is presented in reference to the national development strategy and responses from provincial governments. It also covers the cultural background of the subject in study, as well as the possible damages to the cultural diversity engendered by the said expansion.

29.3 Analysis

29.3.1 Significance of Transportation Network in Yunnan: The Historical and Geographical Context

Tibet, having been a separate kingdom for more than a thousand years, had gone under the rule of the Qing Dynasty in the 1700s and then became part of the People's Republic of China in 1951. Situated on an impenetrable highland plateau and acting as the border between India and China, Tibet has become known for its strategic importance with its proximity to China's south-eastern neighbours.

The provinces of Qinghai, Sichuan, and Yunnan which landlocked the borders of the Tibet Autonomous Region, have always acted as entry points to the region. In the 1600s, the Tea-Horse Road [茶馬古道] gradually expanded in between the provinces for tea trading purposes, as the name of the road, 'by horse for tea' implies. Since then, major urban settlements have developed along the road at the provincial landscape due to their demands en route (Figs. 29.4 and 29.5).

Since the establishment of the People's Republic of China, the roads network was further developed in Yunnan for military purposes, such as the campaign to suppress the remainders of the Chinese Nationalist Party in the Chinese Civil War (1950–1953) and to control the settlement of the border dispute between Tibet and India in the Sino-Indian War (1962).

Following the current era of the 14th Five-Year Plan of China, the high-speed network is expected to expand into a web of passageways corresponding with the idea of 'Eight Vertical and Eight Horizontal' [八縱八橫] networks across the nation. The plan strives to complete these expressways and railways with an operating mileage of 45,000 km that covers cities with a gross population of over 5 million people. The three Western terminus of these lines, Kunming, Chengdu, and Xining, will be further connected into Tibet with ongoing construction of local train lines. This results in the anticipation for these major nodes within the network, especially those townships with new railway stations to flourish and expand further.

Aside from its historical context, Yunnan's magnificent landscapes and historical heritage has its tourism flourished since the late 1990s when heritage sites such as Li-Jiang City and Shangri-La were declared as UNESCO World Heritage sites. In 2016, the province welcomed over 470 million domestic and foreign tourists

Fig. 29.2 **a** 'Open road' condition over time: (left) local residents seen selling produce off a truck by the roadside; (right) trucks emerging into stalls, markets and villages. **b** An 'open road' condition over time: (left) individual houses built on a slope next to the road; (right) more houses are built progressively, thus forming a settlement and gradually transforming the topography

and aimed to attract 1 billion arrivals under the 14th Five-Year Tourism Development Plan.

The growth of the tourism industry was identified as a means to alleviate poverty and urbanise rural villages. Concurrently, there is a growing demand for infrastructure and supporting facilities which further drives this development orientation towards tourism. In the past decade, high-speed networks began developing with priorities for target groups of tourists to drive the local economy in Yunnan, thus shaping tourism as a new priority within urban and transportation development (Fig. 29.6).

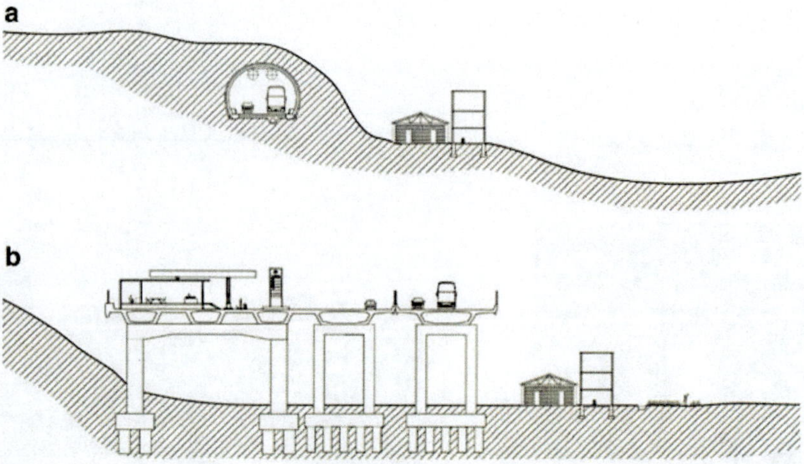

Fig. 29.3 **a** 'Closed road' condition: a tunnel built underground allows no interaction with the local life to occur. **b** A 'closed road' condition: a highway service area providing services such as petrol refilling, rest areas as well as food and beverage exclusively for the vehicles and drivers within the closed network. This system weakens the connection between the road users and local community, thus reducing opportunities for organic growth as planning had been blueprinted

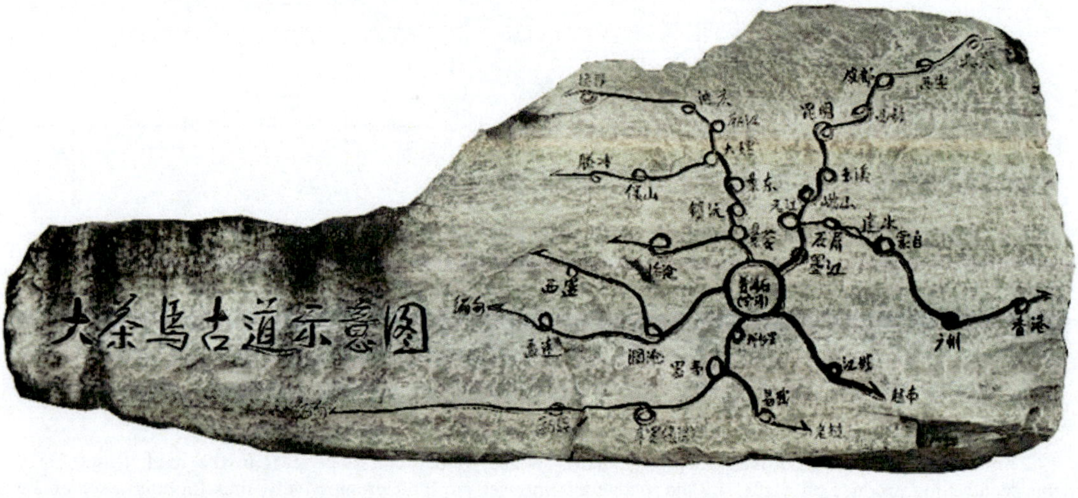

Fig. 29.4 Rough cartographic diagram of the network of the Tea-Horse Road on a stone carving in Pu-Er [普洱] Township

29.3.2 National and Local Infrastructure Policies: Image Construction Propaganda

Since the twentieth century, the transportation network of China has developed unprecedentedly, exceeding 4.24 million km of roads and 121,000 km of railways in 2013, second only to the USA in the total coverage of the transportation network. China's 2030 Road Master Plan aims to achieve 5.8 million km of total road network coverage that includes 400,000 km of national highways, 180,000 km of expressways, and 274,000 km of railways (Fig. 29.7). This national development policy highly corresponds with the 'Chinese understanding of globalisation' (Zhou 2012) where

Fig. 29.5 Since the 1600s, the network of the Tea-Horse Road had grown out from Yunnan towards its neighbouring provinces for tea trading

the Chinese fail to foresee the shortcomings of the rapidly developing infrastructure network, although those disadvantages are widely discussed and recognised globally.

The rapid process of infrastructure development in China hinders it from reaching full maturity. Significant amount of basic infrastructure networks especially in rural areas was left behind with an absence of 'modern infrastructural ideals' (Graham and Marvin 2001). Despite the challenges, China, still determined to catch up with centuries of infrastructure development and her Western counterparts, has relied on the strategies termed in *Speculative Urbanism* discussed by Goldman (2010) in his article on Bangalore, India. China aims to progress to the path of production of 'World Cities'. Although its benefits are merely speculative, it has shown clearly that the magnitude of advancement in infrastructure development becomes propaganda tools for the reputation of China.

This national position on infrastructure development reflects the same vision and policies as Yunnan's provincial government to a more local scale. China's aspiration of maintaining Yunnan's strategic location at the border by road construction is apparent as observed in the case of Teng-Mi Road, which seemingly connects Tengchong, Yunnan, and Myitkyina, Myanmar (Fig. 29.8). Despite the efforts, the results may not be clear and often baffles the locals as the constructed infrastructure ultimately leads to an unreliable and underdeveloped network in Myanmar (Zhou 2012). This competitive infrastructure development among townships to claim its strategic location at the border becomes the crucial objective of China's planning. Therefore, the above argument is clear when the already limited resources are being prioritised to projects that boost the reputation and speculative goals of the country instead of the essential needs of the locals.

Consequently, with the combined effort of the national and local provincial government, the development of transportation networks, especially high-speed transportation, is marketed as the mindful way of prospering the hinterland. However, these new networks prioritise connecting the Yunnan province to other inland cities first rather than individual counties, townships, or villages. Therefore, the focus on new networks left behind the potential to functionally replace the old but

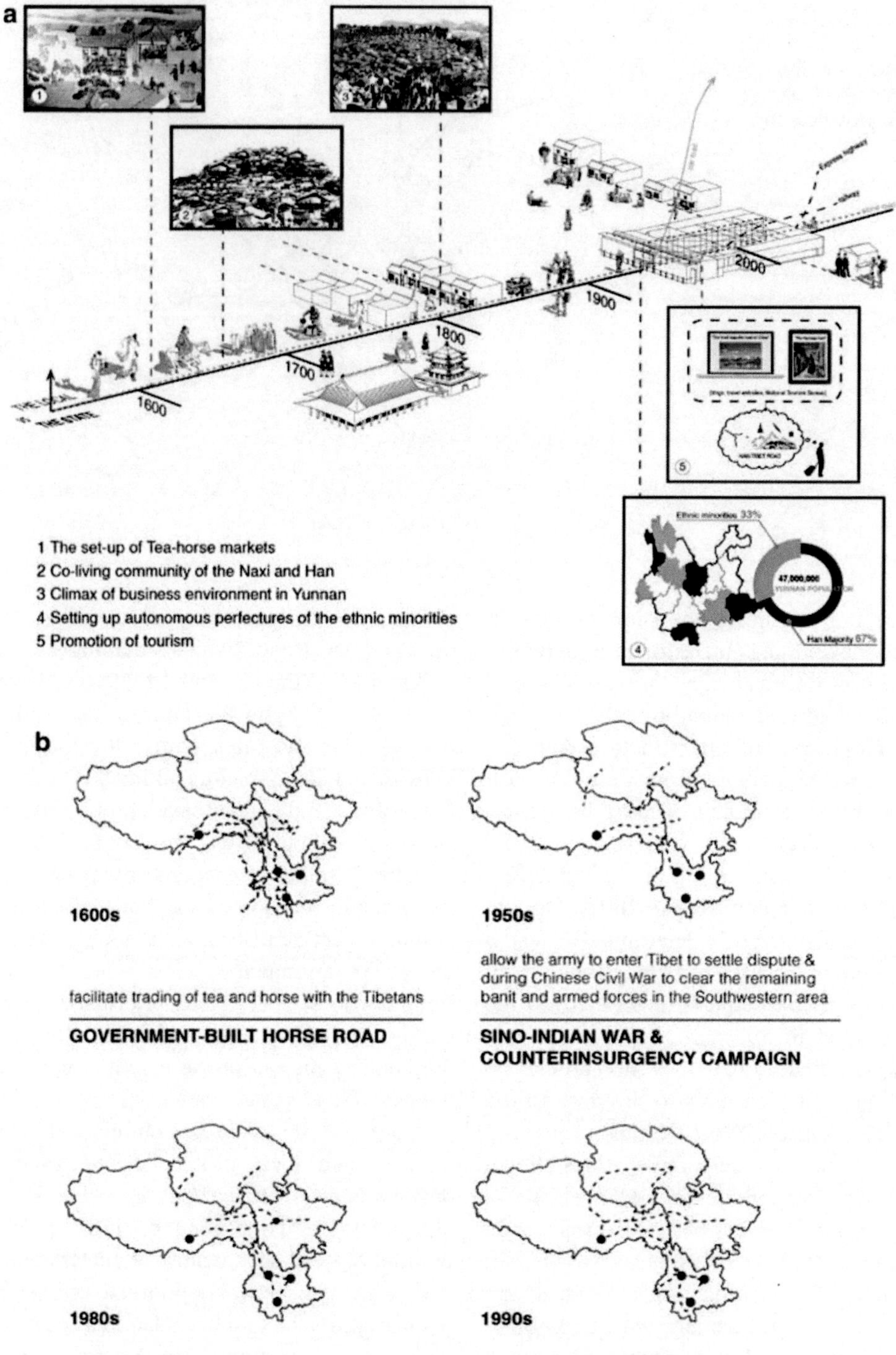

Fig. 29.6 **a** Historical timeline on the factors driving the construction and development of roads between Yunnan and Tibet. **b** The diagrams of the road construction network in relation to their purposes of construction in its own period

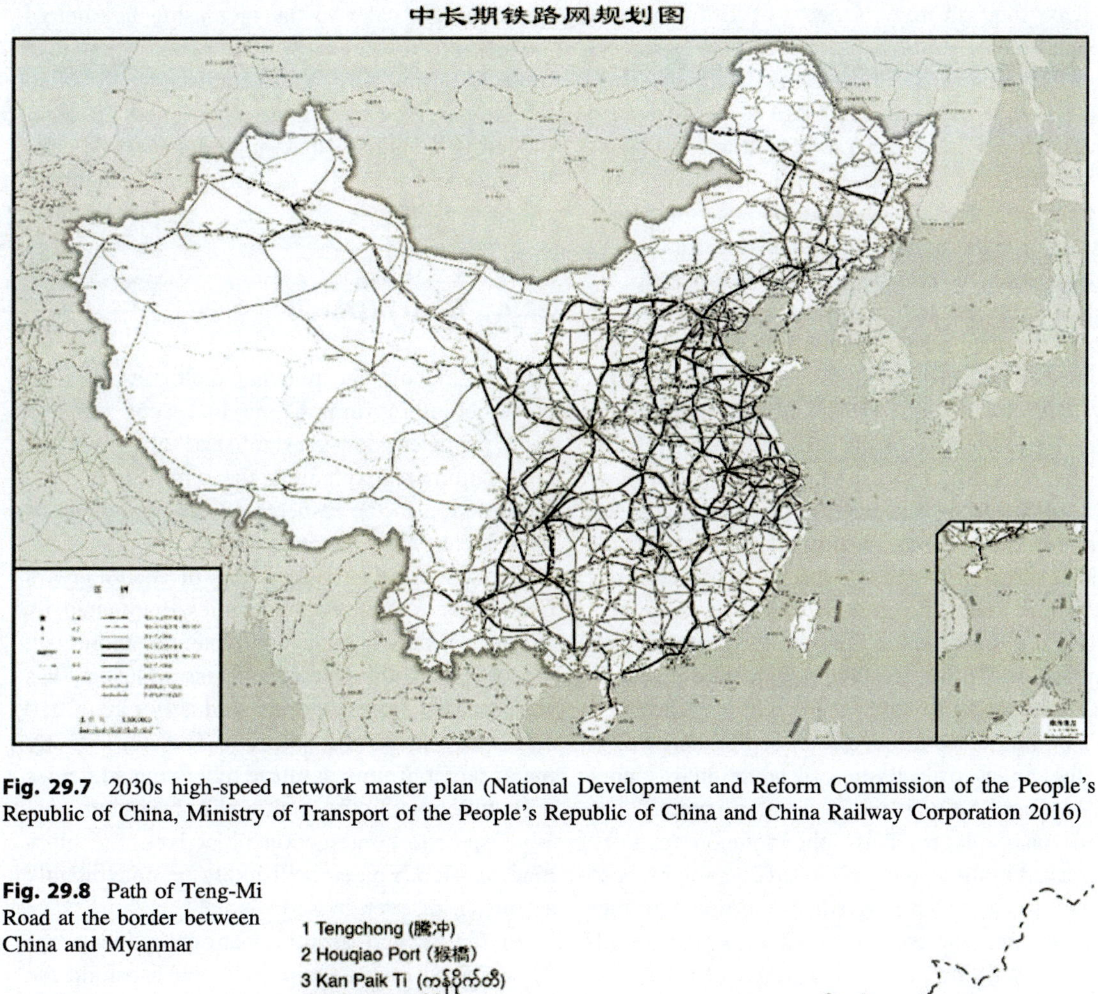

Fig. 29.7 2030s high-speed network master plan (National Development and Reform Commission of the People's Republic of China, Ministry of Transport of the People's Republic of China and China Railway Corporation 2016)

Fig. 29.8 Path of Teng-Mi Road at the border between China and Myanmar

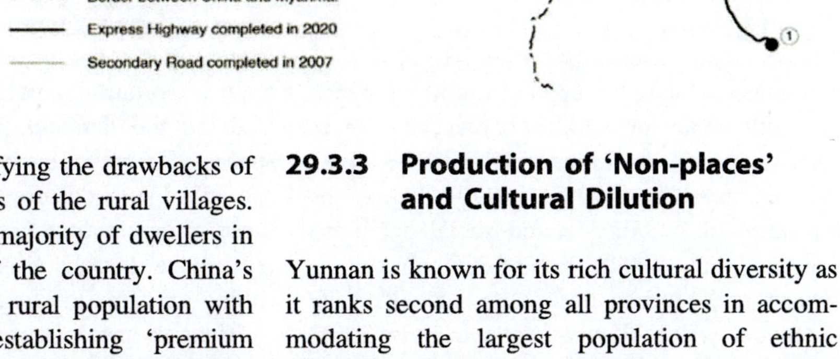

bustling rural roads, amplifying the drawbacks of bypassing the communities of the rural villages. This also leaves behind a majority of dwellers in the developing portion of the country. China's false claims of aiding the rural population with extra connectivity and establishing 'premium networked spaces' (Graham and Marvin 2001) becomes prominent when it instead segregated the market, cultural, and economic interaction between the authentic inhabitants and the tourists.

29.3.3 Production of 'Non-places' and Cultural Dilution

Yunnan is known for its rich cultural diversity as it ranks second among all provinces in accommodating the largest population of ethnic minorities in China. Consisting of 25 ethnic minorities of various cultural traditions, they display an abundance of diversity that also attracts more visitors into the province.

On the positive aspect of infrastructure development, it shortens transit time and connects regions across the provinces which brings convenience to locals and tourists. As a result, it promotes an interchangeable cultural exchange over sceneries that were once-obscured. However, the gradual standardisation to accommodate for such development results in the dilution of the diverse lifestyle. Developed stations do not provide the means for genuine interactions to occur between locals and tourists. In conjunction with the cultural limitations of the 'closed road system', it threatens the genuine living experiences and lifestyle of the affected area.

As established by Lefebvre (1991), 'Production of Space' is associated with human activity. As observed in the historical context of the ancient Tea-Horse Road, the heart of economic activities aided the growth of urban settlement along the roads which then evolved into today's townships. This corresponds with the interpretation of 'market towns'.

As the nucleation of urban settlements grows at the nodes of activities, production of 'non-places' and homogeneous 'abstract spaces' can also take place. This phenomenon can be observed in the system of closed-loop high-speed networks, in which it sacrifices cultural activities in exchange for efficiency. China's new terminals, now acting as the new nodes of urban settlement, focus on accommodating for passers-by tourists which results in the lack of depth of interaction between the locals, ultimately replacing the majority of the cultural lifestyle at each town centre.

This new experience is further heightened by the enhancement of speed and mobility, which is directly relevant to the contest of forces towards various conditions manifested as time–space compression (Harvey 1990), hence the occurrence of mobility-related social exclusion (Kenyon et al. 2002). When the connections between major urban nodes are intensified by the 'tunnel effect' (Graham and Marvin 2001) with point-to-point contact, it leaves behind an immense amount of cultural diversity away from the scene. In addition, the local lifestyle at these developing urban centres is forced to modernise to cater to the increasing number of tourists from the city, who seek alternative lifestyles. Ironically, these tourists would often find themselves in similar experiences to other modernised nodes that underwent the same development priorities of infrastructure network and urban lifestyle.

29.4 Conclusion

Having laid out the relevant dialogues on road construction in rural China in recent years, a basis for discussion is established.

A summary explaining the purposes of road construction in the respective time periods helps us to identify the target groups of those projects. It also covers the background of roadology in China where the existing studies dominated by Western scholars have insufficiently addressed. The term 'sustainable smokeless industry' has surfaced in China's tourism industry circa 2010 and is now under the spotlight. The study of any major target groups of future bypassing strategies will undoubtedly be a thought-provoking one. With specific construction objectives, the planning of such projects will likely be unavoidably atomistic, or even biased.

In the second section, national policies are discussed to reveal the primary push behind the spontaneous growth of the transportation network in the country. Instead of migrating from a natural monopoly to privatised services, it is the rise in the national strength and ability to influence bordering countries, as well as a drastic change in the composition of the population in the transportation network of the case area which leads to the demand for 'premium networked spaces'. Spaces of such cater only to those who can afford such customised services. Regretfully, these transportation network development projects have largely been embarked before the setting up of basic common infrastructure.

The pros and cons of cultural exchange brought by the elevated means of travel are undermined by the propaganda-oriented planning (i.e. the national policies) simply because the motivation behind such initiatives lack genuineness.

Next, attention is drawn to the impact on spatial production of a smaller scale brought by the planning of the vast network. The twofold impact of the construction of the high-speed network brings to cultural diversity has also been discussed in terms of the types of spaces produced. This is with an objective to raise awareness about the proportion of the production of 'non-places' as a by-product of the creation of closed road transportation networks, and a culprit of a monotonous lifestyle created through the process of modernisation. More precisely, the objective is to provide an insight into the negotiation and articulation between the construction of transportation infrastructure (i.e. solely for transit) and spaces produced for activity (e.g. accommodation, social gathering, etc.).

29.5 Discussion

In the course of modernisation, unlike China's single-fold strategy, a multipronged approach should be taken, especially in a region that comprises a complex composition of population.

Henri Mendras in *The Vanishing Peasants* (1970) analysed the discourse surrounding the rural society in Europe which underwent restructuring after World War II. He discussed the then diminishing traditional peasant class as they modernised and pointed out the low percentage of peasants which had dropped to just 4% of the total population by the 1990s. On the other hand, in China, the agricultural population still makes up almost 50% of the total population with a 'redline policy' that sets the country's total arable land to be no less than 1.8 billion mu (120 million ha). It is thus impractical to adopt an undifferentiated measure in achieving the goals of modernisation to match the interest of stakeholders.

Meanwhile, Chinese scholars have begun to propose alternative 'self-help measures' that can be carried out by the rural population, instead of having them rely on the urbanisation policy implemented on a macroscale. These scholars have also been carrying out field work for over half a century in the villages to examine the potential modes of development suitable.

Some scholars of the study have proposed that China can only progress if the agricultural population prospers in their local economy and emphasises on experiences based on interpersonal connections (He 2007). The formation of 'market towns' [集鎮] then came under close scrutiny, and it was revealed that this urban model may be regarded as an intermediate form of a 'settlement system' [聚落系統] which has a significant link with road development. These 'market towns' undergo an evolution from being markets on the road in the past to establishing their own kind of ecology and organisation, as a process of urbanisation independent of the transportation network. Their operation structure and mechanism does not limit the whole population to a specific industry. It in fact promotes a kind of organic integration among cultures. This branch of study is focused on scouting for the rural population who neither want to restrict themselves to an agrarian lifestyle nor leave the villages completely.

This paper has attempted to set off the scrutinisation on spatial composition in transportation architecture from the viewpoint of its service to a regional population, as opposed to exclusively the immediate users of a closed-loop population or even those who are simply serving as a component within the off-scale network. Similar to the discourse on networked infrastructure, transportation architecture has often been studied in silo in the areas of structural advancement, environmental response, and public space integration. It is seldom studied in a holistic view on an urban scale. Moving forward, this research may be extended to a smaller spatial scale whose articulation is centred on evaluating existing hubs along the transportation network—the historical transformation from 'open entities' (which attracted the formation of nucleated settlements centred around economic activities) to 'closed entities' (which exclusively serve as functional 'non-spaces' within the larger 'secessionary network') (Graham and Marvin 2001).

References

Bonsu D (2014) Road transport and agriculture: a comparative study of the implications of road access for subsistence agriculture in the Northern Ghana

Coutard O (2008) Placing splintering urbanism: introduction. Geoforum [online] 39(6). Available at: https://doi.org/10.1016/j.geoforum.2008.10.008

Goldman M (2010) Speculative urbanism and the making of the next world city. Int J Urban Reg Res 35(3):555–581. https://doi.org/10.1111/j.1468-2427.2010.01001.x

Graham S, Marvin S (2001) Splintering urbanism: networked infrastructures, technological mobilities and the urban condition. Routledge, London

Harvey D (1989) The condition of postmodernity: an enquiry into the origins of cultural change. Wiley-Blackwell, Cambridge

Harvey D (1990) Between space and time: reflections on the geographical imagination. Ann Assoc Am Geogr 80(3):418–434

He X (2007) 鄉村的前途 [Future of villages]. Shandong People's Publishing House

Kenyon S, Lyons G, Rafferty J (2002) Transport and social exclusion: investigating the possibility of promoting inclusion through virtual mobility. J Transp Geogr 10(3):207–219. https://doi.org/10.1016/S0966-6923%2802%2900012-1

Lefebvre H (1991) The production of space (trans: Nicholson-Smith D). Blackwell, Oxford

30 Shattering the Narrative

Matxalen Acasuso Atutxa, Mª Angeles López Amado, Natalia Brener Maceiras, Sonia Puente Landázuri, Aurea Soto Vázquez, M. Elvira Carregado Pazos, and Teresa Táboas Veleiro

Abstract

What if we could go back in time with the knowledge we have today? What if we distance ourselves and try to view our territory as if from childhood? Would we think the same way? More importantly still, would we behave the same way? The following narrative endeavors to serve as a manifesto for reflection and a call to action—individual action, yes, but mainly collective. It is a manifesto with more questions than answers, because as a method, we believe in the power of asking the right questions, as a kind of brainstorming to find solutions that will solve problems. In fact, it is a manifesto viewed through a conversation between Marvin, a boy who lives on the moon after the Earth's collapse, hastened by nuclear disaster, and Heidi, a girl who lives in the rural Swiss mountains. It is important to shatter a prevailing narrative in which, in cities, there is always an air of superiority toward the countryside… and to be able to construct an urban–rural model for the future by blurring borders and improving the better part of our territorial planning model so that solutions are not divorced from the origin of the problem. For it is an erroneous premise that breeds our inability to question the true causes of the environmental degradation we face. The climate emergency obliges us, without excuse or delay, to refashion this city-country blueprint, to dismantle the borders between rural and urban, because we continue to propose solutions to the challenges of our time from and for cities, as if they were autonomous entities independent of other territories. Hence, there is a need to reflect on the right to territorial solidarity and dignified conditions for all people with an integrated vision for our future survival, reduced uncertainty, and people at the center of all decision-making. These issues might seem obvious, but we have forgotten them in the last decades. That's why Heidi is here to remind Marvin.

Keywords

reversas · Paradigm shift · Putting people at the center · Degrowth · Transect

M. A. Atutxa · M. Angeles López Amado · S. P. Landázuri · A. S. Vázquez · M. Elvira Carregado Pazos · T. T. Veleiro (✉)
Reversas, Ourense, Spain
e-mail: ttaboas@coag.es

N. B. Maceiras
Reversas, Montevideo, Uruguay

S. Hilal et al. (eds.), *Design for Partnerships for Change*, Sustainable Development Goals Series,
https://doi.org/10.1007/978-3-031-36993-3_30

30.1 Preface

Over the centuries, the city has been conceived as complex system, as humanity's most important challenge in every historical epoch, as at the center of all public policy, and as the raison d'être of most business strategies. Nonetheless, since industrialization, it has also been viewed as alien to and independent of the broader territory, of the rural environment, of natural space. It is seen as indifferent to the ecosystems that it requires for survival—to energy, food, water, and countless resources—and yet in need of them, in need even of a recipient for its own waste.

The climate emergency obliges us, without excuse or delay, to refashion this city-country blueprint, to evaporate the borders between rural and urban, because we continue to propose solutions to the challenges of our time from and for cities, as if they were autonomous entities independent of other territories.

Nowadays, the territorial challenge is always addressed from a series of simplifications that, in their very approach, are removed from an observably true situation. There resides the first source of error in resolving the problem. Therefore, to build a future urban–rural model, there is a need to fine-tune an appropriate territorial planning model such that the solution not deviate from the problem's root (for it is an erroneous premise that breeds an inability to question the true causes of environmental degradation). Again and again, the result reflects the internal contradictions of an economic system that exploits territory in order to obtain the resources it needs for its reproduction. Economic growth and urban expansion develop in lockstep, feeding one another. Solutions for cities do not then resolve the problems inherent to them because of a faulty approach. Such is the case with 15-min tours; the exclusive proximity of facilities or the reduction of private transportation is not enough, but rather the integration of territorial implications into that quarter of an hour. Why not also reduce food, energy, and waste exchanges? It is obviously time to rewrite the rules.

The greater the city—and the less contact it has with the non-urban territory it relies on to survive—the poorer the ecological awareness of the people who inhabit it.

Several international bodies have recognized this issue, including the IEF, a professional non-governmental organization founded in 1997. "People who live in rural areas are usually more aware of the limitations of their natural environment because they live so close to it, and it is easier to see the damage that human activities may do to resources that are important to them," they say.

Or according to the new report, published by the European Environment Agency (EEA) headquartered in Copenhagen, fast, uncontrolled and uninterrupted urban expansion threatens the environmental, social, and economic equilibrium of Europe. "*Cities promote and symbolize the dissociation of humanity to nature, since they function much better than rural economies at effectively satisfying material needs and at the same time reducing environmental impact.*"

We should perhaps ask ourselves to what extent the current, almost irreversible situation of climate emergency that we find ourselves in is the consequence of fixing a skewed gaze between territories, of the ignorance of the ecological, social, and identity footprint upheld by some over others.

30.1.1 What if Marvin Had Met Heidi?

Imagine recovering planet Earth through a collective effort, as dreamed by Marvin, the child protagonist of If *I Forget Thee, Oh Earth* by Arthur C. Clark (1951).

From their lunar habitat to which they embarked on a pilgrimage after nuclear disaster, Marvin understood that it was possible to restore the cradle of humanity. Like him, let us imagine that things can and must be done better. And, though he was sure to never return, Marvin knew that if he passed on to his descendants the mistakes of the past for them to learn from, that they could perhaps one day use the space ships lying in wait to return and recover Earth. Imagine a dialogue outside of time between two children bound by a situation caused by human ambition and blunder: one of them from beyond the planet, and the other, Heidi, from her rural world and understanding of the relationship to territory and of the need to build bridges between city and country. Heidi explains to Marvin how to take care of the environment, offering solutions to repair the damage done.

30.2 A Territorial Web

How ought this spatial conformation be?

Marvin, look at the stars from where you are and observe the workings of the universe. The universe is a constellation formed by stars from different dimensions, themselves different in nature, yet close to one other.

Territory has no concept of the political borders we have created, the countries, autonomous communities, municipalities, towns, villages… In our globalized world, territory is a web of interconnecting cities and towns, a network whose work multiplies rather than compounds.

Shattering the constantly competing rural–urban dichotomy must be part of our political agenda. There is but one territory, a matter of scales of labor, hence the need for integral and cross-cutting views. Both fora are needed—and not only from a purely extractive view of resources. One must look at their habitable environment and overcome the industrial revolution's legacy of defining places exclusively by their productivity.

Economic activity is important, but so too is the preservation of territorial patrimony as part and parcel of sustaining people's lives, of natural patrimony to preserve biodiversity, and of cultural patrimony to maintain our identity.

And all this territorial preservation and activity as a whole must be part of the transfer of knowledge among all brokers to build a better society. It is in this way that accessible and inclusive territory can be realized, minimizing inequalities.

30.3 A Holistic Vision

What ought this new vision be to get us there?

One with an integral sensibility, replacing or perfecting our tools to achieve it.

With the instrumentation of better legal guarantees and the exercise of our legitimate rights, territorial and urban planning have turned into a sort of legal regulation hodgepodge, unintelligible to the common person. This has led to the alienation of citizens who do not comprehend why there should be broad interest in these instruments. Citizens have become concerned only with matters of individual interest because that is what planning has become: a commercial enterprise and, in many cases, pure speculation.

The model of rural and urban territorial planning must be rewritten to shift it from the problem that it is today to a true solution, with a holistic view of territory, prioritizing a strategic vision over a legal one (without detracting from the latter's importance), with an ecosystemic analysis of territory fed by different disciplines and perspectives. It ought to be accessible to all citizens to bring visibility to the importance of the subjects addressed, such as natural resources to sustain life, the preservation of the natural environment and biodiversity, access to dignified housing, and the right to quality public space and to safe and accessible mobility, etc.

It ought to be an integrative and holistic vision, with the aim of putting territory at the service of new egalitarian modes of relating, working, inhabiting, and using under the criteria of the three measures of sustainability: economic revitalization, environmental responsibility, and social cohesion.

Territorial development should integrate a multi-sectoral spatial vision of our desired future and be backed by strategic investments in physical infrastructure and environmental management. Territorial equity must be considered in the development of specific portions of the territory: not only for urban, metropolitan, regional, and rural jurisdictions, but also for watersheds, coastlines, mountains, and borders. This new methodology will be able to measure the interaction between urban, peri-urban, and rural areas in terms of their territorial relations.

Under this system-based blueprint, governance agreements will be the foundation to achieve it. From it springs the necessary effectiveness and coordination for management among different sectors, stakeholders, and interests. There ought to exist shared responsibility for decision-making between the planners and administrators involved in different levels of governance and the political arena.

Territorial planning should influence and foster beneficial relationships between rural, peri-urban, and urban areas. And though different interpretations exist within the discipline, different kinds of spatial planning should base their objectives and approaches on the harmonization of economic development, social inclusion, and environmental protection. The existing challenge is a common one: to rewrite methodologies and tools to promote more balanced and territorially integrated development.

30.4 Plurality: Other Perspectives

How can we acknowledge and integrate diversity?

From a new approach, Marvin, one would have to rethink the new world by placing importance on encounters and cooperation among people.

We would have to abandon the model of the first half of the twentieth century of theoreticians who understood all human needs to be the same, thus proposing a sole, partitioned and ordered model, and zoning by different human activities, a homogeneous model for all, which resolved dwelling, working, and resting, and forgot other issues.

We do best is legal growth. Urban planning legislation in the twentieth century was written for the city, to transform soils into urban expanse, to maximize production and consumption, and for indefinite growth on a finite planet. We created laws, regulations, techniques, and instruments for a growth model. We need redesign laws, regulations, techniques, and instruments to build a new urban, social, and economic planning model: one that is proximate and local. Let us abandon expansion and focus on taking better stock. Let us draw a picture of our vulnerabilities and critical supplies. Let us pursue a replacement rate and energy and food self-sufficiency, without delay and from our existing homes, neighborhoods, cities, municipalities, and territories. In sum, let us legislate degrowth.

If we are to make inroads toward other models, separate from the most rapacious capitalistic models of power and production, we will have to recover mixed and juxtaposed uses in which all of the daily life activities of people of all kinds can be situated. Such needs have historically been ignored by those who planned and built the spaces and settings where life unfolds.

Society will then understand that territory is inhabited by a diverse collective.

We know that this is the challenge of spatial and territorial planning in the twenty-first century—a return to apparent disorder and placing people's lives at the center. We must plan, design, and create a better territory, based on the true, day-to-day, and diverse needs of the human being. And we must eschew the uniform treatment of zoning of traditional planning, adapted for the abstract and standardized necessities of a typified, masculine, middle-aged, healthy citizen who goes to work in his vehicle.

30.5 Pride in Belonging and the Right to Dignity

How can I recover the Earth so that it may provide dignity for all people?

From a rights perspective, Marvin, by avoiding territorial and socioeconomic disparities.

The territorial expression of globalization is the major city. Globalization has favored processes of urban dispersion because its morphology is expansive. It colonizes everything, eliminating all form of rural life, and therefore destroying natural and social ecosystems. The necessities of inhabitants in almost any territory are identical.

Territory therefore becomes a site of conflict between two realities. Because the rural reality has largely been undervalued in its capacity for society-building. Representing for the urban person a site of extraction that is exclusively economic in nature, it is deprived of its social function and relegated to the margins of importance.

Humanity has dedicated all its time, all its knowledge potential to city-planning processes to the city, recall the words of Carolyn Steel in her book Hungry Cities:

> Think about it: only six generations ago, the vast majority lived, literally, in the country. So everyone knew very well from where their food came, because the majority were self-sufficient on what they produced. But when the masses began to move to cities, which have continued to grow, this direct connection with the countryside and farmers began to be lost, especially in the developed world. Living in the urbanised West, the fact that the cities we inhabit are part of an organic continuum can be hard to grasp. Since we appear to live independently of nature, worrying about our waste can seem irrelevant; cranky, even. In Britain, we have never been overly fussed by the matter in any case. Food and raw materials have come too easily to us; and the sea has always been there to dump stuff into when we're done with it.

Global territorial ordering and planning have forgotten the complexity and potential of intricate rural dynamics. We have inherited an urbanism that today and yesterday has amounted to the plunder of territory. What has been done to preserve territory, or to improve it?

Moreover, we have too often conceptualized and planned from the same premises as those made by the Modern Movement in cities, trying to tame and control, to impose rational order on chaos and on the random diversity of people and situations. Rural areas that have lost their agricultural character have been transformed and made to assume urban characteristics, changing the intensity and use of land, and with it small-scale and proximate productivity.

The rural as a way of life is disappearing on the planet, and the space that has not been devoured by the megacity has oft been abandoned. It is necessary, therefore, to dignify it again so that a new rurality can emerge within the territory of the twenty-first century. Direct investment will have to be directed at creating value in the primary sector, and new market rules made to benefit all social actors and sectors, without the environmental cost.

We need territories that are socially just, economically robust, environmentally resilient, and people-focused. We require disruptive strategic initiatives with innovative tools that connect and educate professionals, politicians and, above all, citizens.

30.6 Relocalize, Consume Less

How will we generate a new way of producing and consuming that eludes precariousness and unsustainability?

It's simple, Marvin—by simply understanding the degree of intensity of an exchange.

Shatter the narrative of a productive model based on unlimited consumption. Put an end to the age of abundance/waste, because oil, gas, and coal have been pushed to the limits of exploitation and the materials we need for technology have exceeded the limits of demand. Stop promoting cities as consumption hubs, divorced in their planning from the centers of extraction and production, and therefore dependent on supplies from afar, needy of global transport networks linked to fossil energy. Abandon the citizen-consumer formula and a quality of life defined by consumption.

To get there, our premises must change, and so, too, the indicators and methodologies we employ—which ought to be a far cry from those established by a neoliberal system based on unlimited growth, the outcome of which is the problem we face today. This blueprint has altered and strained the relationship of communities to their territories. We must avoid an environmental discourse predicated on the same economic and financial terms, because therein lies the danger, which is none other than propping up a narrative of a dominant economic model that deliberately hides the environmental problem. This predatory model sees an opportunity to accumulate financial capital time and again, generating problems rather than solutions.

We must abandon expansion, improve stock; trace critical vulnerabilities and supplies in crisis; pursue a replacement rate and energy and food self-sufficiency. We must do so without delay—from our existing homes, neighborhoods, cities, municipalities, and territories.

We must draft new laws and technical regulations and create and use new instruments for better governance in alliance with nature. In this way, we can evaluate the traceability of public resources and the wealth they generate to discover their end use and outcome.

We should not rely on growth indicators alone if it is societal development that we seek. In fact, the exclusive application of these parameters did not create territorial equilibrium, but greater concentration of assets in few hands. Housing, civic engagement, health, wages, education, and security are parameters that need to be set in the future toward societal growth, rather than exclusively GDP.

30.7 Food Geographies for Territorial Intelligence

How will we avoid the repeated breakdown of the production and supply chains that made us more vulnerable?

There is a need to reconnect production, consumption, and space, Marvin, as well as for spatial and territorial planning to help relocalize the food system.

There is obvious need for a disruptive strategy of territorial planning because the global challenges laid bare by the pandemic and the climate emergency forced us to also consider how the food demands of large urban areas are produced, processed, distributed, and ultimately consumed.

In this context, we must become aware of the resources that these large cities require and turn their management into one of the main vehicles of planning. Every day, millions of tons of food are transported, sold, preserved, cooked, eaten, discarded, and made into leftovers in these large cities.

A transformation toward more efficient, inclusive, resilient, sustainable, and proximate agri-food systems is urgent. And it's not just about incorporating food gardens in urban and peri-urban areas: The narrative of a dividing line between the rural and urban must be broken, though some try to maintain this dissociation at all costs despite its suppression of responsible territorial, social, and ecological planning. In depth analysis is needed of the food system from a planning perspective, as well as of the considerable footprint generated by the demand of these large concentrations in population. The challenge will be to adopt territorial classifications that exceed the current model and cross-cutting management actions. Because in the context of resource scarcity, food production also becomes a commodity to speculate over, in competition with other sectors like housing or energy. To this we must add the issue of food insecurity, which is related to extensive industrial processes and long-distance transport. We must also look more deeply into other aspects related to food production, such as adapting to the climate, job creation, health, and well-being.

If this problem is not tackled by territorial planning, we will end up overpowered, for unsustainability already reigns today. And though the very structure of large cities has revolved around food value chains, no one has ever looked out for the value of the territories that produce for them. Mindful we must be then of the true impact of what we eat and its sustainability for the natural world.

30.8 Care and Putting People at the Center

What has happened to us? Why have we lost ancestral practices of care, a human perspective, the protection of the group, collective health, and respect for differences?

Because, Marvin, we have neglected people's rights and fundamental capabilities. We have overlooked people's importance and the value of their talent.

Spatial planning must create places of encounter that are democratic, friendly, and respectful of all people, regardless of their life cycle, ability, or gender. It should also integrate vulnerable groups from an alternative economic, social, and psychosocial perspective.

Our philosophy should be like medicine, the true social function of which is to contribute to individual and collective well-being in a healthful territory.

Understanding each of the spaces that make up a rural area is important to its survival. We will need to focus on building a model of sustainable rural development, based on four core pillars—an environmental ethic that values ecosystems for the survival of biological species; an economic ethic around combined economic activities and the source of raw materials; a cultural ethic, encompassing all the material and intangible patrimony ensconced in development. Above all, there is one bare necessity: people, those who contribute to the creation of social value. Therefore, we must be aware that people's well-being depends in large part on maintenance of the urban environment. Therefore, territorial management, if it is to be effective, must aim to establish a model by and for all people, because it is human beings who have the intrinsic need to organize the physical space in which they live.

We must promote humanization, improved universal accessibility, and quality spaces with criteria that respect diversity and promote equality and solidarity with current and future generations. We must leverage the value of people in building a more human and humanizing world through networks of thought and action that guarantee emotional memory. We must become mindful and encourage the search for synergies between the environment and people's health as we embark on a new journey laden with opportunities.

30.9 Territorial Solidarity

Heidi, how can we build territorial exchange and balance?

By revitalizing relations of cooperation, solidarity, reciprocity, and transparency. The links between the territory's realities must be strengthened and understood.

Inhabiting territory from an integral perspective means carrying out the necessary transition and seeing the complementarity of the rural and the urban. The rural should not be understood only for its support of the city system, but above all in order to re-signify territorial solidarity.

We have to analyze at various scales—regional, national, supranational, and global—the indispensable relationship of each portion of the territory within the whole.

We have to reject the classic conception of winning and losing territories, of beneficiary and subservient territories.

It is essential that territorial action be taken with an endogenous and participatory development approach, taking into account conditions of access to opportunities, so as to mitigate territorial imbalances and social inequalities.

For there to be territorial solidarity, we will reinvent the conception of the relationship between country/city, suppliers/consumers, reflecting on its limitations and potential, and seeking development that can be harmonious in terms of economic growth, social equity, and environmental sustainability.

The articulation between the social and solidarity economy and territory must be developed through careful observation of complex local configurations in which public, private, associative, and cooperative issues and actors contend with understanding of territorial dynamics. And while contexts vary from country to country, innovative contributions to local, social, economic, and sustainable development are needed. Such initiatives must offer a disruptive response to changes in socioeconomic and political regulations, promoting new forms of territorial cooperation, as well as mechanisms that help contribute to a balanced approach to encourage mobilization and social participation in the development of new territorial policies.

The idea of a city-countryside fusion is the common denominator of utopic thought. Today, the possibilities offered us by new technologies again reveal an opportunity for new geographies to inhabit our territories differently.

If we have to suggest for example, for a midsize or large cities in India, as in the case of the proposal for the city of Dongtan, China, as a new urban paradigm for the twenty-first century, we must point out that despite expecting food and energy self-sufficiency, reducing its ecological footprint to 2.4 ha per inhabitant, it will problematically continue to be connected to a global network that completely destroys ecosystems.

While the precise expanse that every inhabitant of a conventional city should have is 5.8 ha for their energy, water, and food needs, we have 2.8 ha available at this moment for the food, water, energy, etc. If growth projections to 2050 materialize, we will be nearly 3 billion more inhabitants on the planet, mostly concentrated in cities. The question is… how will we manage?

30.10 Knowledge Transfer and Quality of Information

Will we be able to transfer knowledge differently to listen and pay attention to our environment?

Yes, Marvin, but with dialogue and cooperation, because knowledge and creativity are different from information and communication. Such skills must allow for constant learning, unlearning and relearning.

The National Institute of Information and Communications Technology in Japan transfers data at a rate of 319 terabits per second over a distance of 3000 km. To give you an idea, that is the equivalent of transferring 80,000 movies per second. On the other hand, a digital divide generates inequality among people according to their access and use of information. 2.9 billion inhabitants on the planet do not even have Internet access; surely a few million more, even if they do have Internet access, discount its use because they do not know how to navigate it or lack the time to do so. Don't you wonder what kind of information is being transmitted at such high speed beyond the reach of nearly half the world's population?

In addition to the knowledge that passes through a virtual network, and that which is formally transmitted through education (an important challenge for society writ large), there is another social knowledge created by the collective that is transmitted through observation and imitation. This knowledge is endowed with great power and wealth. However, given that it does not use the same channels of dissemination, it runs the enormous risk of disappearing, precisely because the other kind of information is transmitted more readily, even if it is of lower quality.

How much consolidated knowledge is out there, created on the basis of doing, experimenting, and feeling in a fundamentally organic way, allowing us to develop as a society, even when such powerful and rapid means of communication did not exist? This knowledge must be preserved and enhanced.

Compared with high-speed data exchange in cities, the communication in rural environments is slower, and sometimes more valuable. Communication is often made with the eyes, touch, smell… and through contact with one's own environment, the territory, and with the beings that inhabit it. Hold onto all the values that industrial and technological advancement are erasing from your knowledge! Because dataism won't help us imagine… Pay attention to those other sources and channels of information so as not to forget what speed leaves by the wayside.

As night falls and Heidi bids farewell to Marvin, she reminds him:

> There is no time to lose! The need is urgent to shatter the current narrative in order to recover and rebuild the planet.

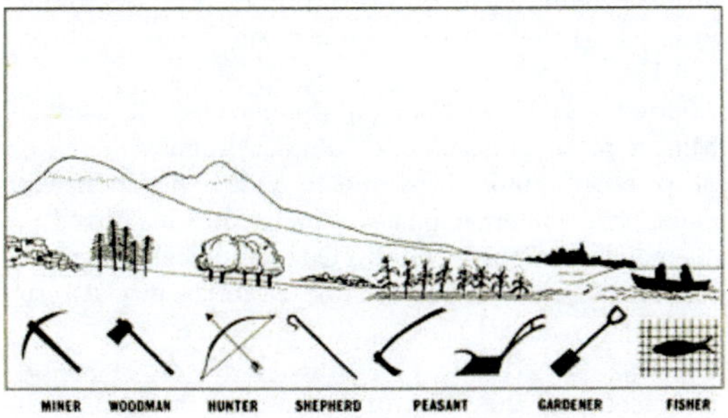

Patrick Geddes, Valley Section 1909

Propose a multi-scale and multi-directional analysis to begin a spatial and temporal transect design process. A rural–urban transect can be the unifying theory and framework for the diverse fields of territorial planning, with the objective of breaking through current divisions. Elaborated as such, the objective is to develop a series of codes to carry out ecological measures at regular intervals as a way of collecting simpler data and not having to study urban footprints. Of course, we must involve scientists in monitoring key indicators of environmental health and use new as well as already existing data with the objective of improving understanding of the environmental problems that affect ecosystems as a consequence of unrestrained urbanization.

Marvin, you would do well to reflect on the ideas that I have transmitted to you, and to put them into practice, because we live in a state of emergency. To do so, we will have to establish a common goal, and draw a map to serve as our guide. I have no doubt that this will make you as happy as I have been in the "Alps."

Bibliography

Ahumada M (2022) El principio de solidaridad territorial y la regionalización. https://www.diarioconstitucional.cl/articulos/el-principio-de-solidaridad-territorial-y-la-regionalizacion/

Christopher A (1965) A city is not a tree. Sustasis Foundation

COL-ECTIU PUNT 6 (2019) Urbanismo feminista, por una transformación radical de los espacios de la vida. Virus Editorial

Corrado M, Mori I (2007) Waste and resources action programme understanding food waste. WRAP

Duran MªÁ (2012) El tiempo no remunerado en la económica global. Fundación BBVA

Echeverri R, Rivero MP (2002) Nueva ruralidad visión del territorio en América Latina y el Caribe. Instituto Interamericano de Cooperación para la Agricultura (IICA)

Fernández M (2021) Tejiendo la Calle. Rua Ediciones

Franke S, Vorstemans A (2011) Compendium for the civic economy. 00 [zero zero] in association with Nesta & Design Concial CABE

Geddes P (1915) Cities in evolution: an introduction to the town planning movement and to the study of civics

Guitton J (2006) Aprender a vivir y a pensar. Encuentro

Junta de Extremadura (2021) Filosofía y Valores de la Accesibilidad universal. https://saludextremadura.ses.es/web/accesibilidad-universal

Junta de Extremadura, Conserjería de Medio Ambiente y Rural, Políticas agrarias y Territorio Natura (2000) An opportunity for everyone. Junta de Extremadura. ISBN 978-84-8107-091-0

Han BC (2022) Infocracy: digitization and the crisis of democracy

Harvey D (2000) Spaces of hope. University of California Press

Huxley A (1944) Time must have a stop. Harper & Brothers, US

Huxley A (1946) Science, liberty and peace. Harper & Brothers, US

Hysler-Rubin N (2011) Patrick Geddes and town planning. A critical view. Routledge

International Telecommunication Union (ITU) (2021) Measuring digital development Facts and figures. ITU Publications. https://www.itu.int/en/ITU-D/Statistics/Documents/facts/FactsFigures2021.pdf

Latour A (2003) Louis I. Kahn. Escritos, Conferencias y Entrevistas. El Escorial

Lefebvre H (1968) The right to the city. Editions Anthropos

Lucas Verdú P, Lucas Murillo De La Cueva P (1998) La solidaridad interterritorial. In: Alzaga Villaamil Ó (dir) Comentarios a la constitución española de 1978. Edersa, Madrid

Mattei U (2013) Bienes comunes. Un manifiesto. Trotta

Meadows DH, Meadows DL, Randers L, Behrens W (1972) The limits to growth. Report to the club of Rome on the predicament of humanity. Universe Publisher

Moro T (1515) Utopía. Ariel 2016

Mumford L (1961) The city in history. Harcourt, Brace & World

Puttnam BJ, Luís RS, Rademacher G, Awaji Y, Furukawa H (2021) 319 Tb/s transmission over 3001 km with S, C and L band signals over > 120 nm bandwidth in 125 μm wide 4-core fiber. In: Optical fiber communications conference and exhibition

Riecchmann J (2019, 2020) Proyecto Medeas "¿Somos demasiado? Reflexiones sobre la cuestión demográfica". Ensayo del númerto.148 de la revista Papeles de Relaciones Ecosociales y Cambio Global https://medeas.eu/

Smith A, Watkiss P, Tweddle G, Campbell A, Mackinnon (2005) The validity of food miles as an indicator of sustainable development. Final report for DEFRA. AEA Technology Environment

Steel C (2020a) Hungry city: how food shapes our lives. Capitan Swing

Steel C (2020b) Sitopia: how food can save the world. Capitan Swing

Thoreau HD (1854) Walden or life in the woods. Ticknor and Fields Boston

Vaz Ferreira C (1953) Sobre la propiedad de la Tierra. Ministerio de Educación y Cultura, Uruguay

Vidal A et al (2003) Movimiento vida independiente, experiencias internacionales. Fundación Luis Vives

Wackernagel M, Rees W (1996) Our ecological footprint: reducing human impact on the earth. New Catalyst Books

Winstanley G (1649) The true levellers standard advanced: or, the state of community opened, and presented to the sons of men

31 The Capillary Boundary: The Critical Redevelopment of the Suburban Area of China's Megacities

Keqing Tang

Abstract

How to maintain vitality in the context of slowing economic development and increasing global uncertainty has become a realistic issue for China's megacities. The urgency of environmental protection has forced the state to reform its planning system to prohibit developers from encroaching on ecological and agricultural space. Under a series of state restrictions on real estate, the traditional means of developers to urbanize large areas of suburban land can no longer balance the cost. The adjustment of the industrial structure and the profound uncertainty caused by the COVID-19 forced the government to constantly explore new urban drives. After decades of rapid development, part of the suburbs surrounded by urban fragments stopped shrinking and formed a unique urban form, which has the dual characteristics of grid and diversification. Many critical practices have taken place on the urban fringes, which have responded to the problems of land use, ecology, fairness, locality, sustainability, and stakeholder participation in the new era. These practices contain a number of strategies that differ from previous ones, and together they constitute the "capillary boundary" effect of the suburbs of megacities: Large tracts of land are divided into micro-plots, centralized frameworks give way to derivative structures, and concrete guidelines give way to ambiguous interfaces. Through typological innovation, architects keep critical in complex, uncertain and changing contexts, helping urban suburbs achieve better ecological quality while slowing urban sprawl and increasing the evolving dynamics of megacities over time.

Keywords

Capillary boundary · Critical redevelopment · Megacities

31.1 Introduction

After 40 years of rapid development through reform and opening up, China's urbanization rate has skyrocketed from 20% to more than 60%, forming a series of megacities with a population of more than 10 million. After the 1990s, land finance stimulated the expansion of megacities. The border between cities and suburbs is moving rapidly. In the process of urbanization of suburbs, from residential areas and high-tech zones to industrial parks, tourist towns and TOD

K. Tang (✉)
Tongji Architectural Design (Group) Co., Ltd., Shanghai, China
e-mail: 30910971@qq.com

S. Hilal et al. (eds.), *Design for Partnerships for Change*, Sustainable Development Goals Series,
https://doi.org/10.1007/978-3-031-36993-3_31

complexes, China's megacities have witnessed a huge transformation from the literal urbanization of suburban land to the urbanization of suburban population. After the rapid expansion, slowing expansion and even stagnation of China's megacities, the relationship between urban fragmentation and suburbs becomes clear.

As economic growth tends to slow down, the significance of land for the development of China's megacities has changed profoundly. In order to achieve the GDP growth target, rural land was once easily converted into urban land by local governments, and now it is strictly protected. The motivation of governments and developers to acquire land around cities has also changed. Once in China, urbanization meant a change in the physical properties of rural land. Farmers turns into "migrant workers." Today, the goal of urbanization is not to change the nature of land, but to focus more on social progress: Farmers do not just transform into migrant workers, but settle as city dwellers.

At the same time, the planning methodology has changed. In the early stage of rapid development, "planning before implementation" was the principle of rapid expansion of China's megacities, with the aim of avoiding disorderly expansion. The content of planning involves land property, development intensity, population, industrial guidance, functions, detailed guidelines, architectural forms, etc. The choice of planning strategies depends on the location of the planned area in the urban framework (Jenks 1996).

In a low-growth environment, land development becomes cautious. In the vicinity of megacities, the new projects show a similar "capillary" characteristics. Grandiose structures give way to derivative structures, short-term sales give way to long-term holding, centralized control gives way to decentralized diversity, and certainty gives way to uncertainty. The demand for large-scale migrant workers gives way to a cautious attraction to new urbanites to achieve a work–life balance. The urban form on the outskirts of the city is taking an interesting transition. In speaking of the factors that determine the success of land use, the importance of the framework gradually gives way to the typological innovation. New strategies are being invented, and architects have the opportunity to participate in the renewal, conservation and decision-making of urban and rural land with a more active role (Fig. 31.1).

31.2 Urban Fragments and Suburbs

31.2.1 Suburbs as the Media Among Urban Fragments

In the traditional Chinese natural economic society, cities have clear geometric boundaries. There is a priori understanding of the division and management of land. In the topology of traditional Chinese cities, land is framed and divided with rational grids that determine

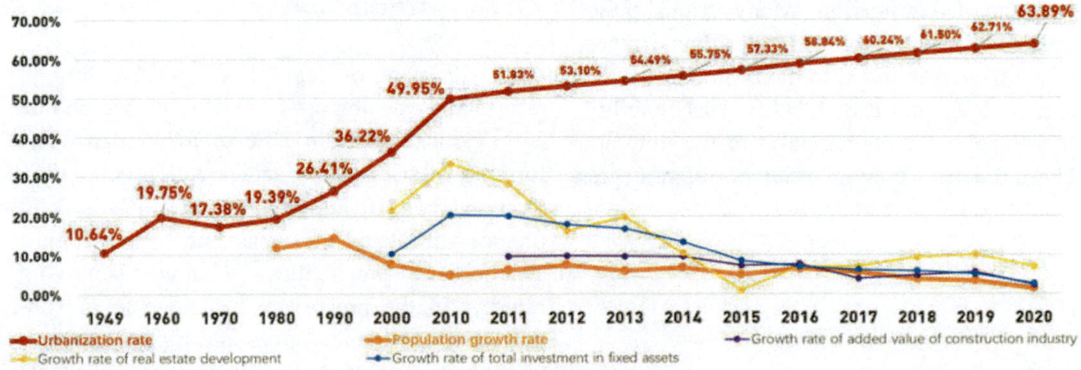

Fig. 31.1 Context of the urbanization of China. *Data source* National Institute of Statistics

hierarchy and resource allocation. In this framework, different contents such as temples, residences, commerce, and agriculture are realized. Settlements of various sizes use such a framework system as a basic model. Architects, on the other hand, as engineers, inherited the macroscopic ritual order, linking it with the building regulations and materials to form a modular system from the city to the human body. The architect is not responsible for interpreting or modifying the system (Li 2021).

Modern China's megacities are formed based on the continuous expansion and extrusion of gridded urban fragments. Agricultural spaces are gradually excluded from the grid of urban fragments. Urban fragments are close to each other, while large tracts of natural and agricultural land between them gradually shrink. Some fragments merge with each other, while others become urban voids due to irreconcilable problems (Xu and Semsroth 2013).

In the evolution of China's megacities, distributed villages and towns gradually expand, integrate, and absorb each other to reframe a new structure. This phenomenon is also reflected in the naming of many urban communities that maintain the names of traditional Chinese settlements, such as towns, villages, settlements, and bays. Formally, sub-districts are traditional settlements within the city, and towns are settlements outside the city. In modern China's megacities, as administrative divisions at the same level, sub-districts and towns are parallel and adjacent. As an administrative division with a scale of tens of square c, sub-districts and towns can be regarded as the primary fragments of China's megacities. At the same time, suburbs (different from outer suburbs) become the contact space among urban fragments. With the advancement of urbanization, suburban land has been shrinking and gradually redefined as urban territory through expropriation, urbanization of the rural population, and homestead retreat, and so on (Figs. 31.2, 31.3, and 31.4).

31.2.2 Spatial Characteristics of the Suburb

The outskirts of modern megacities also have their own "ecology" that should be effectively protected.

The development approach of suburban space shapes its uniqueness. At the edge of the traditional urban fragment, the suburban area represents the end of the expansion of urban space. There may be natural "dividers" in suburban area, such as natural elements (rivers, farmland),

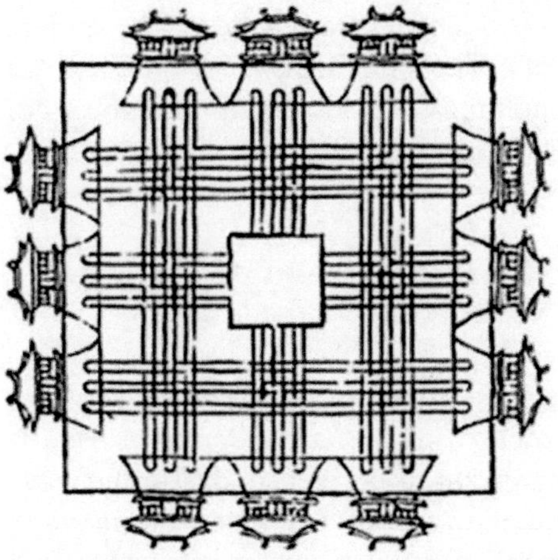

The model of traditional Chinese capital

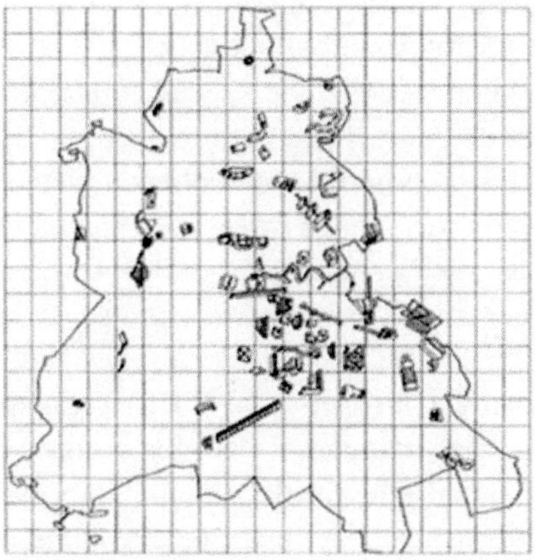

The concept of urban islands of post-war Berlin

Fig. 31.2 Model of traditional Chinese capital (left) and the diagram of urban islands of post-war Berlin (right) (Ungars 1978). *Source* Left: The Rite of Zhou—The Artificers' Record; Right: Ungars, O., M. Cities within the City, Lotus 19, 1978, 82–87

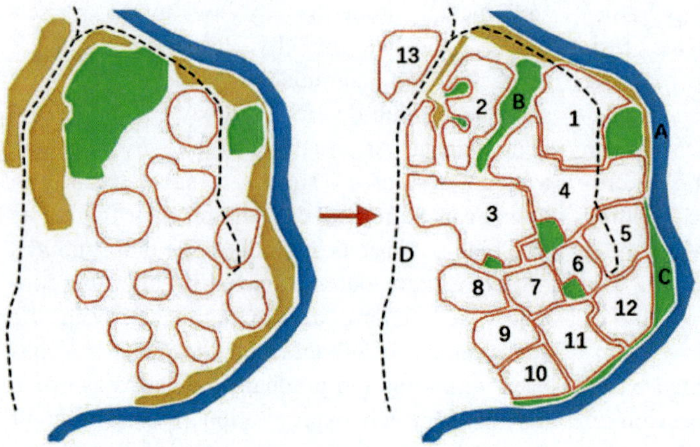

Fig. 31.3 In north Shanghai, the evolution of sub-districts or towns as urban fragments leads to the shrinking of the suburbs as the media space among urban fragments. *Source* Author

infrastructures (railways, expressways), and industrial relics (factories, warehouses). These "dividers" maintain their spatial characteristics in the process of "urbanization." Indeed, in the early stages of rapid urbanization, different administrative divisions had to compromise with each other and work together in a single and complex urban framework, so the integration of urban fragments was so thorough that the boundaries between many administrative divisions disappeared. However, today the space between urban fragments, as a void left by the super-urbanization process, has become a particular urban zone that is not assimilated by surrounding urban territories and is characterized by suburban identity.

Based on a study of the special suburban area among urban fragments in Birmingham, Michael. W. Hopkins proposed a series of ecological characteristics for this unique urban void named urban fringe belts, such as having a significant amount of urban green space, no important protected areas, less disturbance, more plant species, and less rural characteristics (Michael et al 2012). This study illustrates the uniqueness of the rural space between urban fragments in the process of expansion, stagnation, and redevelopment of megacities. This uniqueness belongs neither to the traditional city nor to the traditional countryside.

In contrast, China has similarities, but also its own stories: The similarity is that the void space between urban fragments is often accompanied by special resources (natural ecology, agricultural land, factories, infrastructure, large institutions, etc.). There are opportunities for these resources to be properly protected in the current context. However, even void spaces in China's megacities are still framed based on traditional Chinese urban grid and land ownership. Even in the context of slowing urban sprawl, the large population base still drives the redevelopment of cities. On the boundary of the urban void, there is still land reframing for the purpose of active commercial development. This characteristic of the current suburbs of China's megacities may raise a series of new questions and answers in the process of urbanization in modern China.

31.3 The Intervention of Multiple Resources

In megacities, some special suburbs serve as "blank zones" in the urban territory have a unique attraction to developers due to their less constraints, low cost of demolition and renewal, strong spatial continuity, close connection with urban resources, and urgent need for change. The attraction of suburbs can be understood by the following three phenomena.

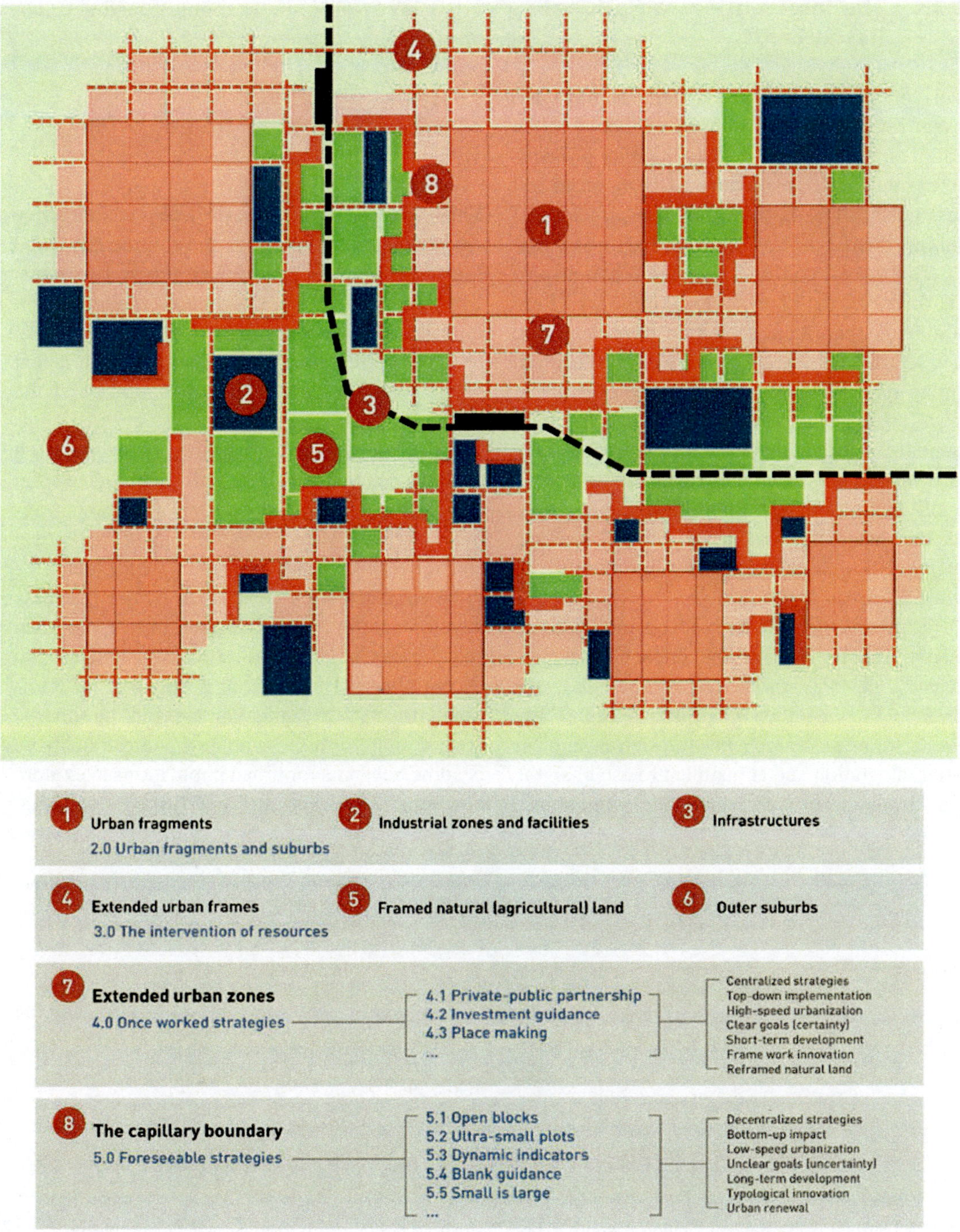

Fig. 31.4 Land composition map of China's megacities and the structure of this paper. *Source* Author

31.3.1 The Intervention of National Resources

As a continuous fringe space, the suburbs are closely intertwined with the urban area. It is an ideal open place for national or intercity projects. For example, the positioning of high-speed rail lines and stations is based on the principle of minimal interference to the original urban fabrics. Ten years ago, high-speed rail station projects tended to choose locations far from traditional urban areas to obtain open land and fewer restrictions; in recent years, high-speed rail stations tend to be anchored at the outskirt of central urban areas, aiming to achieve the integration of stations and local urban spaces in a short period. For example, Suzhou North Railway Station, Chengdu South Railway Station, and Baoshan Railway Station are all seeking integrated development in the suburbs close to traditional urban areas.

The involvement of high-speed rail has also raised questions such as the impact of megastructures on the urban landscape and the increased fragmentation of urban spaces. These issues require a comprehensive urban design for spatial integration and adjustment to establish the sophisticated framework and code, which also inevitably weaken the freedom of land development to some extent (Fig. 31.5).

31.3.2 The Intervention of Urban Resources

The suburban area also provides an attractive environment for urban development. Good logistics infrastructure, fewer urban contextual constraints, and a more open policy environment offer a good place for innovative industries and resource spillovers (Li and Sun 2017). In recent years, with the control of urban scale and the reform of the urban planning system, the definition of ecological and agricultural space has taken precedence over urban territories. Urban land that can be used becomes the remained and fragmented space after the definition of ecological and agricultural zone. As a result, megacities have to be compact, even contracted. Unlike the mass development in the past, the reframe of suburban space today tends to be more marginalized and decentralized. Therefore, as a "divider" of traditional urban space, industrial areas in the suburbs are constantly infused with new content and have become hot spots for urban populations to explore new jobs and lifestyles (Fig. 31.6).

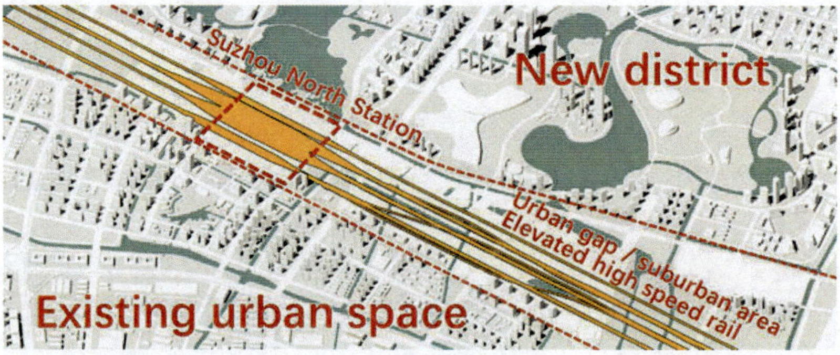

Fig. 31.5 Suzhou North Railway Station is going to be a high-speed rail hub with regional radiation. High-speed rail lines connect to cities through the gap area between urban fragments, trying to minimize the impact on existing urban fabrics while inevitably exacerbating the barriers between urban fragments. It poses a challenge for urban design for this particular area. *Source* Author

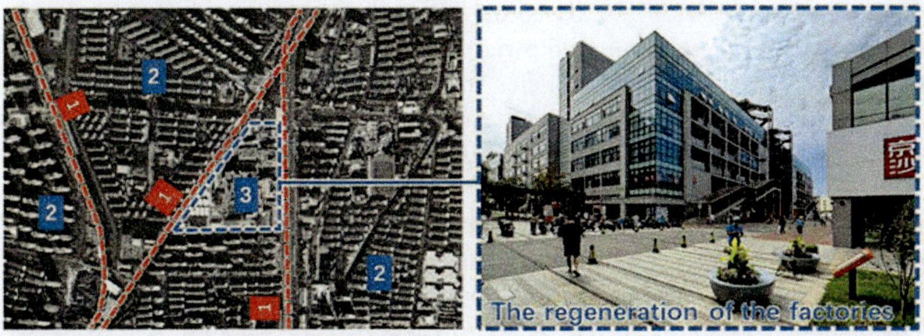

Fig. 31.6 Xinyefang, which once served as an industrial zone between urban fragments along with the railway, is the epitome of Shanghai's suburban development. With the expansion of the surrounding metropolitan area, Xinyefang has been fully integrated with the surrounding residential areas and has been renewed to become a community center for the creative park and surrounding communities. *Source* Author

31.3.3 The Intervention of Suburban Resources

As the urban population continues to spill over, some suburbs have developed a new vitality different from traditional urban areas. With the implementation of the national land spatial plan, the important top-down control, the strict protection of ecological space and arable land has made it difficult to add new construction land in the suburbs. The revitalization of farmer-owned land protected by law has also made it a significant feature of suburban spaces, a unique "rural within the city" landscape. Pushed by the rural local resources, the suburban area is actively or passively exploring a new development model that is different from urban area and embodies ecological and agricultural features (Fig. 31.7).

31.4 Once Worked Strategies

Over the past few decades, although undefined suburban areas are easy to be organized and planed as a whole, there are risks of insufficient sustained investment and management. The government has introduced a series of successful countermeasures centralized framed and controlled by large developers or platforms. From a planning and urban design perspective, it can be summarized into three basic strategies.

Fig. 31.7 In the northeastern suburbs of Chengdu, the giant panda breeding base coordinates farmer-owned and state-owned land with the support of the local government to protect farmland and increase local employment while improving the quality of the park. Images show that one of the panda pavilions avoids farmland and the natural environment through a unique curved form. *Source* Author

31.4.1 Public–Private Partnership: Centralized Planning Plus Phased Implementation

The government works with companies to reframe large suburban areas with systematic master plan and urban design, and enterprises implement complete control, continuous investment, and phased implementation. This approach will help establish sustainable control over the area, improve the efficiency of land acquisition, balance overall interests, and avoid fragmentation. However, in the face of today's long suburban development cycle, this "framework-first" model will reduce the initiative of individual plots and increase investment risk. In fact, it is difficult to obtain a comprehensive and measurable understanding in the initial stages of a project, and post-project evaluations also need to be conducted over a longer period of time. The longer development cycle increases the uncertainty of the suburban space, and the construction conditions of the later developed properties may change due to the developed ones, which will harm the fairness of subsequent developments. Therefore, in development projects with large time spans, it is inevitable to constantly modify the original urban design. However, the current planning and project approval reporting mechanism does not encourage frequent adjustments, as it will undoubtedly increase top-down management costs.

31.4.2 Investment Guidance: Top-Down Industrial Planning

Another strategy to avoid fragmentation and short-sightedness in suburban development is top-level planning. In the feasibility study stage of the regional development project, by determining the appropriate industries, investment objects and spatial characteristics, a detailed industrial plan is established, and each plot is functionally fixed, thus forming a reference basis for urban design. In an era of rapid development, based on this "blueprint" methodology, detailed industrial planning can focus on goal-oriented policies in a relatively short period of time to quickly seize opportunities. This strategy standardizes the development mode of the plot and avoids disorderly urban sprawl. For an urban space that is full of uncertainty and requires continuous investment, this approach may reduce the enthusiasm and risk-taking spirit of entrepreneurs to some extent.

31.4.3 Place Making: Iconic Events

In suburban areas, introducing famous brand and events, such as World Expos, major sporting events, and famous parks, was once a catalyst for creating places of collective memory. As cities expand and integrate, sites may eventually integrate into urban spaces. As a "booster" for development, this "placemaking" strategy can create a fast, robust image that is conducive to sustainable operations. Today, the marginal effect of introducing new events has diminished. The fragmented suburban land and the increase in personnel costs may lead to a continuous increase in investment and operating risk (Tang and Pan 2008).

31.5 Foreseeable Strategies: The "Capillary Boundary"

Under pressure to protect agricultural or ecological spaces in the future, the usable land in the suburbs will trend toward fragmentation and miniaturization. China's megacities could face stagnation or slow contraction. Agricultural, ecological, farmer-owned, and state-owned land may remain intertwined for a long time (Fig. 31.8).

With the rising cost of renewal in the central urban area and a series of rural problems to be solved, suburban areas still have urgency and potential for development (Zhou 2022).

In addition to land ownership and typology, which are the foundations of modern urban forms, another important foundation of China's

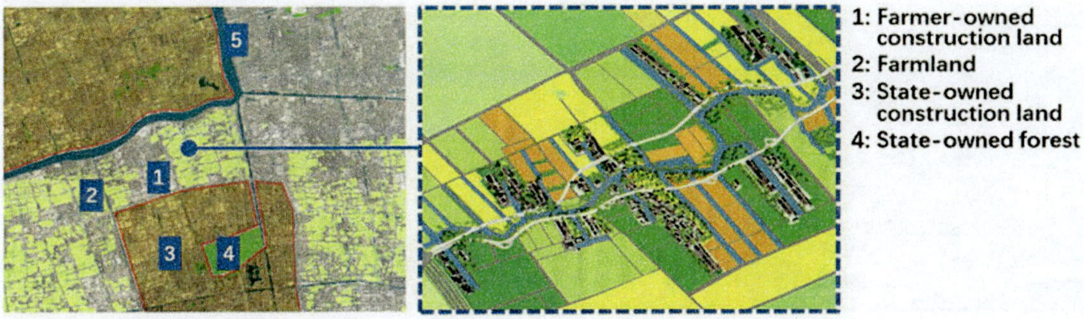

Fig. 31.8 Intertwining of state-owned construction land, farmer-owned construction land, farmland, and state-owned parks in Fengxian District, Shanghai. *Source* Author

modern megacities is a priori framework system with the characteristics of traditional Chinese culture (whether in ancient or modern China, land is not privately owned). At a time when urban expansion slows, subtle changes also take place at the fringe of the traditional urban framework. Some fragmented, de-framed typological features emerge and influence the logic of land division (Zhou and Jin 2021). This characteristic consists of a series of development and design strategies as a whole, which in this paper is called "capillary boundaries." A growing number of meaningful cases support the following five predictable strategies.

31.5.1 Open Blocks: Increase the Public Interface and Do Not Imitate the Urban Space

With the slowing down of urbanization, the evolution of suburban areas has shown a diversified trend. Suburban spaces may be in a state between rural and urban areas for a long time, while ecological and agricultural spaces are unique resources here. This phenomenon will lead to long-term irregularity and imbalance of buildable land. Natural, urban, rural, municipal, or national infrastructure has a lasting impact on land development. Therefore, maintaining the spatial permeability of blocks and increasing the accessibility of each plot and surrounding elements, rather than deliberately pursuing the integrity of urban space, may provide another possible way to improve the fairness of each property. As a result, the traditional urban block featured by "street wall" may give way to the open block, a de-framed spatial form. It provides the various projects the opportunities to show the individuality, independence, and diversity. For the development of plots in the undefined suburban area, openness is supposed to be the more important indicator than the near-line (Fig. 31.9).

31.5.2 Ultra-small Plots: The Opportunity for Fairness, Self-Organization, Universality, and Diversity

The "ultra-small plot strategy," which is closely related to the "open block strategy," provides good adaptability and avoids massive undefined open spaces due to insufficient near-line rate. The "ultra-small plot" effectively lowers the development threshold of each single project, so that the overall development can spontaneously adapt to the construction conditions of different historical periods. The architectural design of each plot can also be adjusted over time. At the same time, the ultra-small plots strategy is conducive to breaking down large development projects into multiple small projects, providing a stage for different design ideas (Fig. 31.10).

Fig. 31.9 Nanjing Mufu Hi-tech Park is located in the gap between the urban space and ecological reserves. The urban design developed a system of open blocks to introduce green space into each plot and keep the dialogue among buildings, courtyards, and the natural land. *Source* Author

Fig. 31.10 In the project "Jiading Mini-Block," each block is defined by the 41.2 m * 41.2 m grid. Architect: Zhang Yonghe. *Source*: Left: Author; Right: Tian Fangfang

The ultra-small plot strategy also poses challenges for coordination between adjacent plots. A series of problems, such as the connection between the ground and underground space, the division of ownership, the coordination of the construction period, the avoidance of mutual interference, and the control of urban interface, make it difficult to predict and control the urban form with traditional urban design methods. Therefore, compared with a detailed masterplan, the development rules, and architectural language generally followed by each plot are particularly fundamental, aiming to provide fair opportunities and promote self-organization and coordination between properties. In the process of urbanization in the past, these factors were often underestimated, and the general layout, the masterplan, was overemphasized.

31.5.3 Dynamic Indicators: Incentives by Plot

Strict and static indicator control of suburban spaces is not the best way to deal with long-term development risks. Dynamic indicator control helps to cope with the changes in policy, market, and industrial development that suburban spaces may face in the process of development. To make the urban design adaptable and fair in the long term and avoid repeated planning modifications, dynamic indicators can be used as incentives or levers to effectively balance the unpredictable negative impacts generated by previous development activities. Private-owned public spaces, flexible green space rates, and flexible coverage are all potential urban design tools. As more and more projects may face a development cycle of more than ten years, it is necessary to introduce a standardized and dynamic evaluation system to scientifically divide the scope of rights and responsibilities. At the same time, it is important to avoid abuse and simplification of power in the implementation process (Fig. 31.11).

31.5.4 Blank Guidance: Stay Open for Additional Possibilities

In the history of modern China's urbanization, industrial planning has played a supporting role in urban design and land-use planning, aiming to avoid the imbalance and short-sightedness of land finance. With the gradual maturity of the global industrial chain and the relative stability of each city's industrial advantages and models, it is becoming increasingly challenging to propose the effective industrial planning for the fragmented suburban land. At the same time, newly emerging industries and innovative ideas often deviate from the traditional scope. In fact, in fragmented suburbs, a controlled blank space for industrial planning is unlikely to lead to disorderly urban sprawl, but may stimulate new content of the unknown. Therefore, the establishment of industry blocklists and safelists to avoid the formation of the unfair tilt is conducive to stimulating the vitality against the uncertainty. Indeed, the goal of urban design is not to limit the content and function of each plot. Still, more importantly, it is necessary to provide future-oriented ingress and exit mechanisms for additional possibilities.

31.5.5 Small is Large: The Irreplaceable Social Effect of Individual Architects and Their Works

Suburban areas provide the stage for clustered or decentralized small projects. Companies known for their flat management structures and quick judgment show their advantages in suburban development compared to developers with over-sophisticated institutions. Products based on rigid market research aimed at all groups of people have also given way to customized products with unique characteristics. The shift from mass production to personalized customization has attracted more individual architects or small design studios to explore new urban concepts and values on smaller suburban sites, thus making meaningful explorations and accumulations for future urban renewal. The suburban area has the potential to be the exhibition site for open innovation and individual values of the megacity (Fig. 31.12).

31.6 Conclusion: The Urban Form of the Suburban Area

Before China's urbanization rate reached 60%, the rapid integration of suburbs and cities had taken place under the government's guidance by

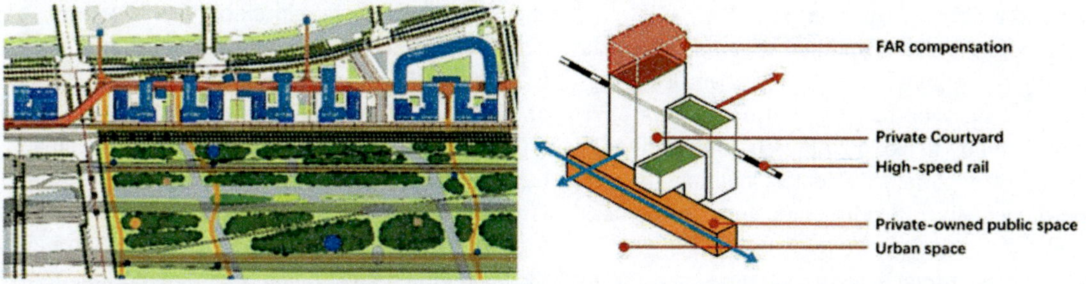

Fig. 31.11 In the urban design competition of Suzhou North Station, designers developed a typological solution for the rail-front plots to protect the properties from rail noise and enhance the publicity and continuity between blocks. By introducing the dynamic FAR policy, the guideline encourages each property to provide more private-owned public space in the opposite direction of the railway. *Source* Author

Fig. 31.12 In the suburb of east Shanghai, along Longdong Elevated Highway, Zhangjiang Riverfront Harbor, the famous integrated circuit industrial park in Shanghai, invited multiple groups of architects to introduce diversity, vitality, and new ideas to the park. Architects: Atelier Liu Yuyang and Atelier Archmixing. *Source* Left: Author; Right: Chen Hao

learning from the experience and cases of western urban planning. In the process of urbanization from 60 to 80%, suburban development may present a different picture. In this stage, many endogenous problems of China's megacities are gradually being exposed. In 2014, the state proposed a new urbanization goal aimed at changing the phenomenon of "separation of urban and rural areas" through the "urban–rural integration" strategy. Rather than trying to turn suburbs into traditional urban spaces, the suburbs are incorporated into the overall urban planning as a unique urban form, giving full play to the advantages of suburban industries, landscapes, and populations. The form of China's megacities is facing a period of relative stability after rapid expansion, and suburban area may remain fragmented for a long time. Unlike past urbanization stories in China, the suburbs of future megacities gradually lose the motivation to be assimilated by the urban fabric, but instead present and maintain their own characteristics.

Recognizing and adapting to fragmentation is a constructive attitude. Architects, urban designers, managers, decision-makers, and developers have the opportunity to explore different tools and build a series of exploratory-oriented, continuous, and environmentally friendly strategies (Koolhaas 1989).

Based on urban design practices, we try to predict some anomaly strategies: "Open blocks" means improving the freedom and fairness; "ultra-small plots" means lower innovation costs; "dynamic indicators" implies policy flexibility; "blank guidance" refers to friendliness to the unknown; "small is large" implies the importance of personal and diversifying thinking. In the face of suburban spaces that are no longer changing rapidly, the development model has gradually shifted from systematic and detailed control to typological intervention. This intervention model takes projects located in the suburbs of megacities as part of the capillary boundary, which is a unique urban form of the suburbs.

The "capillary boundary" is a phenomenon in the late period of rapid urban development in China, which can be understood as another relationship between settlements and natural (agricultural) land in China's megacities. This relationship is based on the slowdown in development growth, the shortage of land resources, long-term development, strict cost control, and full consideration of uncertainty. "Capillary boundaries" provide an attempt to get out of the box of a priori grid. In this attempt, there is a certain ambiguity between the definition of natural land and settlements, contributing to the unique urban form of the suburbs of the city. Here, critical redevelopment and typological innovation is more valuable than hierarchical frameworks (Alexander 1965). Unprecedented, more and more Chinese architects together are influencing the value of cities through their individual creations in this new period, aiming to make it possible for megacities to maintain the vitality of their urban boundaries while no longer claiming land from the nature.

References

Alexander C (1965) A city is not a tree. Archit Forum 122 (58–62):58–61

Jenks M (1996) Compact city: a sustainable urban form? Routledge

Koolhaas R (1989) Towards the contemporary city. Design Works Rev 17(winter)

Li X (2021) From the city to the construction: a new dimension in the architecture of the city 2021(10): 17–22

Li Z, Sun M (2017) An introduction to the study on technical system of community-oriented regeneration of "downtown factories" in the Yangtze River Delta. Archit J 8:82–88

Michael IW et al (2012) The ecological significance of urban fringe belts. Urban Morphol 16(1):41–54

Tang K, Pan J (2008) City: mega events and event cities. Time Archit 4:6–10

Ungars OM (1978) Cities within the city. Lotus 19:82–87

Xu K, Semsroth K (2013) Fall and revitalization of "publicness": Chinese urban space in comparison with European ones. Urban Plann Forum 3:61–69

Zhou X (2022) Fengxian New City practice for the development goal of an "Independent and Comprehensive Node City." Shanghai Urban Plann Rev 1:134–140

Zhou Y, Jin Y (2021) Spatial planning of city and county level green space system: Fengxian District, Shanghai 2021(4):36–43

32 Is Sea the New "Land"? Re-thinking Land Use and Re-framing Sustainable Urban Development in the Context of Marine Urbanization

Luis Carlos Mestrinho

Abstract

The introduction of built infrastructure into the marine environment has long followed human occupation in coastal or estuarine areas. However, the projection of an increase in its use, the artificialization of the coasts and marine areas and the changes brought to the marine ecosystem suggest that interventions in the marine environment can no longer be considered one-off activities, but rather a phenomenon called ocean sprawl. This argumentative essay posits that the idea of land use is outdated, and we must consider the sea also becoming the new "land". It suggests that professionals in the built environment should open their minds to reflect on how to manage this reality of "land use conversion" in a different environment, the sea. This framework has implications for how we deal with this coupled urban-marine system. Like its counterpart on land, marine urban sprawl also brings a rupture in the physical environment and the functioning of ecosystems, commonly with a loss for marine biodiversity. It is advocated that interventions in aquatic ecosystems require a reset in our thinking and consider the needs that may be at stake on both land and sea. It is also argued that the concept of sustainable urban development has to be reframed to account for the particularities of this territory, which may differ from the characteristics found on land. The management of artificial light at night is highlighted as an example of these particularities. Finally, the role of architects in reconciling the built environment and aquatic ecosystems is briefly explored.

Keywords

Marine urbanization · Extended urbanization · Ocean sprawl · Sustainable urban development

32.1 Introduction

The introduction of built structures[1] into the marine environment adjacent to cities is commonplace in history and records of construction in this space to support maritime traffic and protect shallow coasts date back to at least 2000 BC (Bugnot et al. 2021). However, the increase in the use of the marine environment to support urban activities (Evans et al. 2019), the strong

L. C. Mestrinho (✉)
Escola de Arquitetura, Arte e Design/Lab2PT,
University of Minho, Braga, Portugal
e-mail: luismestrinho@hotmail.com

[1] The term "built infrastructure" is here defined as "all products of human design and construction that are intended to deliver services in support of human quality of life" (Wilbanks et al. 2012).

S. Hilal et al. (eds.), *Design for Partnerships for Change*, Sustainable Development Goals Series,
https://doi.org/10.1007/978-3-031-36993-3_32

artificialization and shielding of the coasts (Floerl et al. 2021) and the changes brought to the marine ecosystem because of the introduction of marine infrastructure (Bulleri and Chapman 2010) suggest that the pattern, scale and impact of such interventions has reached a new critical level, a distinguishing phenomenon known as "ocean sprawl" (Duarte et al. 2013; Firth et al. 2016).

When defining the process of urbanization Fadigas (2010) remarks:

> (urbanization) is a process of land use transformation that occurs whenever agricultural and forestry use is replaced by another, predominantly urban use. It is a cultural and economic process, an expression of social dynamics and not just a building process.

Although the definition might be correct, in the contemporary world, the field of incidence of this process is no longer restricted to the aforementioned areas. The conversion of land use, which until then took place commonly in agricultural and forestry environments, nowadays also occurs in the marine environment where urbanization advances, no longer limited to the ground as usual, but also over the intertidal zone and the nearshore estuarine and marine waters (Bulleri and Chapman 2010).

32.2 Is "Sea" the New "Land"? Marine Urbanization as a Form of Extended Urbanization

The ubiquity of hard structures to serve urban purposes and the diversity of uses frequently noticed in waters adjacent to urban areas, as well as the increase in land reclamation and artificial islands development t (Fig. 32.1), may lead us to question if the sea is not becoming the new land.

We could see before that the process is not just about building, so the answer might not be straightforward. What can be noticed is that, increasingly, the urban fabric, the manifestation of dominance of the city over the rural area (Lefevbre 1970), is extrapolating its power over marine areas in a sort of extended urbanization of the sea. As already noted by Couling (2020), the ocean has become a site of spatial and environmental convergence, a kind of "hinterland" for urbanized territories.

Thus, professionals of the built environment, such as urbanists and architects, can no longer take the concept of land use as something that happens strictly on land. They should reflect on how to manage this reality of conversion in a very different domain, the sea. Indeed, Brenner and Schimd (2012) advocate for the acknowledgement of new urbanization processes that are reshaping the world. It seems to be what is occurring in the interface between land and sea in several parts of the world. On the one hand, the proliferation of urban structures seaward is expanding this limit, on the other, the artificialization and hardening of the coast to establish a border against sea level rise fixes artificially a limit where impermanence was the rule. As a result, the social-ecological balance of this relationship may be altered, especially in territories where the border between land and sea is tenuous, as in the mangroves, or where the influence of the tides goes far beyond the limits of the coastline, as in the case of deltas and estuaries.

32.3 Re-framing Sustainable Urban Development in the Context of Urbanization on Sea

Bugnot et al. (2021) noted that the growth of urban areas has already been identified as one of the main causes of biodiversity loss in terrestrial ecosystems but posit that urbanization is not just a concern on land, as the pattern of occupation and expansion of coastal cities are fuelling a boom in maritime construction, a process that strongly impacts the marine ecosystem and ecological processes. Therefore, like its counterpart on land, marine urban sprawl, a sea use conversion process, also brings changes to the physical environment and the functioning of ecosystems (Airoldi et al. 2005), including in the ecological connectivity at large scales (Bishop et al. 2017), commonly with a loss for marine biodiversity.

Elmqvist et al. (2013) recall that in the history of urbanism, gradually, the socio-ecological

Fig. 32.1 Satellite overview of the city of Tokyo by Copernicus Sentinel-2 showing land reclamation in Tokyo Bay area

dimension of urbanization has been neglected. The commonly noticed dichotomy between the built environment and natural processes may also be reproducing in the urban-marine interface with implications not only for biodiversity but also affecting the social dimension. For example, the allocation of monofunctional infrastructures such as ports in urban areas often gave rise to the rupture of the people's relationship with the seascape, a separation capable of leading to an emotional disconnect between citizens and aquatic ecosystems as far as this latter becomes "out of sight, out of mind" (European Commission 2021) compromising the appreciation of the multiple opportunities and services offered by urban-marine systems.

Climate change-induced sea level rise and extreme events add an extra layer of complexity. Solutions designed and implemented to tackle such contemporary issues are not exempt from social controversy. For instance, environmental organizations have already expressed concern about the consequences that the artificial island project of Lynetteholm (Copenhagen-DK) may create to the functioning of the whole Baltic Sea environment. For them, the project, promoted as a flood protection and climate change adaptation measure, may obstruct one of three deep channels in the Øresund strait and have regional impacts by changing saltwater flow regime to the Baltic Sea (Coalition Clean Baltic 2021). In Japan, where urban areas were hit by the 2011 tsunami, building new defence structures against extreme events has been continually criticized by locals, despite being designed to protect them. Local communities fear the adverse effects that massive shielding of the coast can bring to themselves (cultural loss) and local economic activities, such as tourism, which depends on the landscape, and fisheries whose fish stocks depend on the cycle of nutrients transported in the flow of the rivers (Hardy 2019).

Therefore, interventions of this type require a reset in our thinking and force us to consider the needs that may be at stake both on land and at sea. It is crucial to recognize the permeability between their limits and manage the different aspects that govern each specific realm. This is why we shall take with precaution the pledge that a marine project is sustainable because "it saves resources on the land". Without looking at the marine world itself, this statement about sustainability may sound incomplete or misleading.

Given the amount of people living in coastal areas, it is unlikely that it will be possible to alleviate the demand for infrastructure and marine resources to meet the multiple needs of stakeholders. In fact, the United Nations estimates an increasing pressure on marine resources and marine biodiversity in the coming years due to the increase in the number of coastal populations (UN 2017; UN-HABITAT 2018). Thus, in ecological terms, the same challenge that arises for the sustainability of urban development on land, namely "to strike the optimal balance between conserving the diversity of nature and advancing human sustainable living (UNEP 1995, p. 923) is also reproduced in the context of marine urban development. However, admitting that they share the same problem does not imply that it can be solved in the same way. Terrestrial processes may differ from processes that occur in marine environments, hence, to achieve sustainable urban development at sea, some adjustments in adaptive solutions may be paramount to fit with the configuration and functioning of the urban-marine ecology.

The issue of artificial light at night—ALAN—illustrates how one same environmental stressor could be managed differently within the land-sea interface. Artificial lighting goes hand in hand with urbanization, and it has been argued that growth in artificial light worldwide represents a potentially perturbation to the natural cycles of light and darkness (Gaston et al. 2013). This brings implications for living species which, through evolution, have adapted their physical, behaviour and interaction profiles according to natural light availability and characteristics (Longcore and Rich 2004 cited in Tamir et al. 2017). While the effects of light pollution on terrestrial ecosystems are increasingly established, relatively less attention is paid to the marine habitats (Depledge et al. 2010 cited in Davies et al. 2014; Longcore and Rich 2004), albeit the latter is also affected.

According to the literature, various mitigation measures can prevent or limit ALAN and simultaneously fulfil people's lighting needs. Some of them, such as shielding luminaires and limiting the extent of the illuminated area, aim to regulate light spillage, while others aim to limit light intensity and manipulate light spectrum (Falchi et al. 2011). All solutions can bring benefits to species (both humans and non-humans) and the environment. Nonetheless, because underwater lightscapes are different from terrestrial ones, lighting spectrum can have greater significance in the former lightscapes becoming especially important when discussing ALAN. Underwater, spectral information facilitates various colour-guided behaviours (Tidau et al. 2020) and variation of wavelengths drives a range of photobiological responses. Therefore, spectral sensitivity is seen as crucial to understanding and mitigating ALAN effects (Tidau et al. 2020). Seawater properties control to what extent the different lengths of light penetrate the water column and attenuate long wavelengths (red light at the end of the spectrum) a phenomenon called spectral narrowing (Fig. 32.2). To put in simply, short wavelengths (blue) are

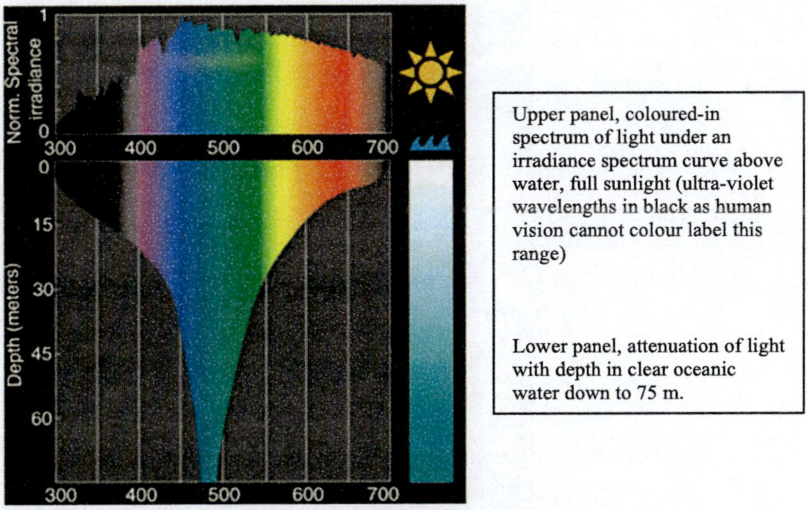

Graph 32.1 The legend stands as Representation of light underwater adapted by Marshall (2017) © Justin Marshall

minimal attenuated and goes deeper towards the bottom (Graph 32.1).

As a result, it has been noticed that at sea not only cycle, intensity and spectra may influence species (both on land and at sea) but also that blue light, the same kind strongly emitted by modern luminaries such as Metal Halide and white LED lighting (Fig. 32.2), is potentially more derelict to marine biodiversity. Blue light is likely to reach deeper levels of the water column and interact more easily with living organisms[2] compared with light composed of longer wavelengths (Tamir et al. 2017).

For this reason, besides spatial coverage, availability and intensity, mitigation of ALAN in marine areas may require special attention to the management of the light spectrum emitted by lighting devices. As suggested by Dominoni and Nelson (2018, p. 5), "caution should be taken when choosing a particular colour for a new light installation".

32.4 What Architects Have to Do with All This? Design Strategies for the Reconciliation of the Built Environment and Aquatic Ecosystems

So far, it has been explained why urbanization should no longer be considered a process that occurs exclusively on land, but also at sea, and the challenges that this phenomenon may imply for urban sustainability. We can ask ourselves what architects have to do with all this and what is the role of architecture in promoting the reconciliation of the built environment with aquatic ecosystems.

It is worth remembering the appeal made by Allen (1999) when he defends the rescue of the role of the architect in different fields, such as local ecology, which was lost with the advent of professional specializations. The same author suggests a reconceptualization of architecture that is less about how things look and more about what they can accomplish (Allen 1999, p. 52).

Now, it is known that, in addition to the pollutants brought by the process of urbanization on land, the introduction of infrastructure into the marine environment causes changes in marine habitat and modification of ecosystem functions and services. To change in this scenario, it has

[2] Technically, "interaction between the intensity and spectral composition of artificial light and the adaptation of an organism's eyes will affect whether visual perception is enhanced, disrupted or unaffected by light pollution" (Gaston et al. 2013, p.921).

Fig. 32.2 ALAN pollution at Victoria Harbour, Hong Kong © Nadine Marfurt on Unsplash

been argued that it is necessary to design the infrastructure differently from the conventional model (Dyson and Yocom 2015). Indeed, Strain et al. (2017) posit that, just as in terrestrial environments solutions were designed to increase biodiversity, restore faunal connectivity and enhance desired ecosystem functions, the same similar approaches can be applied to the design of urban infrastructure in marine environments. This suggestion is in line with Allen's argument in defending an architecture concerned with performance. Designing an urban-marine infrastructure differently means one capable of accomplishing multiple functions that together can satisfy the needs of human life and marine biodiversity.

Of course, as already experienced with terrestrial systems, conserving biodiversity in a novel ecosystem, such as an urban environment, is not a trivial task nor can a one-size-fits-all solution be used. However, as mentioned by Norberg-Shultz (1961), one of the main tasks of the architect is, precisely, "to formulate the problems on the basis of the various and often contradictory needs which are brought forth" and affirms to be "always possible to define socially determined building tasks which should be translated into architectural solutions" (Norberg-Shultz 1961, p. 16).

Living in an era of massive biodiversity loss, partially due to human-induced changes in land and sea use (European Commission 2020), a critical socially determined building task is not only to reduce the ecological footprint of the urbanization process but to reconcile the built environment with the (modified) natural system providing a net gain for biodiversity. Thus, when dealing with building tasks in the urban-marine setting, architectural solutions should respond to this call for action by translating into the project this contemporary requirement of reconciling the built environment with the aquatic ecosystem.

In fact, the role of architects in the matter of wild-life inclusive design to promote biodiversity in built-up urban areas has already been recognized by ecologists when they assume the need for interdisciplinary collaboration from an early stage of design procedures (Apfelbecket al. 2020). However, as noted by some authors (Nilon et al. 2017 cited in Apfelbecket al. 2020; Garrard et al. 2018) guidance on project-specific level for biodiversity enhancement is still scarce. Although investigation in this field has improved for terrestrial systems, this knowledge gap may be even more pronounced when it comes for urban-marine biodiversity.

Despite the gaps that still exist, it is possible to foresee architectural contributions coming from different design strategies. First, by reconciling society with the marine environment (a relationship that was profoundly broken with the predominance of industrial uses) a design approach based on experience which Ruddick (2016) calls "reinvention". Second, by providing the conditions for built-up areas (whether made of "hard" or "soft" structures[3]) to serve as urban habitats for species or allow the continuity of ecological processes, a design approach based on the species' critical requirements to thrive.

Examples of these strategies are appearing, some of them based on the human experience of marine biodiversity such as the new pier for the city of Saint-Petersburg (USA), by Michael Maltzan Architecture and Tom Leader Studio, or to promote an active engagement with the marine environment like Faaborg Harbour Bath (Denmark) by JDS Architects. Others are grounded on non-human needs like Afsluitdijk Fish Migration River (Netherlands) by Benthem Crouwel Architects and West 8 or, still, in natural processes like the nature-based coastal defence with absorbent borders at Dragør municipality (Denmark) by Arkitema. Albeit these are promising examples to show that reconciliation is not only a desirable but feasible task, we cannot say that a socio-ecological focus has already become a new paradigm for interventions taking place in urban-marine settings. The urgency of tackling the problem of biodiversity loss and the dimension of human influence on the urban-marine ecosystem leads us to consider that architects should look more closely at the issue, as they can influence interactions and, in doing so, have a role to play in the results.

[3] Respectively Grey Structures solutions and Nature Based Solutions as defined in IPCC- CZMS (1990).

References

Airoldi L et al (2005) An ecological perspective on the deployment and design of low-crested and other hard coastal defence structures. Coastal Eng 52:1073–1087; 1076–1077

Allen S (1999) Infrastructural urbanism, points + lines: diagrams and projects for the city, pp 48–57; p 52

Apfelbeck B et al (2020) Designing wildlife-inclusive cities that support human-animal co-existence. Landscape Urban Plann 200:103817; 11; 5

Bishop MJ et al (2017) Effects of ocean sprawl on ecological connectivity: impacts and solutions. J Exp Marine Biol Ecol 492:7–30; 7

Brenner N, Schmid C (2012) Planetary urbanization. In: Gandy M (ed) Urban constellations. Jovis, pp 10–13; 13

Bugnot AB et al (2021) Current and projected global extent of marine built structures. Nat Sustain 4:33–41; 38

Bulleri F, Chapman MG (2010) The introduction of coastal infrastructure as a driver of change in marine environments. J Appl Ecol 47:26–35; 28–29

Coalition Clean Baltic (2021) Adverse and large-scale environmental impacts of the Lynetteholm project. Letter to the Denmark Transportation Committee. Available at https://www.ft.dk/samling/20201/lovforslag/L220/bilag/39/2407870/index.htm, Accessed August 2022

Couling N (2020) Ocean space and urbanisation: the case of two seas. In: Couling N, Carola H (2020) The urbanisation of the sea—from concepts and analysis to design. nai010 Publishers, Rotterdam, p 316; p 20

Depledge MH, Codding CA, Bowen RE (2010) Light pollution in the sea. Marine Pollut Bull 60:1383–85 cited in Davies TW (2014) The nature, extent, and ecological implications of marine light pollution. Frontiers Ecol Environ 12(6):347–35, 347

Dominoni DM, Nelson R (2018) Artificial light at night as an environmental pollutant: an integrative approach across taxa, biological functions, and scientific disciplines. J Exp Zool A Ecol Integr Physiol 329(8–9):387–393. https://doi.org/10.1002/jez.2241

Duarte CM et al (2013) Is global ocean sprawl a cause of jellyfish blooms? Front Ecol Environ 11:91–97; 95

Dyson K, Yocom K (2015) Ecological design for urban waterfronts. Urban Ecosyst 18:189–208; 192

Elmqvist T, Redman CL, Barthel S, Constanza R (2013) History of urbanization and the missing ecology. In: Urbanization, biodiversity and ecosystem services: challenges and opportunities. Springer, Dordrecht, p 755; 19

European Commission (2020) EU Biodiversity Strategy for 2030 Bringing nature back into our lives, Brussels, Publications Office of the European Union, p 22; p 2

European Commission (2021) Emotional disconnect with Europe's aquatic environments: Report for the European Commission's Mission Board for Healthy Oceans, Seas, Coastal and Inland Waters, Brussels, Publications Office of the European Union, p 110; p 18

Evans AJ et al (2019) From ocean sprawl to blue-green infrastructure—a UK perspective on an issue of global significance. Environ Sci Policy 91:60–69; 60

Fadigas L (2010) Urbanismo e Natureza—Os desafios. Edições Sílabo, p 152; p 29

Falchi F et al (2011) Limiting the impact of light pollution on human health, environment and stellar visibility. J Environ Manage 92:2714–2722; 2715–2716

Firth LB et al (2016) Ocean sprawl: challenges and opportunities for biodiversity management in a changing world. Oceanogr Marine Biol Annu Rev 54:193–269; 194

Floerl O, Atalah J, Bugnot AB, Chandler M, Dafforn KA, Floerl L, Zaiko A, Major R (2021) A global model to forecast coastal hardening and mitigate associated socioecological risks. Nat Sustain 4(12):1060–1067

Garrard GE et al (2018) Biodiversity sensitive urban design. Conserv Lett 11(2):1–10; 2

Gaston KJ et al (2013) The ecological impacts of nighttime light pollution: a mechanistic appraisal. Biol Rev 88:912–927; 923

Hardy M (2019) Ominous views of Japan's new concrete seawalls. Available at https://www.wired.com/story/photo-gallery-japan-seawalls/. Accessed August 2022

Hauck TE, Weisser WW (eds) (2019) Animal-aided design in the living environment—integrating the needs of animal species into the planning and design of urban open spaces. Kassel Universitat/Technishe Universitat München, p 60

IPCC-CZMS (1990) IPCC Response Strategies Working Group Reports, pp 129–159; p 150

Lefebvre H (1970) Translated by Bonnono (2003) The urban revolution. University of Minnesota Press, Minneapolis, p 189; pp 3–4

Longcore T, Rich C (2004) Ecological light pollution. Front Ecol Environ 2:191–198 cited in Tamir R et al (2017) The spectral and spatial distribution of light pollution in the waters of the northern Gulf of Aqaba (Eilat). Sci Rep 7(42329):1–10; 2

Longcore T, Rich C (2004b) Ibidem, p 197

Marshall J (2017) Vision and lack of vision in the ocean. Curr Biol 27(11):494–502; 495

Nilon CH, Aronson MFJ, Cilliers SS, Dobbs C, Frazee LJ, Goddard MA et al (2017) Planning for the future of urban biodiversity: A global review of city-scale initiatives. BioSci 67:332–342

Norberg-Shultz C (1961) Intentions in architecture. The MIT Press, Cambridge, p 242; p 16

Perini et al (2021) ECOLOPES a multi-species design approach to building envelope design for regenerative urban ecosystems. In: Markoupoulou F, Marengo (eds) Responsive cities: design with nature symposium proceedings, pp 368–381

Ruddick M (2016) Wild by design strategies for creating life-enhancing landscapes. Island Press, Washington, p 243; p 45

Strain EMA, Olabarria C, Mayer-Pinto M et al (2017) Eco-engineering urban infrastructure for marine and coastal biodiversity: which interventions have the greatest ecological benefit? J Appl Ecol 00:1–16

Tamir et al (2017) The spectral and spatial distribution of light pollution in the waters of the northern Gulf of Aqaba (Eilat). Sci Rep 7(42329):1–10; 8. https://doi.org/10.1038/srep42329

Tidau S et al (2021) Marine artificial light at night: an empirical and technical guide. Methods Ecol Evol 12:1588–1601. https://doi.org/10.1111/2041-210X.13653p.1596

Tidau S et al (2020) Ibidem, p 1592

United Nations Environment Programme (1995) Global biodiversity assessment. Cambridge University Press, Nairobi, p 1125; p 923

United Nations Human Settlements Programme (2018) UN-Habitat background paper on Blue Economy and Cities. Nairobi, p 44; p 13

United Nations (2017) Factsheet: people and oceans. In: The UN ocean conference, New York, p 7; p 5

Wilbanks T, Fernandez S et al (2012) Climate change and infrastructure, urban systems and vulnerabilities. U.S. Department of Energy, Washington, p 86; p 4

33 Strategies for the Reclamation of Mining Lands—Framework for an Architectural Study with "Influencing Factors" and "Global Perspectives"

Akshatha Ravi Kumar and Irena Fialová

Abstract

This research is mainly concentrated on the post-mining lands of Ústecký Kraj (Ústí nad Labem Region)—Districts of Ústí nad Labem, Teplice, Most, Chomutov in the Czech Republic. The region of Northwest Bohemia (Ústecký Kraj) has always attracted the attention of the extractive industries in all their various guises over the centuries as a result of the rich mineral deposits found there. It is no coincidence that the hills bordering the region are known collectively as the Ore Mountains. The discovery of silver and tin in these mountains in the twelfth century triggered the slow but steady settlement of the area. However, the discovery of coal lying alluringly close to the land surface finally unleashed the full panoply of extractive technology on the region. In the century and a half since the Industrial Revolution, this region's physical, cultural, and social landscape has been irreparably altered in the scramble to uncover the vast deposits of brown coal in the North Bohemian basin stretching roughly from Kadaň to Ústí nad Labem. Each city affected by the mining activities has significant characteristics of its own in terms of history, culture, and land. Even though post-mining lands are a problem that affects the entire Ústecký Kraj, reclamation solutions for these lands must be specifically adapted to the cities and inhabitants that are local. Additionally, to ensure sustainable solutions and a sense of belonging for the living communities. Therefore, this paper proposes an *architectural study framework* to recognize the *global perspectives*, strategies, and policies in line with mining lands reclamation and systematically analyses the *influencing factors* of urban innovation and development in Ústecký Kraj.

Keywords

Reclamation strategies · Post-mining lands · Shrinking cities · Social–economic–environmental factors · Global perspective · Sustainability

33.1 Introduction

33.1.1 The History of the Czech Coal-Mining Sector

The Czechoslovak Communist Party and government strategies of the 1950s and 60s emphasized the development of heavy industry and

A. Ravi Kumar (✉) · I. Fialová
Department of Urban Design, Architecture and Urbanism, Czech Technical University, Prague, Czech Republic
e-mail: ravikumar.akshatha07@gmail.com

I. Fialová
e-mail: ravikumar.akshatha07@gmail.com

S. Hilal et al. (eds.), *Design for Partnerships for Change*, Sustainable Development Goals Series,
https://doi.org/10.1007/978-3-031-36993-3_33

energy, dependent almost exclusively on brown coal. The largest coal deposits are located in the basins of the foothills of the Ore Mountains (Fig. 33.1). These areas were developed exclusively based on coal mining at the expense of other economic activities, the natural environment, the existing built environment, social structures, and public health. Everything had to make way for coal mining as coal was considered the "lifeblood of industry." Mining executives, mining projection auxiliary operations, and especially Communist party functionaries were rewarded for ever-increasing the quantities of coal mined and the excavation and relocation of as much overburden as possible. In 1979, the craze for coal was in full swing as villages, one after another, were swallowed up. Not even the royal city of Most was spared devastation (Říha et al. 2011).

At the crossroads of several-related modernizing projects in the twentieth century, the sprawling surface mines of the north Bohemian brown coal basin expanded voraciously, swallowing 116 villages and parts of several larger cities by 1980. An area of over 1100 km^2 was heavily mined from Kadaň to Ústí nad Labem for brown coal burning in many thermal power stations, electrical power stations, and factories (Fig. 33.2). In the 1970s and 1980s, mining increased on a massive scale, and because of the expansion of mining operations, whole villages, towns, and even cities (Most) were demolished to extract the coal that lay beneath; their inhabitants were rehoused in large-scale new prefabricated apartment buildings which were poor in quality (Glassheim 2007). The combination of a regime hell-bent on mining brown coal by whatever means possible and at any cost and a militantly atheistic philosophy proved to be ruinous for the region's religious structures: between 1945 and 1989, a total of 536 churches, chapels, monasteries, synagogues and Jewish cemeteries were destroyed in the Ústí region. The 104 churches

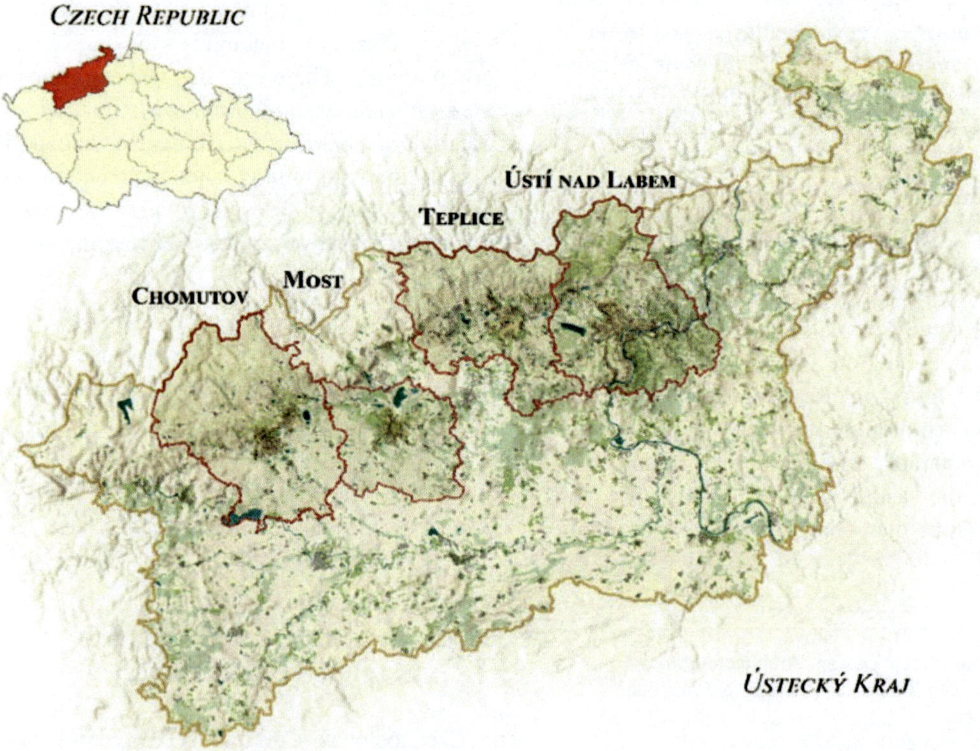

Fig. 33.1 Region of study: Ústecký Kraj—Ústí nad Labem, Teplice, Most, Chomutov, Kadaň

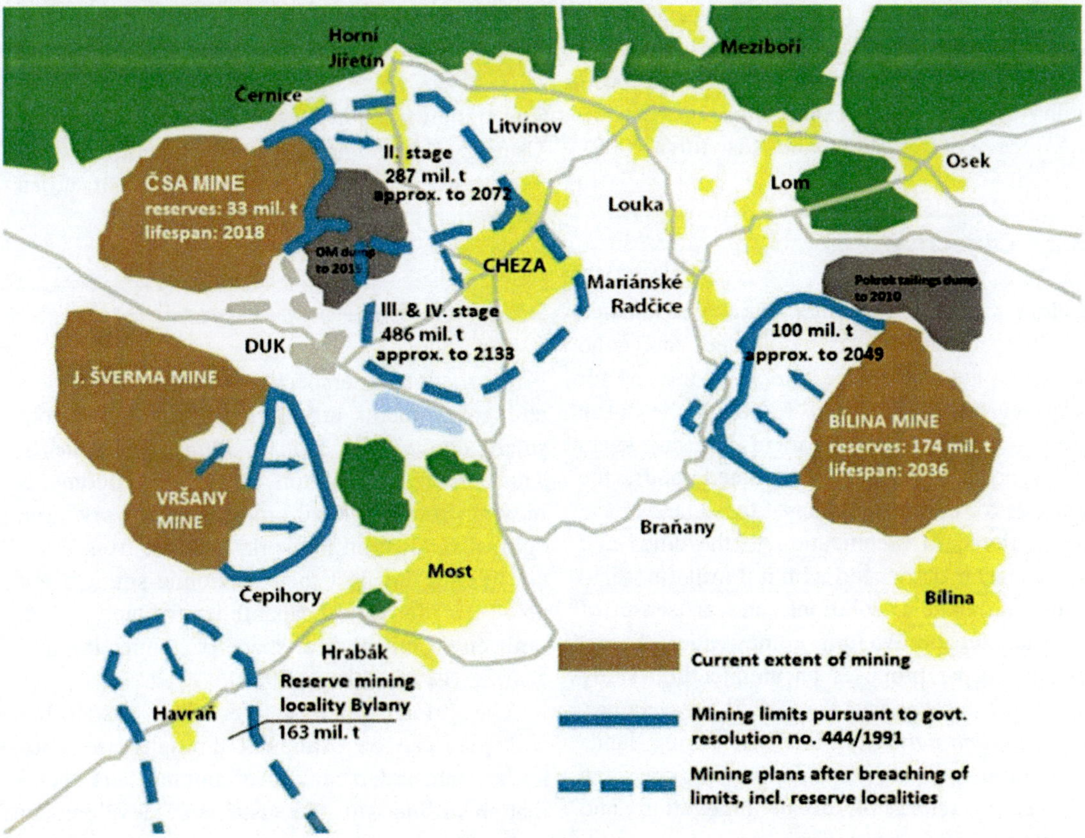

Fig. 33.2 The North Bohemian brown coal-mining limits as they apply to the Most region and plans for continued mining should the limits be repealed; yellow = built-up areas, green = forested areas. *Source* Kořeny o.s. (translated into English)

that were razed over that time represent nearly half of all the churches that disappeared throughout Czechoslovakia (Stehlík et al. n.d.).

33.1.2 The Impact of Mining on the Land, Environment, and Local Community

All types of opencast mining and its technology seriously impact urbanistic components and functions, significantly altering the original spatial development. Chomutov, Most, Teplice, and Ústí nad Labem districts involve a region that accounts for more than 76% of all coal mining and produces more than 35% of electricity in the Czech Republic (Vrablikova et al. 2016).

Once mining operations have been carried out on land, cities shrink—affecting the social and cultural aspects of the city, causing the loss of historical heritage and local communities due to migration. The landscape development is disturbed, manifested by, e.g., contaminated soil, water, the atmosphere, a drop in water tables and sources for drinking water, and a decrease in the diversity of plant and animal life. The original ecosystems are removed, the topography is significantly altered, the fundamental ecological relations are unchangeably disrupted, and biodiversity is decreased. These factors consequently lead to total ecological destabilization, elimination of aesthetic values, and decrease in the recreational potential of the city/town associated with the mining lands. Therefore, post-mining cities are often called "*landscapes without a memory*" (Sklenicka and Kasparova 2008).

33.2 Objective

Ideologies and design interventions associated with the reclamation of exploited land and shrinking cities have been a topic of interest for many years in this region. Although the entire Ústecký kraj (Ústí nad Labem Region) share similar problems concerning mined lands, the strategies for the reclamation of these lands have to be tailored in specification to the cities and people that are associated with it. Furthermore, to ensure sustainable solutions and a sense of belonging for the dwelling communities. Therefore, this paper proposes an architectural study framework to recognize the *global perspectives, strategies, and policies* in line with mining lands reclamation and systematically analyses the *influencing factors* of urban innovation and development in Ústecký Kraj.

33.3 Framework for Architectural Study

The framework for the architecture study of the Ústecký Kraj comprises two segments—*global perspectives* and *influencing factors* of urban innovation and development (Fig. 33.3).

33.4 Global Perspectives

From a global perspective, many stakeholders and governments are shifting their focus from solely *capital provision* to *innovation systems*. The core feature of this emerging economy is moving from a "global production network" to a "global innovation network," that is, from cities considering not just their economic strength but also their innovation capabilities for land and its built environment as a space of connecting and sharing (Li and Zhang 2020).

The global perspectives for sustainable strategies can be examined through case studies and an understanding of mining land reclamation in line with the sustainable development goals.

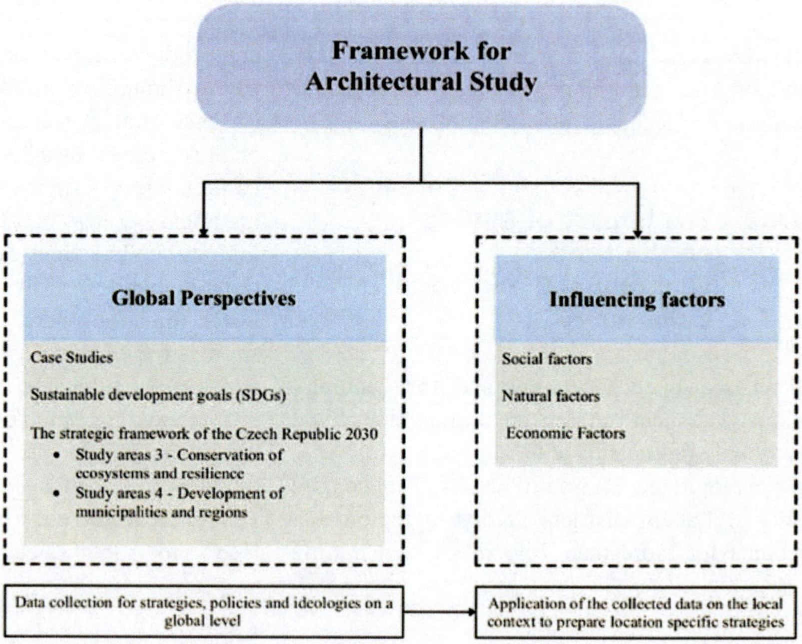

Fig. 33.3 Framework for architecture study and the steps involved in this paper

33.4.1 Case Studies

This research utilizes and builds upon secondary sources, including published, peer-reviewed literature, journal articles and reliable sources of information, and verified media sources regarding mining lands. This paper's methods rely on data collection and narrative assessment of existing case studies relevant to mining lands and their settlements. The case study selection is subject to several criteria, which are as follows:

1. Case study concerns itself with the reclamation process, sustainable strategies, and policies of lands subjected to degradation due to mining activities.
2. Case study incorporates one or more uses of technological interventions and data collection, which could be a new understanding.
3. Case study is based on the location of mining lands from all over the world (outside and within Europe) to understand its impact on different social, ecological, and economic backgrounds.
4. The case study was conducted during the past 10 years and was documented in the English language.

Based on the criteria mentioned above, the selected case studies will be analyzed to assess the impact of mined lands on cities and local environments, as well as the impact on the livelihood and well-being of individuals as a direct result of investigating the connection of land with the community. Following the assessment, a framework of suggested recommendations for reclamation strategies will be extracted and synthesized to lay a foundation for future research.

33.4.2 Sustainable Development Goals (SDGs)

"Achieving sustainable development is challenging, and the mining industry must ramp up its engagement, partnership and dialogue with other industry sectors, government, civil society and local communities' (World Economic Forum 2016).

In 2016, World Economic Forum published a report—*"Mapping Mining to the Sustainable Development Goals: An Atlas."* This report explains that the linkages between mining and the SDGs aim to encourage mining companies of all sizes to incorporate relevant SDGs into their businesses and operations, validate their current efforts and spark new ideas. Success will also require a substantial and ongoing partnership between governments, the private sector, communities, and civil society. They hoped to spur action that would leverage the transformative power of collaboration and partnership between the mining industry and other stakeholders.

This report structures the guidelines and encourages mining companies to align their ongoing operations with the SDGs, educating the mining industry on reducing the risk of extensive land degradation in the future. However, the limiting factor is that no straightforward policy or strategy has been developed to restore the already degraded land and form alliances needed for its transition.

33.4.3 The Strategic Framework of the Czech Republic 2030

To improve the quality of life, the Strategic Framework of the Czech Republic 2030, which integrates the Sustainable Development Goals (SDGs) into national priorities, aims to focus on essential values that matter to people rather than economic factors. Taking a new approach to connecting two key concepts—*sustainable development and quality of life*—the Government of the Czech Republic believes that enhancing the quality of life means improving people's access to good healthcare and well-being, quality education, work–life balance, sustainable environment and also personal safety in the community. "Quality of life" is to become one of the core indicators of measuring sustainability. This strategic framework is also an advancement toward achieving Goal 17—Partnership for the goals of the SDGs (Department of Sustainable development 2017). This document consists of strategies and policies that acknowledge the more specific problems in the country.

The strategic framework (Fig. 33.4) builds on six priority areas: (1) people and society; (2) economic growth; (3) conservation of ecosystems and resilience; (4) development of municipalities and regions; (5) global development and; (6) good governance.

Study areas *3—Conservation of ecosystems and resilience* and *4—Development of municipalities and regions* addresses the issues of transforming the post-mining landscape in the area of the closed mining complexes and keeps its impact on the environment and shrinking cities in the loop since it is a more region-specific problem (Fig. 33.5).

33.4.4 Study Areas 3—Conservation of Ecosystems and Resilience

Vision—"Agriculture, forestry and water management respect natural limits and global climate change; they improve soil quality, slow water drainage from the landscape and help maintain biodiversity. The development of settlements and technical infrastructure, especially transport infrastructure, takes place with the utmost regard to maintaining and strengthening ecosystem services provided by the landscape" (Department of Sustainable development 2017).

According to the Sustainable development department of the Czech Republic, the key area:

Fig. 33.4 Strategic framework of the Czech Republic 2030 (Department of Sustainable development 2017)

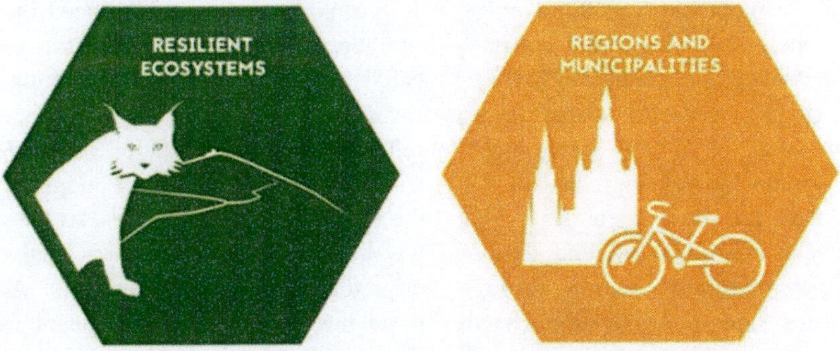

Fig. 33.5 Strategic framework of the Czech Republic 2030 showing study areas 3—conservation of ecosystems and resilience and 4—development of municipalities and regions (Department of Sustainable development 2017)

Resilient Ecosystems, focuses primarily on the goals and targets in the Planet category of the 2030 Agenda (SDG 6, 13, 14, and 15). With the exception of climate action, the Czech Republic is ahead of or in line with the OECD average results for every underlying goal. The results regarding protecting biodiversity and creating favorable conditions for terrestrial ecosystems are particularly close to the target levels. Besides the Planet goals, SDGs 2, 11, and 12 are also addressed in this key area (Department of Sustainable development 2017).

The policies in this key area state that the ecosystems of the Czech Republic are under significant pressure from intensive agriculture, forestry, and mining activities, which contribute to soil degradation and decrease its ability to retain water hence making the land unfit for further use. There is more landscape fragmentation and less biological diversity. A significant part of the surface water is contaminated by discharge from fields, wastewater, and toxic release due to mining. There can be a synergy between improvement in soil fertility, carbon capture, and storage in the form of organic matter, and retaining water in the landscape. Suitably adjusted farming conditions may provide the restoration of ecosystem services; distribution of subsidies will depend on the fulfillment of these conditions. The synergy between the necessary landscape restoration and creating jobs in the country in landscape maintenance and tourism will also contribute to balanced regional development (Department of Sustainable development 2017).

33.4.5 Study Areas 4—Development of Municipalities and Regions

Vision—"Responsible use of land creates the conditions for a balanced and harmonious development of municipalities and regions, improves spatial cohesion, directs the suburbanization trend and limits forced mobility. Cities and towns create preconditions for maintaining and improving the quality of life of their population. Competent public administration communicates openly with citizens and integrates them systematically into decision-making and planning. Housing is adapted to climate change" (Department of Sustainable development 2017).

According to the Sustainable development department of the Czech Republic, the key area: *Regions and Municipalities* brings a subnational perspective to implementing the SDGs and provides a framework for mainstreaming sustainable development to regional and local policies. This key area touches upon achieving its vision and contributes to implementing SDGs 6, 7, 9, 10, 11, 12, 13, 16, and 17. The Czech Republic scores above the Organization for Economic

Co-operation and Development (OECD) average, particularly in water, sanitation, and equality-related targets; however, the country also faces challenges arising from regional disparities in income and the quality of the environment (Department of Sustainable development 2017).

The policies in this key area state that inequalities between municipalities and regions are growing; they manifest mainly through internal peripheries and poor availability of public services and amenities, as well as a result of the outflow of the younger and more educated population to cities. At the same time, municipalities and regions have been dealing with the consequences of gradual suburbanization, affecting smaller settlements. The gradual urban sprawl also leads to poorer availability of public services, increases forced mobility, and related negative phenomena, for example, health-threatening factors induced by an increase in traffic. Cities and municipalities will also have to start accepting measures for adaptation to climate change and prevent its impact by 2030. Climate change is a typical example that, by its cross-sectional nature, imposes complex demands on municipalities, their representatives, and expert bodies. However, some cities and municipalities have yet to face those demands because they are still struggling with various problems, such as an unclear situation in terms of competition between the national administration and municipal administration, shortcomings in governance effectiveness, quality, and legitimacy, including involvement of the public. By 2030, it is necessary to support municipalities and regions in planning spaces smaller than regions and transcending the territory of a single municipality that will consider functional relations. There will also be a need to develop a network of services in the territory and co-operation among municipalities. Support emission-free transport, provide new uses for brownfield sites, degraded post-mining land, re-urbanize city centers, reduce greenhouse gas emissions and take adaptation measures, such as better care for greenery in cities, or mitigation measures, e.g., in the form of passive energy civil engineering (Department of Sustainable development 2017).

Shrinking cities will need to adapt to sociodemographic changes. Cities that economically active residents are leaving are called shrinking cities. It is a negative process, which is reflected in the degradation of the physical part of the city, the reduction of budget revenues, and the inefficient use of services. At the same time, it has a negative effect on the attractiveness of the city, for example, for investors. Although this trend has not been as significant in the Czech Republic as in other European regions, it is a trend that cannot be avoided. Almost one-sixth of cities are already shrinking (Schmeidler et al. 2011). This situation raises new demands for the management of affected cities; responding only to enhancing economic performance is insufficient (Rink et al. 2014). In addition to the new approach to spatial planning, local authorities and the state must also modify management and improve and prepare public services. They must be prepared for the new trend and not leave the cities to change based on ad hoc decisions (Department of Sustainable development 2017).

33.5 Influencing Factors

The strategies and policies created by the Department of Sustainable development of the Czech Republic can be adapted to the region of Ústecký Kraj. In order to see effective results of these policies in the cities associated with the degraded mining lands—Chomutov, Most, Teplice, and Ústí nad Labem, it is necessary to understand the significant influencing factors that are associated with each of these cities.

Based on a multidimensional perspective, this paper suggests that *natural, economic, and social factors* are three significant factors that condition urban innovation and development (Zhang et al. 2022).

The natural factors—including environmental quality, geographic location, and city scale—are prerequisites for conditioning urban innovation and development. Economic factors are also crucial, including economic level, industrial structure, industrial agglomeration, and technological innovation. Social factors are guarantee factors, including administrative hierarchy, cultural environment, population structure, and government management and services; i.e., they are essential for cities to adapt to the current dynamic situation. This segment of the architectural study provides theoretical support and practical directions for formulating and adapting policies for urban innovation development (Anttiroiko 2016; Lauer and Liefner 2019; Deng and Chen 2020).

Research into the influencing factors of urban innovation and development is relatively fragmented and lacks comprehensive and systematic study. Researchers from various disciplines have different opinions concerning the influencing factors of urban innovation and development, leading to significant differences in how this phenomenon is understood. When it is challenging to identify and determine the scope of influencing variables, it is more difficult to design suitable measurement scales or even use quantitative analysis. Hence, an exploratory and deductive investigation enquiry, such as a qualitative research method, is needed.

33.5.1 Social Factors

Social factors are the influencing factors included in the social system of urban innovation and development. The social system is restricted by population, policy, and social structure. Culture, scientific level, and traditional habits are all factors that must be considered when analyzing the relationship between social organizations and human activities. These factors are closely related to the innovation and development of cities and urban centers (Ma and Wang 1984; Yigitcanlar et al. 2008; Barrado-Timón et al. 2020). The qualitative analysis results identify the social factors affecting cities' innovation and development: cultural environment, administrative levels, government management, service measures, and population structure (Zhang et al. 2022) (Fig. 33.6).

33.5.2 Natural Factors

Although natural science was originally a discipline dealing with the relationship between plants and animals and their habitats or environment, it has also been applied in the fields of physiology, mathematics, and economics, as well as urban planning (Lenicki and Lewandowska 2016; Wang et al. 2019). The applied research dimension of natural sciences in urban development focuses on urban ecology, which has introduced the ecological viewpoints of plants and animals into human communities. Related concepts such as "environment," "natural process," "competition," and "symbiosis" can be used to analyze cities or communities. According to the qualitative analysis results, the natural factors affecting the innovation and development of cities mainly include the ecological environment, city scale, and geographic location (Zhang et al. 2022) (Fig. 33.7).

33.5.3 Economic Factors

Economic activity is associated with a series of activities people pursue to organize production, circulation, and consumption for their survival and development. Materials and energy are collected, processed, and produced in economic activities to become economic products, some of which become consumer goods, and some used to expand production. In this respect, people and societies turn natural materials and energy into products they need to meet immediate and long-term development needs. Thereby, industries are formed (Yao et al. 2015). The essence of the complex system's internal interaction is material flow and transformation. Therefore, when considering economic factors for urban innovation and development, this study mainly analyzes the material form of economic input and output rather than their monetary form. Based on the qualitative analysis results, the economic factors affecting the development of urban innovation

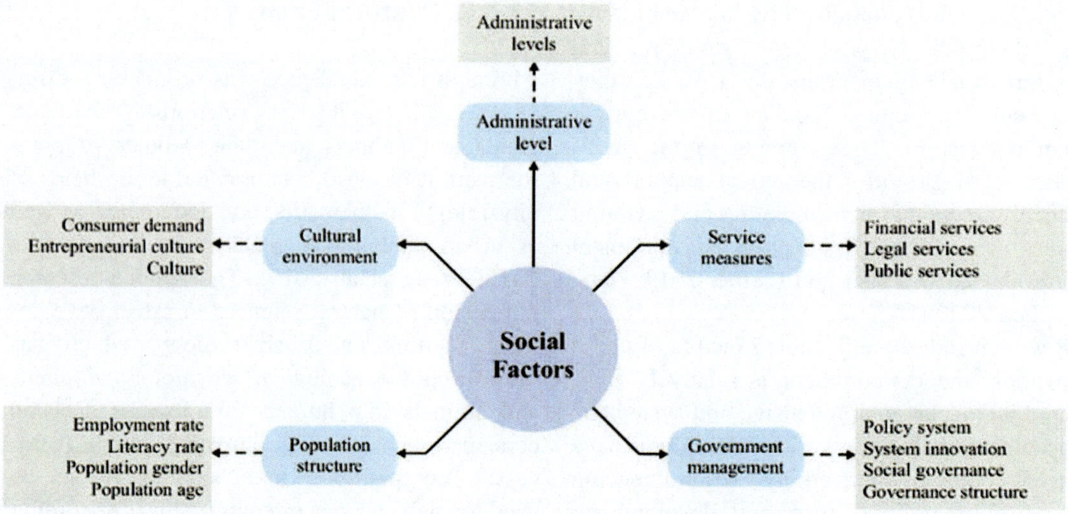

Fig. 33.6 Social factors affecting urban innovation and development (Zhang et al. 2022)

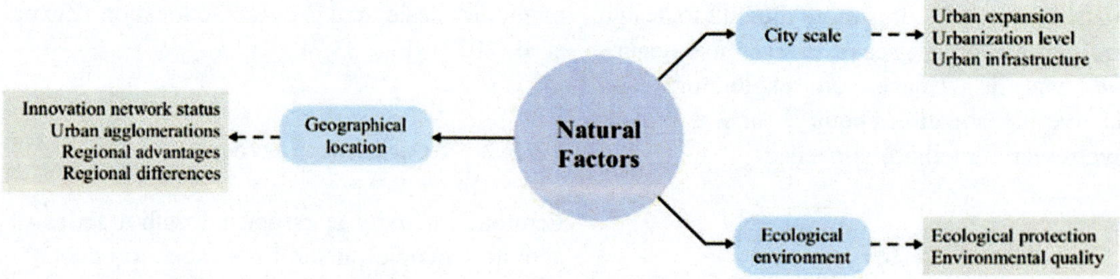

Fig. 33.7 Natural factors affecting urban innovation and development (Zhang et al. 2022)

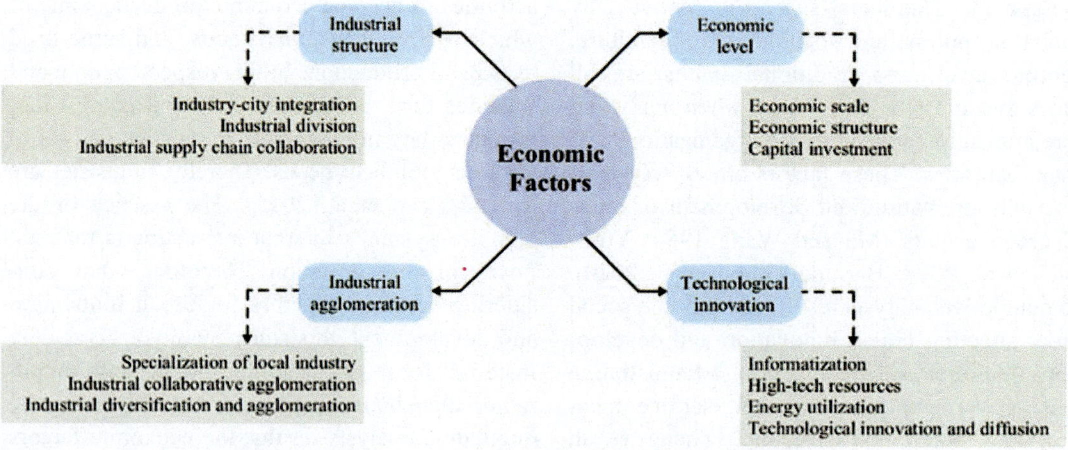

Fig. 33.8 Economic factors affecting urban innovation and development (Zhang et al. 2022)

mainly comprise economic level, technological innovation, industrial agglomeration, and industrial structure (Zhang et al. 2022) (Fig. 33.8).

Natural, economic, and social factors have mutual influences and interdependencies. The impact of natural factors on social factors is mainly due to the accumulation of human resources caused by large-scale cities, increasing the population's education level. Government policies are inclined toward cities with high administrative levels. In the case of the Czech Republic, cities in the central region have accumulated many innovative elements and have advantages in terms of economic level and services, whereas cities in the northwest (Ústecký Kraj) still have ample scope for development. Natural factors are counteracted by the performance of such social factors as the guidance of government policies making the city's environmental protection atmosphere stronger, social governance and services improving the ecological appearance of the city, and increasing the level of economic development, expanding the city's scale. Economic factors, as critical factors in urban innovation and development, are based on natural factors and are protected by social factors: these, in turn, have a direct impact on the development of urban innovation.

33.6 Conclusion

A good environment is an essential condition for the existence of life on Earth in any form. Humankind intervenes in the environment with activities associated with securing their everyday existence, mainly industrial (mining) and agricultural production, transportation, and energy. These interventions are primarily harmful when they are connected with the production of emissions, wastewater, various waste products, deforestation, land grabbing, degradation of land, etc.

In line with state of the art, transforming the post-mining landscape in the area of the closed mining complexes in the Ústecký Kraj region requires a systematic series of strategies. The Department of Sustainable development in the Czech Republic introduced a framework with definitions, policies, and iterative processes requiring periodic reviews and revisions. However, several aspects of the framework's limitations stem from the fact that these policies and strategies will hold good in reclaiming the Ústecký Kraj post-mining lands and communities only if they are explicitly tailored to suit the distinctive characteristics of the cities associated with the land (Chomutov, Most, Teplice, and Ústí nad Labem districts).

This paper sets up an initial Architectural study framework that connects how to approach the problem from a global to a specific local level. However, the framework only establishes a more general model of the interaction of global perspectives, strategies, policies, and influencing factors in the innovation and development of cities and degraded post-mining land. Hence, it lays the foundation for further research. Additionally, huge differences between cities (at the administrative level and/or geographical location, for example) mean that this theoretical model still has room for further specialization. Therefore, future research might benefit from this proposed theoretical framework system to conduct more detailed research on specific cities and enrich this first model with a combination of qualitative and quantitative research methods.

References

Anttiroiko A-V (2016) City-as-a-platform: the rise of participatory innovation platforms in Finnish cities. Sustainability 8(9):922. Available at https://doi.org/10.3390/su8090922

Barrado-Timón D, Palacios A, Hidalgo-Giralt C (2020) Medium and small cities, culture and the economy of culture. A review of the approach to the case of Spain in light of international scientific scholarship. Sustainability 12(18):7321. Available at https://doi.org/10.3390/su12187321

Barton AG (2013) Environmental mining limits in the North Bohemian Lignite Region. Envigogika 8(4). Available at https://doi.org/10.14712/18023061.418

Deng Z, Chen Y (2020) Research on place-making in innovation districts. City Plann Rev (in Chinese)

Glassheim E (2007) Most, the town that moved: coal, communists and the 'gypsy question' in post-war Czechoslovakia. Environ Hist 13(4):447–476. Available at https://doi.org/10.3197/096734007x243168

Kašparová I, Sklenicka P (2008b) Restoration of visual values in a post-mining lands cape. J Landscape Stud 1–10
Lauer J, Liefner I (2019) State-led innovation at the city level: policy measures to promote new energy vehicles in Shenzhen, China. Geogr Rev 109(3):436–456. Available at https://doi.org/10.1111/gere.12320
Leźnicki M, Lewandowska A (2016b) Contemporary concepts of a city in the context of sustainable development: perspective of humanities and natural sciences. Problemy Ekorozwoju 11–45–54. Available at http://yadda.icm.edu.pl/yadda/element/bwmeta1.element.baztech-bc129d7c-823f-4aac-9a6a-c8b935e796dc
Li L, Zhang X (2020) Spatial evolution and critical factors of urban innovation: evidence from Shanghai, China. Sustainability 12(3):938. Available at https://doi.org/10.3390/su12030938
Ma SJ, Wang SR (1984) The social-economic-natural complex ecosystem. Acta Ecol Sin 4(1):1–9 (in Chinese)
Office of the Government of the Czech Republic, Department of Sustainable Development (2015) Strategic Framework Czech Republic 2030. Polygrafie of the Office of the Government of the Czech Republic in Prague. Available at https://www.vlada.cz/assets/ppov/udrzitelny-rozvoj/projekt-OPZ/Strategic_Framework_CZ2030.pdf
Říha M et al (2011) The environmental mining limits in the North Bohemian Lignite Region. Společnost pro krajinu. Available at https://frontiers-of-solitude.org/sites/default/files/file-uploads/limitsreport.pdf
Rink D et al (2014) The governance of urban shrinkage in cities of post-socialist Europe: policies, strategies and actors. Urban Res Prac 7(3):258–277. Available at https://doi.org/10.1080/17535069.2014.966511
Schmeidler K, Jiříčková H, Zámečník P (2011) Výzva shrinking cities u nás, v Evropě i ve světě. Urbanismus a územní rozvoj 14(6):21–27 (in Czech)
Stehlík M, Hlaváček T, Plachý J, Kuča K, Macek P (n.d.) Destroyed churches of Northern Bohemia 1945–1989. Society for the Restoration of Ústeck Monuments. Available at http://www.znicenkostely.cz/
Vrablikova J, Wildova E, Vrablik P (2016) Sustainable development and restoring the landscape after coal mining in the northern part of the Czech Republic. J Environ Prot 07(11):1483–1496. Available at https://doi.org/10.4236/jep.2016.711125
Wang M-H, Ho Y-S, Fu H-Z (2019) Global performance and development on sustainable city based on natural science and social science research: a bibliometric analysis. Sci Total Environ 666:1245–1254. Available at https://doi.org/10.1016/j.scitotenv.2019.02.139
World Economic Forum, UN Sustainable Development Solutions Network, and Columbia Center on Sustainable Investment (2016) Mapping Mining to the Sustainable Empowered lives. Resilient nations. Development Goals: An Atlas, https://www.undp.org/publications/mapping-mining-sdgs-atlas. World Economic Forum. Available at https://irp-cdn.multiscreensite.com/be6d1d56/files/uploaded/Mapping_Mining_SDGs_An_Atlas.pdf
Yao L, Liu J, Wang R, Yin K, Han B (2015) A qualitative network model for understanding regional metabolism in the context of social–economic–natural complex ecosystem theory. Ecol Inform 26(1):29–34. http://linkinghub.elsevier.com/retrieve/pii/S1574954114000612
Yigitcanlar T, O'Connor K, Westerman C (2008) The making of knowledge cities: Melbourne's knowledge-based urban development experience. Cities 25(2):63–72. Available at https://doi.org/10.1016/j.cities.2008.01.001
Zhang J-X et al (2022) Influencing factors of urban innovation and development: a grounded theory analysis. Environ Dev Sustain. [Preprint]. Available at https://doi.org/10.1007/s10668-022-02151-7

34 Refusing Land's Capture: A New Status for a Finite Resource

Francisco Díaz and Camillo Boano

Abstract

Through this argumentative essay, we seek to frame a new status for land. First, with the help of companion intellectuals we analyze the construction of land as stable property that ultimately became financial security, to the point that nation-states ended up providing the apparatus to secure land property. Then, based on different alternatives of refusal we aim to destabilize the 'secure' position of land as property. This opens the space to propose a new status for land: an infrastructure for coexistence, a scarce resource that escapes capture. For if private property is at the root of inequality, then we may start to think of strategies that run away from that condition to ensure the right of future generations to have a place to live with dignity.

Keywords

Commons · Property · Ownership · Soil · Ecosystem

F. Díaz (✉)
Pontifical Catholic University of Chile, Santiago, Chile
e-mail: fdiazp@uc.cl

F. Díaz · C. Boano
Politecnico DI Torino, Torino, Italy

C. Boano
UCL, London, UK

34.1 Introduction

In 1797, Thomas Paine wrote a manifesto called *Agrarian Justice*. For him, poverty did not exist before civilization and, in its natural state, the earth "was, and always would have continued to be, the common property of the human race" (Paine 1797). Acknowledging the progress made by agriculture, the arts, sciences, and manufacturing, the US-born English philosopher argued that improvements made by those who work over the land may generate profit, but these developments do not entitle them to claim ownership over that land. Thus, he proposed that those who cultivate lands should compensate the community for their use through a system—that he devised and calculated—in which each citizen would receive an annual payment from age 21. This payment was not philanthropy or charity. It was not a tax either. It reflected, instead, the right every citizen has to receive compensation for being excluded from portions of the earth—those removed from the realm of the common to pass them to private dominion.

Paine's argument contradicted what John Locke proposed in 1689: that the work over land was the reason that justified its privatization as property. Whoever took the wild land and added his work to it, Locke argued, could appropriate it to improve it and make it more productive (Locke 1689). For Locke, the mix between human labor and the earth's resources would result in the right

S. Hilal et al. (eds.), *Design for Partnerships for Change*, Sustainable Development Goals Series,
https://doi.org/10.1007/978-3-031-36993-3_34

to own land. The appropriation and cultivation of it implied a journey from a primitive state to a civilized one. The legitimacy of land capture relied on a productivist idea of improvement.

By basing property on the work on it, Locke understood that unproductive land—according to capitalist precepts—could be considered wasteland (Bhandar 2018:36). The common ownership of land was understood as an almost wild state. Paine, however, skillfully used this same argument for the opposite end. For if Locke's story were true, "it follows (...) that the idea of land ownership began with cultivation and that there was no such thing as land ownership before that." That is, there was nothing natural about land ownership. It was a cultural construct (Paine 1797). As such, it could be contradicted, challenged, and, why not, dismissed.

34.2 Stabilizing Land

Anthropologist Tania Li observed that although you can subdivide land, it can never be eliminated: land does not move from its position. Thus, to enjoy its property as a private good, the owner needs to expel those who do not want to share that place with. The forms of this exclusion can be physical (fences or weapons), legal (rules of use), or economic (discarding those who cannot pay). Li argues that those forms always include "an attempt to defend exclusion in terms of its legitimacy" (Li 2014:591).

The possibility of excluding others from it, is, for Canadian geographer Blomley (2004:6), the key for the right to own land. In *Unsettling the City*, he reminds us that when we sell or buy land we think we are transferring a piece of it, but we don't realize that we are actually exchanging the rights to exclude from that land. Implicit is the possibility of evicting someone who occupies property without authorization, mobilizing the state apparatus in favor of whoever owns the property (Bhandar 2018:189). It is the threat of eviction that sustains the connection between property and exclusion. The meaning of property, as law scholar Bhandar (2018:20) reminds us, "translates into an absolute right to do what you want for the purpose you own." Having property on land means understanding it as something appropriable, as a thing at our service, thus establishing a direct relationship between property and possession. At the end of the day, the question of ownership is one of access: access to acquire the land, to occupy it and the chance to deny access. Ownership is exclusive precisely because it grants the right to exclude.

As Blomley (2004:xiv) points out, there is an assumption that "property brings certainty" and that "certainty brings peace and prosperity." In fact, property rights are one of the cornerstones of modern states. As land ownership does not protect itself but requires a legal body that recognizes it and a physical force that defends it, the existence of a power superior to the owner—the state—becomes imperative. This implies the establishment of a set of institutions to enact them: a police force to enforce the law, a bureaucratic system to register titles, and a citizenship that legitimizes and respects all these procedures. Respect and care for private property are what states ensure to their citizens, making citizenship worthwhile for the upper classes. Hence the historical connection between property and citizenship, and therefore between private property and the public sphere, as it was known in ancient Greece. In Hobbes' *Leviathan*, the state emerges as the force that prevents the barbarity of "war of all against all." In Bhandar's argument, "the fear of losing one's property functions as an expression of the fear of losing civilization altogether," to the point that "savagery, defined by disrespect for the law of property, is that against which the law of property must be protected" (Bhandar 2018:101). This assembly of laws, institutions, and people is the evidence that the care for land ownership implies the protection of social structures. At the end of the day, property is not something fixed, acquired, and forgotten but has to be constantly settled (Blomley 2004:xi).

The imposition of a state apparatus for registering and safeguarding such titles means excluding pre-existing property formats. In the processes of colonization, for example, everything that did not have a document was considered vacant land, thus ready to be appropriated by a settler. This argument underlies practically the entire occupation of indigenous territories in the Americas. But this state apparatus is not the only thing that imposes itself on the ground. There is also a vision of it as an abstract territory, without history, memory, or affect, which could be treated as an object and, thus, finally, be able to subdivide and alienate it (Blomley 2004:55).

The idea that the territory in its natural state was a waste and therefore should be subject to improvements, explains the need to measure and quantify the land to distribute it among those responsible for 'improving' it. Through this process the land becomes a bargaining chip, a form of retribution for services to the colonizing state. In fact, it was the British crown's debt to the army what drove both the surveying and the valuation of the land; thus, the land was delivered to pay that debt, which made it necessary to "measure, map and appraise all the appropriate lands" (Bhandar 2018:44).

Carried out through topographic surveys, these measurements established a shared basis to enforce property. The representations resulting from these surveys are descriptions of a new world, one in which land becomes an abstract, ownable, measurable, and objective entity. Cadaster plans allowed the geometric, precise identification of what was only described in words. These drawings created a new scenario, an abstract reality that materializes on the ground through boundaries and fences that limit property. With a space that is already abstract, it is possible to "treat it as the objectified and alienable 'object' of property" (Blomley 2004:55). In coordination with the land titles, these representations became part of the legal body of the state. Given that the assembly between property titles and these plans allowed the states greater efficiency in collecting taxes (Bhandar 2018:80), this apparatus was key to the economic valuation of the land and its subsequent transformation into tradable merchandise.

Confined and determined by cadaster plans, the ground gets fixed: it does not move from its position nor can it be removed from the map. That persistence over time transformed the land into a security. Measured, appraised, drawn, and recorded, land gave access to credit to those who owned it, that is, to the credibility necessary to borrow money. That is why, for the English utilitarian philosopher Jeremy Bentham, the meaning of property did not depend so much on the work that could be put on it, as Locke argued, but rather on its implications for the system's functioning. Land ownership was the source of all certainties and all expectations; that is why it had to be maintained at all costs. "If the property were reorganized with the direct intention of an equality of possessions," Bentham (1871:120) says, "evil would be irreparable. There would be no security, no effort, no abundance!" This opposition between property and equality explains, conversely, why the property is such a determining factor in economic inequality. As Bhandar (2018:19) argues, private property is at the root of inequality. Land doesn't just sustain life. It also upholds the status quo.

As security, land ownership made possible the emergence of a system of financial speculation (Bhandar 2018:86). For classical economic theory, land value does not necessarily depend on its present condition but rather on the expectation of a future value, under the assumption that land will develop its full potential for profitability (Blomley 2004:84). Yet, even considering the lack of transparency in what is known as the 'land market,' the protection of property as security is a structural element of the global economic system. Not by chance the World Bank itself stipulated that "secure and well-defined rights to land are key to the ownership of household assets, productive development and the functioning of the market" (Deininger 2003: xix). The state apparatus is asked to secure these rights over the land, so it can secure the credit system. Thus, the state stabilizes the land. However, since nation-states are coincident with a specific piece of land of the planet, then the state ends up stabilizing what was supposed to give it stability (Fig. 34.1).

Fig. 34.1 Private property sign affixed to a fence's gate in Ogunquit, Maine, USA. 22 July 2017, 15:57:59. Picture by Mark Buckawicki. CC License

34.3 Destabilizing Land

But, what if land ownership were not that stable ground on which our lives and the state apparatus lay? How can we think of destabilizing the construction of land as private property? To counter such a solid arrangement, we will examine several forms of refusal—*no*, abolition, *désœuvrement*, and use—in order to explore different ways to find the cracks in the structure of land property.

In *A Feminist Theory of Refusal*, Bonnie Honig suggest reading Bartleby not as a figure of suspension but of intensification. His refusal to comply with his employer's demands and requests, while also refusing to not comply with them, intensifies a situation. Bartleby refuses to give reasons for his noncompliance; he is silent or only repeats his formula "I prefer not to" (Honig 2021). Eventually, the law office is relocated in response to Bartleby's occupation of the workplace, and Bartleby is finally evicted and arrested. Bartleby's "I prefer not to," responds to the concern that giving reasons and making demands just enters would be dissidents into the tradeoffs of their opponents and compromises projects of radical transformation. Following Honig, our first step would be to intensify the opposition by saying no to the logics of private property. We prefer not to sustain such structure.

Kennan Ferguson, in *No Politics*, reflects that the "presumption that political action is always positive—that must lead to something programmatic and organized—obscures the potentials of outright refusal for many kinds of thinking and action" suggesting that the *no* "is pregnant with possibilities, promises, and dangers" (Ferguson 2021:xi). Fergusson maps a taxonomy of *nos*, in which the first is the "No of resistance," "the *no* of both the two-year-old and the subversive. It says, simply, that the current situation does not work… [it recognizes] the compulsory power of the system she or he confronts but stand against it, nonetheless" (Ferguson 2021:xi). A *no* that opens a space of "the subversive and the saboteur, those who find that the best way to use their wooden shoe is to throw it in the machine" (Ferguson 2021:xi). For Ferguson (2021:xii) "this constitutes the negativity of 'enough, already'" and therefore, "it is immediately accessible and […] easily recognized as political." It's a *no* that "does not overthrow extant power but rather attempts to interrupt it, depriving it of its totality" and in some radical form is a "*no* of revolution, of the complete (or at least as far as the imagination can reach) overthrow of the existing social order" (Ferguson 2021:xi) that has taken specific historical forms of unionization in Marx's, or the rejection of Nation-State in Fanon's for example. A second type of *no* is the one that "puts in its place a whole set of alternative possibilities and effects [...] Rejecting one or more preconditions, such a *no* replaces it with another condition and traces the results and implications of this alterity" (Ferguson 2021:xiii). These are the *nos* of science fiction, or of counterfactual history proper of practices as Forensic Architecture, and non-philosophy, to use the words of Laurelle's attempt to open alternative ways of thinking and being. A third *no* "is the *no* of absolute refusal, of confrontational negation, of abolition" (Ferguson 2021: xiv), surely it "draws from the first two forms of *no*, in that it rejects the status quo and looks elsewhere. But it differs from them by its very existence standing against the thing it opposes. To archaism, such a naysayer posits anarchism; to consensus, dissent; to order, laughter" (Ferguson 2021:xvi). In this line of though, our aim here is to reject the status quo and look beyond: against property we postulate commonality.

On another level, following Ruth Wilson Gilmore, refusal can be understood as abolition. For the US geographer, "Abolition is a totality and it is ontological. It is the context and content of struggle, the site where culture recouples with the political; but it is not struggle's form" (Wilson Gilmore 2022). Without a form, it has no repair. It's a *no* that fails to reward. However, there is something prophetic about abolition; some element that moves the imagination to think beyond what is abolished. The experiences of abolitionist movements show the complex challenges of simultaneously opposing oppressive institutions while proposing new ways of making the world. As Stefano Harney and Fred Moten said, "is not so much the abolition of prisons but the abolition of a society that could have prisons, that could have slavery, that could have the wage, and therefore not abolition as the elimination of anything but abolition as the founding of a new society" (Harney and Moten 2013:42). For us, abolishing land property does not imply a new form. Abolition means undoing property and showing its cracks.

Refusal can also take the form of *désœuvrement*. This is a complex term that, in its very simplistic kernel, points to the need to resist the productive ends of property. *Désœuvrement* has a central role in Agamben's thought translated as *inoperosità*—inoperativity or inoperativeness—a resistive gesture central to other thinkers as Bataille. For Bataille, *désœuvrement* is a form of pure negativity, a *négativité sans emploi* that, on a philosophical level, escapes the Hegelian dialectic of progress and, on a social level, is related to excess and refusal to work that radically reject the utilitarian aims of modern society (De la Durantaye 2009:18–19). Such understandings of *désœuvrement* draw a clear line between what is work and what is not but suffer from what Agamben called the incapacity of modernity to conceive of *désœuvrement* other than as a refusal of labor (De la Durantaye 2009). In Agamben's words: "the only coherent way to understand inoperativeness is to think of it as a generic mode of potentiality that is not exhausted" (Agamben 1998:62). To the understanding of politics as a means-end schema, Agamben opposes the one of *désœuvrement*, a politics of pure means capable of opening a terrain upon which human potentiality is not exhausted by any one identity or vocation. On our side, we think of the chance of making land property inoperative, that is, as a permanent potential that is never exhausted.

In the *Highest Poverty Monastic Rules and Form-of-Life,* Agamben (2013) interprets the Franciscan theory of poverty and use, suggesting that the very political task of the present is to think of a form-of-life entirely removed from the grasp of the law, and a use of bodies and of the world that would never be substantiated into an appropriation. Asking "what is so outrageous and deeply political in the *highest poverty* practiced by the Franciscans?," the Italian philosopher discovers that the Franciscan rule is a life that coincide with his own form, a life that is a form-of-life. From a legal point of view, that form-of-life can be achieved only through the *abdicatio omnis iuris*, or waiver to any form of law. The only practice that can be maintained—as it allows the survival of the individual—is the use of things. Use here is contrasted with property not only because it represents a "different way of owning," but also because it theorizes a relationship with the world that is independent from the paradigm of appropriation. For the Franciscans there is not a form of economic life or a form of legal life, rather there is simply a form-of-life that resists appropriation. It only allows use. Tracing the different emergences of *usus pauper*—a poor use where "the perfection of the rule consists in the renunciation of ownership and not in the scarcity of use" (Agamben 2013:56) *usus facti*, the simple act of using something, and *simplex usus as a de fato use* separated by a legal *usus* (property)—Agamben understand they signify the relationships of non-appropriation of the human with the world. In other words, Franciscans simply define and characterize use as opposed and the right to property. How would it be to detach use from property in the case of land?

In a sort of summary of these ideas, Pierre Bélanger, argues for "the unmapping of settler urbanism," which means "destroying the dispossessive categories that sanction exclusion, exploitation, extraction, and erasure. Dismantling the structures that obviate the legal landscape of treaties and that are constructed to sever relations between lands, waters, beings, cycles, and communities. Unplanning oppressive policies. Unnaming colonial place names. Debasing base maps. Debunking benchmarks. Redrawing legends. Retroceding lands" (Belanger 2020:127). Part of that attitude is what we want to recover here, by redefining and rethinking the logics of land property. We seek to understand land not as something protected or separated for the private use but as something resisting capture, a sort of "fugitive" resource, using Fred Moten's words (Fig. 34.2).

34.4 Thinking Beyond

Why do we think of destabilizing land property regimes? Land is a finite, non-renewable resource. The amount of soil in our planet is limited and has to feed and host us all. Ownership over something so crucial for the lives of human and no-human beings is hard to sustain. In fact, if we extrapolate the current rates of population growth, soon it will be impossible for all humanity to have access to owning land. But even if it were, that would mean excluding thousands of other non-human species. Furthermore, access to property, including land, is a crucial predictor of the position within the social hierarchy, to the point that there are obvious

Fig. 34.2 Deforestation in Tierras Bajas, Bolivia. This digital photograph of deforestation associated with the Tierras Bajas project in eastern Bolivia was taken by astronauts from the International Space Station on April 16, 2001./NASA

social differences in access to real estate (Blomley 2004:38). Merging these two aspects, we can visualize the following scenario: if each increase in population brings with it an increase in demand for land, but at the same time the land stays finite, its price can only go up. But if the price rises so high that it becomes unattainable for that growing population, at some point there will be a concentration of land ownership in a few hands, and access to it will be impossible. Have we ever wondered what will happen when no one can afford land ownership? Will we continue to defend the right to exclude access to land even though it leaves beings of our species without a place to live?

One possibility could be that land ownership would not depend on a title that guarantees it—which allows accumulation—but rather on the work and care put on it. This argument justifies land grabs and squatting (Boer et al. 2019). For if someone who has no land occupies an unoccupied property as a place to live, "shouldn't their interests in the property prevail over those of a genuinely absentee owner?" (Bhandar 2018:33).

A pragmatic vision could answer that land can be vertically multiplied to house population's growth. But land is not only a place to dwell. It is actually the basis of all our biological interactions. Our lives depend on land's ability to perform its ecosystem processes. These properties of land as soil can be enhanced but not multiplied. Land is finite. We can't make it grow. It needs to be protected and cared for. The right to own land does not ensure that soil will be allowed to perform its ecosystemic role. The exploitation of native forests, monocultures, and urbanization processes show that a resource so crucial as land can't be left in private hands.

Since the refusal of land property can have multiple forms, it would be naïve to prescribe a recipe. What matters is to start thinking of land as a fugitive resource—one that, by resisting capture, ceases to be a resource and instead becomes an infrastructure for coexistence. This doesn't mean we can't use it. It means that its organization, management, and care over time do not depend on the will of its owners. Thus, no matter the form of this refusal of land property, we need to imagine worlds in which land can become an infrastructure that ensures the right of future generations to have a place to live with dignity. For, if the concentration of land ownership continues over time, we already intuit the worst scenario: following the current trends, there will be no land left for the common good. And in a world without access to land, many species—including ours—will end up being evicted from their own planet.

References

Agamben G (1998) Homo sacer: sovereign power and bare life. Stanford University Press, Stanford

Agamben G (2013) The highest poverty: monastic rules and form-of-life. Stanford University Press, Stanford

Belanger P (2020) No design on stolen land: dismantling design's dehumanizing white supremacy. Arch Des 90

Bentham J (1871) Theory of legislation. Vertheimer, Lea and Co, London

Bhandar B (2018) The colonial lives of property: law, land, and racial regimes of ownership. Duke University Press, Durham

Blomley N (2004) Unsettling the city: urban land and the politics of property. Routledge, London

Boer R, Otero-Verzier M, Truijen K (2019) Architecture of appropriation. On squatting as spatial practice. Het Nieuwe Instituut, Rotterdam

Deininger K (2003) Land policies for growth and poverty reduction. World Bank and Oxford University Press, Washington DC

De La Durantaye L (2009) Giorgio Agamben: a critical introduction. Stanford University Press, Stanford

Ferguson K (2021) The big no. University of Minnesota Press, Minneapolis

Wilson Gilmore R (2022) Abolition geography: essays towards liberation. Verso, London, New York

Harney S, Moten F (2013) The undercommons: fugitive planning and black study. Minor Compositions, New York

Honig B (2021) A feminist theory of refusal. Harvard University Press, Cambridge, MA

Li T (2014) What is land? Assembling a resource for global investment. Trans Inst Br Geogr 39(4):593

Locke J (1689) Second Treatise of Government. Chapter 5 "Of Property". Accessible in https://press-pubs.uchicago.edu/founders/documents/v1ch16s3.html

Paine T (1817) Agrarian justice. W.T. Sherwin, London

35 Vernacular Eco-lodges for the Protection of the Bolivian Amazon Rainforest

Laura Otero Pedraza, Joanna Maria Lesna, Andrea Terceros Barron, and Kasper Martlev

Abstract

To keep our forests alive, dead trees should not have more value than living trees. A paradigm change within the global economic system could regenerate wilderness and restore communities. Bolivia has extensive rainforest territory, but it is in great danger. The lack of opportunities in rural areas pushes people to migrate to big cities abandoning their communities and leaving behind a great source of indigenous knowledge. This, and the cattle industry, illegal logging and drug trafficking are the main factors in destroying thousands of rainforest hectares yearly. The private investment in lands for conservation is crucial to protect these areas. The project proposes a scalable eco-lodge concept in the lands surrounding the Amboró Park in Bolivia, designed in collaboration with communities living near natural territories and using local materials. This proposal envisions a symbiotic cooperative where the eco-lodge depends on a healthy natural ecosystem and vice versa. An exchange of knowledge between locals and professionals is meant to happen in all areas: from architectural design to management structure. The proposal aims to evolve as a non-extractive business model to generate sources of income for the communities and increase awareness of the park's conservation. This proposal is a model to be copied in any rural area near green bodies. Its success should slowly create a belt of protection connecting trails for explorers and routes for wildlife. Based on qualitative interviews and on-site observations of four eco-lodges in Bolivia, this paper intends to respond to a current social and ecological demand for alternatives.

Keywords

Amazon forest · Deforestation · Rainforest · Bolivia · Eco-lodge · Communities · Conservation · Ecotourism

35.1 Introduction

Bolivia is one of the most biodiverse countries on Earth due to its extensive territory of rainforest (Killeen 2021), and together with the other nine countries sharing the Pan Amazon area, a great

L. O. Pedraza
Atelier Madera, Atelier Madera, Santa Cruz, Bolivia
e-mail: lauraoteropedraza@gmail.com

J. M. Lesna
London, UK
e-mail: joanna.maria.lesna@outlook.com

A. T. Barron (✉)
A.TE Arquitectura, Santa Cruz de La Sierra, Santa Cruz, Bolivia
e-mail: arquitectura.ate@gmail.com

K. Martlev
Copenhaguen, Denmark

S. Hilal et al. (eds.), *Design for Partnerships for Change*, Sustainable Development Goals Series,
https://doi.org/10.1007/978-3-031-36993-3_35

source of resources. As the global demand for commodities grows, the region's economic activity is progressively expanding the footprint of modern society. In the words of Timothy Killeen (2021): R*ampant deforestation driven by multiple social and economic phenomena threatened to transform its landscapes, degrade its aquatic resources and overwhelm its indigenous communities*. The environmental degradation in the Amazon is a complex global matter that requires meaningful reforms with solid solutions.

Knowing that the heart of the problem is socio-economic (Killeen 2021), we started our research by analysing the different realities on the ground. To better understand how the monetisation of the ecosystem could find a realistic alternative to face conventional business models, corruption and illegal activities, we observed the ongoing sustainable projects in the region. It has become evident that businesses that provide a stable income and small investments are not sustainable (Navia 2022). This has been replicated merely for economic reasons and, in sum, because no other effective economic models are suited for the context. A sustainable model must challenge this to become naturally replicable.

The primary inspirational literature for this paper is "A Perfect Storm in the Amazon Wilderness: Success and Failure in the Fight to Save an Ecosystem of Critical Importance to the Planet" by Killeen (2021).

This book is about a search for sustainability. It is to be considered a complete guide to understanding the complex ecological and socio-political scenario of the largest tropical forest on the planet. As an advocate for the conservation of the Amazon, Killeen writes based on decades of experience living and working in the region. His analysis of his book is objective and clear-sighted, reminding the reader that the realities in the Pan Amazon region show the different actors shaping the future of how humans can coexist with Nature on Earth.

We determined that privately owned land at the border of a National Park is the first key component, and the second is to find a community that can become a potential ally.

The social component is elemental in this equation, and locals are as important as the environment. Communities are disappearing because young people migrate to big cities to avoid social exclusion (Sören Gigler 2015), placing their cultural heritage in great danger. Other communities are the ones clearing the forest, as Killeen (2021) mentions in his book: *More often, people clear forest to generate wealth by selling timber, cultivating a crop or raising livestock. They are acting in their own self-interest to create wealth for their families and jobs for their communities*. As the pursuit of economic growth around the Amazonian Rainforest has multiple faces, consumers remain isolated from producers by a host of intermediaries contributing to the commercial forces driving deforestation. It is not a novelty that the responsibility for environmental degradation belongs to everyone around the globe (Killeen 2021).

This research proposes a combination of the actors involved in this global equation, placing into the scene all the stakeholders with legitimate interests in the activities, to integrate approaches. We take ecotourism as a flexible medium that allows exploring different practices simultaneously within the philosophy of sustainability and eco-lodges as the centre of the strategy. Borrowing the definition of an eco-lodge from The International Finance Corporation (2004): *Eco-lodges are of particular interest to the sustainable development community because they are small, medium and microenterprises that can generate a variety of positive economic development impacts in highly rural, biodiverse areas, where other types of development underway or under consideration are frequently damaging to the environment.*

As ecotourism increases worldwide with a new definition of luxury, Bolivia presents its potential to set the standards of a fresh development paradigm. A strong structure for this business must combine the strength of conservation experts, investors, local communities and the beauty of Bolivia's rainforest.

We see an opportunity for a successful business model: *How can an eco-lodge be the solution to protect the Bolivian rainforest from degradation?*

35.2 Context

With 714,834 m^2, Bolivia has 8.4% of the total area of the Amazonia. It is the country with the biggest area, following Brazil and Peru (RAISG 2020). However, in only 20 years (from the year 2001 until 2021), Bolivia lost circa 10% of its tree cover. This accounts for the loss of 6.67 million hectares of tree cover (Global Forest Watch 2021).

Since the 1980s, the annual deforestation rates have kept increasing, and the leading causes are wildfires, shifting agriculture, forestry and commodity-driven deforestation. In 2021, a total of 556 thousand hectares of tree cover were lost. 85% of this loss was caused by commodity-driven deforestation (Global Forest Watch 2021).

In 2022, 15% of all fires in the Amazonia happened in Bolivia (Finer et al. 2022). Furthermore, these fires are not occurring naturally: they are usually ignited intentionally by humans trying to clear land for farming, illegal logging or other motives (Greenberg 2022).

A controversial governmental decree signed in 2019 expanded the demarcated land for agricultural production and allowed for "controlled fires" in the forests (Mendez and Mercado 2019). That same year, more than 4 million hectares of forest were burnt in Bolivia, nearly 2.9 million hectares in the Chiquitania region (Saccardi 2019) (Fig. 35.1). The highest number of the past two decades.

Deforestation is still a growing issue despite having protected areas that account for 16% of the entire Bolivian territory (Arteaga Zambrana 2023).

The Bolivian Government has a precedent of violating its protected areas: the TIPNIS, which stands for Territory and National Park Isiboro Secure, is a clear example of this. The TIPNIS is an indigenous territory and a protected area (Fundación Solón 2017). However, in 2008, the Government of Evo Morales permitted the construction of a major highway which would divide this territory into two parts and, because of this, drastically increase the deforestation rates, affect the biodiversity and lead to an expansion of the already existing coca leaf plantations and illegal settlements. The indigenous people of the low-lands marched to defend their territory and opposed the road construction, yet they were intimidated and victimised (International Rights of Nature Tribunal 2023). The TIPNIS case is an example of the controversies in Bolivia regarding the extractivist interests of the people in power.

Moreover, even though Bolivia has 22 protected areas (Arteaga Zambrana 2023), there is always the risk of becoming a "paper park". Governments tend to lack the infrastructure and money to control the boundaries of such extensive land areas effectively. Consequently, many protected areas become "paper parks" that exist solely in formal maps (Ostrom and Nagendra 2007). However, even if the boundaries of a protected area can be controlled efficiently, deforestation tends to happen in its surrounding areas (SESMAD 2014).

The traditional "fortress" conservation model isolates the protected area from human disturbance to best preserve the place's biodiversity (Robbins 2007). However, the traditional "fortress approach" to governing protected areas has limitations (Boillat et al. 2010) because the local people who depend on the resources for their livelihood are excluded from critical decision-making. As a result, the locals' living conditions will likely decline, together with their interest in protecting the area (SESMAD 2014).

Some alternative models recognise the importance of the local community's involvement in conserving protected areas (Boillat et al. 2010). Ostrom and Nagendra (2007) state that protected areas managed by the communities can be equally effective, and under certain circumstances even more effective, than public "strictly protected areas". Although it is challenging to reconcile nature conservation with development, new alternatives must be implemented.

Fig. 35.1 Fires in the Chiquitania region in the year 2019. Photograph by Andres McLean

35.3 Methodology

We start the research understanding that over the last three decades, many have already explored variants of ecotourism and lodging in Bolivia. The different examples of existing projects show an interest in finding the formula for their success. These experiences provide valuable information about the methods implemented and key factors of their achievements and failures.

This research relies on four case studies that each use different methods related to ecotourism to protect the Bolivian rainforest from degradation. Data is gathered from on-site observations and qualitative interviews with people working closely with the community-based eco-lodges. As part of the research, we visited 3 out of the 4 case studies and interviewed the owners and community members. All interviewees gave their consent for the interview to be used for research purposes and gave consent for their names to be used. In addition, we maintained an ongoing conversation with Roberto Navia, an environmental journalist and Alex Villca, an indigenous Leader.

Analysing the diverse business models of these existing organisations allows us to evaluate the efficiency of the different approaches and thus provide our solution to how an eco-lodge can protect the Bolivian rainforest.

1. **Eco-lodges in Madidi run by members of the Community of San Jose de Uchupiamona**

Multiple eco-lodges—Chalalan Ecolodge, Sadiri Lodge, Madidi Jungle Ecolodge—inside the Madidi National Park are owned by individuals of the indigenous community of San Jose de Uchupiamona. The data comes from a documented three-day visit in November 2021 in Madidi. It uses on-site observations and an interview with Alex Villca, Co-founder of Madidi Jungle Eco-lodge and Defender of indigenous Rights and Nature in Bolivia. The follow-up interviews with Alex Villca took place on September 2022 and January 2023 via telephone.

2. **Refugio Los Volcanes**

It is a lodge in a natural reserve on the edge of the Amboró National Park. The data gathered for Refugio Los Volcanes is based on well-documented multiple on-site visits, interviews with guides, cooks and drivers and interviews with Canadian owner and manager Warren McCain between September 2021 and November 2022.

3. **Tsimane Lodge**

Three lodges deep inside the Tipnis National Park. The interview took place via telephone on 20 September 2022 with Daniel Coimbra, photographer of the lodge. It was conducted through video conferencing and email.

4. **Alta Vista**

Alta Vista is a research centre that helps understand and protect the environment. The interview took place via telephone on 22 September 2022 with Hermes Justiniano, the Strategic Advisor in Alta Vista.

The interviews aimed to collect qualitative information to understand current methods and the challenges of Bolivian eco-lodges. The case studies were evaluated based on the following criteria:

1. **Location**

This aspect analyses whether the location of the eco-lodges should be inside or outside a park and determines how it is possible to access it.

2. **Ownership**

The aim of this is to question who owns the land. Whether private investors, NGOs, local communities, or the government.

3. **Involvement of community**

This aspect questions the community's role and whether the community is involved in management and ownership.

4. **Production**

This criterion analyses what type of production is used, how much of the preserved area is combined with production and if the production is for their use or to sell. In addition, it analyses how it could be sustainable and if the community is self-sustained.

5. **Offers**

This criterion questions the type of services the community provides tourists.

6. **Architecture**

This aspect investigates the typology of the lodges, the kind of materials that are used and the local construction techniques.

7. **Forest protection**

This aspect analyses how the evaluation of the environmental impact is conducted, the implemented methods of forest protection and which area of the forest communities can protect.

Ultimately, based on analysis and evaluation of the precedents, we propose a scalable model of eco-lodges, which includes community, sustains the cultural heritage, is self-sustained and plays a role in preserving the Amazonian Forest. In order to achieve scalability, the proposal consists of (1) a bespoke project of eco-lodges in Amboro National Park with a potential for expansion; (2) an open-source project for indigenous communities which can be applied in various locations of the Amazonia.

35.4 Case Studies

35.4.1 Community of San Jose de Uchupiamona

In one of the most biodiverse areas on Earth, The Madidi National Park, an indigenous community in Bolivia, pioneered the eco-lodge business with Chalalan, their first ecotourism venue. Aware of the potential of their territory, a group of young community leaders opened the doors of their territory by creating a sustainable business and attracting customers from around the world. In the 80s, an explorer named Yossi Ghinsberg, who got lost in their territory and established a friendship with them, played a crucial role (Ghinsberg 2005). They offer an exceptional experience in wildlife viewing that, together with their eco-lodge structure, provides many work opportunities to different community members. Alex Villca, Co-founder of Madidi Jungle Eco-lodge and Defender of indigenous Rights and Nature in Bolivia, mentions that they experience less migratory problem than other communities. The facilities are very functional, modest and safe. The classic jungle bungalows are elevated from the humid ground. Walls and floors are made with rustic wood and tall roofs of thatched palm trees. Just as the houses in their community. These little bungalows are far from each other, so visitors can enjoy a more profound experience of the jungle, having the natural feeling of being alone in the middle of nowhere. All the materials are obtained from the wilderness, except for modern comfort.

The experience is focused on exploring the surroundings to spot a jaguar or other animals in their natural habitat. Experienced guides and community members always make these expeditions. The time spent in the private bungalow is only for resting at night. There is the main bungalow, where the kitchen and dining area are, run by a group of ladies from the community. That is the only place with electricity powered by a diesel motor.

The excellence of their service is linked to how personal the project is for them. The amount of knowledge of the jungle they share with the visitors is invaluable and admirable. Their native language is Quechua. Generally, they know Spanish, but some also speak English. Their commitment to the project is easily perceived as one experiences the place.

Nevertheless, as this community ultimately depends on this project, their main predator, far from being the jaguars, has started hunting them: the government of Bolivia (Villca 2022). Their success as a conservation project means power and freedom. It also represents an alternative model away from extractive and away from ample business opportunities for the government. This marked the beginning of an unequal war for defending the protected territories and natural resources. Since then, indigenous communities have awakened and involvement in political matters has been their most direct way to protect Nature. Though the eco-lodge business in Madidi is still ongoing, these problems have drastically affected the influx of people and their independence, now they are on a survival mode (Villca 2022).

The analysis of this example exposed a policy problem. At the same time, it proves that indigenous communities can be the best protectors of Nature while running a successful sustainable business.

Although we will not focus this research on the government's role in territory management, it is a subject to consider while choosing the area to create a project.

35.4.2 Refugio los Volcanes

Refugio los Volcanes is a private natural reserve of 300 ha next to a National Park owned by foreigners who believe in conservation. Their business focuses on a relaxed experience of wildlife viewing, combined with various hikes inside the property. The lodges are unpretentious and very comfortable, without any specific style and use solar energy. Although their cabins are made with conventional materials and techniques, they use the property's wood and rocks

Precedence	Chalalán Ecolodge-Community of San Jose de Uchupiamona	Refugio los Volcanes	Tsimane Lodge	Alta Vista
Location	in the Madidi National Park	next to a National Park	Isiboro Sécure Indigenous Territory & National Park	Outskirts of Santa Cruz
Organization ownership	Individuals of Community of San Jose (indigenous community)	Private owners	Private owners	Alta Vista (NGO)
Land ownership	Community of San Jose (indigenous community)	Private owners	Land owned by indigenous community, business model owned by owners	Alta Vista (NGO)
Size	210.000 hectares	300 hectares	12 million hectares	3.360 hectares
Community	management, guides, cooking, cleaning and maintance, sharing knowledge about jungle with visitors	35 people, not involved in management; guides, cooking, cleaning and maintance, sharing knowledge about jungle with visitors,farming	300 indigenous and 100 locals;	group of researchers and professionals
Production	None	vegetables and coffee	None	Cattle, vegetables, honey
Service and offers	accomodation; guided tours of the forest; animal spotting; rainforest interpretation; sailing, canoe and boat rides; handcrafting workshop; learning about rituals of the Uchupiamonas Indigenous people; learning ecological processes and primary forest habitats; dining;	hikes, guided tours and experience of wildlife viewing; wellbeing; sports	fishing trips	None. It is primary a research center for regenerative agriculture
Architecture	elevated jungle bungalows	typical modern, solar-powered	elevated jungle bungalows and tents	Colonial Architecture
Materials	in-situ materials: wooden floor and walls, thatched palm tree roofs	concrete, bricks and wood	in-situ materials for construction: wood; equipment and furniture shipped from the city	Concrete, bricks, terracotta tiles
Forest protection	community surveils the perimeter of the forest, to protect against illegal settlements	indirect protection through their presence	indigenous community understands the value of their land and protects it	By showing alternative and sustainable ways of working with the land
Problems	goverment because of their lack of support for these initiatives and attempt to sabbotage them	lack of involvement of community in the management	land scarcity (organization depends on land owners)	not profitable as a business model yet

Fig. 35.2 Comparison table of the four case studies

for construction. Even though Refugio is not considered an eco-lodge because their villas are made with cement and bricks, they operate as an ecotourism business.

They have a daily flow of visitors worldwide and a growing local market, this is due to their successful marketing strategy and access to technology. The elasticity that ecotourism can have allows them to add to their offer of well-being activities and sports. This business provided many job opportunities for a community nearby for more than 25 years. Up to 35 persons have an income that relies on this business.

The fact that the project depends on this community to function and vice versa can be perceived as a fantastic relationship, but it is their biggest weakness. There is a lack of awareness among the staff regarding the importance of the project for them. While several strategies have been implemented over the past couple of years to improve the experience, those that keep failing are related to the staff members. We observed that the non-existent involvement of the community in the decision-making at a managing level could cause this. Which is a problem that Chalalan at Madidi does not have. Keeping the community members away from the management responsibilities can be dangerous in a more existential way in case of a potential risk of abandonment of the property by the investors. It would shape a significantly different reality for them.

On the other hand, beyond the business, Refugio los Volcanes has recently embarked on a new mission that can shape the future of many people in the community. On the path to becoming a more self-sustained place, they have started producing vegetables and coffee, providing the staff with new knowledge.

35.4.3 Tsimane Lodge

Deep within the Isiboro Sécure indigenous Territory and National Park, where the Amazon jungle meets the Andes mountains, the 3 Tsimane eco-lodges await. The Tsimane Program provides exclusive fishing trips for people from around the world. Prices are high, and the capacity of trips is small. This means that even when fully booked, the business will have minor damage to Nature, bringing in significant revenue. The lodges are fully built with materials from the forest. However, furniture and equipment had to be shipped from the city to fulfil the high standards of comfort.

To be allowed to bring people to the National Park, Tsimane Lodge, owned by foreigners, has partnered with the indigenous population who owns the land by right. Tsimane Lodge pays a fee to the local community. This fee is approximately similar to the projected profit of the Tsimané program. However, without business risks and represents their profit as partners of the Tsimane project. According to the Tsimane Lodge, they provide more economic benefits for indigenous peoples than any other tourism project in Latin America, helping to improve the quality of life and build a sustainable future while protecting their cultural heritage. Although they provide the natives with job opportunities as guides, maintenance and river transportation mainly, they are not included at a managing level. The members are not used to full-time work and might not be reached when needed. The Tsimané Eco-Lodge operates on a rotational principle that allows the community members to take turns. Daniel Coimbra (2022) explained that it is a system that has been implemented in various experiences around the Amazonian rainforest with different indigenous communities as a progressive understanding that they have a different way of managing their time and seeing the world.

Daniel believes this model has pros and cons. From his experience, ecotourism needs to take the financial aspects seriously. He has seen many NGOs trying to work with local communities but forgetting the finances, and those programs have disappeared (Coimbra 2022). The finances work well for the Tsimané program. They have a solid non-extractive business model that brings in significant money to both the owners and the local community—they share 50/50. The funds from the fee are primarily used for social projects, infrastructure development, transportation, and medical and education projects.

The downside of Tsimané's business model is that the local communities can decide that they no longer wish to allow Tsimané Eco-Lodge to do business in their territory whenever they want. It is an investment in land that investors can never buy. This aside, the Tsimane Eco-Lodge managing team meets with their indigenous partners and national park authorities each February for an annual meeting to assess the financial results of the project, conduct comprehensive accountability and celebrate the year's success. They provide annual employment to more than 300 indigenous and 100 non-indigenous locals. Their project is a means to help the natives preserve their millenary culture and tradition while preserving their territory. Thanks to this close collaboration, a constant exchange of information on the matter keeps the indigenous aware of the richness of their territory.

As part of their ecotourism mission, the Trimane Eco-Lodge, also partners with the Wilderness Conservation Society for biological research on dorado, studying various strains of the species in the jungle environment. These studies are meant to be more integrated into the project and, in the process, involve more local communities in the subject.

35.4.4 Alta Vista

In a property owned by an NGO called Alta Vista, around 3360 ha are being tested as part of a research program on regenerative production. Approximately 75% of the land is in a perfect state of conservation. We decided to speak to Hermes Justiniano, the strategic advisor of the program. With a long list of conservation projects and more than 40 years in the field, Hermes knows the Bolivian context and its challenges. We wanted to share our ideas and hear his own opinions on the subject.

In the interview, he mentioned that, on the one hand, in the Bolivian context, private property plays a substantial role (Justiniano 2022). His insights on the subject are like those of Killeen (2021). He explained how a realistic proposal should be considered as a business model. On the other hand, he said production is a crucial aspect. Alta Vista is known for its success as a model of regenerative production in Bolivia. He said the way to know this is because it is being copied. Big property owners, with a reputation for having little interest in Nature, have heard about how much more productive the land can be with sustainable practices and are now implementing the system and spreading it.

Speaking with Hermes was crucial for understanding that the whole of a sustainable project must involve different fields, and production is an exceptionally important one for a long-term impact on the practices of locals (Fig. 35.2).

35.5 Results and Discussion

Based on the conclusions from the interviews, we propose an eco-lodge project that considers the different criteria analysed in the former case studies.

35.5.1 Location

On our mission to find a perfect equation, we focused on the bespoke project on the outskirts of Santa Cruz. At the Amboro National Park borders, an extensive protected area of around 63.700 km^2 (SERNAP 2020) combines all ecosystems and generic social conditions. The Amboro National Park is one of Bolivia's 22 national protected areas (Fig. 35.3). It is home to a great variety of species and ecosystems. It is estimated that there might be over 4000 species of plants, and there are 1236 registered animal species (SERNAP 2020). The primary threat to the Amboro ecosystem is small-scale farming from colonists, who clear the areas for agriculture (Killeen 2021).

The mission of the future project is to expand the frontier of the National Park and protect it from deforestation. In addition, being next to a National Park will provide a more intense experience for visitors. Therefore, we propose the eco-lodges be in the northern area of Buffer Zones of the Amboro National Park—Integrated Management Natural Area (ANMI).

The area is easily accessible. Its north part is located only approximately 150 km away from Santa Cruz, the largest city in Bolivia. It allows transport for visitors as well as necessary equipment for the project. The ANP is one of Bolivia's most important protected areas, and its biodiversity and accessibility make it an excellent potential for ecotourism (Soria 2008).

As the community plays a crucial role in the model of eco-lodges, it is essential to choose a site near the existing community facing the problem of migration to the city. The project has a potential for expansion within the area of Amboro National Park but also in other regions of the Amazon Rainforest. In the long term, the multiple eco-lodges protecting different National Parks should serve as corridors for wildlife and explorers.

35.5.2 Ownership

As The Amboro National Park is almost exclusively public land, in the ANMI area, most of the land is privately owned (Bucklin 2010). The responsibility of owning land for conservation purposes must not be underestimated. The land

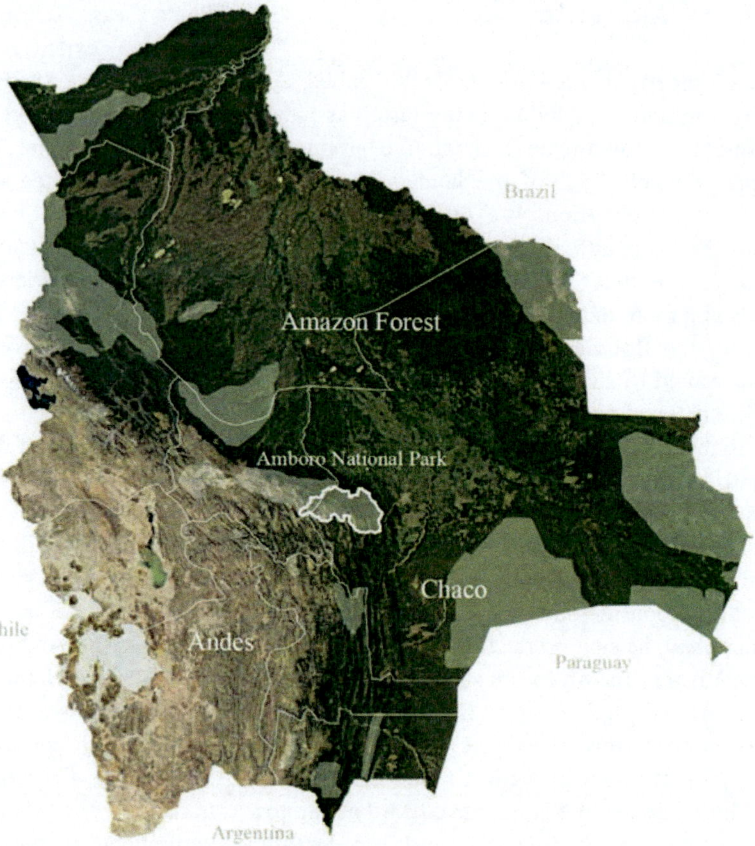

Fig. 35.3 Map of Bolivia and National Parks. Amboro National Park is located at the encounter between three phytogeographic regions: the Andes, the Chaco and the Amazon

ownership strategy we propose is to acquire land surrounding protected areas (Fig. 35.4). Over time, the community members working on the eco-lodge will have the opportunity to become landowners.

As learned from the example of the Tsimane Lodge, including the indigenous and local community in the organisation's profit results in improved efficiency and investment quality. Our proposal goes further than that. The long-term strategy is to bring partners into the business within those in the community that show interest. Integrating community members into a partnership for the board, directors and management teams must be implemented after a series of environmental management and sustainability training. It is essential to maintain the sovereignty of the property to guarantee its protection (Fig. 35.5).

35.5.3 Production

A sustainable plan for resource management and monitoring must generate a side business for the project and can become a robust economic activity that benefits nearby communities. This should be seen not only for internal consumption but also as part of research for sustainable practices within a community. Part of this is to recover knowledge in management and resources and explore technologies in the field.

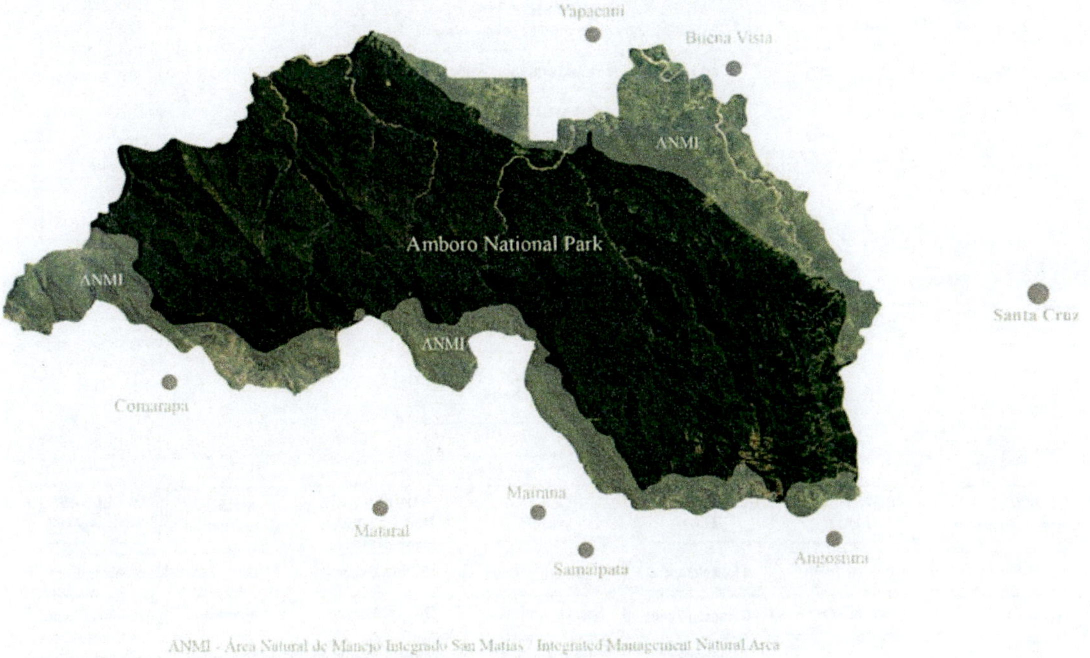

Fig. 35.4 Satellite image of the Amboro National Park and its surroundings. Possible site locations are located in the buffer zone surrounding the park (ANMI)

The production should not negatively impact the environment and the ecosystem and be carbon emission negative. The type of production should be chosen together with the community members to maintain their interest, cultural heritage and knowledge, for instance, vegetable, coffee cultivation and crafts. Furthermore, providing training opportunities could improve the education level within the communities. Respectful production can add value to the location. The cross-collaboration of different professionals and community members can potentiate innovation.

35.5.4 Involvement of Community

The real guardians of the territory are those who call it home. Close communities must be included as much as possible to promote a permanent exchange of knowledge and to engage them in a shared commitment. As mentioned in the introduction, rural communities are as endangered as the environment and must not be left out of a strategy. A sustainable project is a path to integrate education and provide people with the right tools for a bright future (Fig. 35.6).

35.5.5 Offers

The project proposes a model that grows organically based on site-specific demands and is built in phases. The first phase should supply the basic needs for housing, water, energy, food production and waste management. With time, the eco-lodge will keep developing based on the interests.

Initially, a good combination of wildlife viewing, sports, wellness, culinary experience and architecture has the power to attract a significant number of people while providing work opportunities and knowledge exchange between the community and experts in each field (Figs. 35.7 and 35.8).

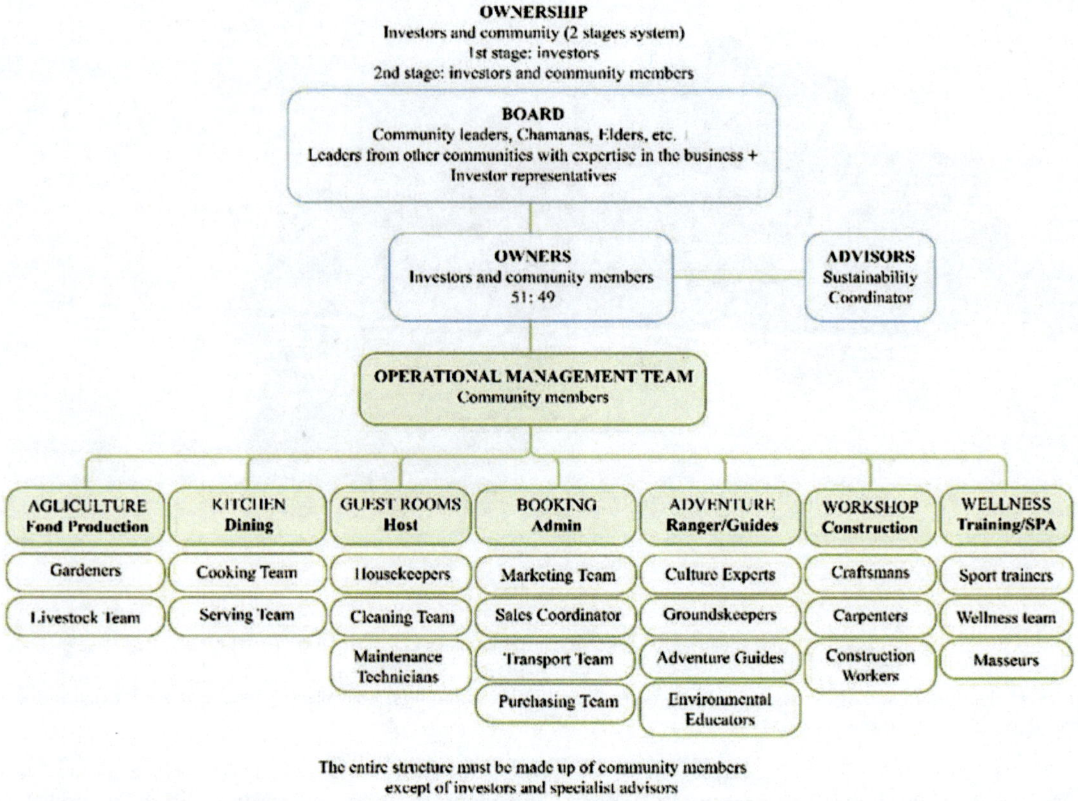

Fig. 35.5 Investors and community ownership diagram

35.5.6 Architecture

The facilities must reflect the principles of what the project aims to achieve. The selection of materials and techniques must result from the abundance of local resources and be carbon negative. All rural communities guard ancient techniques of construction that have the potential to become the main attraction of the eco-lodges. As the project must authentically protect the area, this architectural experience must be understood as the moment when locals value their knowledge and bring that into the scene. The architectural proposal aims to use vernacular architecture and technology. The eco-lodge could be an example of sustainable and rural architecture for others (Fig. 35.9).

35.6 Future Research

The scope of the project's future research for the short term is to analyse potential locations for the eco-lodges parallel to interviewing community members located in the Amboro National Park. The objective would be to plan a pilot and structure the business model in all possible scopes, including finding potential investors, management, training the employees, partnerships with local communities and forest monitoring.

Architectonically, future research would go in two directions. The first one proposes a modular and scalable project that would be open-sourced. It will reach a wider audience and create a more

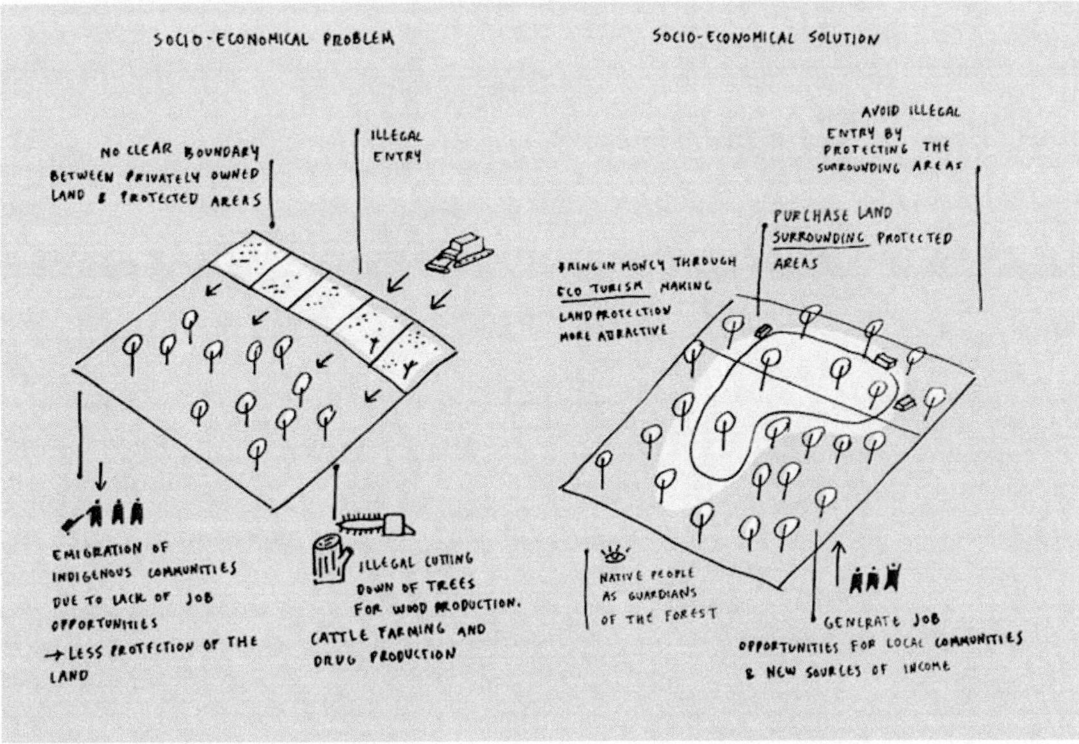

Fig. 35.6 An illustration explaining the socio-economic problem and proposed solution to protect the Amazon Rainforest and National Parks from illegal entry for wood cutting and setting fires with the community's support

significant impact on the protection of forests through ecotourism.

The other proposal would consist of a bespoke eco-lodge design and a structured business plan. The architects would serve as the intermediaries for potential investors, guiding them through the land acquisition process, the business structure, the employees' training, and the eco-lodge's construction and design. Long-term research aims to measure the impact of an eco-lodge on deforestation and life quality within rural communities.

35.7 Conclusion

This research aims to understand if and how an eco-lodge in Bolivia can successfully help protect the rainforest from degradation. The paper analyses the existent eco-lodges and ecotourism. The eco-lodges offer wildlife watching, hikes, retreats and experiences within local communities and the Amazon Rainforest, which will attract tourists who are conscious and responsible and want to experience Nature under the guidance of the locals. The primary purpose of the eco-lodges is to generate income and to protect the environment.

The case studies were analysed based on seven criteria: location, ownership, community involvement, production, offers, architecture and forest protection. The current examples of eco-lodges typically follow SGD principles, which include sustainable architecture, locally grown food, fighting poverty by providing job opportunities, using clean energy, partnerships with local communities and protecting Nature.

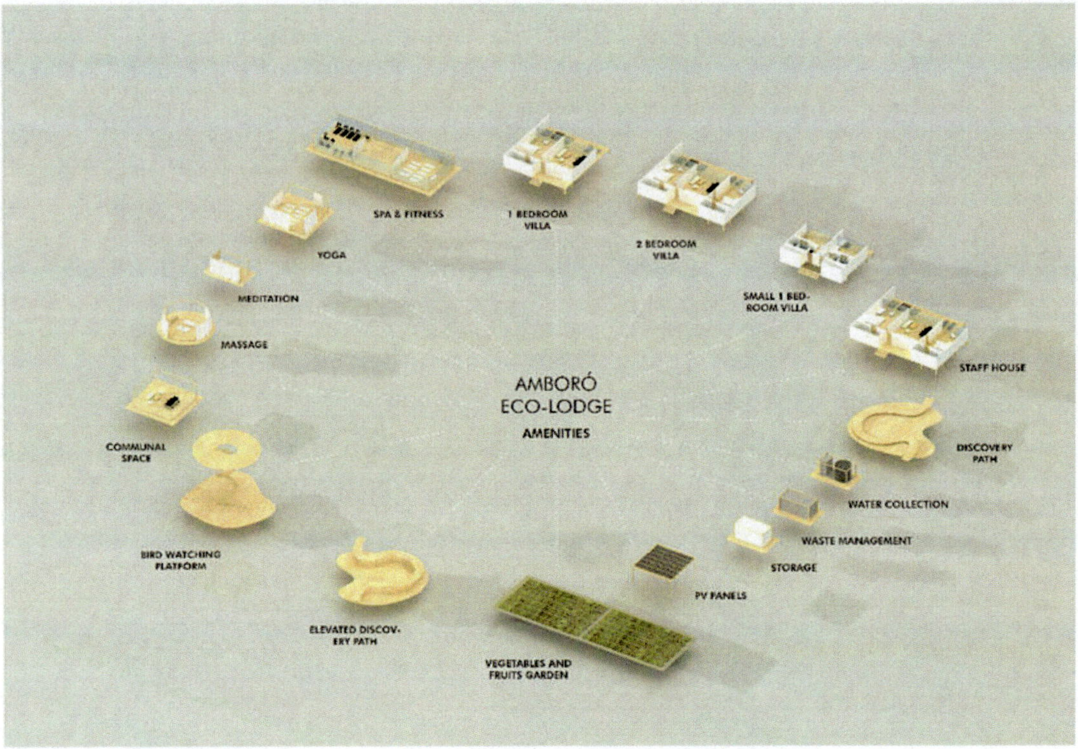

Fig. 35.7 Possible eco-lodge amenities constructed based on the site-specific demand

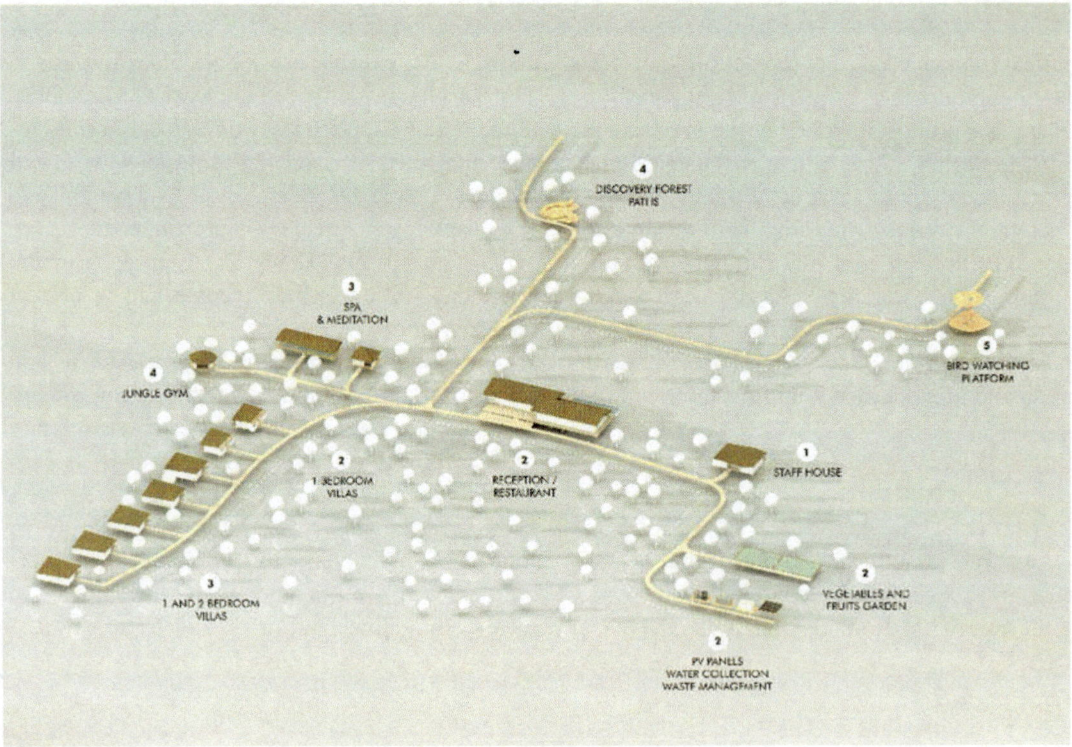

Fig. 35.8 Eco-lodge master plan. The numbers represent the construction phases of the amenities, which expand organically over time, and the initial investment costs and risks are kept low

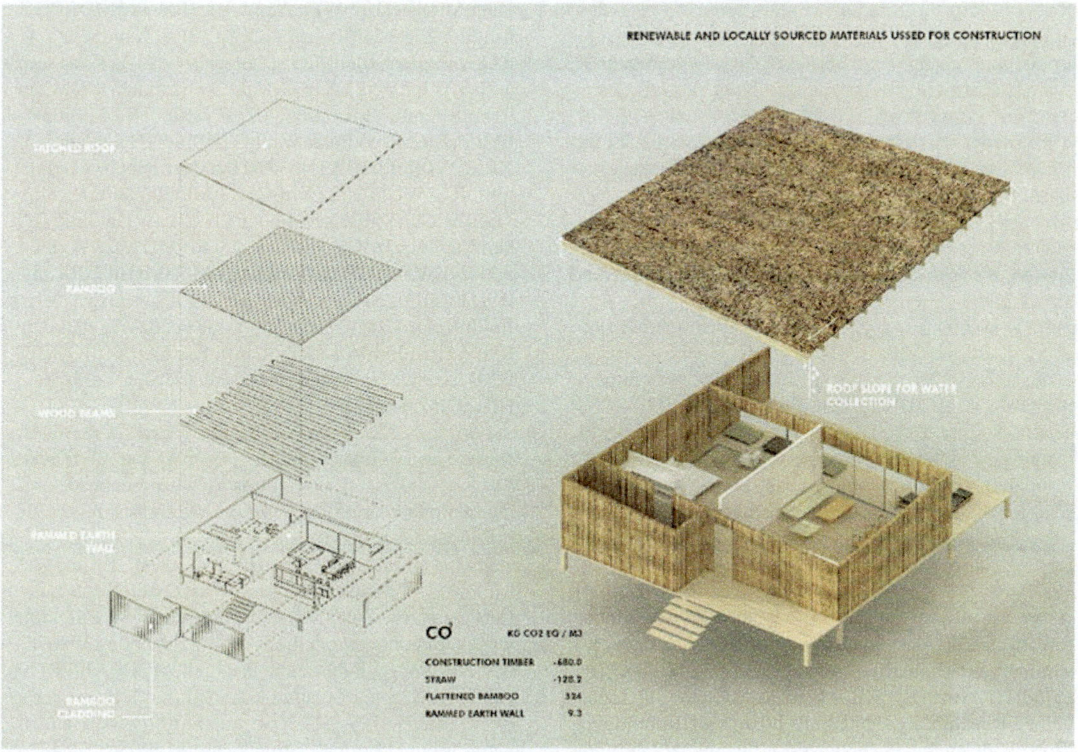

Fig. 35.9 Prototype example of the Villas. Built with traditional construction methods and local materials (such as earth, wood and straw)

Eco-lodges thrive when the community members own the property or organisation and are involved in the management and direct profit. Local ownership ensures more motivated employees and communities that take part in making the place as good as possible. A crucial aspect lies in involving the indigenous and local communities, as they have a particular interest in the biodiverse rainforest and are the most affected by the frequently damaged environment.

The result of the paper is an example framework of a successful eco-lodge by the border of the Amboro National Park, which would serve as a hub for the protection of the vulnerable rainforest. The location in the buffer zone of the Park is proposed as it can protect the forest from outsiders who cut down the trees around the Park to expand their land for agricultural purposes. The proposal includes community members in eco-lodge management. The profit from eco-lodges should be split between the owners, private investors and local communities. This collaboration, where knowledge about the area, crafts, business, education, Nature and sustainability can be shared, is seen as the most sustainable ownership system. Many private investors are motivated to invest in sustainable projects, so they can protect Nature while generating money. The future research includes interviewing community members in the Amboro Park Area and deep analyses of potential locations which could be the most beneficial for communities and strategic to protect the Amazon from illegal deforestation.

References

Arteaga Zambrana J (n.d.) Áreas protegidas de Bolivia. Academia Riquezas de Bolivia. Retrieved 9 Jan 2023, from https://riquezasdebolivia.com/areas-protegidas-de-bolivia

Boillat S, Alca J, Álvarez A, Bottazzi P, Camacho DP, Serrano E, Biffi V, Mathez-Stiefel SL, Larsen PB,

Rist S (2010) Protected areas and indigenous peoples in Bolivia and Peru: dilemnas, conflicts, and ways out. In: Global change and sustainable development: a synthesis of regional experiences from research partnerships. Perspectives of the Swiss National Centre of Competence in Research (NCCR) North-South, University of Bern, vol 5. Geographica Bernensia, Bern, Switzerland, pp 501–515

Bucklin DN (2010) Protected-area effectiveness near dynamic colonization zones: forest clearance in and around Amboró National Park. Oregon State University, Bolivia

Coimbra D (September 20, 2022) Personal communication

Finer M, Costa H, Villa L (2022) Amazon Fire Season 2022. MAAP: 168. Retrieved 9 Jan 2023, from https://maaproject.org/2022/amazon-fires-2022

Fundación Solón (2017) TIPNIS: the saga for the rights of nature and indigenous people. Retrieved 9 Jan 2023, from https://fundacionsolon.org/2017/10/24/tipnis-the-saga-for-the-rights-of-nature-and-indigenous-people

Ghinsberg Y (2005) Jungle: a harrowing true story of survival. Boomerang New Media, Austin

Global Forest Watch (2021) World Resources Institute. Retrieved 9 Jan 2023, from https://www.globalforestwatch.org/dashboards/country/BOL/?category=summary&dashboardPrompts=eyJzaG93UHJvbXB0cyI6dHJ1ZSwicHJvbXB0c1ZpZXdlZCI6WyJkb3dubG9hZERhc2hib2FyZFN0YXRzIl0sInNldHRpbmdzIjp7InNob3dQcm9tcHRzIjp0cnVlLCJwcm9tcHRzVmlld2VkIjpbXSwic2V0dGluZ3MiOnsic2hvd1Byb21wdHMiOnRydWUsInByb21wdHNWaWV3ZWQiOltdLCJzZXR0aW5ncyI6eyJzaG93UHJvbXB0cyI6dHJ1ZSwicHJvbXB0c1ZpZXdlZCI6W10sInNldHRpbmdzIjp7InNob3dQcm9tcHRzIjp0cnVlLCJwcm9tcHRzVmlld2VkIjpbInNoYXJlV2lkZ2V0IiwiZG93bmxvYWREYXNoYm9hcmRTdGF0cyJdLCJzZXR0aW5ncyI6eyJzaG93UHJvbXB0cyI6dHJ1ZSwicHJvbXB0c1ZpZXdlZCI6WyJzaGFyZVdpZGdldCJdLCJzZXR0aW5ncyI6eyJzaG93UHJvbXB0cyI6dHJ1ZSwicHJvbXB0c1ZpZXdlZCI6W10sInNldHRpbmdzIjp7Im9wZW4iOmZhbHNlLCJzdGVwSW5kZXgiOjAsInN0ZXBzS2V5IjoiIn0sIm9wZW4iOnRydWUsInN0ZXBJbmRleCI6MCwic3RlcHNLZXkiOiJzaGFyZVdpZGdldCJ9LCJvcGVuIjp0cnVlLCJzdGVwSW5kZXgiOjAsInN0ZXBzS2V5IjoiZG93bmxvYWREYXNoYm9hcmRTdGF0cyJ9LCJvcGVuIjp0cnVlLCJzdGVwc0tleSI6IndpZGdldFNldHRpbmdzIn0sInN0ZXBzS2V5IjoiIiwib3BlbiI6ZmFsc2UsInN0ZXBJbmRleCI6MCwiZm9yY2UiOnRydWV9LCJzdGVwc0tleSI6IiIsIm9wZW4iOmZhbHNlLCJzdGVwSW5kZXgiOjAsImZvcmNlIjp0cnVlfSwib3BlbiI6dHJ1ZSwic3RlcEluZGV4IjowLCJzdGVwc0tleSI6ImRvd25sb2FkRGFzaGJvYXJkU3RhdHMifSwib3BlbiI6dHJ1ZSwic3RlcHNLZXkiOiJzaGFyZVdpZGdldCJ9&location=WyJjb3VudHJ5IiwiQk9MIl0%3D&map=eyJjZW50ZXIiOnsibGF0IjotMTYuMzk0ODY5NTQ4NzA1ODc0LCJsbmciOi02My41NDY0MzYzMTAwMjY2OH0sInpvb20iOjQuNjU5NTg1MzIyMzE4OTczLCJjYW5Cb3VuZCI6ZmFsc2UsImRhdGFzZXRzIjpbeyJvcGFjaXR5IjowLjcsInZpc2liaWxpdHkiOnRydWUsImRhdGFzZXQiOiJwcmltYXJ5LWZvcmVzdHMiLCJsYXllcnMiOlsicHJpbWFyeS1mb3Jlc3RzLTIwMDEiXX0seyJkYXRhc2V0IjoicG9saXRpY2FsLWJvdW5kYXJpZXMiLCJsYXllcnMiOlsiZGlzcHV0ZWQtcG9saXRpY2FsLWJvdW5kYXJpZXMiLCJwb2xpdGljYWwtYm91bmRhcmllcyJdLCJib3VuZGFyeSI6dHJ1ZSwib3BhY2l0eSI6MSwidmlzaWJpbGl0eSI6dHJ1ZX0seyJkYXRhc2V0IjoidHJlZS1jb3Zlci1sb3NzIiwibGF5ZXJzIjpbInRyZWUtY292ZXItbG9zcyJdLCJvcGFjaXR5IjoxLCJ2aXNpYmlsaXR5Ijp0cnVlLCJ0aW1lbGluZVBhcmFtcyI6eyJzdGFydERhdGUiOiIyMDAyLTAxLTAxIiwiZW5kRGF0ZSI6IjIwMjItMTItMzEiLCJ0cmltRW5kRGF0ZSI6IjIwMjItMTItMzEifSwicGFyYW1zIjp7InRocmVzaG9sZCI6MzAsInZpc2liaWxpdHkiOnRydWUsImFkbV9sZXZlbCI6ImFkbTAifX1dfQ%3D%3D&showMap=true&treeLossPct=eyJoaWdobGlnaHRlZCI6ZmFsc2V9

Greenberg C (2022) Amazon rainforest fires 2022: facts, causes, and climate impacts. Greenpeace. Retrieved 9 Jan 2023, from https://www.greenpeace.org/international/story/55533/amazon-rainforest-fires-2022-brazil-causes-climate

International Finance Corporation (2004) Ecolodges: exploring opportunities for sustainable business. Retrieved 18 Dec 2023, from https://www.ifc.org/wps/wcm/connect/da53b1ff-7064-45bf-bd0a-f392075bf3b7/Ecolodge_Publication_Part1.pdf?MOD=AJPERES

International Rights of Nature Tribunal (n.d.) Tipnis case. Retrieved 9 Jan 2023, from https://www.rightsofnaturetribunal.org/cases/tipnis-case/#1590541198488-1be5e24e-f794

Justiniano H (September 22, 2022) Personal communication

Killeen T (2021) A perfect storm in the amazon wilderness: success and failure in the fight to save an ecosystem of critical importance to the planet. The White Horse Press, Cambridgeshire

Mendez A, Mercado I (2019) Disaster strikes in Bolivia as fires lay waste to unique forests. Mongabay. Retrieved 9 Jan 2023, from https://news.mongabay.com/2019/09/disaster-strikes-in-bolivia-as-fires-devastate-unique-forests

Navia R (2022) Personal communication

Ostrom E, Nagendra H (2007) Insights on linking forests, trees, and people from the air, on the ground, and in the laboratory. Proc Natl Acad Sci USA. https://doi.org/10.1073/pnas.0607962103

RAISG (2020) Amazonia Under Pressure, p 68. www.amazoniasocioambiental.org. Accessed 6 Jan 2023

Robbins P (2007) Encyclopedia of environment and society. Sage Publications, Thousand Oaks

Saccardi S (2019) Forest fires in Chiquitania and the aftermath. Bolivian Express Magazine. Retrieved 9 Jan 2023, from https://bolivianexpress.org/blog/posts/forest-fires-in-chiquitania-and-the-aftermath

SERNAP (2020) Bolivia. Parque Nacional y área natural de manejo integrado Amboró. Retrieved 9 Jan 2023, from http://sernap.gob.bo/amboro/#sc-tabs-1673298928923

SESMAD (2014) Social-ecological systems meta-analysis database: theory. Critique of fortress conservation.

Retrieved 9 Jan 2023, from https://sesmad.dartmouth.edu/theories/85

Sören Gigler B (2015) Poverty, inequality, and human development of indigenous peoples in Bolivia. Retrieved 9 Jan 2023, from https://doi.org/10.1596/978-1-4648-0420-5_ch3

Soria F (2008) The impact of community-based ecotourism projects in Amboró National Park. Institute for Advanced Development Studies (INESAD), La Paz

Springer Nature (n.d.) Instructions for Authors: Manuscript Guidelines | Springer—International Publisher. Retrieved 18 Dec 2022, from https://www.springer.com/gp/authors-editors/book-authors-editors/your-publication-journey/manuscript-preparation

Villca A (November, 2021) Personal communication

Check for updates

36 A Designed Emancipation: An Urbanist Re-framing

Kweku Addo-Atuah

Abstract

Much of the current literature in architectural and urban planning practice extols the use of partnerships and traditional techniques in public projects, especially in the Global South. However, these studies often fail to acknowledge the precarity of indigenous or local populations in these contexts. Often caught in the crosshairs of domestic and geopolitical entanglements, these populations are at constant risk of eviction from their lands, and with that, a loss of critical lifelines. Through narrative and argumentative storytelling, this essay examined the land and water grabbing threats that undermine the territorial sovereignty of African nations by both internal operators and by external, often monied foreign interlopers. Utilizing a speculative study component, the essay examined the tensions at play in shaping the built-natural environment interface, with a particular focus on the western Sahel, and specifically, Niger. The study covered several issues, including collective land ownership and stewardship, national and regional stability, traditional knowledge and techniques, and the nuances of modernity across the rural–urban spectrum in facilitating a community-led national development framework. The essay builds upon the wealth of studies by conservation scholars and specialists in how we engage with indigenous communities and reframe our relationship with land and water resources from one of unchecked extraction to one of mutual benefit. Therefore, the essay seeks to remind all of us of the inherent links between developmental interventions, indigenous communities, their lands and knowledge-rich contributions to achieving resilient "inclusive futures" as outlined in the Sustainable Development Goals.

Keywords

Indigenous land sovereignty · Community-led development · African urbanism · Sahelian cities · Geopolitics of land · Peri-urban interface · Local knowledge · Fluid modernity

36.1 Historical Grounding

'Kaka, (grandfather in Hausa) tell me our country's story once more.' 'Again! exclaimed Boubacar. How many times must I retell this story?'. Relenting due to his grandson Idris' pleading eyes, Boubacar launched into character complete with his gurmi, a lute instrument…Our story begins in the Sahel. Stretching across the African continent, its western point lies upon the Atlantic Ocean while its eastern edge forms an enclave that rests on the Red Sea. As Sahelian peoples,

K. Addo-Atuah (✉)
Providence, USA
e-mail: kaddoatu@gmail.com

S. Hilal et al. (eds.), *Design for Partnerships for Change*, Sustainable Development Goals Series,
https://doi.org/10.1007/978-3-031-36993-3_36

we occupy a transitional cushion between the Sahara Desert to our north, and to the south, where the savanna and forested terrains dominate the African landscape (see Fig. 36.1). Modern-day Sahel occupies nine semi-arid countries—Senegal to Eritrea—is home to various ethnic groups, and serviced regionally by the Niger and Nile Rivers, and Lake Chad.

Our lands sit within the Western Sahel, comprising six countries, including Senegal, Mali, Nigeria's northernmost edges and our own, Niger (see Fig. 36.2). At its height, the Western Sahel birthed the Mali and Songhai Empires, served as an epicentre of intense learning and knowledge production as well as enshrined upon its lands, the Sudano-Sahelian architectural style evident throughout many of our cities and towns, including Timbuktu and Agadez. Following a shift into smaller empires and kingdoms (Kanem-Bornu and Hausa) post 16th-century, our lands eventually fell under French administration in 1901. The French colonial government established the Nigerien capital at Niamey in 1904 and shaped a settlement strategy along the Niger River's banks. It was a spatial arrangement steeped in racial ideology and preference; the colonial administration placed European settlers upstream and the African indigenes downstream,

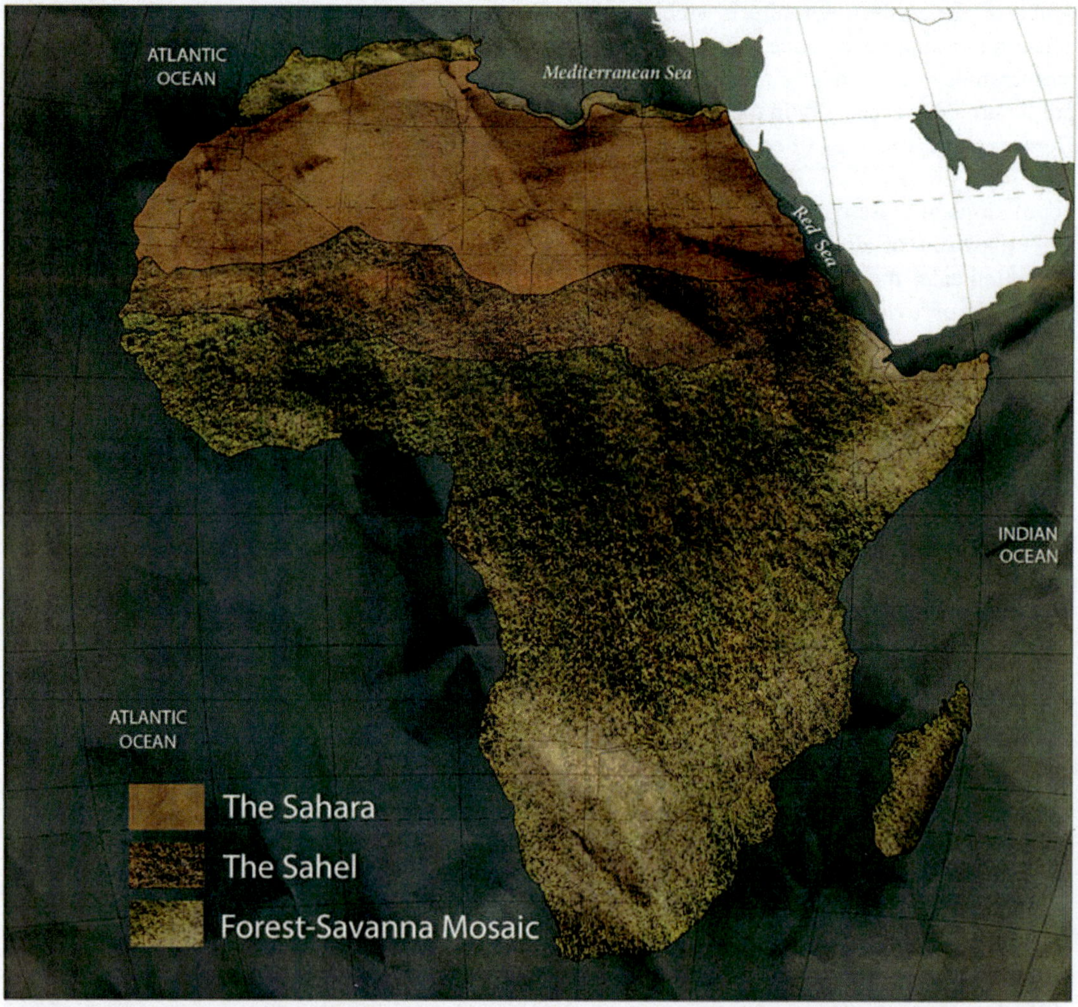

Fig. 36.1 Africa Biomes (by Author)

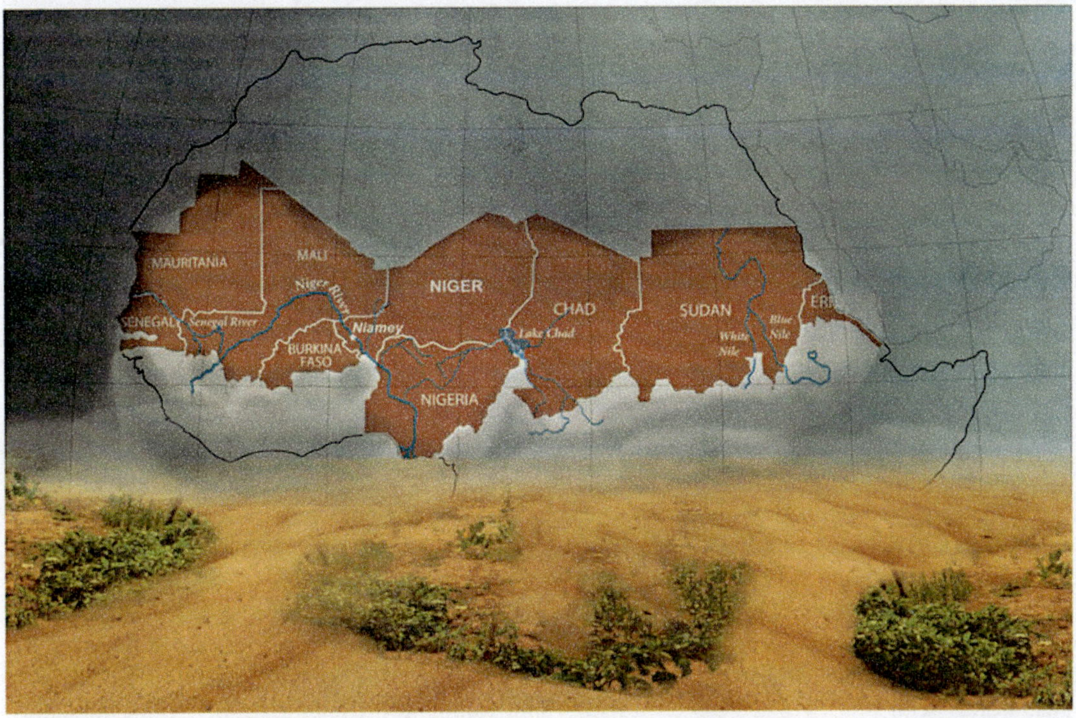

Fig. 36.2 Sahelian Societies (by Author)

with planted, recreational buffers between the two… "Idris, Idris, didn't I tell you to go get ready!"

36.2 Formative Years

That story was one that Idris often replayed. It was an evolutionary journey of Niger and its 27 million+ people, bound together by a pastoralist tradition as key drivers of identity, culture, political and economic capital. As a young boy, Idris repeatedly probed his grandfather, Boubacar, a noted griot about these stories while perched at the elder's feet during Boubacar's monthly visits. Raised alongside the Niamey elite, Idris and his siblings grew up firmly separated from the rural existence of their forebears, a sheltered reality that saw them chauffeured from one stately courtyard home to the another while visiting friends. As a living repository of knowledge, Boubacar instilled in Idris a fuller vision of Niger, helping demystify the nuances of its people by highlighting key distinctions and parallels of cultural, religious, economic and leisure practices. Particularly enthralling for young Idris was when Boubacar would puncture the air with his fists as he wove tale after tale of the eminent myths of their people and its guiding principles as realized through real-world translations.

But nothing thrilled Idris more than listening as Boubacar described how their people settled the Sahelian landscape, deftly navigating between its steppes and occasional plateaus alongside grasslands, tall perennials and trees to increase sheltering comfort. Boubacar would speak of the evocative building forms, material choices, layering of public and private space, and of course, the communal aspect of building production. For Idris, these stories awoke the innumerable possibilities and fluidity of human–nature interactions and that, when necessary, humans would constantly work to reshape and reframe environments to suit their most pressing needs. That awakening sparked his intention to study architecture and landscape architecture at university, where he was introduced to the material, sociological and ecological implications of the people, place and environment interface.

Raised in less grandiose fashion was Yasmin, who was born into a family of mid-scale farmers and ranchers in a thriving agricultural village some 25 km outside of Niamey. Her upbringing was one rooted in concrete, experiential encounters, rather than an imagined, conceptual existence borne from borrowed memories of an old man. Actively involved from childhood, Yasmin spent much of her free time toiling alongside her family on the multigenerational property and cheerily moving from one family dwelling to the other. From loading bales of hay to feed the horses and maintaining inventories to assisting her older cousins and ranch hands with the cattle and goats, Yasmin began to understand communality as a fundamental aspect of life. Her favourite memories were times spent with grandparents as they regaled her and their other grandchildren with stories and songs of their family lineage as they planted, pruned and harvested a rotation of crops on their land. However difficult or unpleasant the cultivation tasks were, it began to sprout an awareness of land, air, soil and water synergy as the drivers to her family's stability in times of both lack and abundance, either due to human error or biophysical stresses.

On Yasmin's walks to-and-from school, she would frequently pass her neighbours' properties, waving almost nonstop to both young and old. When she could steal some time, she would often stop to chat to friends at the large grazing lands and lake the community used, especially in times of extreme dryness and shrinking water levels. Immersed so deeply in this context, Yasmin was heartbroken when her parents sent her to the capital for high school, believing that qualifications from a good city school would ensure her a place at the best universities. In Niamey, life for Yasmin was different, no less joyous than her home village, but faster paced and mobility there much more car-centric than the walking and biking culture she had left behind. She found neighbours more closed-off until repeat encounters, little to no open or cultivated spaces, a sense of individualism and of course, she could always sense a slight impatience and irritation from her classmates when she heralded traditional knowledge. It was these colliding realities of her upbringing, from firmly rural to urban, and often feeling and seeing herself adjust as she moved between these worlds that led Yasmin to study ecology and urban planning at university. Yasmin felt she needed to deepen her understanding of urban and natural systems, the impact of land use policy and physical implications on human and non-human quality of life, including such matters as scale, and connections vs. disconnections.

36.3 Contemporary Realities

With these potent mutual interests in the natural and constructed world symbiosis, it is no surprise that Idris and Yasmin became close at school, cultivating a gradual trajectory from frequent collaborators, courtship, and shortly after graduation, marriage. As Idris and Yasmin continued to evolve over the years, oscillating between rapid speed and tepid contraction, so too did Niger. While land proved a viable economic and political capital tool historically, a decentralized zoning process sparked a boom in land speculation and brisk transformation of Niger's land holdings. As zoning shifted from the colonial era of state power and consolidated in the post-colonial era by furthering processes of socio-spatial marginalization, zoning power expanded to include both municipal and private sector actors (Körling and Moussa 2019). The City of Niamey, under a process called lotissement, (See Fig. 36.3) began to absorb hectares upon hectares of agricultural lands near its immediate limits and transformed them into urban land lots for "housing, industry, commerce and public facilities" (Belko-Maïga 1985, 61). While the decentralized zoning process fuelled the social mobility aspirations of would-be urban dwellers, the process also ran afoul of the customary or collective land tenure system that has been a mainstay for centuries (Körling and Moussa 2019).

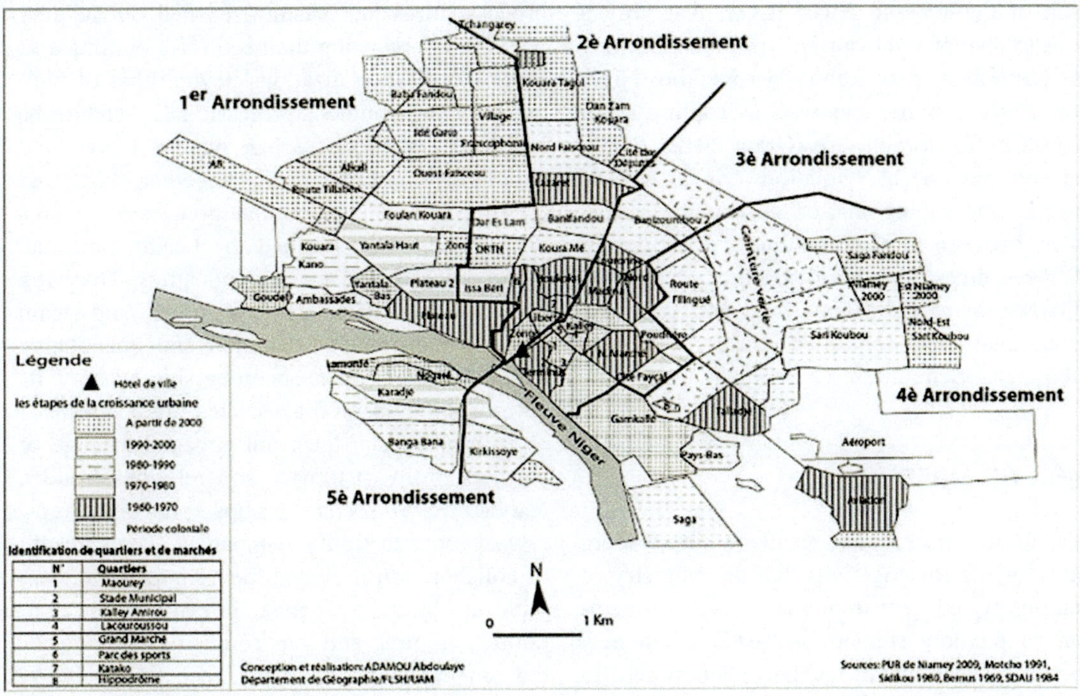

Fig. 36.3 Lotissement process: Niamey's spatial growth from 1937–2010. Source: Abdoulaye ADAMOU (2012)

When predicated on due diligence, decentralized zoning produced several benefits from shoring up neighbourhoods' integration into the larger city and national water, electricity, and transportation networks as well as bolstered sheltering and economic prospects. However, due to the complex land registration regulations, it undercut customary land holders in favour of local private elites and in-house public servants, who used their insider knowledge to acquire large tracts of land. In response, traditional leaders turned to private sales by auctioning off parcelled lots to protect themselves from both public and private sector encroachment, while rural families looked to their educated youngsters to facilitate land registrations and convert their lands into variable economic uses (Körling and Moussa 2019).

Concurrent with Niger's intense urbanization and its capital Niamey in particular, and a return to state-controlled zoning is the increasing incidences of land and water grabbing both within Niger and the Sahel region (Koigi 2022). These incidences stem from the misinformed consensus of industrialized nations that Africa's vast arable lands and water resources are underutilized and in perpetual abundance. This consensus conveniently forget that these lands hold value outside of agricultural use and that a third of Africa's population faces water scarcity due to declining volumetric depths of regional water sources like the Niger and Nile Rivers (Koigi 2022). Often in cahoots with sitting governments, these land and water grabs by individual foreign investors, governments and multinational corporations often fall under the guise of conservation or protected areas. These protected areas drive out local populations to prop up foreign economic interests for such ventures as glamping, trophy hunting, mining and crop production for export to feed industrialized populations (Sène 2022; Aryeetey and Lewis 2021).

Large-scale grabs include Office du Niger located in neighbouring Mali, and precariously sited in the Inner Niger Delta. Alongside disrupting the social, economic and hydrological health of the stressed Niger River, that project stands to disrupt and displace 1 million+ farming and pastoralist populations across the Niger River Basin countries, as well as capture some 470,000 ha for foreign use (Grain 2012). These land and water rights violations trigger several domestic and subregional crises, including skirmishes between local communities over diminishing resources, and deadly conflicts due to the militarization of indigenous land and water on international NGO-controlled conserved or protected areas (Sène 2022).

36.4 An Open Call

It is these crises, and resultant diminishing quality of life for Nigeriens that the Ministry of Planning placed intense pressure on the government to publicly support and establish a new national development agenda. Determined to quell rumours of partisanship, and opaque bidding and contract awarding processes, the Ministry of Planning with full presidential backing launched an open call for urban visioning strategy proposals for Niger. The open call encouraged proposals to consider multiple concerns, from the socio-economic to the ecological, public health, drivers of peace, crime and conflict, national and regional sovereignty alongside practices and processes of urban–rural place-making in this semi-arid, pastoral-dominant context. In short, the Ministry was cognizant of the looming UN Sustainable Development Goals (SDGs) deadline and sought to find actionable solutions that would fast-track Niger's capacity to dramatically boost quality of life and ensure its self-sufficiency. Each proposal team would have 6 months to calibrate their work and present these findings in multiple arenas around Niger. At long last, the call that Idris and Yasmin had yearned for both in school and their first working years had materialized!

Equal parts terrified and excited, Idris and Yasmin leapt into action, familiarizing themselves with the open call brief, its expectations and range of skills required. Following this process, Idris and Yasmin reached out to their network, recognizing the need for a multitude of professional specializations beyond that of typical built environment professionals—architects, landscape architects, urban planners, civil and structural engineers, etc. Altogether, Idris and Yasmin assembled a competition team of some 20 members that spanned the health, financial, environmental and energy industries. Divvying up the task list, Idris, Yasmin, and team embarked on an ethnographic and quantitative examination of Niger, encompassing Niamey, its seven subregions and associated prefectures.

In each locale, the team engaged a range of partners, from customary and religious leaders, activists, civil society groups, prefecture reps, townspeople to utility companies. That breadth of collaboration provided their competition team with an intimate spatial, infrastructural, economic, cultural and environmental framework that an analytical approach focused solely on Niamey or generalized assumptions of other Nigerien realities would have failed to acknowledge, or address with targeted action. The team's cross-country canvassing took them through Niger's northern edge where the nomadic Tuareg reside, the Nigerien south where the sedentary agriculturalists hold court and to the centre where much of the Hausa reside, alongside semi-nomadic groups like the Fula.

The insights the team gleaned were expansive, helping them reflect upon and truthfully respond to Niger's intricacies. Sparse school lectures were swapped for in-depth exposure through workshop visits to various artisans and in-field analyses of the distinctive and overlapping architectural typologies regarding form-making, material choice, and construction techniques and detailing. For Idris, these travels served as clarifying opportunities of Boubacar's stories. Listening to and mulling over perspectives and motivations different from his own helped place the myths, origins, historical truths and outright lies of Niger and its peoples in precise context. Yasmin dove into the particularities and nuances concerning nomadic, semi-nomadic and sedentary lifestyles of Niger's major ethnic groups, their struggles amidst resource shortages and how their developmental ambitions differed from their urban counterparts? How do people use land and water, their response to encroachment or competition, how do they define public space, what it contains and what does it look like? Conversations with farmers and agriculturalists also helped re-situate Yasmin's appreciation of the planting and harvesting methods of the people, and the multifaceted benefits of indigenous crop and flora species.

While there were certainly inklings of what their proposal could look like, after 3 months spent deconstructing Niger, its peoples and landscapes, Idris, Yasmin and the team realized that it would not be an exercise of harshly drawn lines and demarcated spatial planning. How could it be when Niger finds itself occupied by peoples living such dynamic lives, influenced by the flows and patterns of rain, wind, water, land, seasonal migration and food! Whereas prevailing architectural constructs in industrialized countries espouse dense arrangements and often coerce a highly urbanized lifestyle, such an approach may incur massive resistance from Nigeriens, with many from cultures rooted in rurality and a preference for a slower life. With this exposure to the daily practices, realities and rituals of Nigeriens, Idris and Yasmin's team spent the next 3 months developing a proposal steeped in fluidity, predicated on adapting, anticipating and accommodating changes, both seasonal and non-seasonal.

To accentuate this fluidity, Idris and Yasmin's team relied on speculative futures to inspire and inform their audience across the public, private and civil society partner spectrum. Eschewing hard-lined drawings and laser-cut or CNC-built models, the team drew upon multiple media, from photographs, simulated walkthroughs via AI, VR/AR technologies, loose sketches to other visualization and text strategies that spoke to a poetic and collaborative process of city, town and village placemaking. Traditional techniques, materiality and practices informed much of their urban design approach. Reinvigorated with centuries-long knowledge of agricultural practices, Idris and Yasmin's team emphasized various agroforestry processes endemic to the Sahel, which combined everything from selective pruning of naturally-sprouting trees to zai pits—half-moon contours (UN News 2022). Together, these techniques work to restore previously degraded lands, mitigate drought threats, encourage water-harvesting and soil permeability and aid high-quality native crop yields due to improved microclimatic conditions. Working from the mappings generated in coordination with local communities, the team produced highly accurate spatial readings of typical livelihoods, as well as its relationship to Niamey. These maps supported scenario explorations, on such issues as migration—both internal and across the Sahel—towns and villages most vulnerable to annexation into Niamey as well as agricultural and water assets susceptible to internal conflict and external take-overs.

Understandably, for Idris and Yasmin's team, Niger's developmental strategy would elevate an urban–rural or peri-urban interface, rather than an urban morphology-heavy focus. The team promoted multidimensional strategies that provided rural, peri-urban and urban residents accessibility, connectivity and continuity. From schools and community centres to sanitation, transportation and energy infrastructure, the architectural and landscape architectural response experimented with layers of communality and privacy, and identified underutilized spaces as opportunities for both liminal and sustained connections. The team regarded such spaces as linchpins for enriching and establishing more secure transitions and interactions between people, goods and circulation networks within and outside of national borders. The team produced scenarios of buildings and landscapes composed of locally sourced earthen brick and adobe infused with varying cement concentrations—thus crafting energy-efficient structures—alongside a broad spectrum of native vegetation selected for multi-use, including drainage capacity and drought hardiness (Kéré and Morris 2020; Snyder 2020).

36.5 Principles, Partnerships and Practices in African Urbanism(s)

And so it was at the Ministry of Planning auditorium where amidst boisterous crowds, candidates presented their open call proposals and were eventually culled to three finalist teams. Project Team 1 rather briskly, introduced their proposal, a scheme so universalist in its scope and response, that it could fit anywhere in the world. It was clear that serving Niger and its people would simply not do. So resolute in its vision to divorce Niger from its own origins that it championed protected areas and reafforestation schemes, forgetting the value of robust ecosystems comprising deserts, grasslands, savannahs and forests. Project Team 2's proposal fought valiantly to straddle the Nigerien reality and aspiration through extensive cross-country infrastructure and mixed-use development linkages. However, the proposal platformed urban areas as the only drivers to a "modern" Niger, failing to appreciate the benefits of peri-urban and rural areas in shaping a prosperous nation. Idris and Yasmin's team clocked in as the third finalist with a proposal upholding the soul and traditions of the Nigerien people (Fig. 36.4).

Theirs was a proposal steeped in promoting national and global peace and prosperity for both people and planet. Its highlights include reorganizing public land and water—*the commons*—back into local populations' purview through negotiated tenure rights with national, municipal and local governments rather than the protected areas approach advocated by organizations like African Parks (Sène 2022). Urbanism-wise, Proposal 3 sought to situate its buildings and constructed landscapes in service and in community with Niger's vast natural systems, thereby inscribing its value in the national consciousness, and help protect against foreign takeovers. Additional efforts included adapting centuries-old agricultural and land restoration methods alongside polyculture strategies to boost forest and food species diversity, nutritional robustness and yield, rather than the monoculture and genetically modified practices often pushed onto African societies (Ahmed 2022).

Therefore, which urban visioning proposal will the Ministry of Planning endorse? Which proposal will better enshrine the sovereignty of Niger and Nigeriens, while recognizing its regional and global contributions to achieving the SDGs? Would the Ministry act boldly and champion a difficult but necessary community-led national development framework or cower towards a short-term foreign investment and interests-focused approach shrouded in secrecy and typically a destabilizing venture? The latter would mean that organizations like African Parks armed with Global North financing, can continue to ink transnational conservation management deals at the expense of indigenous populations—those who have long preserved these ecological jewels—and therefore, will endure the most harm (Sène 2022). It upturns their lives, entrenches insecurity and conflict that reverberates far beyond these areas, and festers carcerality when these protected areas wield state power to punish locals for accessing their own land and water resources (Sène 2022).

As urbanists, like Idris and Yasmin, it is incumbent that our visioning and strategic interventions that shape land, especially in the Global

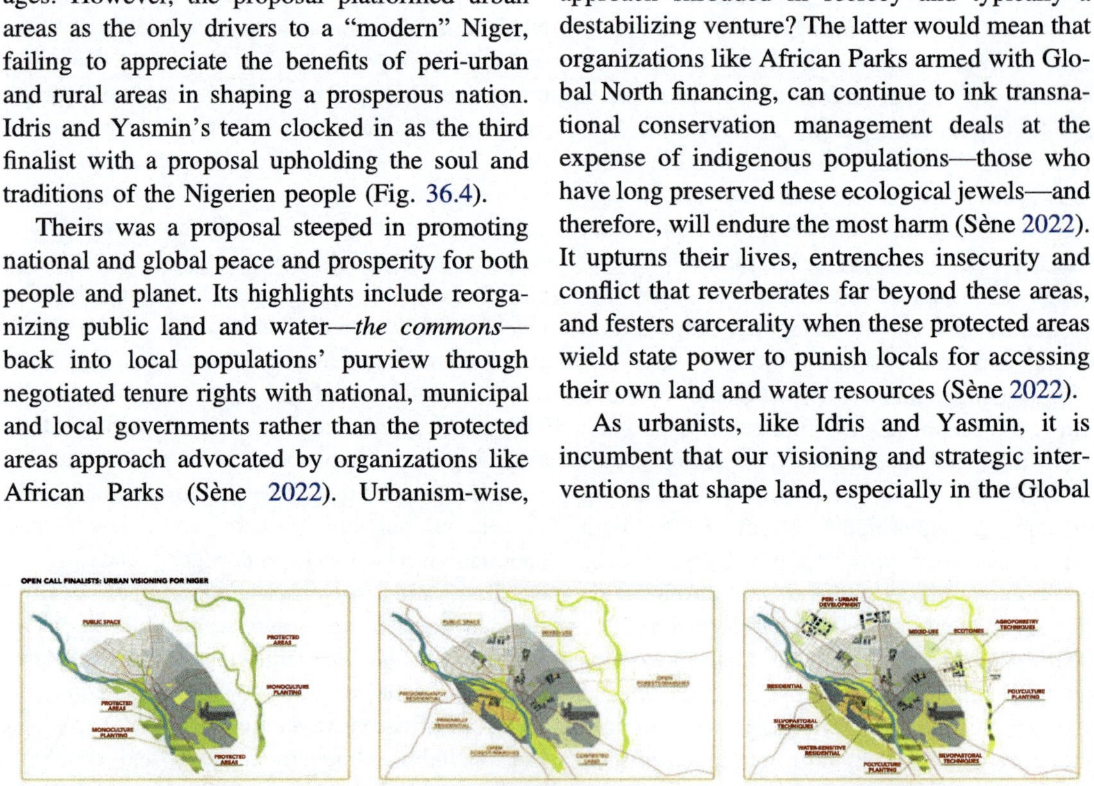

Fig. 36.4 Finalist proposals for Niger—Phasing (Image by Author)

South, do not reinforce deleterious colonial land practices, where local populations were pushed off prime lands for the express benefit of settler populations and a cabal of affiliated local elites (Ogada 2022). Thus, an urbanist re-framing conscripts us to lead with humanist principles of upholding health, safety and welfare in our work. Do the projects municipal and national governments sponsor endanger already vulnerable populations and/or concentrate resources, public goods and infrastructure connectivity and continuity to only a few sections of a city, town or village? An urbanist re-framing compels us to reconsider the conflicting interests on public projects, which partners are present, who and what ideas or needs are assigned the greatest value, and which are ignored or obfuscated.

Beyond this ethical provocation, this is simply a principle of stability. As built environment professionals know intimately, to build on shrinking soils or disturbed lands yields disastrous results unless we undertake appropriate structural reinforcement measures. It has frankly become untenable to continue advancing socio-spatial policies that further embed covert and overt injustices in national and global developmental agendas. It tasks urbanists to examine and confront the root causes of puzzling resistance to steady, transformative outcomes in countries globally, from the rural to urban to peri-urban, and work to design new systems to prevent such imbalances from self-replication.

References

Ahmed K (2022) Fonio just grows naturally: could ancient indigenous crops ensure food security for Africa? The Guardian. https://www.theguardian.com/global-development/2022/jul/07/fonio-indigenous-crops-africa-food-security. Accessed 12 Sept 2022

Aryeetey E, Lewis Z (2021) African land grabbing: whose interests are served? The Guardian Nigeria. https://guardian.ng/sunday-magazine/newsfeature/african-land-grabbing-whose-interests-are-served/. Accessed 10 Sept 2022

Belko-Maïga G (1985) La politique foncière au Niger: Causes et conséquences des affrontements entre urbanisme périphérique et système foncier traditionnel à travers l'exemple de Niamey. Doctoral thesis, Institut de Géographie, Université de Paris IV

Everyculture (2022) Niger. Everyculture.com. https://www.everyculture.com/Ma-Ni/Niger.html. Accessed 20 Aug 2022

Grain (2012) Squeezing Africa dry: behind every land grab is a water grab. Grain.org. https://grain.org/article/entries/4516-squeezing-africa-dry-behind-every-land-grab-is-a-water-grab#Stop%20the%20water%20grab. Accessed 15 Sept 2022

Kéré F, Morris J (2020) Building on architectural traditions of the Sahel. MetMuseum.org. https://www.metmuseum.org/perspectives/articles/2020/7/francis-kere-james-morris-sahel-architecture. Accessed 8 Sept 2022

Koigi B (2022) The plague of water grabbing and its consequences. https://www.fairplanet.org/story/the-plague-of-water-grabbing-and-its-consequences/. Accessed 20 Sept 2022

Körling G, Moussa IH (2019) 'Tout a été loti!': decentralisation, land speculation and urban expansion in Niamey, Niger. Kritisk Etnografi Swed J Anthropol 2 (1–2):67–79. http://urn.kb.se/resolve?urn=urn:nbn:se:uu:diva-409762

Ogada M (2022) The stealthy, surreptitious second coming of Western Colonialism in Africa. https://ogada.co.ke/2022/07/29/the-stealthy-surreptitious-second-coming-of-western-colonialism-in-africa/. Accessed 1 Oct 2022

Sène AL (2022) Western nonprofits are trampling over Africans' rights and land. ForeignPolicy. https://foreignpolicy.com/2022/07/01/western-nonprofits-african-rights-land/. Accessed 21 Sept 2022

Sène AL (2022) Land grabs and conservation propaganda. AfricaIsACountry. https://africasacountry.com/2022/06/the-propaganda-of-biodiversity-conservation. Accessed 21 Sept 2022

Snyder M (2020) In Niger, an architect looking to the country's design's traditions. The New York Times Style Design Magazine. https://www.nytimes.com/2020/08/10/t-magazine/mariam-kamara-architect-design.html. Accessed 8 Sept 2022

United Nations (2022) Bringing dry land in the Sahel back to life. UNNews. https://news.un.org/en/story/2022/01/1110322. Accessed 24 Sept 2022